HANDBOOK OF INSECT REARING VOL. II

HANDBOOK OF INSECT REARING VOL. II

Edited by

PRITAM SINGH

Entomology Division, Department of Scientific and Industrial Research, Auckland, New Zealand

and

R.F. MOORE

United States Department of Agriculture, Cotton Production Research Unit, Florence, SC 29503, U.S.A.

ELSEVIER
Amsterdam – Oxford – New York – Tokyo 1985

ELSEVIER SCIENCE PUBLISHERS B.V.
Molenwerf 1
P.O. Box 211, 1000 AE Amsterdam, The Netherlands

Distributors for the United States and Canada:

ELSEVIER SCIENCE PUBLISHING COMPANY INC.
52, Vanderbilt Avenue
New York, NY 10017, U.S.A.

Library of Congress Cataloging in Publication Data
Main entry under title:

Handbook of insect rearing.

Includes bibliographies and indexes.
1. Insect rearing--Handbooks, manuals, etc.
I. Singh, Pritam, 1935- . II. Moore, R. F. (Ray F.)
SF518.H36 1985 638 85-1609
ISBN 0-444-42467-9 (U.S. : set)
ISBN 0-444-42465-2 (U.S. : v. 1)
ISBN 0-444-42466-0 (U.S. : v. 2)

ISBN 0-444-42466-0 (Vol. II)
ISBN 0-444-42467-9 (Set)

Printed in The Netherlands

PREFACE

The concept of the Handbook of Insect Rearing was developed by the editors in 1980. However, the original idea was conceived by the senior editor about 25 years ago when he was a graduate student. He was dismayed that there was no standard book available providing guidelines for rearing insects, or describing specific procedures to follow in order to produce insects for research in the laboratory and field. He then and there decided that there was a need for a 'cookbook-style' text to deal with standard methods for rearing insects.

In March 1980 the senior editor participated in a conference, "Advances and Challenges in Insect Rearing", jointly sponsored by the Agricultural Research Service of the United States Department of Agriculture and the Insect Rearing Group, held at Capital Airport Inn, Atlanta, GA, U.S.A. He presented a paper, "Insect Diets: Historical Developments, Recent Advances and Future Prospects." His recommendation that rearing techniques for pest species, beneficial species, common laboratory species, and other significant species should be compiled and published in the form of a technical handbook was received with enthusiasm by the conference participants and the conference organizer, Dr R.F. Moore.

Our concept of an instruction handbook on insect rearing, featuring practical step-by-step methods which could be easily followed by entomologists and technicians, was announced in 1981 in an Insect Rearing Group newsletter called FRASS. The response was enthusiastic and encouraging. Following this announcement, invitations were sent to rearing specialists with the request that they provide contributions to the handbook. Most accepted, and many favorable comments on the merits of the idea were received.

The original list of contributors was compiled from the 1981 publication "Arthropod Species in Culture", by W.A. Dickerson and associates. This list was expanded as contributors and other interested persons gave their input and suggested additional species. All the methods described in the handbook have been well established and successfully used for several years in the laboratories of the respective authors.

We proposed, developed and adopted a 'cookbook-style' writing procedure for the handbook — a new venture in the presentation and dissemination of information on rearing aspects. The first model using the present format was developed by the senior editor and was sent to contributors for their guidance in preparing contributions. The original model highlighted each step in rearing, first into various sections followed by changes in spacing and indentation within each section. This model was later revised to accommodate publication requirements to conserve space. With most contributors typing their own manuscripts on CRC forms, some unavoidable variations in format have occurred. However, this should not detract from the value of the procedures that are given.

We believe that this format has significant advantages in comparison with the research report style of presentation, in that it systematically instructs individuals in how to rear an unfamiliar insect. The format provides in sequence information on the equipment, materials and supplies required for the various operations to be carried out. There is no need to search through a narrative to compile the information necessary to rear the insect. Then, the specific instructions in a step-by-step manner are given for each operation in the rearing procedure. In total, the format provides a checklist for both the contributor and the technician to ensure that no essential item is lacking nor vital procedure overlooked. It thus covers those minute details and precautions which are absolutely essential in any successful rearing method. Its merit has become evident to us as we have worked with the format and explained it to the contributors. We hope that this format will be recognized as an effective method of presenting information on rearing insects.

Because of several intrinsic limitations, editing the manuscripts — though enjoyable and rewarding — has been a difficult task. The large geographical distances involved — the publisher in Europe, one editor in New Zealand and one in the United States, and contributors throughout the world — resulted in some long delays. The task of keeping up-to-date records on the status of each of approximately 100 different manuscripts was a challenging one. We are deeply indebted to the contributors, whose patience especially is appreciated, in that a sequence of changes was necessary during the text development stages to establish reasonable uniformity in style, and to ensure that all material was properly collated in the final camera-ready copy. We have made every effort to maintain high standards and prepare a handbook that is technically accurate, useful, legible and attractive in its final type and form, and accept sole responsibility where this has not been accomplished.

It is recognised that other competent researchers who are not contributors also rear some species which are listed in the handbook. However, because there are so many other species reared around the world we were unable to accommodate all the possible contributions. A selection process was therefore used in inviting specialists with many years experience in rearing a particular species. Also some important contributions were delayed, others were left out through oversight or lack of specialist contributors, and some were withdrawn at the last moment. Pending the judgment of the scientific community on the usefulness of this handbook and the need to publish rearing methods for additional species, we believe it possible to prepare another volume on additional species and on other methods for the same species that may be unique or of widespread interest. Any input and comments from users would encourage us to take up the preparation of the third volume.

We thank our respective organizations — the New Zealand Department of Scientific and Industrial Research and the Agricultural Research Service of the United States Department of Agriculture — for the use of facilities and for permission to edit the handbook. Special thanks are due Mrs Edna Starling, who maintained the files, ensured that manuscripts were not lost and were routed to the proper person, and, with the assistance of Mrs Sharon Herndon, typed many of the final CRC drafts. Appreciation is also expressed to the many contributors, who in addition to their expertise, also prepared the CRC manuscripts. This handbook would have not been possible without their valuable assistance.

Pritam Singh
R.F. Moore

September 1984

CONTENTS

INTRODUCTION

PRITAM SINGH, Entomology Division,
Department of Scientific & Industrial Research,
Auckland, New Zealand
and
R.F. MOORE, USDA-ARS, Cotton Production Research Unit,
P.O. Box 2131, Florence, SC 29503, U.S.A.

WHY THE HANDBOOK?

Progress in entomological research and the success of pest management programmes depend on our ability to rear insects and establish colonies in the laboratory. Ultimate success depends on the quality and performance of laboratory-reared insects. Most laboratories devise their own rearing procedures, and information on these procedures may not be readily available to other workers. Sometimes, conversely, there are many published procedures for rearing a species, and then a worker does not know which is the best for his situation. The result is duplication of effort, with consequent waste of both time and money on re-inventing procedures. More importantly, results cannot be compared from one laboratory to another owing to differences in rearing procedures.

There is no standard textbook on rearing procedures which one could follow. This handbook is intended to provide a practical guide for those who wish to rear insects for the first time, whether they be specialists or non-specialists. It is anticipated that the subject matter may be useful not only to entomologists but also to biologists, zoologists, geneticists, ecologists, biochemists, microbiologists, pathologists, toxicologists, plant breeders, State and Federal Government officials dealing with agriculture or the environment and private chemical industry, including biotechnologists.

This is a 'do-it-yourself' book. A unique feature is its new writing style: the advantage of a 'cook-book' style is that it is a simple abbreviated text that is easy and logical to follow. The text layout first lists the equipment, materials and ingredients required and then gives specific step-by-step rearing instructions including precautions. We hope that publication of this handbook is a first step towards standardizing rearing methods for insects in the laboratory.

ARRANGEMENT OF THE HANDBOOK

The handbook is in two volumes. In the beginning of Volume I there are ten chapters dealing with some fundamental information relating to selected areas of insect rearing. It also includes species-specific rearing procedures for the orders - Coleoptera, Collembola, Dictyoptera, Hemiptera, Hymenoptera, Neuroptera and Orthoptera. Volume II deals with rearing of species of Diptera and Lepidoptera. Both volumes have the indexes.

The orders are arranged alphabetically in the two volumes. Within each order, the genera and species are arranged alphabetically. Thus, in Volume

I for example, Coleoptera precede Collembola and Dictyoptera; and within the Coleoptera, rearing methods for *Anthonomus grandis grandis* are to be found earlier than those for *Hypera postica* and *Leptinotarsa decemlineata*. Where several species can be reared by the same method or by slight modification of the method, they are listed under a separate section at the end of the chapter. The scientific and common names cited are those approved by the Entomological Society of America (Werner, 1983) and "Scientific and Common names of Insects and Allied Forms Occurring in Australia" (Carne *et al.*, 1980).

LAYOUT OF REARING INSTRUCTIONS

The information on rearing each species is generally presented in 8 major sections (but with some exceptions), as follows.

1. INTRODUCTION - Provides a brief overview of the species, with its primary characteristics and unique features. Economic importance, distribution, damage, hosts and life cycle information are covered.

2. FACILITIES AND EQUIPMENT - Lists rearing conditions and facilities, design of cages, and a checklist (supplemented with photographs, if necessary) of the equipment needed for different stages of development and phases of the procedure.

3. DIET OR FOOD - Gives the composition of artificial diet, with details of preparation and dispensing. Also deals with the natural host, when used, and precautions to be observed. Sources of ingredients and brand names are given.

4. REARING AND COLONY MAINTENANCE - Includes methods of egg collection and sterilization, diet inoculation, holding larvae, collecting pupae and adults, sex determination, oviposition and mating, plus daily/weekly/monthly rearing schedules. Insect quality traits and important specific problems are discussed.

5. INSECT HOLDING - Gives information on how to manipulate development of the colony and (for weekend work) the insect stages which can be held at lower or higher temperatures. Precautions to be taken in handling are given.

6. LIFE CYCLE - Data on the growth and development of various life stages is provided, generally in tabular form. Includes larval period, number of larval instars, pupal period and weights, adult lifespan, fecundity and oviposition period; also survival data and other biological information (diapause, cannibalism, etc.) needed for rearing management.

7. SUPPLY PROCEDURE - Deals with insect yields, insect orders and methods of packaging, shipping and supplying various stages.

8. REFERENCES - Only selected and important references are listed.

INDEXES

For ready reference, three categories of indexes are provided at the end of both volumes. Indexes locate species by volume and page number. The page numbers for those species for which detailed instructions are given are identified by **bold** type. The page numbers for species referred to in the chapters on multiple species rearing diets and entomophagous insects are listed in regular Gothic type.

INDEX OF TAXONOMIC ORDER - This provides a rapid method of locating groups of genera within their representative families.

INDEX OF COMMON NAMES - The species are listed alphabetically by their common names, where these exist.

INDEX OF SCIENTIFIC NAMES - The species reared are arranged in alphabetical order of scientific names.

THE PAST, PRESENT AND FUTURE OF INSECT REARING

PAST HISTORY (5000 B.C. to A.D. 1900)

Historical records show that insect rearing has been practised for at least 7,000 years, as evidenced by the recent discovery of a Neolithic relic site in Zhejiang Province in China, where silk material was radio-carbon dated (Anonymous 1980, Chou 1980). The Chinese wrote many books on sericulture and silk technology, and took extreme care in rearing their insects. The attendant, called the silkworm mother, was to have no bad or sexual odour, she was not to smoke or wear make-up, and not to eat or even touch garlic or chicory. While tending the silkworms she was to wear simple, clean clothes and sandals, and not stir up the air in such a way as to disturb them.

Other insects have been cultured for a very long time. For instance the lac insect, *Laccifer lacca* Kerr, which formed the basis of the shellac industry, has been reared in India and China for several thousand years. The use of insects in medicine dates from 31 B.C. to A.D. 1578, and a total of 73 species are listed in "Compendium Materia Medica" (Tsai 1982). These insects are still reared today, basically in the same manner, for the benefit of mankind. The history of insect rearing has been traced by Miller (1952), Smith *et al.* (1973), Southwood (1977), Free (1982) and Tsai (1982).

TWENTIETH CENTURY

The era of contemporary rearing began at the turn of this century when Bogdanow (1908) reared the blowfly *Calliphora vomitoria* from egg to adult on an artificial diet. Since then many species have been successfully reared and colonised in the laboratory (Needham *et al.* 1937, Anonymous 1964, Smith 1966, Singh 1977). However, it was not until 1936 that the screwworm, *Cochyliomyia hominivorax*, was mass-produced in a factory. This laid the foundation of the Sterile Insect Technique (SIT), and 10 billion flies have since been produced for release in Mexico and Texas (Bush 1978).

Today a wide variety of insects are produced in millions and used in various pest control programmes. Some examples are the boll weevil, *Anthonomus grandis grandis*; bollworm, *Heliothis* complex; pink bollworm, *Pectinophora gossypiella*; codling moth, *Cydia pomonella*; housefly, *Musca domestica*; cabbage looper, *Trichoplusia ni*; fall armyworm, *Spodoptera frugiperda*; European cornborer, *Ostrinia nubilalis*; tropical fruit flies; mosquitoes; ladybird beetles; and several parasites.

The progress made in contemporary rearing, and the contribution it has made to entomological research programmes and in developing pest control strategies, has recently been documented (Knipling 1984, Singh 1984). Progress is also being made as more difficult problems are investigated, as evidenced by some unique adaptations to rearing given in this handbook. For example, in **Vol. I** of this handbook: a unique substrate for rearing Collembola (Tomlin, p. 317), which may be useful for other soil insects that require similar conditions; an artificial diet and parafilm sachet for feeding and oviposition by *Lygus hesperus* (Debolt, p. 329); development of methods for rearing and handling the difficult group of social wasps (Matthews, p. 401); use of a lyophilized powder of honey bee drones for rearing predaceous coccinellids (Matsuka and Niijima, p. 265; rearing *Xyleborus ferrugineus* (a wood-boring scolytid beetle) on oligidic and meridic diets (Norris and Chu, p. 303); and in **Vol. II**: gnotobiotic rearing of blowflies (Greenberg and George, p. 25); storage of diapause stages, as for pupae of *Delia* spp. (Tolman *et al.*, p. 49, Whistlecraft *et al.*, p. 59); and larvae of *Choristoneura* spp. (Robertson,

p. 227) and *Diatraea grandiosella* (Chippendale, p. 257), to name a few; feeding of adult *Glossina* spp. through a silicon membrane (Langley, p. 97); rearing larvae of *Simulium* spp. in a tank that provides the rapid water flow required by these insects (Edman and Simmons, p. 145); rearing *Lixophaga diatraeae* on its host and also larvae of the greater wax moth, *Galleria mellonella,* by applying maggots recovered from the adult females (King and Hartley, p. 301); use of Gelcarin® as a substitute for agar (Patana, p. 329, Guy *et al.*, p. 487) and sodium alginate (Navon, p. 469). These examples illustrate the diversity of approach required to rear insects and the imagination applied by many individuals to the problems encountered.

The current status of insect rearing (up to 1980) is well documented in a recently published USDA bulletin, "Advances and Challenges in Insect Rearing", edited by King and Leppla (1984). This excellent publication results from a Conference held in Atlanta, Ga, on 4-6 March 1980 under the joint sponsorship of the Agricultural Research Service of the United States Department of Agriculture and Insect Rearing Group. It comprises 36 articles arranged in six sections, four dealing with basic insect-rearing concepts, one with the state of the art, including descriptions of several rearing systems, and one discussing various topics important in the management of insect rearing systems. Recommendations which emerged from the conference focus on priorities and future research areas in insect rearing.

THE FUTURE

The future of insect rearing as a scientific discipline depends on progress in seven major areas:

1. Establishment and development of standard rearing methods, including production of special strains of insects.
2. Standardization of artificial diets which are nutritionally superior and can be economically prepared.
3. Understanding basic behaviour, biology, nutritional requirements, and the chemical and physical stimuli controlling feeding, mating and oviposition.
4. Control of microorganisms in the laboratory and in diets, and overcoming the diseases caused by them.
5. Development of quality control procedures and performance fitness tests.
6. Engineering aspects, including application of automated systems for inoculation and extracting various stages of insects from diets to reduce labour and production costs.
7. Application of insect rearing management (IRM) systems for efficient production.

Out of 1 million described species, only about 1400 have been reared in the laboratory, and of these about three dozen successfully and continuously on a large scale. Most work has been done with plant-feeding species of Lepidoptera, Coleoptera, and Diptera. Relatively little has been achieved with the colonization of predators, parasites and blood-feeding insects. It is apparent that the field of insect rearing is completely open, and there is a tremendous opportunity for workers with an innovative approach.

Insect rearing has today become a complex and specialized field. It encompasses various scientific disciplines and requires the co-operation of entomologists, microbiologists, nutritionists, biochemists, geneticists, ecologists and engineers. In addition, business management skills are a vital element in successful large-scale rearing programmes.

Problems arising in insect rearing such as those involving in-breeding, subtle nutritional deficiencies, diseases and conditioning are not insurmountable. Once the cause of the problem is understood, genetic, physiological and nutritional modifications can be made in rearing techniques. In fact, there is no reason why a 'super insect' with desired traits could not be produced for use in the field or laboratory. As the technology of insect rearing moves forward, the use of standard, high-quality, laboratory-reared insects will play an important role in the advancement of applied and experimental entomology, and in the development of future pest control strategies.

Insect rearing will continue to benefit mankind, as it has done in the past by protecting his crops, livestock and health.

REFERENCES

Anonymous 1964. Symposium on Culture Procedures for Arthropod Vectors and their Biological Control Agents. Bull. WHO 31, 622 pp.

Anonymous 1980. General Entomology Vol. 1. Agricultural Publishing House, Beijing, Peoples Republic of China, 425 pp.

Bogdanow, E.A. 1908. Über die Abhängigkeit des Wachstums der Fliegenlarven von Bakterien und Fermenten und über Variabilität und Vererbung bei den Fleischfliegen. Arch. Anat. Physiol. Abt. Suppl., 173-200.

Bush, G.L. 1978. Planning a Rational Quality Control Program for the Screwworm Fly. pp. 37-47 in Richardson, R.H. (Ed.) The Screwworm Problem: Evaluation of Resistance to Ecological Control. University of Texas Press, Austin, Texas.

Carne, P.B.; Crawford, L.D.; Fletcher, M.J.; Galloway, I.D. and Highley, E. (Eds) 1980. Scientific and Common Names of Insects and Allied Forms Occurring in Australia. Commonwealth Scientific and Industrial Research Organization, Australia, 95 pp.

Chou, I. 1980. A History of Chinese Entomology. Entomotaxonomia. Wugong, Shaanxi, Peoples Republic of China, 213 pp.

Free, J.B. 1982. Bees and Mankind. George Allan & Unwin, London, Boston, Sydney, 155 pp.

King, E.G. and Leppla, N.C. (Eds) 1984. Advances and Challenges in Insect Rearing. U.S. Department of Agriculture Technical Bulletin, 306 pp.

Knipling, E.F. 1984. What Colonization of Insects Means to Research and Pest Management. pp. ix-xi in King, E.G. and Leppla, N.C. (Eds) Advances and Challenges in Insect Rearing. U.S. Department of Agriculture Technical Bulletin.

Miller, D. 1952. The Insect People of the Maori. J. Polynesian Soc. 61 (1 & 2): 1-61.

Needham, J.G *et al.* (Eds) 1959. Culture Methods for Invertebrate Animals. Dover Publications Inc., New York, 590 pp.

Singh, P. 1977. Artificial Diets for Insects, Mites and Spiders. Plenum Press, New York, 594 pp.

Singh, P. 1984. Insect Diets: Historical Developments, Recent Advances, and Future Prospects. pp. 32-44 in King, E.G. and Leppla, N.C. (Eds) Advances and Challenges in Insect Rearing. U.S. Department of Agriculture Technical Bulletin.

Smith, C.N. (Ed.) 1966. Insect Colonization and Mass Production. Academic Press, New York and London, 618 pp.

Smith, R.F.; Mittler, T.E.; Smith, C.N. (Eds) 1973. History of Entomology. Annual Review Inc., Palo Alto, CA., 517 pp.
Southwood, T.R.E. 1977. Entomology and Mankind. Am. Sci. 65: 30-39.
Tsai, J.H. 1982. Entomology in the Peoples Republic of China. N.Y. Entomol. Soc. 90: 186-212.
Werner, F.G. (Chairman) 1982. Common Names of Insects and Related Organisms. Entomological Society of America, 132 pp.

AEDES AEGYPTI

LEONARD E. MUNSTERMANN and LORETTA M. WASMUTH

Vector Biology Laboratory, Department of Biology
University of Notre Dame, Notre Dame IN 46556, U.S.A.

THE INSECT

Scientific Name:	*Aedes* (*Stegomyia*) *aegypti* (Linnaeus)
Common Name:	Yellow Fever Mosquito
Order:	Diptera
Family:	Culicidae

1. INTRODUCTION

The genus *Aedes* now contains more than 1000 species and is found throughout the world. Among the medically important members of this genus, *Aedes aegypti* is the most renowned, not only as a disease vector but also as a laboratory research animal. The distribution of *Aedes aegypti* is confined only to the frost-free zone between approximately 45° North and South latitudes (Christophers, 1960). As a "container breeding" mosquito, it is closely associated with man, and, as man's production of throw-away containers and worn tires has proliferated, so have the populations of this species.

As the common name implies, this mosquito has been historically important in transmission of yellow fever among humans. Today, it is much better known as a vector of the viral disease, dengue-hemorrhagic (breakbone) fever. This syndrome is endemic in Southeast Asia, and beginning in the late 1970's, an epidemic has raged in Latin America and the West Indies involving hundreds of thousands of reported cases.

The life cycle of *Aedes aegypti* is continuous, regulated only by the availability of water-filled containers. The females, after taking a bloodmeal, produce drought-resistant eggs and oviposit them singly just above the water line. When the water level increases, the eggs become submerged and hatch. In the laboratory at 27° C, pupation occurs in 5 days, and the life cycle can be completed in 18 days.

As a stenogamous mosquito (i.e., mates in confined spaces), it is easily colonized from field collected material and hence is the most commonly used mosquito for laboratory studies. The ROCK strain, originating from the Rockefeller Institute, has been in continuous colony for nearly 50 years and is truly the "white rat" of *Aedes aegypti*. Our laboratory has reared hundreds of strains collected from around the world. Only in the case of the sylvan subspecies, *Ae. aegypti formosus*, has difficulty been encountered with blood-feeding and oviposition. The rearing techniques now standard in our laboratory have evolved from those detailed by Craig and VandeHey (1962), though numerous alternate diets and procedures have been used successfully (Gerberg, 1970).

A list of other *Aedes* species successfully reared in the Notre Dame Vector Biology Laboratory is appended (Section 9).

2. FACILITIES AND EQUIPMENT REQUIRED

2.1 Facilities

The entire active life cycle is carried out in controlled environment rooms maintained at 27 ±2° C, 75-80% RH and 18:6 LD photoperiod (light intensity not critical).

2.2 Equipment and materials required for insect handling

2.2.1 Adult oviposition
- 450 ml plastic or wax coated paper cups
- Tap water
- Textured paper toweling

2.2.2 Egg collection and storage
- 20 x 30 x 8 cm plastic vegetable crispers
- 150 ml beaker
- Potassium sulfate powder
- Controlled environment chamber maintained at 15° C

2.2.3 Egg hatch
- Tap water
- 450 ml plastic coated paper cup
- Liver powder suspension

2.2.4 Larval rearing
- 14.6 cm Pasteur type pipets
- Rubber pipet bulb
- Hand tally counter
- 15 x 100 mm petri dish
- Dark surface
- Adjustable lamp
- Liver powder suspension
- 30 cm diam enamel dish pans
- Plexiglass® squares 40 x 40 x 3 cm

2.2.5 Pupal collection and holding
- Nalgene® dropping pipettes 10 cm in length
- 450 ml plastic coated paper cups (Nestyle)
- Cotton balls
- Paper toweling
- Neptune straight-sided 450 ml plastic coated paper cups
- Black nylon tulle netting
- Honey pads

2.2.6 Adult cages
- 3.6 or 15.5 liter white plastic buckets and lids
- Black tulle netting
- Razor knife
- Plier stapler
- Stockinette

2.3 Diet preparation

2.3.1 Larval diet
- Liver powder
- Weighing balance
- 1000 ml Erlenmeyer dispensing flask with 10 ml volumetric head
- Tap water

2.3.2 Adult diet

a. Maintenance
- Cellucotton (absorbent wadding)
- Honey
- 30 cm diam enamel dish pan

b. For oviposition
- Standard strain white laboratory mice 25+ g
- 15 x 15 cm pieces wire window screen
- 2.5 cm wide paper tape

2.4 Brand names and source of supplies

Liver Powder (ICN Nutritional Biochemicals, P.O. Box 28050, Cleveland, OH 44128); honey (commercially processed, locally retailed); Nestyle "16 oz" (450 ml), tall container without lid (Sealright Co., Inc., Fulton, NY 13069); Palmer® No.244 Sof-embossed® Handifold® towels (Fort Howard Paper Co., Green Bay, WI 54305); 30 cm diam enamel dish pans (No.32 basin, General Housewares Corp., P.O. Box 4066, Terre Haute, IN 47803); dropping pipets (Nalge Co.,

Division of Sybron Corp., Rochester, NY 14602); "chimney" (Neptune heavy duty pint [450 ml] container with rolled rim, Neptune Paper Products, Inc., Jamaica, Long Island, NY); plastic buckets (Imperial Plastics, Inc., 101 Oakley St., P.O. Box 959, Evansville, IN 47706-0959, 8 or 34 lb [3.6 or 15.5 liter] size plastic container with lid); stockinette (Tomac® tubular stockinette, American Hospital Supply, McGaw Park, IL 60084); tulle netting (purchased from local fabric store); absorbent wadding (Curity®, Kendall Hospital Products Division, Boston, MA 02110); absorbent cotton or rayon balls (purchased from local physicians supply store)

3. PROCEDURES

3.1 Diet preparation

3.1.1 Larval diet

- Weigh 60 g liver powder into 1000 ml flask
- Mix the powder with 250 ml warm tap water
- Bring volume to 1000 ml with cool tap water
- Let stand until foam subsides (note: the resulting suspension must be agitated before each use)
- Store at 5° C; discard after 2 weeks

3.1.2 Adult diet

a. Honey pads for maintenance

- Shred cellucotton (50 x 50 cm) into 5-10 cm strips and place in 30 cm diam enamel dish pan
- Moisten with 200 ml tap water
- Add 1 kg honey (approx. 2 lb) and knead until a uniform consistency is obtained
- Store at 5° C
- Discard if signs of fermentation (odor) appear
- Sugar cube, raisin, or 15% sucrose solution absorbed in a cotton ball can be used as a substitute for the honey mixture

b. Blood supply for oviposition (Fig. 1)

- Place tape around edges of window screen to enclose rough edges
- Fold screen in half to make an "envelope"
- Place mouse inside envelope and fold the taped edges together to secure the mouse and keep it immobile

3.2 Cage construction - (Fig. 2A)

- Using the razor knife, remove the center of the bucket lid leaving 2-3 cm of the rim intact
- Cut a 14 cm diam hole in the side of the bucket
- Stretch a 40 cm long piece of stockinette around the hole
- Secure the stockinette with a continuous row of staples
- Tape a paper towel to the inside of the cage to provide a resting place for the adults
- Cut a piece of nylon tulle several cm larger than the lid
- Place tulle over bucket and secure with lid (rim)

4. COLONY MAINTENANCE

4.1 Egg hatch

- Place egg paper into paper cup
- Add 400 ml tap water
- Add 5 ml liver powder suspension (3.1.1)
- Hatching will begin in 1-4 h
- Separate larvae into pans within 24 h

4.2 Larval separation and rearing

- Place open half of petri dish on dark surface
- Adjust lamp to shine on one side of dish
- Pour water containing 1st instars into dish
- As larvae swim away from light source, pick them out with a Pasteur pipet
- Place 300 larvae in enamel pan containing 1 liter tap water

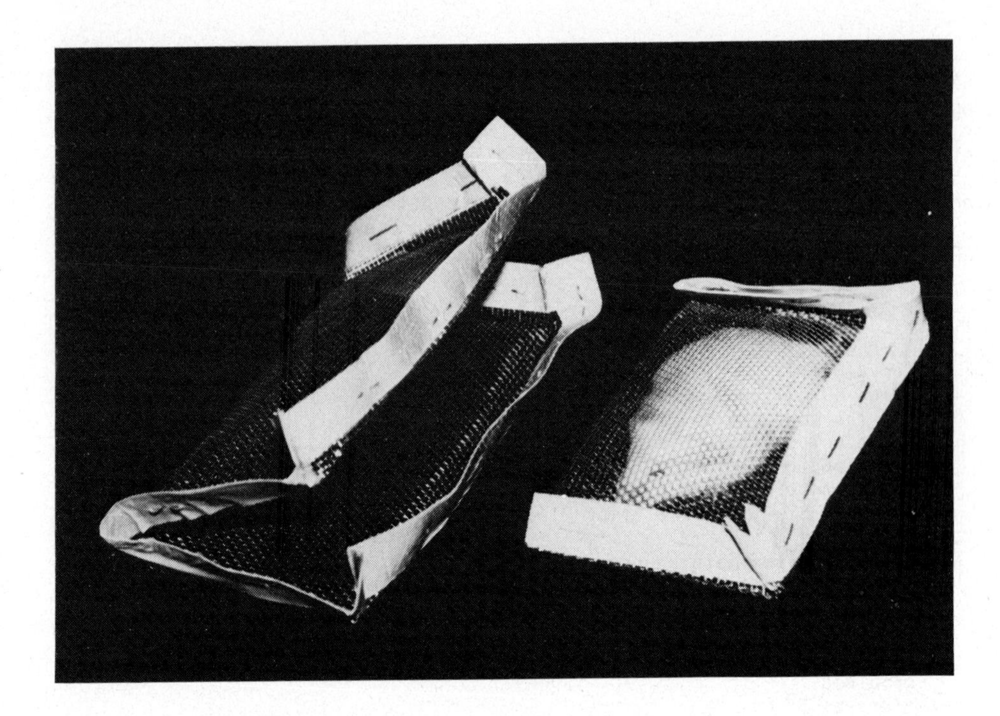

Fig. 1. Mouse "envelope" used to immobilize mouse for bloodfeeding the adult female mosquitoes.

- Add 10 ml liver powder suspension (day 1)
- Cover pan with Plexiglass® square
- Add additional 10 ml liver powder every other day until pupation (if scum forms on water surface, remove by skimming with paper towel)
- Pupation begins on day 5

4.3 Pupal collection

4.3.1 General colony maintenance
- Use razor blade to enlarge hole at tip of plastic pipet allowing pupae to enter without damage
- Remove pupae from pan with pipet
- Place pupae in a paper cup half full of water
- Put cup directly into colony cage (adults will emerge from water surface into cage)
- Maximum of 300 mosquitoes in 8 lb (3.6 liter) size cage; 1500 for the 34 lb (15.5 liter) size
- Pupae must be removed from pans every 48 h

4.3.2 Holding cages

4.3.2.1 Prepare emergence containers
- Place 4 cotton balls in the bottom of a 450 ml paper cup (Nestyle)
- Add 100 ml tap water
- Place a square of paper toweling on top of the cotton balls
- Turn up the edges of the towel to prevent pupae from swimming beneath towel and being caught in the cotton

4.3.2.2 Prepare "chimneys"
- Punch out the top and bottom from the straight-sided cups (Nestyle)
- Place a square of black nylon tulle on top of the cup, securing it

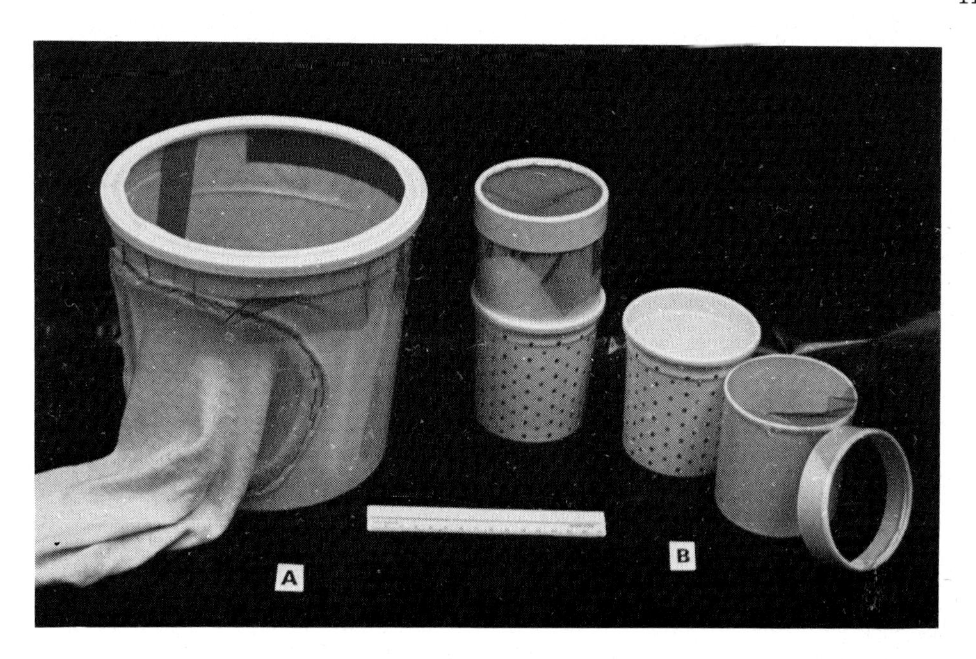

Fig. 2. Cages. A. Gallon-size (3.6 liter) cages for mating and oviposition. B. Pint-size (450 ml) cages for adult emergence and pre-experiment maintenance.

with the rim

4.3.2.3 Assembly - (Fig. 2B)
- Place the pupae on the towel in the emergence container (maximum 50)
- Place chimney firmly on top of cup, wedging the two cups together
- Put honey pad on top
- Adults emerging in container can be held for several weeks
- Change honey pad at least once per week

4.4 Adults

4.4.1 Emergence period is 24-48 h post pupation; sex ratio is approximately 1:1

4.4.2 Mating occurs freely in small cages

4.4.3 Oviposition

4.4.3.1 Blood feeding
- Place mouse in "envelope" on top of cage until most females have fed (observe blood-engorged abdomen)
- Refeed cages every 4th day for colony maintenance
- Females will oviposit 2-3 days post blood meal

4.4.3.2 Oviposition media
- Line inside wall of paper cup with paper towelling
- Fill the cup half full with tap water
- Put cup inside cage with blood fed mosquitoes

4.5 Egg maturation
- Remove egg containers from cage
- Empty all but 1 cm water from cup
- Leave container, loosely covered, in insectary 3 days to complete embryonation
- Allow egg paper to air dry completely
- Put beaker of saturated potassium sulfate solution in vegetable crisper
- Remove egg paper from cup, label with date and strain name, put in

crisper and replace crisper lid
- Store at 15° C for 3-6 mo (note: extreme care must be taken to prevent contamination of strains with stray eggs that may stick to fingers)

4.6 Sex determination
- Female pupae and adults are usually larger than the males
- The plumose antennae of the male adult is densely hairy and conspicuous; that of the female is scarcely visible

4.7 Special handling of adults

4.7.1 Etherizing
- Saturate cotton ball with ethyl ether
- Place on top of cage or chimney
- Cover container with plastic for 1-2 min or until adults drop to bottom of container and appear relaxed
- Mosquito can then be handled by the legs using watchmaker forceps

4.7.2 Mouth aspirator
- Glue a small piece of plastic window screen to cover one end of a 30 cm length of plastic or glass tubing with an inside diam of 10 mm
- Insert the screened end into a length of rubber tubing
- Place the free end of rubber tubing into the mouth and inhale to draw adults into the section of glass tube
- Exhale to remove the adults from the tube and into the receiving container

5. REARING SCHEDULE

a. Weekly
- Bloodfeed females twice a week
- Collect eggs
- Replace oviposition containers
- Replace honey pad
- Hatch eggs as necessary to maintain colony cages

b. Daily
- Check pans for pupae, nutrient level, and scum formation

6. LIFE CYCLE

Life cycle data for developmental stages under optimum rearing conditions of 27° C, 80% RH and 18:6 LD photoperiod are given (Craig and VandeHey, 1962; Crovello and Hacker, 1972)
- Embryonation of egg 72 h
- Maximum hatch 24 h
- No. of larval instars 4
- Pupation begins on day 6
- Adult emergence begins on day 8
- Mean no. of eggs laid per female in one egg batch 85
- Oviposition 4 days post blood meal
- Mean male lifetime 10 days (min 2, max 45)
- Mean female lifetime 21 days (min 2, max 70)
- Life cycle 17-20 days (egg to egg)

7. PROCEDURES FOR SUPPLYING INSECTS
- Since the eggs are drought resistant, embryonated egg papers sealed in a plastic sandwich bag can be placed in an envelope and sent via regular postal service
- Before international orders can be filled, the appropriate government shipping permits must be obtained
- Eggs of all species maintained at the Vector Biology Lab, including those listed in Section 9 may be obtained by

writing the Director, PROF. GEORGE B. CRAIG, JR. at the following address:

VECTOR BIOLOGY LABORATORY
DEPARTMENT OF BIOLOGY
UNIVERSITY OF NOTRE DAME
NOTRE DAME IN 46556, U.S.A.

8. REFERENCES

Christophers, S.R., 1960. _Aedes aegypti_ (L.), the Yellow Fever Mosquito: Its Life History, Bionomics and Structure. London: Cambridge University Press. 739 pp.

Craig, G.B., Jr., VandeHey, R.C., 1962. Genetic variability in _Aedes aegypti_ (Diptera: Culicidae) I. Mutations affecting color pattern. Ann. Entomol. Soc. Amer. 55: 47-58.

Crovello, T.J., Hacker, C.S., 1972. Evolutionary strategies in life table characteristics among feral and urban strains of _Aedes aegypti_ (L.). Evolution 26: 185-196.

Gerberg, E.J., 1970. Manual for mosquito rearing and experimental techniques. Amer. Mosq. Contr. Assoc. Bull. No. 5, 109 pp.

9. APPENDIX

Species reared by _Aedes aegypti_ method at the Vector Biology Laboratory, University of Notre Dame.

Aedes aegypti - mates readily in small cages, oviposits on paper towelling, feeds easily on mice; eggs store 6-12 mo
Aedes albopictus - occasionally shows slight autogeny
Aedes alcasidi
Aedes annandalei
Aedes atropalpus - feeds on mice after first autogenous egg batch. Larvae are sensitive to overfeeding. Usually requires oviposition stimulus of old pupal skins before it will oviposit on paper towelling.
Aedes bahamensis
Aedes cooki - 3 mo maximum egg storage
Aedes epactius - needs oviposition stimulus of old pupal skins
Aedes flavopictus - 3 mo maximum egg storage
Aedes furcifer - oviposits on paper creased accordian style; store eggs 17-20 days before hatch and reflood 4-5 times to hatch eggs; eggs can keep for 1 yr
Aedes gardnerii imitator - 3 mo maximum egg storage
Aedes hebrideus - 3 mo maximum egg storage
Aedes heischi
Aedes kesseli - extremely difficult to maintain
Aedes katherinensis - 3 mo maximum egg storage
Aedes malayensis - 3 mo maximum egg storage
Aedes malikuli - 3 mo maximum egg storage
Aedes marshallensis - 3 mo maximum egg storage
Aedes mascarensis
Aedes mediopunctatus
Aedes mediovittatus - does not cage mate readily; hold adults for 8-10 days before placing oviposition container into cage
Aedes metallicus - must reflood eggs a second time for a good hatch
Aedes muelleri - 3 mo maximum egg storage
Aedes pernotatus - 3 mo maximum egg storage
Aedes polynesiensis
Aedes pseudalbopictus - 3 mo maximum egg storage
Aedes pseudoscutellaris - 3 mo maximum egg storage
Aedes riversi - 3 mo maximum egg storage
Aedes seatoi
Aedes sierrensis - hatch with nutrient broth method

Aedes taeniorhynchus
Aedes terrens
Aedes togoi
Aedes tongae tabu - 3 mo maximum egg storage
Aedes vittatus

ACKNOWLEDGEMENTS

This work was supported by NIH Grant No. AI-02753 to Prof. G. B. Craig, Jr. We thank Dr. J. Freier for rearing information on several of the species listed in the Appendix. Ms. Peggy Hodges and Ms. Ernestine Hodges did the preliminary testing for many of the procedures listed here.

AEDES TRISERIATUS

LEONARD E. MUNSTERMANN and LORETTA M. WASMUTH

Vector Biology Laboratory, Department of Biology
University of Notre Dame, Notre Dame IN 46556, U.S.A.

THE INSECT

Scientific Name: *Aedes* (*Protomacleaya*) *triseriatus* (Say)
Common Name: Eastern Tree Hole Mosquito
Order: Diptera
Family: Culicidae

1. INTRODUCTION

As the common name implies, this mosquito has a larval habitat largely restricted to rot holes in trees. Its geographical range is confined to the continental United States and southern edge of Canada, extending from Florida to Maine in the East and as far west as the wooded regions of the middle Great Plains states. Discarded containers and tires are a facultative larval habitat, particularly if shaded and containing high levels of organic nutrients.

This species came into prominence as a laboratory animal relatively recently. It has been confirmed as a vector of California Encephalitis (La Crosse virus), a disease particularly affecting young children. In wooded rural and surburban areas, this mosquito is also a probable vector of dog heartworm (*Dirofilaria immitis*). It is a persistent summertime pest in woodland recreational areas and parks because of the female's taste for human blood, though usually she feeds on other sylvan creatures (Nasci, 1982).

Unfortunately, *Aedes triseriatus* is not so easily colonized from field-collected specimens. Maintenance of field strains almost always requires insemination of the females by laborious induced copulation techniques. However, one strain (WALTON) was colonized in the early 1970's and now is the strain normally used in laboratory experiments since it can complete its life cycle in enclosed cages (at least 60 x 60 x 60 cm) without difficulty.

In the field, *Aedes triseriatus* seldom undergoes more than 2 generations annually. An egg diapause protects it through the cold and/or dry seasons. In the laboratory, the life cycle is much longer than that of *Aedes aegypti*, primarily because it requires cooler temperatures (21° C) for optimal larval growth and adult survivorship. At this temperature, the 4 larval stages take 14 days to pupation, 4-5 days to adulthood, and at least a week until the first oviposition. The eggs require an additional week or more for embryonation before hatching. As a result, the laboratory life cycle is usually 5-6 weeks duration. Special procedures essential for successful rearing of this mosquito are cool insectary temperatures along with a summer photoperiod, induced (hand-mating) copulation, attractive oviposition substrate and egg storage.

A list of other *Aedes* species reared at the Notre Dame Vector Biology Laboratory by the *Ae. triseriatus* method is appended (Section 9).

2. FACILITIES AND EQUIPMENT REQUIRED

2.1 Facilities

The entire active life cycle is carried out in controlled environment rooms maintained at 21 $\pm 2^{o}$ C, 75-80% RH and 18:6 LD photoperiod (light intensity not critical).

2.2 Equipment and materials required for insect handling

2.2.1 Adult oviposition

2.2.1.1 Group

- 354 ml aluminum beverage container
- Black 100% cotton percale cloth (prewashed and rinsed 3 times in tap water)
- Balsa wood strips, 2 mm thick, soaked in tap water for several days

2.2.1.2 Single female

- Balsa wood strips, 25 mm x 75 mm (2 mm thick), soaked in tap water for several days
- Glass shell vials 22 mm i.d. x 80 mm
- 90 ml wax-coated paper cup
- Tulle netting
- Rubber bands
- Vial rack
- Razor knife

2.2.2 Egg collection and storage

- 20 x 30 x 8 cm plastic vegetable crispers
- Plastic sandwich bags
- Paper towels

2.2.3 Egg hatch

- Nitrogen gas cylinder with pressure regulator
- Nutrient broth
- Weighing balance
- 2 liter flask
- Rubber or Tygon® tubing

2.2.4 Larval rearing

- 14.6 cm Pasteur type pipets
- Rubber pipet bulb
- Hand tally counter
- 15 x 100 mm petri dish
- Dark surface
- Adjustable lamp
- Liver powder suspension
- 30 cm diameter enamel dish pans
- Plexiglass® squares 40 x 40 x 3 cm

2.2.5 Pupal collection and holding

- Nalgene® dropping pipettes 10 cm in length
- 450 ml plastic coated paper cups (Nestyle)
- Cotton balls
- Paper toweling
- Neptune straight-sided 450 ml plastic coated paper cups
- Black nylon tulle netting
- Honey pads

2.2.6 Adult cages

- 3.6 or 15.5 liter white plastic buckets and lids
- Black tulle netting
- Razor knife
- Plier stapler
- Stockinette

2.2.7 Induced copulation

- Cotton balls
- Container of crushed ice
- 23 x 80 mm shell vials
- Stereo dissection microscope
- Wood applicator sticks, plain, 14 cm long, long grain birchwood
- Minutien pins
- Fine watchmaker forceps
- Flat tray
- Glass tubing
- Rubber or Tygon® tubing

- Pinch clamps
- Depression slides
- Glass rods to fit into flat tray
- Nitrogen gas cylinder with pressure regulator

2.3 Diet preparation

2.3.1 Larval diet
- Liver powder
- Weighing balance
- 1000 ml Erlenmeyer dispensing flask with 10 ml volumetric head
- Tap water

2.3.2 Adult diet

a. Maintenance
- Cellucotton (absorbent wadding)
- Honey
- 30 cm diam enamel dish pan

b. For oviposition
- Standard strain white laboratory mice 25+ g
- 15 x 15 cm pieces wire window screen
- 2.5 cm wide paper tape

2.4 Brand names and source of supplies

Liver Powder (ICN Nutritional Biochemicals, P.O. Box 28050, Cleveland, OH 44128); honey (commercially processed, locally retailed); Nestyle "16 oz" (450 ml), tall container without lid (Sealright Co., Inc., Fulton, NY 13069); Palmer® No.244 Sof-embossed® Handifold towels (Fort Howard Paper Co., Green Bay, WI 54305); 30 cm diam enamel dish pans (No.32 basin, General Housewares Corp., P.O. Box 4066, Terre Haute, IN 47803); dropping pipets (Nalge Co., Division of Sybron Corp., Rochester, NY 14602); "chimney" (Neptune heavy duty pint [450 ml] container with rolled rim, Neptune Paper Products, Inc., Jamaica, Long Island, NY); plastic buckets (Imperial Plastics, Inc., 101 Oakley St., P.O. Box 959, Evansville, IN 47706-0959, 8 lb or 34 lb [3.6 or 15.5 liter] size plastic container with lid); stockinette (Tomac® tubular stockinette, American Hospital Supply, McGaw Park, IL 60084); tulle netting (purchased from local fabric store); absorbent wadding (Curity®, Kendall Hospital Products Division, Boston, MA 02110); absorbent cotton or rayon balls (purchased from local physicians supply store); Balsa (Joy Products, 201 3rd Ave, Box 374, Menominee, MI 49858); nutrient broth (Difco Labs, Detroit, MI 48201); forceps (Electron Microscopy Services, Box 251, Fort Washington, PA 19034; No.5 Dumont Tweezers); minutien pins (Clair Armin, 191 W. Palm Ave., Reedley, CA 93654); No.37 Lily® cup (Lily Tulip Division of Owens-Illinois Co., Toledo, OH); Wood applicator sticks (American Scientific Products, 1210 Waukegan Rd., McGaw Park, IL 60085)

3. PROCEDURES

3.1 Diet preparation

3.1.1 Larval diet
- Weigh 60 g liver powder into 1000 ml flask
- Mix the powder with 250 ml warm tap water
- Bring volume to 1000 ml with cool tap water
- Let stand until foam subsides (note: the resulting suspension must be agitated before each use)
- Store at 5° C; discard after 2 weeks

3.1.2 Adult diet

a. Honey pads for maintenance
- Shred cellucotton (50 x 50 cm) into 5-10 cm strips and place in 30 cm diam enamel dish pan
- Moisten with 200 ml tap water
- Add 1 kg honey (approx. 2 lb) and knead until a uniform consistency is obtained
- Store at 5° C
- Discard if signs of fermentation (odor) appear
- Sugar cube, raisin, or 15% sucrose solution absorbed in a cotton ball

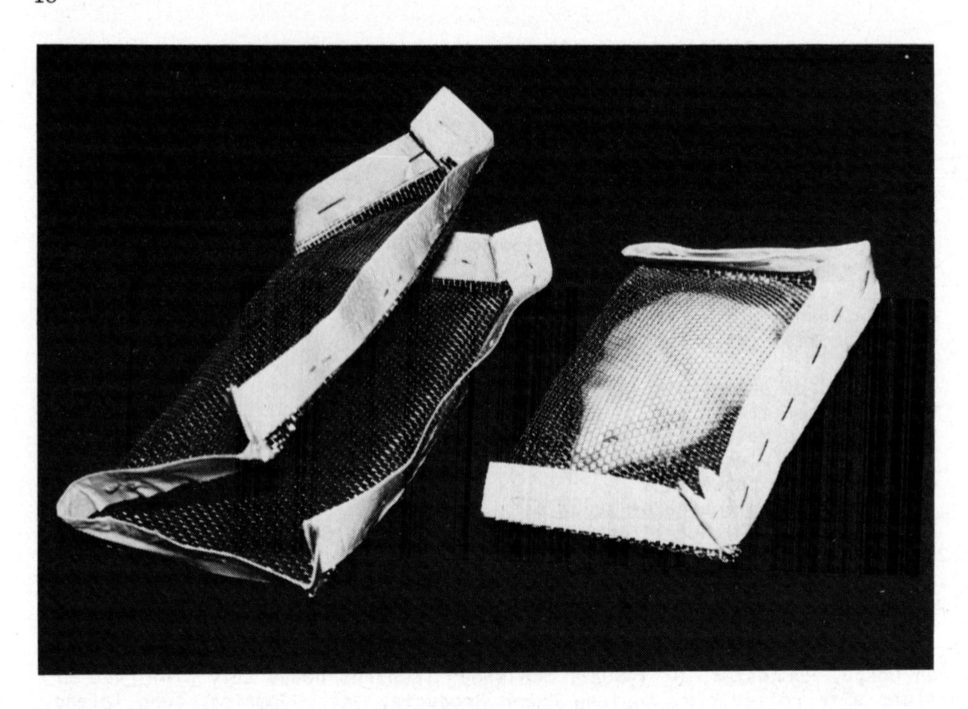

Fig. 1. Mouse "envelope" used to immobilize mouse for bloodfeeding the adult female mosquitoes.

can be used as substitute for the honey mixture

b. Blood supply for oviposition (Fig. 1)

- Place tape around edges of window screen to enclose rough edges
- Fold screen in half to make an "envelope"
- Place mouse inside envelope and fold the taped edges together to secure the mouse and keep it immobile

3.2 Cage construction - (Fig. 2A)

- Using the razor knife, remove the center of the bucket lid, leaving 2-3 cm of the rim intact
- Cut a 14 cm diameter hole in the side of the bucket
- Stretch a 40 cm long piece of stockinette around the hole
- Secure the stockinette with a continuous row of staples
- Tape a paper towel to the inside of the cage to provide a resting place for the adults
- Cut a piece of nylon tulle several cm larger than the lid
- Place tulle over bucket and secure with lid (rim)

3.3 Hatching media preparation

- Dissolve 1 g nutrient broth powder in 1000 ml tap water
- Attach tubing to pressure regulator on nitrogen gas cylinder (Fig. 3)
- Bubble nitrogen into nutrient broth solution for 15 min

3.4 Micro dissection needle construction

- Soak end of applicator stick in tap water several hours to soften
- Using small pliers, insert minutien pin into end of stick
- Allow to dry before use

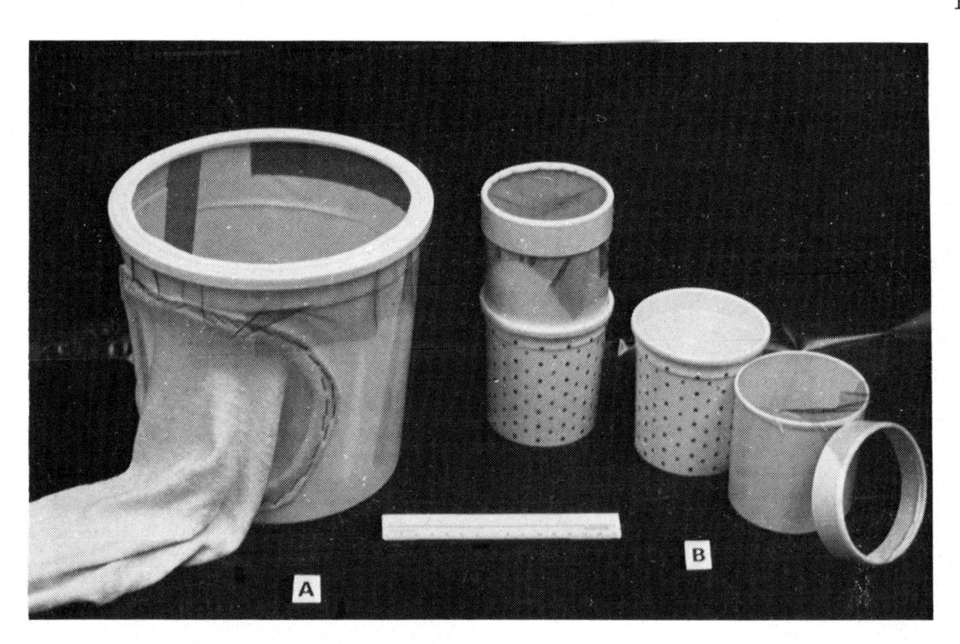

Fig. 2. Cages. A. Gallon-size (3.6 liter) cages for mating and oviposition.
B. Pint-size (450 ml) cages for adult emergence and pre-experiment maintenance.

4. COLONY MAINTENANCE

4.1 Egg hatch
- Place balsa egg strip and/or cloth into paper cup
- Submerse in hatching media (3.3)
- Leave egg strip and larvae in nutrient broth no longer than 24 h

4.2 Larval separation and rearing
- Place open half of petri dish on dark surface
- Adjust lamp to shine on one side of dish
- Pour water containing 1st instars into dish
- As larvae swim away from light source, pick them out with a Pasteur pipet
- Place 300 larvae in enamel pan containing 1 liter tap water
- Add 10 ml liver powder suspension (day 1)
- Cover pan with Plexiglass® square
- Add additional 10 ml liver powder every 5th day until pupation
- Pupation begins on day 14

4.3 Pupal collection

4.3.1 General colony maintenance
- Use razor blade to enlarge hole at tip of plastic pipet allowing pupae to enter without damage
- Remove pupae from pan with pipet
- Place pupae in a paper cup half full of water
- Put cup directly into colony cage (adults will emerge from water surface into cage)
- Maximum of 300 mosquitoes in 8 lb (3.6 liter) size cage; 1500 for the 34 lb (15.5 liter) size
- Pupae must be removed from pans every 72 h

4.3.2 Holding cages

4.3.2.1 Prepare emergence containers

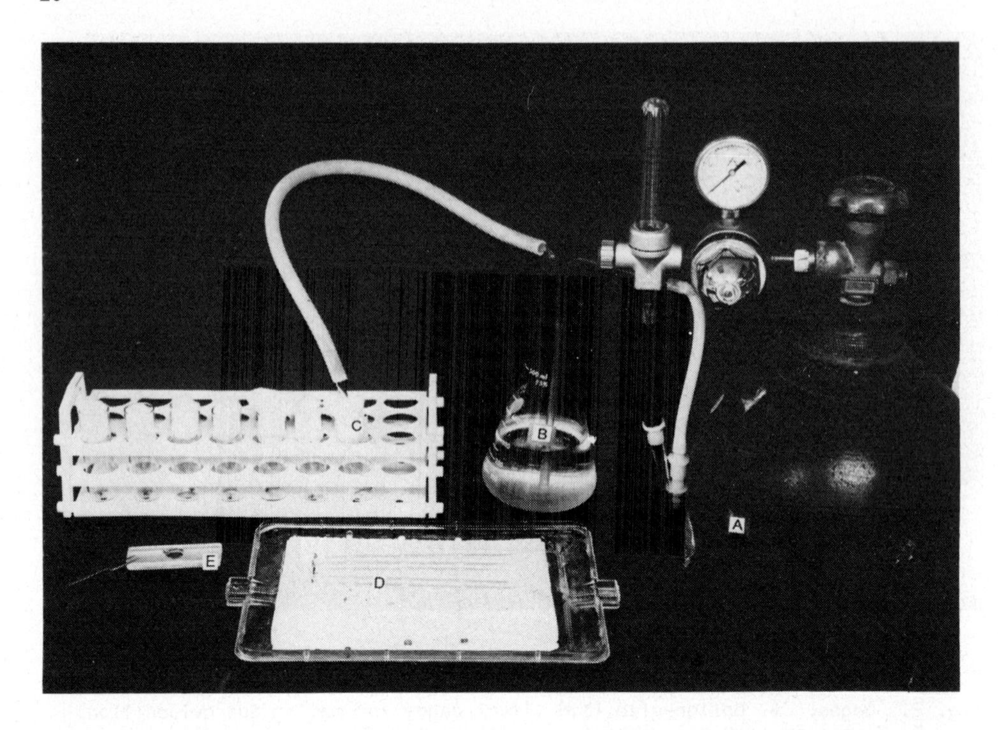

Fig. 3. Setup for induced copulation. A. Nitrogen gas cylinder with regulator B. Bubbling chamber for saturating the gas with water vapor C. Glass vials for anaesthetizing the females D. Tray of males prepared for copulation procedure E. Well slide as the mating couch.

- Place 4 cotton balls in the bottom of a 450 ml paper cup (Nestyle)
- Add 100 ml tap water
- Place a square of paper toweling on top of the c
- Turn up the edges of the towel to prevent pupae from swimming beneath towel and being caught in the cotton

4.3.2.2 Prepare "chimneys"
- Punch out the top and bottom from the straight-sided cups (Neptune)
- Place a square of black nylon tulle on top of the cup, securing it with the rim

4.3.2.3 Assembly - (Fig. 2B)
- Place the pupae on the towel in the emergence container (maximum 50)
- Place chimney firmly on top of cup, wedging the two cups together
- Put honey pad on top
- Adults emerging in container can be held for several weeks
- Change honey pad at least once per week

<u>4.4 Adults</u>

4.4.1 Emergence period is 4-5 days post pupation; sex ratio is approximately 1:1

4.4.2 Induced copulation

4.4.2.1 Blood feed females
- Place mouse in "envelope" on top of cage until most females have fed (observe blood-engorged abdomen) (Fig. 1)
- Place fed females individually in shell vials
- Plug vial opening with cotton

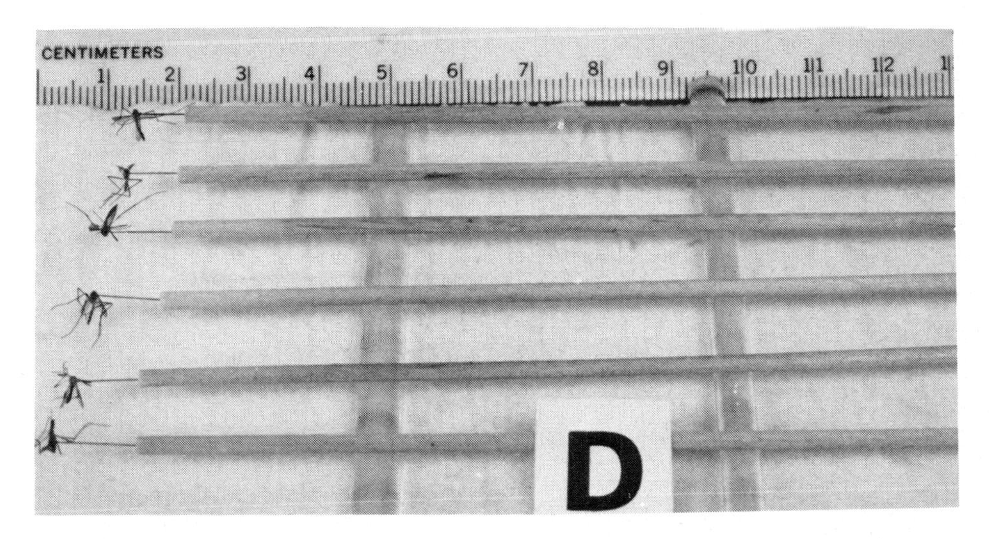

Fig. 4. Enlargement of the tray of males prepared for induced copulation.

4.4.2.2 Prepare males
- Place males in individual shell vials, plugging vial opening with a cotton ball
- Place several vials of males into container of crushed ice until they become immobilized
- Remove the male from vial, and using a stereo dissection microscope, impale the mosquito with the micro dissection needle through the middle of the thorax from the left to right side (for a right-handed person)
- Pull off the head with forceps
- Place the prepared male on glass rods over moist towelling to hold for subsequent copulation (Fig. 3D, 4)

4.4.2.3 Anesthetize females with nitrogen immediately prior to mating
- Fit 500 ml flask containing 300 ml water with a 2-hole rubber stopper
- Insert one piece of glass tubing thru hole in stopper to bottom of flask
- Insert a second piece of glass tubing above the water level
- Attach rubber or Tygon® tubing from the gas regulator to the first glass tube (Fig. 3A)
- Attach another piece of rubber tubing to the second glass tube (Fig. 3B)
- Insert a Pasteur pipet into the free end of tubing
- Adjust gas pressure to give a gentle bubbling effect through the water in the flask
- Insert pipet into a vial containing one female (do not remove cotton plug) (Fig. 3C)
- Female will be anesthetized in 1-2 min

4.4.2.4 Mating
- Place anesthetized female, ventral side up, on depression slide with the genitalia extending over the edge of the depression (Fig. 3E)
- Bring the genitalia of the male and female together, venters facing each other, at a 45° angle until the female genitalia are firmly clasped by the male
- Leave in this position until the female is released
- Put mated females into cage containing an oviposition can

4.4.3 Oviposition

4.4.3.1 Blood feeding
- Place mouse in "envelope" on top of cage until most females have fed

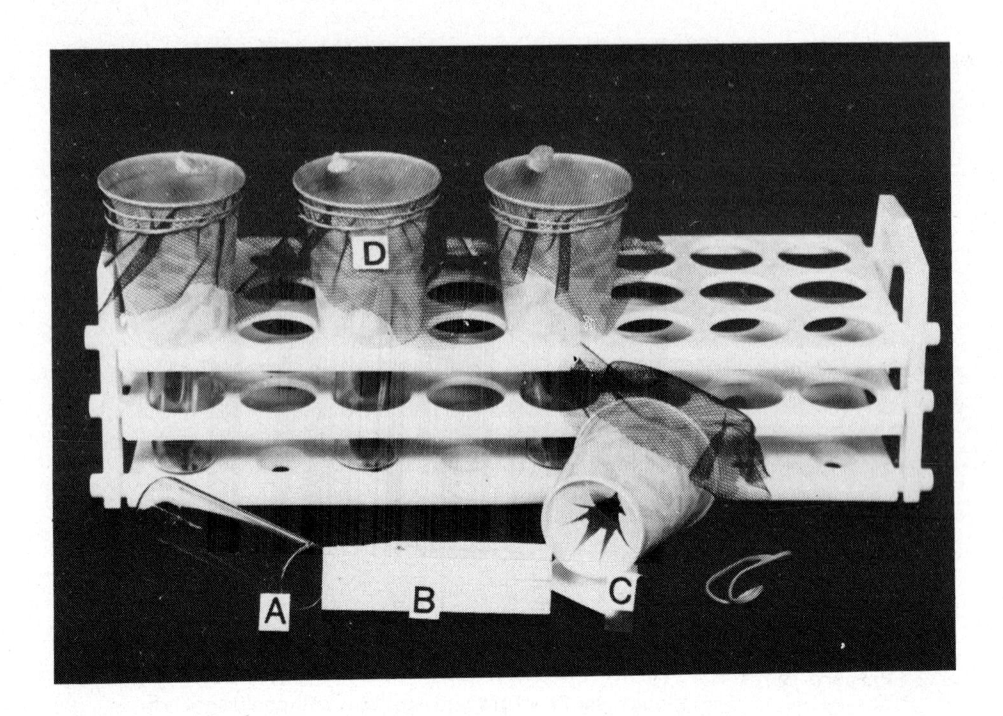

Fig. 5. Individual female oviposition chambers. A. Glass vial B. Balsa wood oviposition substrate C. Dixie cup cover chamber and net D. Assembled chamber.

(observe blood-engorged abdomen)
- Refeed caged every 5th day for colony maintenance
- Females will oviposit 3-4 days post blood meal

4.4.3.2 Oviposition media - group rearing
- Line inside wall of beverage container with black cloth
- Secure with 2 paper clips
- Insert balsa strip (5 cm x 10 cm) and secure with clip
- Fill container half full with tap water
- Place container in cage with mated females

4.4.3.3 Oviposition media - single female
- Use razor knife to slit bottom of 90 ml cup
- Make 2 perpendicular slits across the bottom
- Make 2 additional slits dividing the bottom into 8 pie-shaped sections (Fig. 5C)
- Cut piece of netting to fit over top of cup and secure it with a rubber band
- Place balsa strip in shell vial (Fig. 5A and B)
- Fill vial half full with tap water
- Label vial and place in vial rack
- Place mated female into cup through slits
- Invert cup and place over vial in rack, taking care to leave space between netting and top of vial (Fig. 5D)
- Place small honey pad on top

<u>4.5 Egg maturation</u>
- Remove egg container from cage
- Leave container, loosely covered, in insectary for 5 days to complete

embryonation
- Remove cloth and/or balsa strip from can
- Blot on paper towel to remove excess moisture
- Put cloth and balsa in plastic bag
- Close bag by folding several times and secure with staples
- Label bag with strain name and date
- Put in vegetable crisper and store in insectary as long as 6-12 mo

4.6 Sex determination
- Female pupae and adults are usually larger than the males
- The plumose antennae of the male adult is densely hairy and conspicuous; that of the female is scarcely visible

4.7 Special handling of adults

4.7.1 Etherizing
- Saturate cotton ball with ethyl ether
- Place on top of cage or chimney
- Cover container with plastic for 1-2 min or until adults drop to bottom of container and appear relaxed
- Mosquito can then be handled by the legs using watchmaker forceps

4.7.2 Mouth aspirator
- Glue a small piece of plastic window screen to cover on end of a 30 cm length of plastic or glass tubing with an inside diam of 10 mm
- Insert the screened end into a length of rubber tubing
- Place the free end of rubber tubing into the mouth and inhale to draw adults into the section of glass tube
- Exhale to remove the adults from the tube and into another container

5. REARING SCHEDULE

a. Weekly
- Bloodfeed females twice a week
- Collect eggs
- Replace oviposition containers
- Replace honey pad
- Hatch eggs as necessary to maintain colony cages

b. Daily
- Check pans for pupae, nutrient level, and scum formation

6. LIFE CYCLE

Life cycle data for developmental stages under optimum rearing conditions of 21° C, 80% RH and 18:6 LD photoperiod are given.
- Embryonation of egg 7 days
- Maximum hatch 24 h
- No. of larval instars 4
- Pupation begins on day 14
- Adult emergence begins day 18
- Mean no. of eggs laid per female in one egg batch 80
- Oviposition 5-6 days post blood meal
- Mean adult lifetime 90 days (max 138)
- Life cycle 35-40 days

7. PROCEDURE FOR SUPPLYING INSECTS

- Since the eggs are drought resistant, embryonated egg papers sealed in a plastic sandwich bag can be placed in an envelope and sent via regular postal service
- Before international orders can be filled, the appropriate government shipping permits must be obtained
- Eggs of all species maintained at the Vector Biology Lab, including those listed in Section 9 may be obtained by

writing the Director, PROF. GEORGE B. CRAIG, JR. at the following address:

VECTOR BIOLOGY LABORATORY
DEPARTMENT OF BIOLOGY
UNIVERSITY OF NOTRE DAME
NOTRE DAME IN 46446, U.S.A.

8. REFERENCES

Nasci, R.S., 1982. Differences in host choice between the sibling species of treehole mosquitoes *Aedes triseriatus* and *Aedes hendersoni*. Am. J. Trop. Med. Hyg. 31: 411-415.

9. APPENDIX

Species reared by *Aedes triseriatus* method at the Vector Biology Laboratory, University of Notre Dame.

Aedes triseriatus - hatch with nutrient broth method, copulation must be induced manually, feeds on mice, oviposits on balsa and/or black cotton percale; eggs store 1 yr
Aedes brelandi
Aedes geniculatus - cage mates in 3 m high x 1 m x 1 m cage, using dawn-dusk simulator for lighting conditions
Aedes hendersoni
Aedes tehuantepec -low percentage of eggs hatch
Aedes zoosophus

ACKNOWLEDGEMENTS

This work was supported by NIH Grant No. AI-02753 to Prof. G. B. Craig, Jr.

CALLIPHORA VICINA, PHORMIA REGINA, and PHAENICIA CUPRINA

BERNARD GREENBERG and JEANNETTE GEORGE

Department of Biological Sciences, University of Illinois at Chicago, Chicago, Illinois 60680, USA

THE INSECTS

Scientific Name: *Calliphora vicina* (Robineau-Desvoidy)
Common Name: Bluebottle fly

Scientific Name: *Phormia regina* (Meigen)
Common Name: Black blowfly

Scientific Name: *Phaenicia cuprina* (Wiedemann)
Common Name: Sheep greenbottle fly

Order: Diptera
Family: Calliphoridae

1. INTRODUCTION

Calliphora vicina is commonly found associated with human activities and habitations. It is worldwide, in the subtropics in winter, in the temperate zone mostly in spring and autumn, and in the subpolar zone in summer. Adults are attracted to fruit, decaying meat, and feces; the larvae develop chiefly in carrion.

Phormia regina is also associated with man but to a lesser degree; it is less apt to enter houses or buildings. It occurs mostly in the Northern Hemisphere and is considered to be a spring-fall fly, but many adults are on the wing throughout the summer in temperate regions. They frequent the inflorescences of wild carrot or Queen Anne's Lace. *Phormia* is the principal blowfly found in dead fish on Lake Michigan shores. Larvae develop in carrion but can be facultative myiasis producers in mammals and birds.

C. vicina and *P. regina* are primarily carrion feeders in the wild, being among the initial members of a succession of carrion decomposers. Either may be a factor in wound myiasis of domestic and other animals. Various pathogenic microorganisms (several types of poliovirus, Echo and Coxsackie viruses, and salmonellae) have been isolated from wild flies; thus they are of potential public health importance. These species are widely used laboratory insects and have contributed to studies in physiology, vision, neurology, and nutrition; they are also used as tools in forensic entomology in establishing time of death. *Phormia regina* is occasionally, though rarely, used in maggot therapy in wound treatment.

Phaenicia cuprina is similar to *Phaenicia sericata*. The generic name *Lucilia* is also used. It inhabits the drier and warmer parts of Africa and Asia and its introduction into Australia and the New World is a recent occurrence. Its habits vary depending on locale; in South Africa, it is reported to be a scavenger, common in abbatoirs and butcher shops, although it also causes myiasis in sheep; in Guam, adult flies prefer masses of liquifying garbage to carrion and are attracted to human excrement and privies; in Australia, it is associated with domestic animals, and although it is a carrion breeder in sheepless areas of northern Australia and New Guinea, the adults are strongly attracted to the wool of sheep elsewhere, and by far the greater percentage has been reported to occur in the living animal (Whitten *et al*. 1975), frequently causing the death of the host. Up to 90% of all sheep "strikes" in Australia and South Africa are caused by this species, and thus it is of considerable economic importance.

The life cycles of the three species are similar. A gravid female selects a medium and oviposits a cluster of eggs, which hatch into first instar larvae. When mature, the third instar larvae cease feeding, and wander from the food supply to a suitable pupariation site. They empty their gut contents, acquire deposits of fat, and contract to form the puparium which is the sclerotized larval cuticle. Within this barrel-shaped structure the larva molts into the pupal stage from which the adult fly emerges, usually in a week or more. All of these stages are influenced by temperature and humidity; photoperiod and handling may also be factors. Kamal (1958) reported that longevity increased in several calliphorid and sarcophagid species raised under fluctuating temperature and humidity conditions; also, that larger cage size increased longevity. Bionomics of the three species are listed in Table 3.

2. FACILITIES AND EQUIPMENT REQUIRED

2.1 Facilities

- Ventilated room with temperature control (humidity and photoperiod control optional) maintained at 20-25°C, 50% RH, and LD 12:12
- Bench, working, and storage space (including sink)

PRECAUTION

Small-scale rearing may be done in a fume hood; it is not advisable to rear blowflies without ventilation due to putrefactive odors associated with the colonies.

2.1.1 Adult maintenance equipment

- 25- to 30-liter cage (with muslin or stockinette sleeve)
- Glass or plastic petri dishes, 15 x 100 mm or the equivalent
- Erlenmeyer flasks, 100-1000 ml
- White cotton knit fabric
- Absorbent cotton
- Plastic wrap
- Small whisk broom or brush
- Grill scraper

2.1.2 Oviposition/egg collection

- Larval container set up as described in section 3.2
- Glass or plastic petri dishes, 15 x 100 mm or the equivalent
- Forceps
- Fine brush
- Filter paper (Whatman #1 or equivalent)
- Screen, maximum 60-mesh or fine nylon gauze
- Pipets, 5 or 10 ml blow-out type with tip removed
- Test tubes, 25 x 200 mm
- Distilled water squeeze bottle, 500-ml capacity
- Reagent bottles

2.1.3 Larval maintenance equipment

- Container (plastic, glass, or metal) appropriately sized
- Pupariation medium (see 2.2.1)
- Tray or shallow pan, size not critical
- Forceps
- Nylon mesh (if needed as cover) 10/inch or so apertures
- Rubber bands

2.1.4 Diet preparation equipment

- Storage containers for dry and liquid raw materials
- Scissors, scalpels, razor blades
- Killing container (if necessary for larval diet animals)
- Refrigerator-freezer
- Autoclave (if needed for gnotobiotic rearing)

2.1.5 Gnotobiotic rearing
- Bunsen burner
- Test tubes, 25 x 200 mm
- Erlenmeyer flasks, 100-2000 ml
- Non-absorbent cotton or polyfoam stoppers
- Pipets, 5 or 10 ml blow-out type
- Pupariation medium (see section 6.3.1)

2.2 Raw materials/reagents

2.2.1 Adult/larval maintenance
- Formalin 37%
- Sand or coarse sawdust and wood chips

2.2.2 Natural diet components
- Water
- Sugar
- Milk (dry or skim)
- Carcass, e.g., fresh-killed or thawed mouse, rat, fish; muscle, liver, or other viscera
- Rat fetus (gnotobiotic)
- 20-day-old chick embryo (gnotobiotic)

2.2.3 Artificial diet components
It may be necessary to rear larvae on a diet with known quantities of vitamin, mineral, amino acid, lipid, and carbohydrate components. One suggested formula is listed in Table 1.

2.2.4 Egg sterilization (see Table 2)

2.2.5 Other
- Ether, anhydrous
- Phenol 5%
- Carbon dioxide

3. DIET

3.1 Natural adult diet preparation procedure
- Place dry milk and sugar in separate small containers
- Wick construction: roll a piece of absorbent cotton around a small test tube; around this fold a strip of knit cotton fabric so that it fits snugly on the cotton and hangs free below the test tube, reaching the bottom of the container
- Place liquid ingredients in separate containers (250-500 ml Erlenmeyer flask if water; 25-50 ml Erlenmeyer flask if liquid milk) which have been fitted with wicks

PRECAUTIONS
(i) If the colony does not thrive on dry food components, a switch to their liquid versions (30% sugar syrup, skim milk) may help.
(ii) Replace liquid milk daily to avoid souring and coagulation.
(iii) Replace wicks when they become soiled.

3.2 Natural larval diet preparation procedure
- Fill a container (use size appropriate to scale of rearing) 2/3 full of moistened sand or sawdust
- Partially bury a freshly killed or recently thawed carcass or part of a carcass in this medium

PRECAUTIONS
(i) If the humidity is uncontrolled and variable, improve the microclimate for the larvae by moistening the pupariation medium and partially burying the carcass; larger portions of larval food may not dehydrate as rapidly as smaller amounts (however, see (ii) below). In extreme cases, partially cover the larval container with plastic wrap.

(ii) Limit the mass of larval food supplied to a quantity which is consistent with the expected number of larvae from a given ovipositional period; excess food will either putrify or dry out; dried food will not be available to the larvae, and will attract dermestids.

3.3 Artificial larval diet preparation procedure (see Table 1)

- Put the dry components of the artificial medium into a suitably sized container (60-ml capacity glass jar)
- Add water and stir thoroughly. Adequate larval growth should be gained with either dry yeast or vitamin B mixture; use both if needed
- Add 0.48 ml of the previously prepared vitamin solution, stir, then plug the mouth of the jars with non-absorbent cotton
- Autoclave 20 min at 15 psi, then cool to room temperture and add fly eggs sterilized as in section 6

PRECAUTION

Autoclaving or heat sterilization may destroy nutritive values or render protein unsuitable for utilization or optimum growth of first instar larvae.

Table 1a. Artificial growth medium

Component	Amount	% (wet basis)
Casein	4.0 g	21.27
McCollum-Davis salt mixture No. 85	0.04 g	0.21
Cholesterol	0.04 g	0.21
Agar	0.16 g	0.85
Methyl-p-hydroxybenzoate	0.09 g	0.48
Vitamin B mixture	0.48 ml	2.55
Distilled water	14.0 ml	74.43

[0.4 g dry yeast added will enhance larval growth]
(After Brust and Fraenkel 1955)

Table 1b. B-vitamin mixture

Vitamin	Amount (mg)
Thiamine	60
Riboflavin	30
Nicotinic acid	120
Pyridoxine	30
Pantothenic acid	60
Choline chloride	1200
i-inositol	600
Folic acid	6
Biotin	1

Dissolve in 480 ml of distilled water; 0.48 ml of this solution will contain 2.107 mg total B-vitamin content. (After Brust and Fraenkel 1955)

4. REARING AND COLONY MAINTENANCE

4.1 Colony start-up

4.1.1 Collection of starter specimens

- Place a carcass (rabbit, large rat, large fish) at the site picked for collection (anchor it to the ground if possible - even better, enclose it in a 1/2 in.-mesh (1.3 cm) wire cage anchored to the ground - if the carcass is to be left in the field for more than a few hours)
- Use a "Hands Off!" sign if appropriate; female flies are attracted to exudates from the normal body orifices and will oviposit there
- Sites may be either urban or rural
- Slash the body to obtain additional ovipositional sites
- Check the site at least daily and capture and identify any adult flies found on the bait [detailed descriptions of these three species - sex differences, adult and larval keys, and illustrations of diagnostic

characters - may be found in James (1947), Hall (1948), Zumpt (1965), Anon. (1967), and Greenberg (1971)]
- Starter specimens (any stage, but pupal is probably easiest) may also be obtained from stock colonies maintained by academic or research institutions, or from biological supply houses (P. cuprina specimens may come from struck host animals in the field)

PRECAUTIONS
(i) Examine all specimens, regardless of source, for phoretic mites prior to using them to start or to add to a colony.
(ii) Properly identify all specimens, regardless of source. P. cuprina, for instance, closely resembles P. sericata, and these two species may mate to yield vigorous and fertile hybrids.
(iii) Do not put the bait (or carcass) on or too near an ant bed; the ants may steal eggs and young larvae.

4.1.2 Cage preparation
- Set up the cage (volume 25-30 liters minimum)
- The stockinette sleeve should be long enough to be securely tied; it may be fastened less conveniently with a twist-tie or rubber band
- Put adult and larval diet in the cage as appropriate to the starter specimens used
- Add specimens
- Maintain the colony at 20-25°C and 50% RH; a LD 12:12 photoperiod is acceptable
- Replenish water, milk, and sugar as needed; P. regina spends more time in flying and will require more sugar than C. vicina
- Supply a carcass for oviposition bait and for subsequent larval diet at the proper time (see Table 3)
- Under normal conditions, replace the larval food every 2-3 weeks; check the supply as needed; replace sooner if it is consumed or dries out before the larvae reach the prepupal stage
- Clean cage as needed; grill scrapers and whisk brooms are useful

PRECAUTION
Remove dead flies and dried carcass remnants from the cage to deter any dermestid and cockroach infestations on a weekly or at least biweekly schedule; a thorough cleaning of the entire insect holding area on a regular basis (including vacuuming and washing if feasible) is recommended.

5. QUALITY CONTROL

Evaluate the progress of the colony by noting obvious trends in colony density, larval yield and/or size from a given amount of food, and adult mortality. More specific quality checks may also be made.

5.1 Quantitative evaluations
- Female fecundity (150-200 eggs/clutch; 4-8 or more clutches/female)
- Adult longevity (see section 7)
- Larval length (C. vicina, 3rd instar, approx. 16 mm; somewhat less for the other 2 species)
- Pupal weight (approx. 0.05-0.06 g for C. vicina; 0.04 g for P. regina)
- % adult emergence (see section 7)

6. GNOTOBIOTIC REARING PROCEDURE (for 1000-2000 eggs)

Wipe down the bench-working area with a quaternary ammonium chloride solution before starting this procedure and prepare any sterile solutions or equipment which may be required. See Table 2 for formulae.

6.1 Adult oviposition
- Place a piece of liver or the equivalent in a petri dish; use enough so

that it won't dry out should it be left in the cage 24-48 h
- Vary bait freshness, if necessary, to attract the female flies
- Remove eggs with a small brush, dissecting needle, wire loop, or fine forceps

6.2 Egg sterilization (see Table 2)
- Place the eggs in a 25 x 200 ml test tube containing 35 ml of 1% detergent solution
- Agitate tube and contents so as to separate the eggs; viable eggs sink
- Decant excess liquid
- Repeat twice; this should be sufficient to separate *C. vicina* eggs
- If not, add 35 ml of 1% sodium sulfite solution, agitate, and decant as before (clustered eggs of *P. regina* and *P. cuprina* may have to be rinsed in a sulfite solution to dissolve their adhesive)
- Add 35 ml sterile distilled water (wash down the tube interior as you add the water), agitate, and decant
- Repeat three times, leaving a just-pourable amount of water after the last rinse
- Flame mouth of tube with a Bunsen burner
- Unplug a previously prepared sterile 125-ml Erlenmeyer flask which contains 50 ml sterile distilled water
- Flame mouth and pour the rinsed, separated eggs into the center of flask
- Add 1 ml fresh bleach (e.g. Clorox®) swirling the flask to mix the contents
- Replug and soak with intermittent swirling for 15 minutes
- Remove plug, flame mouth again, carefully decant excess liquid, flame mouth, and replug
- Unplug, flame mouth, and add 50 ml sterile distilled water
- Agitate the flask 2 min, unplug, flame, and decant
- Repeat four times, leaving approx. 1 ml of water after the last rinse
- Using a sterile pipet (5 or 10 ml blow-out type with part of the tip removed), remove the sterilized eggs for subsequent procedures (0.1 ml contains ca 500 eggs)
- Other egg disinfection procedures may also be used (see Table 2)

PRECAUTIONS
(i) Older eggs (more than half developed) may be more heavily contaminated and thus require either longer or stronger disinfection; younger eggs may be less resistant to the toxic effects of the sterilants.
(ii) Disinfectants with free chlorine may adversely affect egg survival, depending on concentration and exposure time.
(iii) To surface-sterilize accidentally-contaminated pupae, soak 10 min in a 5% phenol solution followed by a sterile distilled water rinse. An alcoholic solution may reduce the percentage of successful eclosions.

Table 2. Egg disinfection procedures

A. White's solution method
1. Detergent wash; use 1% sodium sulfite to separate eggs if necessary; rinse 2X with distilled water.
2. Formalin (5%); soak 5 min with intermittent swirling.
3. Sterile distilled water; rinse 2X.
4. Fresh NaOCl, e.g., Clorox, (0.2%) 0.1 ml in 50 ml sterile distilled water; soak 2 min with swirling.
5. Sterile distilled water; rinse 2X.
6. White's solution; soak 60 minutes.
   a. 0.25 g Mercuric chloride
   b. 6.5 g NaCl
   c. 1.25 ml HCl
   d. 250 ml 95% ethanol
   e. 750 ml distilled water
7. Sterile distilled water; rinse 7X.

B. Bleach method
1. Detergent wash; rinse 2X.
2. Fresh NaOCl, e.g., Clorox (2%), 1 ml in 50 ml sterile distilled water; soak 15 min with swirling.
3. Sterile distilled water; rinse 3X.

C. Hyamine method
1. Detergent wash; rinse 2X.
2. Hyamine 10-X (1%) in 70% ethanol; soak 20-30 minutes.
3. Sterile distilled water; rinse 3X.

D. Other proven disinfectants
1. Detergent wash
2. Disinfect in one of the solutions below:
a. 0.1% benzalkonium chloride (25 min)
b. 3% Lysol® (2-3 min)
c. 5% formalin + 1% KOH (30 min)
d. Merthiolate .005% (20 min)
3. Rinse several times in sterile distilled water.

Note: Agitate or swirl the eggs during each wash, soak or rinse step.

6.3 Larval holding for gnotiobiotic rearing
- This procedure is best carried out in a disinfected hood, glove box, or the equivalent

6.3.1 Pupariation medium sterilization
- Place a 2.5-cm layer of sand or sawdust in a 1000 ml Erlenmeyer flask
- Stopper the flask with non-absorbent cotton or a polyfoam plug
- Autoclave for 60 min at 121°C

6.3.2 Chick embryo technique
- Place an egg which has been incubated for 20 days into an egg carton, blunt end up
- Swab the egg with 2% tincture of iodine solution and with scissors and forceps which have been previously sterilized for 15 min at 121°C, cut around the egg shell ca 1/3 down from the top of the egg so as to remove a cap
- Unplug and flame the mouth of the flask with a bunsen burner
- Drop the embryo into the flask
- Flame the mouth of the flask and plug again
- Unplug; flame the flask that contains the disinfected eggs and pipette them directly on the embryo; one embryo will support ca 100 larvae
- Flame and replug the flask

6.3.3 Rat fetus technique
- Sacrifice a gravid, near-term rat in a carbon dioxide atmosphere and immediately sever the cervical vertebrae
- Shave the abdomen and disinfect thoroughly with tincture of iodine, merthiolate, or other standard disinfectant
- Using sterile scalpel, scissors, and forceps, make a median longitudinal incision through the abdominal wall and expose the uterus, being careful not to perforate the intestines (this would contaminate the body cavity)
- Free the uterus from adhering body tissues and place it temporarily into a sterile covered petri dish
- Unplug and flame the mouth of the flask
- Drop the fetal mass into the flask
- Inoculate with fly eggs as in 5.3.2; this biomass will be somewhat less than that of the chick; fewer larvae will be supported

7. HOLDING INSECTS AT LOWER TEMPERATURES

Eggs, larvae, or pupae may be held at 4°C for short periods, provided that they are not allowed to become dehydrated. Duration of these periods for each stage and the relative survival through emergence should be determined if this is a factor important in colony rearing (such as to maintain the colony or to obtain specific stages over technician vacations or holidays).

8. OTHER SPECIES

This rearing method may be used for other species of Calliphoridae, with modifications as required for temperature, humidity, and ovipositional preference.

Table 3. Bionomics of Calliphora vicina, Phormia regina and Phaenicia cuprina

Stage	Duration		
	C. vicina[b]	P. regina[b]	P. cuprina[c,d]
Egg	24 (20-28) h[a]	16 (10-22) h	8-10 h
Instar 1	24 (18-34) h	18 (11-32) h	14 h
Instar 2	20 (16-28) h	11 (8-22) h	32 h
Instar 3	48 (30-68) h	36 (18-54) h	48-68 h
Prepupa	128 (72-290)h	84 (40-168)h	36-48 h
Pupa	11 (9-15) d	6 (4-9) d	5 d
Total immature	18 (14-25) d	11 (10-12) d	15-19 d
% emergence	31-43	87-94	.....
Time to 1st copulation	5-9 d	3-7 d	2-8 d
Time to 1st oviposition	8-15 d	5-8 d	9 d
Time span of:			
mating activity	.....	19 (10-25) d	F usually once
oviposition[e]	6 (2-15) d	22 (18-28) d	43 d
adult life	25 (24-35) d	52 (45-68) d	71 d

[a] h=hours; d=days; F=female.
[b] At 25.5-29°C, 48-52% RH, C. vicina and P. regina, from Kamal (1958).
[c] P. cuprina, largely after Subramanian and Mohan (1980), at 25.6°C.
[d] At 25°C, Whitten et al. (1975) noted 5-6 day feeding stages, a brief time as wandering 3rd instars, 6-7 days as pupae (males), 7-8 days as pupae (females), 2-4 days from eclosion to copulation, and 5-6 days to oviposition.
[e] At 20-22°C, Mackerras (1933) noted somewhat longer ovipositional span.

9. REFERENCES

Anon., 1967. Pictorial keys to arthropods, reptiles, birds, and mammals of public health significance. pp. 121-133 In: National Communicable Disease Center, U.S. Public Health Service.

Brust, M. and Fraenkel, G., 1955. The nutritional requirements of the larvae of the blowfly, Phormia regina (Meig.). Physiol. Zool. 28: 186-204.

Greenberg, B., 1971. Flies and disease, Vol. I. Ecology, classification, and biotic associations. Princeton Univ. Press, Princeton, New Jersey, 856 p.

Hall, D. G., 1948. The blowflies of North America. The Thomas Say Foundation, 477 p.

James, M. T., 1947. The flies that cause myiasis in man. USDA Misc. Publ. No. 631. U.S. Govt. Printing Office, Washington, D. C., 175 p.

Kamal, A. S., 1958. Comparative study of thirteen species of sarcosaprophagous Calliphoridae and Sarcophagidae (Diptera). I. Bionomics. Ann. Entomol. Soc. Amer. 51: 261-271.

Mackerras, M. J., 1933. Observations on the life-histories, nutrition requirements and fecundity of blowflies. Bull. Entomol. Res. 24: 353-362.

Subramanian, H. and Mohan, K. Raja, 1980. Biology of the blowflies Chrysomia megacephala, Chrysomia rufifacies and Lucilia cuprina. Kerala Jour. Vet. Sci. 11:2, 252-261.

Whitten, M. J., Foster, G. G., Arnold, J. T., and Konowalow, C., 1975. The Australian sheep blowfly, Lucilia cuprina. Ch. 16 in: King, R. C. (ed.) Handbook of genetics. Vol. 3. Invertebrates of genetic interest. Plenum Press, New York, 874 p.

Zumpt, F., 1965. Myiasis in man and animals in the Old World. Butterworths, London, 267 p.

10. RECOMMENDED READING

Browne, L. B., 1958. The choice of communal oviposition sites by the Australian sheep blowfly Lucilia cuprina. Aust. Jour. Zool. 6: 241-247.

Evans, A. C., 1936. Studies on the influence of the environment on the sheep blow-fly Lucilia sericata Meig. IV. The indirect effect of temperature and humidity acting through certain competing species of blow-flies. Parasit. 28: 431-439.

Greenberg, B., 1968. Gnotobiotic insects in biomedical research. pp. 410-416 In: Miyakawa, M. and Luckey, T. D. (eds.) Advances in germfree research and gnotobiology. CRC Press, Cleveland, Ohio. 439 p.

Greenberg, B., 1970. Sterilizing procedures and agents, antibiotics and inhibitors in mass rearing of insects. Bull. Entomol. Soc. Amer. 16: 31-66.

Greenberg, B., 1973. Flies and disease, Vol. II. Biology and disease transmission. Princeton Univ. Press, Princeton, New Jersey, 447 p.

Norris, K. R., 1959. The ecology of sheep blowflies in Australia. pp. 514-544 In: Biogeography and ecology in Australia, Series Monographiae Biologicae, Vol. VIII. CSIRO, Canberra, Australia.

Povolný, D. and Rozsypol, J., 1968. Towards the autecology of Lucilia sericata (Meigen, 1826) (Dipt., Call.) and the origin of its synanthropy. Academia, 2: 1-32.

Sedee, P. D. J. W., 1952. Qualitative vitamin requirements for growth of larvae of Calliphora erythrocephala (Meigen). Experientia 9: 142-143.

Sedee, P. D. J. W., 1954. Qualitative amino acid requirements of larvae of Calliphora erythrocephala (Meig.). Acta Physiol. Pharmac. Neerl. 3: 262-269.

Sedee, P. D. J. W., 1958. Qualitative B vitamin requirements of the larvae of a blowfly, Calliphora erythrocephala (Meig.). Physiol. Zool. 31: 310-316.

Webber, L. G., 1958. Nutrition and reproduction in the Australian sheep blowfly, Lucilia cuprina. Aust. Jour. Zool. 6: 139-144.

CULISETA INORNATA

W. G. FRIEND and R. J. TANNER

Department of Zoology, University of Toronto, Toronto, Ontario, Canada M5S 1A1

THE INSECT

Scientific Name: *Culiseta inornata* (Williston)
Common Name: Mosquito
Order: Diptera
Family: Culicidae

1. INTRODUCTION

Culiseta inornata is a pest mosquito in the prairie provinces of Canada, and a known vector of western equine encephalitis, Japanese encephalitis, and St. Louis encephalitis viruses (McLintock, 1964). In the wild, the females bite large mammals, and both sexes take nectar. It is a large mosquito (female: length 5-6 mm, wing span 9-11 mm). Because of its size and relative ease in rearing and handling, it is an excellent experimental animal.

Owen (1942) and McLintock (1952, 1964) were probably the first to successfully rear *C. inornata* continuously. Feir *et. al.* (1961) and Pappas (1973) modified their techniques, and the techniques described herein are a further modification.

It is frequently difficult to get field-collected adult females to take blood meals from small laboratory mammals. Our colony was obtained from the Canadian Department of Agriculture, Lethbridge, Alberta in 1977, where the insects were fed on guinea pigs. In our laboratory, the females fed poorly on guinea pigs. We switched to rabbits, with increasing success. At the present time, 75% of the females feed well on the rabbit; apparently selection has occurred during the life of our colony.

About 5-9 days after blood-feeding, the females lay their eggs on a water surface in a egg raft of 100-200 eggs. The eggs hatch in 3-4 days. The larvae grow through 4 instars, and pupate about 2-3 weeks after hatching. Males pupate first, with smaller pupae than the females. Adults emerge 3-4 days after pupation. The males emerge first and frequently commence mating before the females are fully emerged from the pupal cuticle, even in closely confined quarters. Consequently, no special flight cages are required.

Adult males and females are easily differentiated. The males have feathery antennae and three-part mouth parts; and the females have a one-piece mouthpart (to the naked eye), and are larger and darker. Our colony provides 150-300 females per week for experimental purposes.

2. FACILITIES AND EQUIPMENT REQUIRED

2.1. Facilities

The *Culiseta inornata* colony is maintained in a controlled environment room at 20±1°C illuminated with incandescent light at LD 15:9. The humidity is not controlled.

2.2. Equipment and materials used

2.2.1. Oviposition

- 50-60 laying cages. A laying cage is a 8-dram glass vial (9.5 x 2.5 cm) with a plastic top constructed from a 4-dram pill vial (i.e. Rigo® Plasticlear vials) (5 x 2.5 cm) which is inverted over the 8-dram vial. A 1.5 cm diameter hole is cut through the bottom of the pill vial, and a piece of fiberglass window screening glued over the hole. A 2.5 x 10 cm piece of sheet acetate or other suitable material is wrapped around the bottom of the pill vial and taped, forming a lip to hold a sugar cube (Fig. 4).
- 2 racks for the laying cages, to hold them at a 25-30° angle. A rack is made from galvanized hardware screening (approximately 1 mesh per cm) secured to two pieces of 2 cm plywood, cut as shown in Fig. 5. A few appropriate wires of the screening are then cut and bent upwards to prevent the laying cages from rolling (Fig. 6).
- 1 x 9 cm pieces of bond paper.

2.2.2. Larval maintenance

- 10-12 plastic dish pans, approximately 35 x 30 x 10 cm.
- Bates' Medium S (Bates, 1941), which consists of 1 g $MgSO_4.7H_2O$, 0.5 g $CaSO_4.2H_2O$, and 0.5 g NaCl per liter of distilled water.
- 6-8 air stones with tubing.
- An aquarium air pump or other suitable air supply.
- Blotting paper cut in 3 cm wide strips slightly longer than the width of the pans.

2.2.3. Pupae collection and adult maintenance

- 4-5 glass finger bowls 10.7 cm diameter x 5.5 cm high.
- 2-4 plastic eye droppers (Beral® disposable pipettes, Beral Enterprises, Canoga Park, CA.), with the tip cut on a 45° angle to produce a 3.5 mm diameter hole.
- 4 wooden cages, 38 x 21.5 x 21.5 cm, with a wooden floor, a fabric sleeve on one end, the front a sheet of plexiglass which can be slid out for cleaning and which has two 4 cm holes stoppered with corks, and the other three sides screened (Fig. 1).
- 5 cm petri dishes, or equivalent.
- wettable cotton batten.
- sugar cubes.
- squirt bottle.

2.2.4. Adult feeding

- 6-8 feeding cages, each made from a plexiglass cylinder 5 cm long with an outside diameter of 4.5 cm. A groove is milled close to each end. A square of nylon stocking is secured over one end with a rubber O-ring in the groove and the nylon pulled tight. A sleeve cut from the leg of a nylon stocking is fastened to the other end of the cylinder with an O-ring (Fig. 2). Feeding cages can also be used in feeding experiments (Friend & Hewson, 1978).
- 1 aspirator, consisting of a 40 cm glass tube (11 mm OD) with a 30° bend about 7 cm from the end. A 40 cm piece of flexible tubing is attached to the other end of the glass tube. To prevent inhaling the mosquitoes, a piece of nylon stocking is stretched over the end of the glass tube before the flexible tubing is attached.
- 1 rabbit harness, consisting of a wooden box, inside dimensions 37 x 14 x 14 cm. One end of the box is cut down to the level of the animal's neck, and a semi-circle cut out. When the rabbit is placed in the box, its head is allowed to protrude over this opening, and a piece of wood with a corresponding semi-circle is slid down between two angle brackets to encircle the animal's neck. This piece of wood is secured with a small nail through a hole on each side. A strap of 2 cm wide elastic is attached to each side of the box. These are joined with Velcro® over the back of the rabbit. A 20 cm square platform is attached to each side of the front of the box (Figs. 3, 7).

Fig. 1. Adult cage for *C. inornata*.
Fig. 2. Feeding cage for adult females.

- Two 7 x 10 x 5 cm high pieces of foam rubber.
- Two 20 cm long pieces of 2 cm wide elastic, with 5 cm pieces of Velcro® attached to each end.

3. DIET

3.1. Larval diet

- Tetra-min® Staple Food for tropical fish, ground finely, and mixed to a slurry with water before adding it to the pan.

3.2. Adult diet

- 1 New Zealand White rabbit, maintained in a communal animal facility using standard techniques. It is used weekly, the mosquitoes taking about 150 µl of blood at each feeding session. The rabbit shows no ill effects or mosquito bites; our present rabbit has been used for 3 years. It is not advisable to use the same rabbit for several different insect species. Some rabbits seem to be able to "cool" their ears; the ears become cold to the touch after the cage of insects is put on. Such an animal is useless for mosquito feeding.

4. REARING AND COLONY MAINTENANCE

The following outlines a weekly schedule for a colony producing 100-150 females twice a week for experimental purposes. The time required is less than 1 day per week.

4.1. Larval maintenance

- Fill 2-3 plastic larval pans with 3 liters of Bates (1941) medium.
- Add an air stone on a low pressure air supply to each pan, such that the stone produces 1 or 2 streams of bubbles. More disturbance than this may turn over the egg rafts and prevent hatching.
- Add 1 tablespoon (3-4 g) of ground Tetra-min® fish food, made into a slurry.

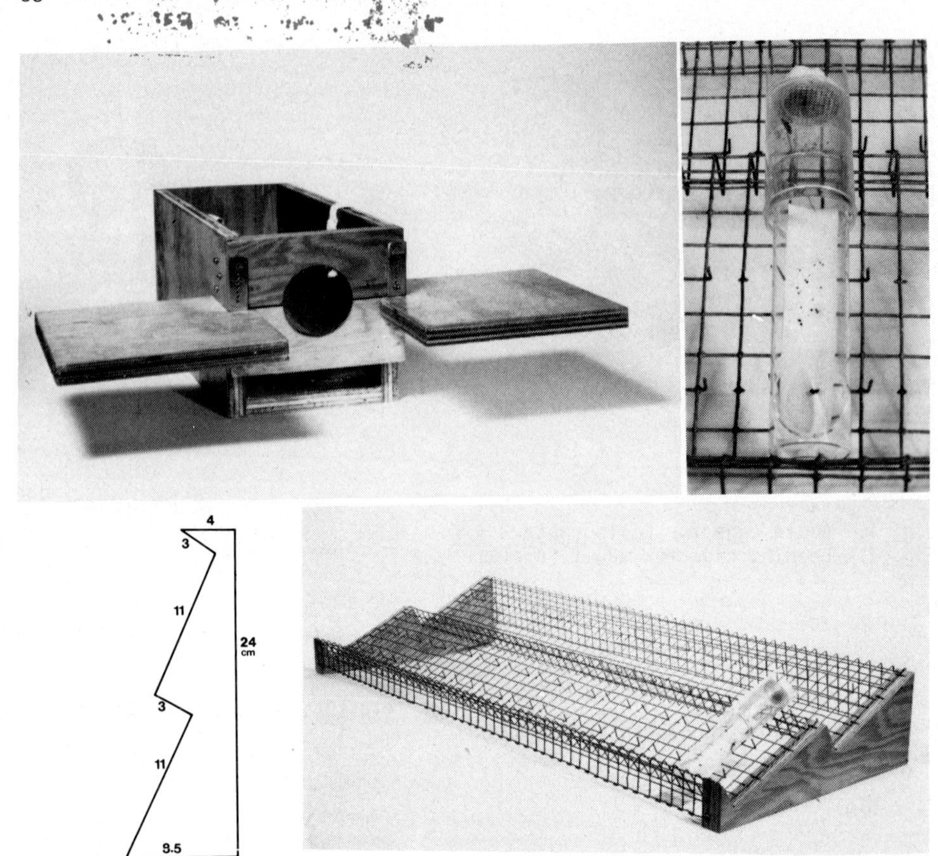

Fig. 3. Rabbit harness for C. inornata.
Fig. 4. Egg-laying cage.
Fig. 5. Template for the ends of the rack.
Fig. 6. Rack for laying cages.

- Skim off any floating food with a piece of the 3 cm wide blotting paper, by holding the paper at each end and drawing it from one end of the pan to the other, and removing any food collected. Otherwise a bacterial pellicle may form which can interfere with larval respiration.
- Float 10-15 egg rafts in each pan.
- Skim the pans with fresh blotting paper whenever bacterial scum is evident.
- Add 2 g of slurried food whenever it appears depleted (about once a week).

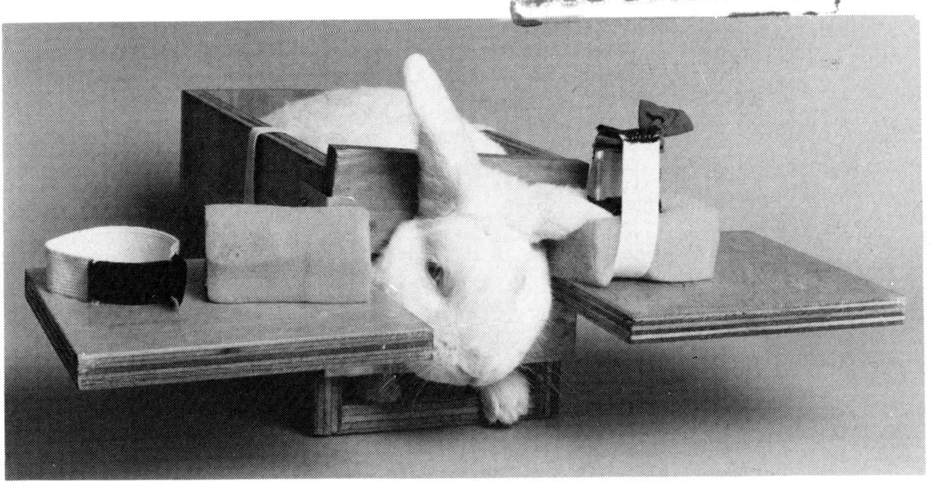

Fig. 7. Rabbit in harness. A feeding cage is on the rabbit's left ear.

4.2. Pupae collection and adult maintenance

After 2-3 weeks, the larvae pupate. The pupae generally stay on the surface of the medium, but they can, and do, swim. There are several methods of collecting pupae; our experiments require females of a specific age, and the dish of pupae must be free of larvae. Pupae moult in 3-4 days, and consequently should be collected frequently enough to prevent colonization of the rearing room!

In order to obtain adults of a known age ± 1 day, we set up cages twice a week. The adults are used for our experiments approximately 5 days after moulting, and any remaining can be fed and allowed to lay eggs. On the weekly feeding day, we have one rearing cage with pupae collected that day, one cage with pupae collected 3-4 days before, and two cages with adults from pupae collected the previous week. Adults from the latter two are used for feeding and the cages then placed in the freezer to kill any remaining adults.

- Collect the pupae with the wide-mouth eye-dropper and put them in Bates' medium in a finger bowl.
- Place the dish of pupae in a rearing cage with a sugar cube and a small dish of water-soaked cotton batten.
- Every few days replenish the water dish with a squirt bottle through one of the holes in the cage front.
- Remove adults for feeding or experimentation with an aspirator through one of the holes in the front of the cage, using cotton to plug the rest of the hole.
- When a cage of insects is no longer required, remove the finger bowl, sugar cube and water dish, and place the cage in the freezer.

4.3. Adult Feeding

- For each week's feeding, take females from at least two rearing cages and females which have previously laid eggs, to ensure a genetic mix.
- Aspirate 8-9 females into each of 6 feeding cages through the feeding cage sleeve.
- Twist the sleeve shut and secure it with an elastic band.

- Place the rabbit in the wooden box with its head secured as described earlier (Fig. 7).
- Lightly blow on a cage of mosquitoes through the stretched nylon (this seems to increase the numbers feeding).
- Place the cage, with the stretched nylon end down, on the rabbit's ear.
- The ear is sandwiched between a piece of foam rubber and a cage of females, and the sandwich is secured with a piece of elastic and Velcro® (Fig. 7).
- Allow the mosquitoes about 10 minutes to feed.
- Remove the feeding cage and replace it with another until all cages have had access to the rabbit. Not all females will feed; we get about 35-40 feeding of 45-50 in the cages.
- Remove 2 fed females from a feeding cage with the aspirator and blow them into the top of a laying cage. To prevent escapees, a tissue is held over the opening of the laying cage top, and the aspirator containing the fed females is introduced between the tissue and the edge of the laying cage top.
- Add 5-8 ml of water to the laying cage bottom, and place a strip of bond paper up one side for the animals to walk on.
- Maneuver the laying cage top (with the females) gently onto the bottom.
- Place the laying cages on the wire rack, paper side down.
- Place a sugar cube on the top of each laying cage, resting on the plastic strip (Fig. 4).
- Leave the rack in the rearing room.

5. REFERENCES

Bates, M., 1941. Studies in the technique of raising anopheline larvae. Amer. J. Trop. Med. 21: 103-122.

Feir, D., Lengy, J.I. and Owen, W.B., 1961. Contact chemoreception in the mosquito, Culiseta inornata (Williston); sensitivity of the tarsi and labella to sucrose and glucose. J. Insect Physiol. 6: 13-20.

Friend, W.G and Hewson, R.J., 1978. A small volume, thermostatically controlled apparatus for feeding radioactive diets to mosquitoes and other sucking arthropods. Mosquito News 38: 536-541.

McLintock, J., 1952. Continuous laboratory rearing of Culiseta inornata (Will.) (Diptera: Culicidae). Mosquito News 12: 195-201.

McLintock, J., 1964. The laboratory cultivation of Culiseta. Bull. Wld Hlth Org. 31: 459-460.

Owen, W.B., 1942. The biology of Theobaldia inornata Williston, in captive colony. J. econ. Ent. 35: 903-907.

Pappas, L.G., 1973. Larval rearing technique for Culiseta inornata (Will.). Mosquito News 33: 604-605.

DACUS TRYONI

N. W. HEATHER and R. J. CORCORAN

Entomology Branch, Department of Primary Industries, Meiers Road, Indooroopilly, Brisbane, Queensland, Australia 4068.

THE INSECT

Scientific Name:	*Dacus tryoni* (Froggatt)
Common Name:	Queensland fruit fly (QFF)
Order:	Diptera
Family:	Tephritidae

1. INTRODUCTION

The Queensland fruit fly *Dacus tryoni* (Froggatt) occurs throughout coastal and sub-coastal districts of eastern Australia. In northern and central regions it regularly occurs some hundreds of km inland but in southern areas records from far inland areas become irregular. Outside of Australia it has been found in Papua New Guinea, New Caledonia, and some other South-Pacific countries, having apparently spread in traded fruit. (Drew *et al* 1978)

It attacks the fruits of a wide range of wild, ornamental and commercial plants. Most favoured are stone and pome fruits, citrus, tomatoes, capsicums, guavas, mulberries, loquats and brazilian cherries. Native hosts are largely rainforest species. (May 1953)

Females pierce the skin of host fruit with their ovipositor and lay a number of eggs within the flesh of the fruit at the site of each sting. After hatching, larvae tunnel through the fruit pulp. When mature they leave the fruit and pupate in the soil. Fruit damage results from both feeding and secondary infections by plant pathogens.

D. tryoni are active all year round in moist tropical areas but in sub-tropical and temperate regions they become active in spring or early summer, stinging fruit after having overwintered as adults. Activity continues through until late autumn. Several generations would normally be undergone each year although individual adults can live for 6 months.

We have maintained strong colonies of *D. tryoni* for more than 10 years for use in investigations of postharvest commodity treatments. In the laboratory, *D. tryoni* can be reared on a number of artificial diets. Our method is a modification of the dehydrated carrot medium of Christenson *et al* (1956) which was developed from the fortified carrot medium of Finney (1956).

A laboratory colony may be established either from infested field-collected fruit or from an already established laboratory colony of which there are currently several in Australia. In the first instance close attention may need to be given to feeding but once established in the laboratory strong colonies can be maintained with little trouble.

2. FACILITIES AND EQUIPMENT REQUIRED

2.1 Facilities

Small colonies of Queensland fruit fly can be maintained without excessive demands on floor or bench space. One or two small cages for the adults together with several plastic dishes for larval rearing are all that is required.

Our adult colonies are maintained in a room with a controlled temperature of 25 ± 2°C and access to natural day light, while larvae are reared in smaller conditioned rooms under artificial light at 27° ± 1°C and 65% ± 5% R.H. In sub-tropical coastal areas, QFF can be reared successfully without temperature or humidity control, though it may be necessary to place an electric radiator heater in the rearing room during winter. High humidity can cause problems in tropical areas since the powdered yeast fraction is deliquescent and flies can become trapped in it. Whilst the generation time can be decreased by increasing the room temperature up to about 30°C, in practice slightly lower temperatures are used to create more comfortable working conditions with a consequent slight increase in development time.

2.2 Equipment and materials required for insect handling

2.2.1 Adult oviposition and egg collection

a. Laboratory reared flies
- fresh, whole apples
- glass or plastic petri dishes
- paraffin wax
- pins (about 21 gauge)
- Zephiran® (alkyldimethylbenzylammonium chlorides in aqueous solution)
- fixed-blade scalpel
- electric hot plate

b. For field collected specimens it is advisable to use ripe, fresh, whole fruit such as stone fruit, papaw, mango or banana for the first one or two generations until the population is built up
- fresh, ripe, whole fruit, as above
- pins (about 21 gauge)
- Zephiran®

2.2.2 Inoculation of diet and larval rearing

- 100 ml beakers
- 25 ml graduated pipettes
- SGP® 502S absorbent polymer
- stereo dissecting microscope
- rearing trays - flat dishes 23 cm diameter, 2 cm deep, available commercially as pot plant bases
- cages with wooden bases and tops and organdie-gauzed sides to hold the above dishes (Fig. 1)
- Zephiran®
- sieve, 1.7 mm mesh, with base
- sieved sawdust (insecticide free)

2.2.3 Pupal collection

- sieve, 1.7 mm mesh with base

2.2.4 Adult holding and rearing

- cages 65 cm^3 with wooden bases and tops, terylene mosquito netting on four sides, sleeve on one face (Fig. 2)
- plastic containers 17 x 12 x 7 cm with air-tight lids
- Wettex® sponges
- cube sugar
- Amber BYF® - autolysed brewers' yeast fraction

2.2.5 Brand names and sources

- Zephiran® - Winthrop Laboratories, Ermington, New South Wales, Australia, 2115
- SGP® 502S absorbent polymer - Henkel Corporation, 4620 West 77th Street, Minneapolis, MN 55435, U.S.A.

2.3 Diet preparation equipment

- top-pan balance (2500 g)
- 2 liter measuring cylinder
- 25 ml measuring cylinder
- domestic plastic crisper boxes 25 x 25 x 12 cm
- petri dishes
- spatula
- commercial blender

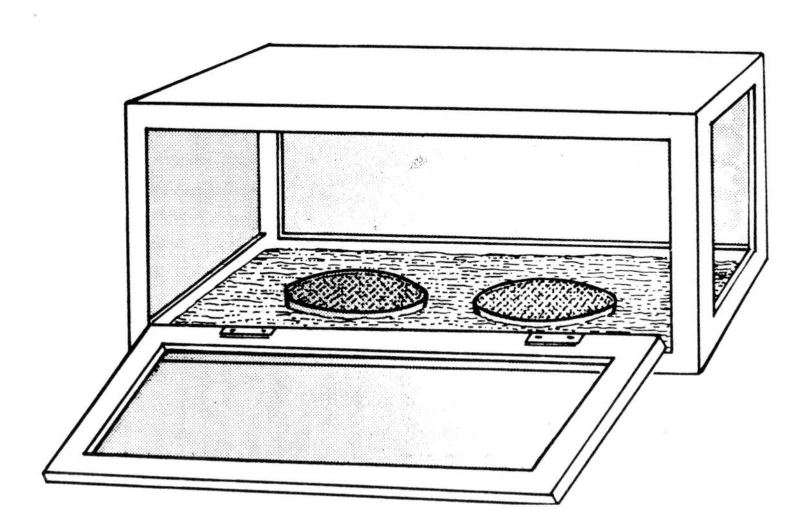

Fig. 1. Larval rearing cage with bottom lined with sawdust, shown holding dishes containing larval diet. (illustration by S. Sands)

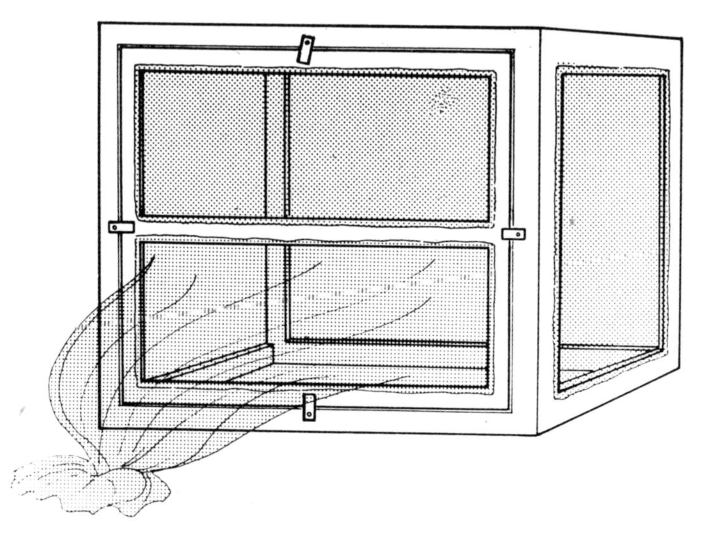

Fig. 2. Adult holding cage showing detachable front frame with gauze sleeve. (illustration by S. Sands)

3. DIET

3.1 Natural

When it is necessary to build up a laboratory colony from small numbers of flies collected from the field as larvae in infested fruit or even as adults it is advisable to use whole fruit for the first oviposition, possibly with larval diet for rearing. Preferred host fruits could be pome or stone fruits or bananas.

3.2 Artificial

3.2.1 Composition

		%
dehydrated, sliced carrot	300 g	12.3
Torula yeast (Sanitarium)	100 g	4.1
Nipagin M® (methyl p-hydroxy benzoate)	10 g	0.4
conc. hydrochloric acid	21 ml	4.8
water	2 000 ml	82.3

3.2.2 Procedure for preparation of diet

- soak carrot, yeast and Nipagin® in 1000 ml of water overnight
- blend ingredients together with the remaining 1000 ml of water using an electric blender
- consistency of the finished product should be that of a thick paste
- medium is placed in flat rearing trays (23 cm diameter, 2 cm deep)
- one batch of medium is sufficient for approximately three trays
- unused medium may be kept for up to one week provided that it is covered and kept refrigerated

3.2.3 Brand names and sources

- Nipagin M® - Nipa Industrial Estate, Llantwit Fardre, Mid Glamorgan, Gt. Britain CF38 2SN
- Sanitarium Health Food Company, Sydney, N.S.W., Australia.

4. INSECT HOLDING

4.1 Cages

Sizes and shape of cages mentioned are not critical. These parameters are largely determined by other factors such as shelf and room size and general convenience in handling e.g. depth of cage not to exceed an arm's length. Both adult and larval cages may require ant-proofing. This can be achieved by supporting the cages on ant-traps (Fig. 3) containing paraffin oil. Ensure that no part of the cage or sleeve contacts the shelving or otherwise bridges the traps.

5. REARING AND COLONY MAINTENANCE

5.1 Egg collection

a. From field collected material. If numbers of adults are low, use whole fruit of a preferred host to maximize egg survival.
- rinse fruit and pin with 0.05% Zephiran® in water
- make pin holes in fruit
- place in fly cages for several hours
- wash fruit again in Zephiran® 0.05%
- this fruit is then held for 2 or 3 days, split and placed, cut side down, on a tray of artificial diet.

b. From an established laboratory colony - a hollowed out half apple has been found to be best for this purpose (Fig. 4).
- cut an apple in half and hollow out each half with a scalpel - in the final stages this is achieved by scraping rather than cutting
- there should be no more than 1 mm of flesh left in the areas for oviposition
- rinse hollow apple halves, pin, and petri dishes in 0.05% Zephiran® in water
- make pin holes in the apple halves; about 20 - 30 holes are sufficient as too many holes allow the eggs to dry out

- allow apple halves and petri dishes to drip dry
  melt paraffin wax in beaker over electric hot plate
- seal the apple-half to the flat surface of the petri dish with molten paraffin wax applied with an art brush
- place overnight in adult fly cages

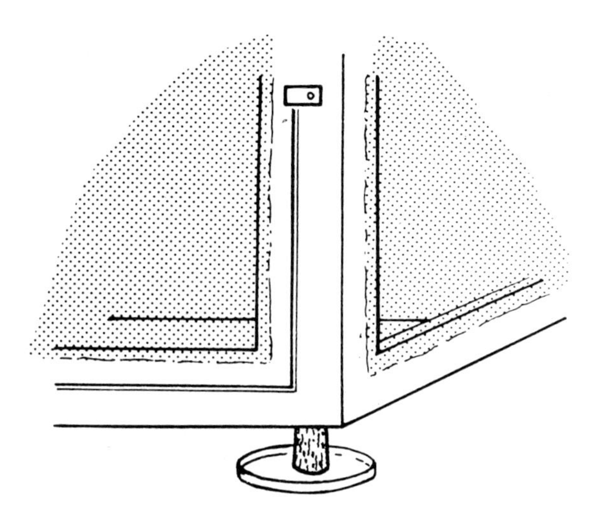

Fig. 3. Ant excluder made from a cork and petri dish containing paraffin oil, shown under adult holding cage. (illustration by S. Sands)

Fig. 4. Hollow apple waxed to upturned petri dish ready for egg collection. (illustration by S. Sands)

5.2 Diet inoculation

Eggs laid into the hollow apple collect on the inner surface or fall down inside the dome. The sealing of the apple with wax prevents desiccation

- make up a solution of 0.6 g SGP absorbent polymer in 200 ml boiling water, allow to cool, add 0.2 ml Zephiran®
- break wax seal and remove apple from petri dish
- scrape eggs gently from the inner surface of the apple and from the petri dish with a fixed-blade scalpel and stir into the cooled SGP gel to produce a uniform suspension
- the eggs in one or more small samples of this gel, e.g. 0.5 ml are then counted under a stereo dissecting microscope and the number of eggs per ml of suspension calculated
- distribute the suspension over the medium in the flat dishes via a 25 ml graduated pipette in such a way that 5 to 6 eggs are applied per gram of diet.

5.3 Larval rearing

a. Natural

If fresh fruit is used as the oviposition medium it should be rinsed in Zephiran® after oviposition and held for two to three days at 27°C. The fruit can then be split and placed on the surface of the artificial diet and held over sawdust in a gauzed cage (Fig. 1). For small colonies of flies where there is sufficient pulp in the oviposition fruit to feed the larvae the fruit can be left whole and merely held over sawdust.

b. Artificial

- Each flat dish of medium, now inoculated with eggs is placed in one crisper box 25 x 25 x 12 cm and covered with a plastic sheet and a sheet of brown paper - these being tied on by a piece of string. The plastic sheet is necessary to prevent the eggs on the surface of the medium drying out before hatching and the brown paper is necessary to inhibit the growth of a yeast which may appear on the surface of the medium in response to light. After three days when all of the eggs have hatched these sheets are removed.
- The dishes are then placed in large holding cages with about 1 cm of sieved dry sawdust in the bottom on the cage. Sawdust is hygroscopic and absorbs enough moisture from the general environment of the open media dishes to prevent desiccation of pupating larvae.

5.4 Pupal collection

- Mature larvae spring out of the medium and pupate in the sawdust usually under the dish. At 27°C, the first larvae are ready to pupate seven days after egg laying
- Pupae and sawdust can be swept up gently with a pan and brush
- Pupae and sawdust are then readily separated by gentle sieving by hand in a 1.7 mm sieve
- Pupae are then spread out not more than 1 cm deep in plastic dishes and placed in the adult fly cages.

5.5 Adults

- First adult emergence begins at 18 days after egg laying at 27°C. Newly emerged adults require sugar (most conveniently supplied in the form of cubes) and water. A water container is made up of a plastic container (17 x 12 x 7 cm) the lid of which is slit to allow a piece of wettex sponge to protrude and lie on the lid whilst the lower part remains in contact with water (Fig. 5)
- To enable flies to reach sexual maturity a protein food source is required. We use Amber BYF®, autolysed brewers' yeast fraction in our laboratory but other autolysed yeasts may be suitable. Protein is not readily utilised by flies unless it is hydrolysed.
- The powdered yeast fraction is placed in the cages in plastic dishes
- Fresh food and water are provided weekly with the water wicks of Wettex® sponge being renewed every two weeks.

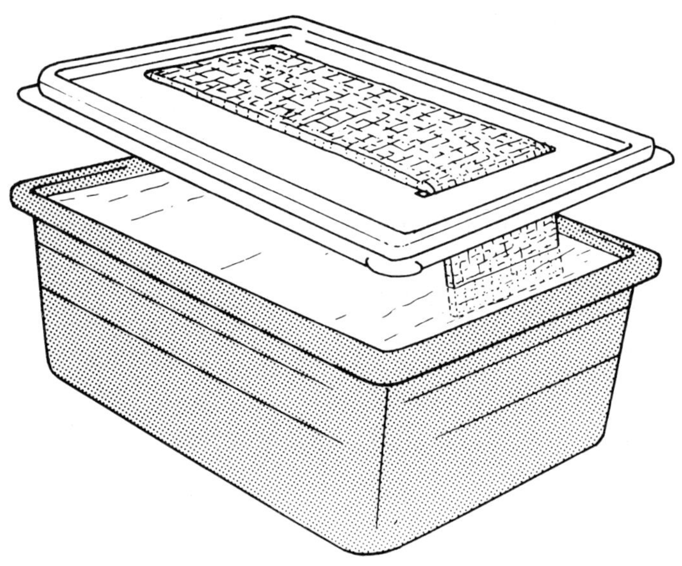

Fig. 5. Water dispenser made from plastic lunch box with sponge wick. (illustration by S. Sands)

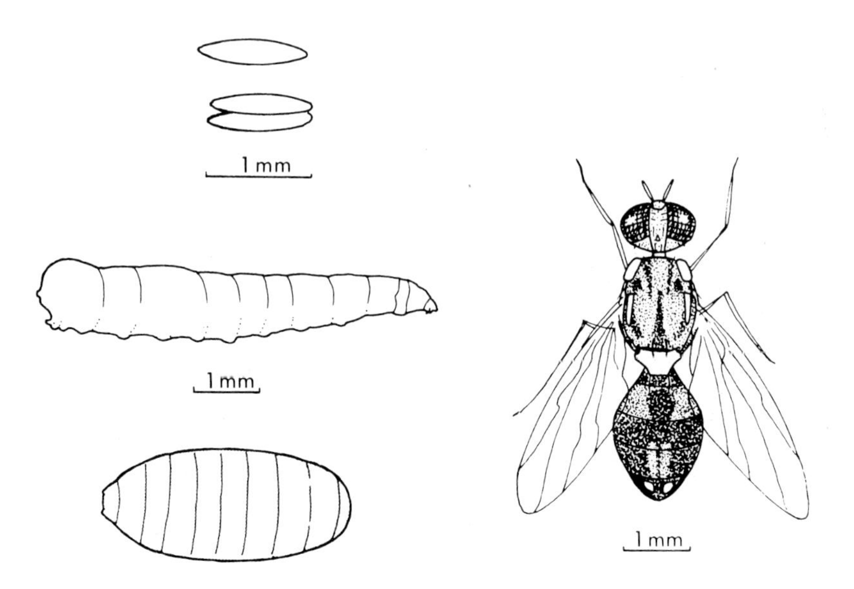

Fig. 6. Stages of the Queensland fruit fly; egg, larva, pupa and adult. (illustration by S. Sands)

- Cages become soiled with time so colonies should be allowed to run down after about 3 months, when the cage frames are stripped, washed and repainted and gauzes replaced.

5.6 Rearing schedule

Day 1 - prepare apple halves, place in adult cage, leave overnight
- feed and water adults
- weigh out ingredients for larval medium, soak overnight

Day 2 - prepare medium
- remove apple from adult cage
- prepare egg suspension in SGP
- inoculate medium with eggs
- cover boxes of newly inoculated medium with plastic and brown paper

Day 5 - remove plastic and brown paper covers from boxes or larval medium inoculated on day 2. Place dishes of medium in cages over sawdust

Day 8 - As for day 1

Day 12 - Gently sieve pupae from sawdust and place in fly cages

Day 15 - Replace wicks in water dispensers

5.7 Special problems

The rearing of this insect is, for the most part, trouble-free provided that the following precautions are observed:

- all instruments, containers, or equipment used in egg collection and inoculation to be rinsed in 0.05% Zephiran® solution
- medium should be protected from light until the larvae begin working through it (at about 3 days).

6. LIFE CYCLE DATA

6.1 Development data (Fig. 6)

a. Eggs
- Eggs show a range of hatching times of 1 to 3 days

b. Larvae
- first pupation 7 days after egg laying } at 27°C
- last pupation 11 days after egg laying }

c. Pupae
- first adult emergence 18 days after egg laying } at 27°C
- last adult emergence 21 days after egg laying }

6.2 Survival data
- Mortality from egg to pupal stage 20 - 25% at 27°C

6.3 General information

6.3.1- Females lay 5 to 12 eggs in each oviposition puncture
- Pre-oviposition period, 15-20 days
- Mean adult longevity, 10 weeks
- Maximum adult longevity, 26 weeks
- The sex ratio is 1:1

REFERENCES

Christenson, L.D., Maeda, S. and Holloway, J.R. 1956. Substitutions of dehydrated for fresh carrot in medium for rearing fruit flies. J. Econ. Ent. 49: 135-136.

Drew, R.A.I., Hooper, G.H.S. and Bateman, M.A. 1978. Economic fruit flies of the South Pacific region. Watson Ferguson & Co. Brisbane, 137 pp.

Finney, G.L. 1956. A fortified carrot medium for mass-culture of the oriental fruit fly and certain other tephritids. J. Econ. Ent. 49: 134.

May, A.W.S. 1953. Queensland host records for the Dacinae (fam. Trypetidae). Queensland, J. Agric. Sci. 10: 36-79.

DELIA ANTIQUA

J.H. TOLMAN, J.W. WHISTLECRAFT, and C.R. HARRIS

Research Centre, Agriculture Canada, University Sub Post Office, London, Ontario, Canada. N6A 5B7

THE INSECT

Scientific Name: *Delia antiqua* (Meigen)
Common Name: Onion maggot
Order: Diptera
Family: Anthomyiidae

1. INTRODUCTION

The onion maggot (OM), *Delia antiqua* (Meigen), is a major pest on several *Allium spp.* across Europe, Asia and North America (Loosjes, 1976). OM undergo at least 2 complete generations per year, surviving extremes of climate either in pupal diapause during winter or in pupal aestivation during hot periods in the summer (Perron and Lafrance, 1961; Loosjes, 1976). Within 1 week of adult emergence, females mate and begin to lay eggs either in the soil surrounding the onion plants or in the leaf junctions of the host plant. Immediately after hatching, larvae generally migrate to the basal portion of the bulb to feed. Up to 50 larvae may be found in large onions during periods of heavy infestation. Larvae develop through 3 instars and then leave the host plant to pupate in the soil.

Several procedures for small scale culture and/or mass rearing of OM have been developed using either artificial diets or natural food sources for both adults and larvae (Rawlins, 1953; Niemczyk, 1964; Allen and Askew, 1970; Ticheler, 1971). The following technique has been developed over the past 20 years at this laboratory and is now successfully used to maintain an insecticide susceptible strain of OM, as well as other strains resistant to cyclodiene, organophosphorus and carbamate insecticides.

2. FACILITIES AND EQUIPMENT REQUIRED

2.1 Facilities (also applies to cabbage maggot (CM))

Rearing is carried out in controlled environment rooms or cabinets. Adults and continuously reared larvae are maintained at 22±1°C, 60±5% RH with LD 16:8. Light at 250-350 foot candles is provided by fluorescent tubes positioned vertically on the walls of the rooms or cabinets. Larvae scheduled for diapause induction and extended cold storage require fluctuating conditions of 8 h light at 17±1°C followed by 16 h dark at 7±1°C. Diapause pupae are stored in the dark at 1±0.5°C.

2.2 Equipment and materials

2.2.1 Adult emergence (also applies to CM and seed corn maggot (SCM))

- 2000 OM pupae (maximum)(volumetric measure, 1000 pupae = 30 ml)
- 450 ml plastic container
- 10 cm plastic petri dish
- "standard soil mixture" = sterilized muck:Plainfield sand mixture (2:1 v/v) - used as a substrate for pupal emergence
- distilled water

Fig. 1. Cage for holding adults. A-water source; B-dry food; C-honey source; D-oviposition dish.

2.2.2 Adult rearing and egg collection (also applies to CM and SCM)

- cages (30 cm^3) consisting of a plywood floor and rigid wooden frame covered with fibreglass screening (1 mm mesh) on 3 sides and having a glass rear panel. A removable cloth sleeve front provides easy access to the cage interior (Fig. 1).
- 450 ml plastic container with tight-fitting lid from which one or more paper towel wicks protrude for use as a water supply (Fig. 1).
- 10 cm filter paper with honey applied to one side is placed on the screened roof of each cage as a carbohydrate source
- 10 cm petri dish containing 2-3 g of dry food mixture
- "standard soil mixture" used as a medium for oviposition
- oviposition dishes constructed from a 15 cm plastic petri dish (platform) glued to a 10 cm petri dish bottom (reservoir). A dental wick (3.75 x 1.0 cm), inserted through a 1 cm hole drilled in the platform, into the water-filled reservoir provides sufficient moisture for approximately 300 cc of soil mixture for at least 2 days (Fig. 2).
- fresh onion quarters placed on the soil to provide an oviposition stimulant for gravid females
- distilled water
- aspirator

2.2.3 Larval rearing

- pesticide residue-free onion supply (either dry onions grown from seed, *e.g.*, cv. Rocket, Autumn Spice, or onions grown from sets, *e.g.*, cv. Yellow Ebenezer, Stuttgarter); cold storage facilities are needed to insure adequate storage duration and quality for the onions after harvest
- clean brick sand for use as a substrate for cut onions and as a pupation medium

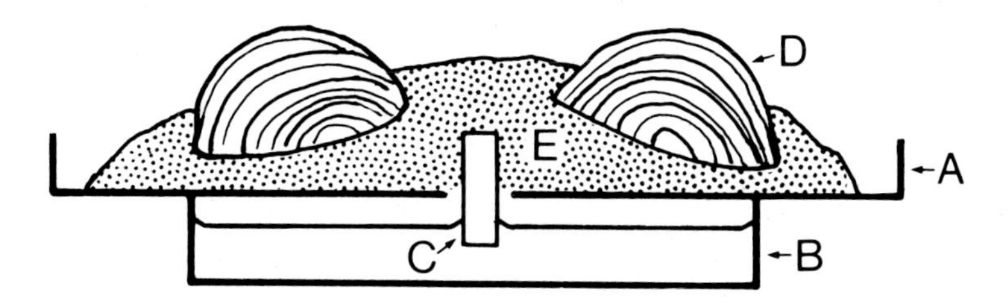

Fig. 2. Complete oviposition dish. A-platform; B-reservoir; C-dental wick; D-onion quarter attractant; E-"soil mixture".

- plastic tubs (40 x 30 x 15 cm) having 4 (1 cm) drainage holes in the bottom for use as rearing tubs
- plastic sheeting (45 x 35 cm) to cover rearing tubs for 7-10 days to prevent desiccation of eggs and young larvae
- clay tile chips to loosely cover drainage holes in each tub, retaining the sand substrate while allowing good drainage of effluent produced by onion breakdown during larval feeding
- metal trays to contain effluent draining from rearing tubs

2.2.4 Pupal collection

- flotation device for pupae recovery consisting of plastic tub (40 x 30 x 15 cm) with an exit tube 2.5 cm below the tub rim; water source and kitchen sieve to collect the pupae that have floated from the sand and onion mixtures (Fig. 3).

2.2.5 Pupal storage (also applies to CM)

- plastic 15 cm petri dishes
- 450 ml plastic containers with tightly fitting lids
- "standard soil mixture"
- 0.1% potassium permanganate ($KMnO_4$) solution
- spray mister for application of $KMnO_4$

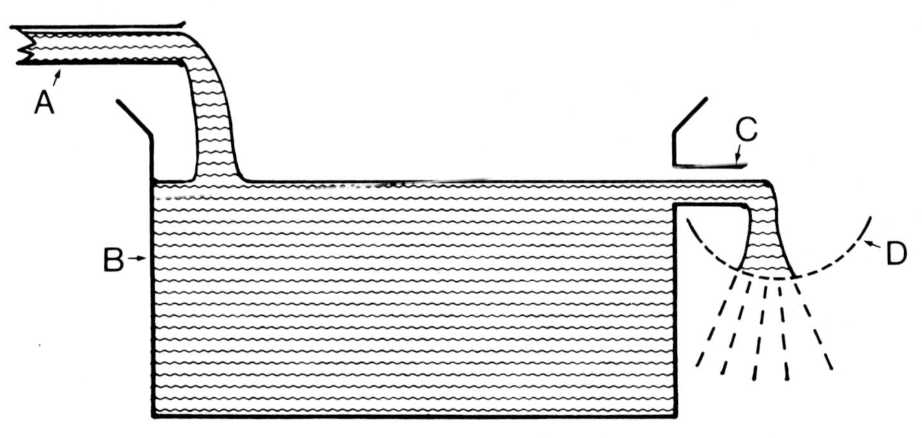

Fig. 3. Diagramatic representation of pupal recovery. A-water source; B-plastic tub; C-exit tube; D-kitchen sieve.

3. ADULT DIET

3.1 Composition (also applies to CM and SCM)

Ingredients	% by volume
brewers' yeast	50
yeast hydrolysate	33
soybean flour	17

3.2 Preparation (also applies to CM and SCM)
- place desired quantity of ingredients in a 4 liter glass jar
- thoroughly mix together with a spoon or rubber spatula
- seal tightly for prolonged storage at room temperature

4. REARING PROCEDURE

The technique can be used for field collected or laboratory reared OM. Although not an axenic technique, cleanliness is essential for continued productive rearing.

4.1 Adult emergence (also applies to CM)
- mix 2000 pupae with 250-300 cc of moist "standard soil mixture"
- place pupae:soil mixture in 450 ml plastic tub, lay 10 cm plastic petri dish on soil surface to minimize desiccation and insert complete emergence container into adult cage (Fig. 1).
- emergence begins within 10-12 days and continues for 3-4 days (Table 6.1)
- males generally emerge before females

4.2 Adult maintenance (also applies to CM and SCM)
- hold no more than 2000 flies (approximately 1:1 ratio ♀:♂ in each screened cage (Fig. 1)
- replace or replenish all food, available *ad libitum*, at time of each egg collection. Heaviest feeding occurs during the first 5 days after emergence
- when handling the positively phototropic flies, place a strong light at the rear of a large (1 m^3) box covered with a black cloth. Place the cage in the box. Most flies migrate immediately towards the side of the cage closest to the light source; gently brush or blow any stragglers from food containers or oviposition dishes

PRECAUTIONS (also applies to CM and SCM)

(i) Isolate field collections from other Dipteran cultures for at least 1 complete generation to reduce the possibility of disease, *e.g.*, *Entomophthora spp.* which results in numerous dead flies with swollen abdomens clinging to the top and sides of the cage.

(ii) Overcrowding of adult cages decreases egg production and can also increase the potential for disease outbreaks such as microsporidian infections, *e.g.*, *Nosema spp.*, which result in sudden, premature decline in egg production.

(iii) If cages become dirty before useful egg production ends, gently aspirate surviving flies into a collection bottle and release into a clean cage.

(iv) Avoid high humidity (over 70% RH) which can lead to fungal growth on food and water containers, increasing the potential for disease outbreaks.

(v) Avoid both high and low temperature extremes which decrease adult longevity and/or fecundity (Tables 6.2.1 and 6.2.3).

(vi) Evenly distributed light sources in the environmental cabinet or insectary reduce the problem of crowding of flies in the cage corners due to positive phototropism.

(vii) Wash and thoroughly rinse all cages and equipment immediately after use.

4.3 Egg collection

- at 22±1°C, oviposition begins 6-7 days after adult emergence, continuing for up to 3 weeks, depending on adult density and the strain under culture
- add oviposition dish (Figs. 1 and 2) to each cage when flies are 6 days old
- change oviposition dishes every 2-3 days
- ensure that each reservoir is filled with distilled water and that the soil is kept moist to prevent desiccation of the eggs
- soil should be "mounded" (Fig. 2), allowing a moisture gradient to develop within the substrate
- washing of eggs in a surface sterilizing solution is not necessary for general rearing on a natural food source
- divide eggs collected from 1 cage during the peak oviposition period (i.e., adults 10-20 days old) between 2 rearing tubs to prevent over-crowding

4.4 Continuous larval rearing and pupal collection

- cover each of 4 drainage holes in the rearing tub with clay chips
- add clean, moist, brick sand to each tub to a depth of 6 cm
- tightly pack freshly cut onion halves, root end down, in the sand (Fig. 4)
- scatter onion quarters and soil containing OM eggs over the onions and water in with 250 ml of water
- cover each tub with a plastic sheet during the critical early days of egg hatch and larval establishment
- holding rearing tubs at 22±1°C with an L:D 16:8 photoperiod produces the maximum number of pupae while maintaining optimum rate of development (Tables 6.2.1 & 6.2.2)
- to obtain approximately 2000 large (15 mg+) pupae per tub, fresh onions must be added at least once after the initial onion supply is 75%-80%

Fig. 4. Complete rearing tub (minus plastic cover) showing onion food source covered by soil containing eggs.

consumed. Remove the plastic sheet, old onions and OM larvae, add freshly cut onion halves, cut side upward, and replace the original onions and maggots on top of the fresh food source.
- do not replace the plastic sheet after replenishing the onion food supply
- recover pupae by flotation (Fig. 3) 3 weeks after the tub is set with 0-48 h old eggs
- count collected pupae volumetrically and set up as in Section 4.1 to begin the cycle anew
- population increases by this method average 10x per generation

PRECAUTIONS
(i) Do not use red onions as a food source for larval rearing.
(ii) Avoid extreme wetness in sand; dry sand can be added during feeding to adsorb excess moisture.
(iii) Do not wash out pupae during the first 48 h after pupation as newly formed pupae do not float.

4.5 Insect quality
- pupal weight - at least 15 mg/pupa
- at 22±1°C, adult flies should live approximately 4 weeks, producing significant numbers of fertile eggs for at least 2 weeks

5. REARING OM FOR DIAPAUSE INDUCTION AND LONG TERM STORAGE

The tendency of larvae exposed to low temperatures to enter pupal diapause can be used to advantage if one desires to accumulate large numbers of the insect for uses such as dispersal in a sterile insect release program or as a food source for production of predators and/or parasites. Alternatively, small cultures may be placed in diapause to minimize labour requirements. Diapausing pupae can be stored at least 1 year at 1°C (Fig. 5) with little effect on emerging flies; non-diapausing pupae should be stored no longer than 8-10 weeks.

5.1 Larval rearing conditions
- when diapausing pupae are desired, set up rearing tubs and hold for 5 days as described in Section 4.4, to allow rapid egg hatch and establishment of larvae in the food source

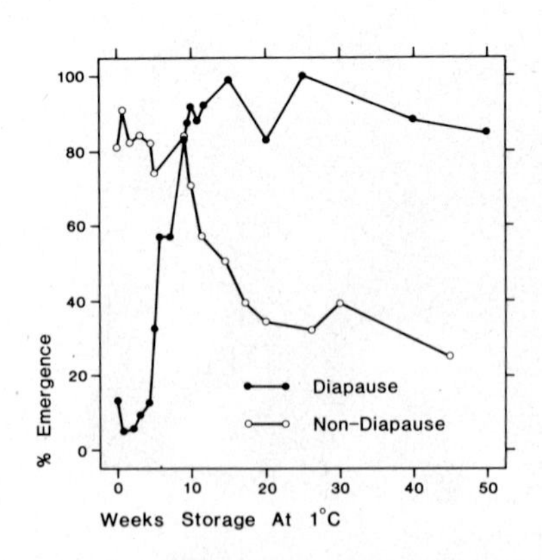

Fig. 5. OM adult emergence from pupae produced in diapause inducing and continuous rearing conditions. Samples of the pupae reared under both conditions, removed from storage at indicated intervals, and numbers of adults emerging within 15 days recorded.

- transfer rearing tubs to diapause inducing conditions of 8 h light at 17±1°C followed by 16 h dark at 7±1°C (10.3°C average daily temperature)
- as the food source is consumed, replace onions as in Section 4.4. Cooler temperatures and improved food utilization increase production to 2500 pupae/tub.
- recover pupae by flotation approximately 6 weeks after rearing tub is set with 0-48 h old eggs

PRECAUTIONS

as in Section 4.4 PRECAUTIONS

5.2 Pupal storage

5.2.1 Small cultures (also applies to CM and SCM)

- volumetrically count 2000 pupae, mix with 200 cc of moist "standard soil mixture" and place in a 15 cm plastic petri dish
- before covering, apply a light spray of 0.1% $KMnO_4$ solution to the soil surface to inhibit fungal growth
- store in dark at 1±0.5°C

5.2.2 Mass cultures (also applies to CM)

- volumetrically count 13,000 pupae and place in a 450 ml plastic container
- pierce 5 (2 mm) holes in the bottom of the container for drainage and 5-10 (1 mm) holes around the top of the container for air exchange
- cover with a tightly fitting lid
- store in dark at 1±0.5°C

PRECAUTION

Diapausing pupae must be stored at 1±0.5°C for at least 8 weeks to fulfill "chilling" requirements of diapause (Fig. 5).

6. LIFE CYCLE DATA

6.1 Standard rearing conditions

Table 6.1
Life cycle data under standard rearing conditions (22±1°C; 60±5% RH; LD 16:8)

LIFE STAGE	CONTINUOUS CONDITIONS Duration (days)	DIAPAUSE INDUCING CONDITIONS Duration (days)
ADULT		
preoviposition	6 - 7	6 - 7
oviposition	21	21
EGG	3	3
LARVA	12 - 15	5
		35 - 40[1]
PUPA	10 - 16	in diapause[2]

[1] 8 h light at 17±1°C followed by 16 h dark at 7±1°C.
[2] pupae must be chilled at 1±0.5°C for at least 8 weeks to fulfill diapause requirements.

6.2 Other conditions

Experience over many years has shown that temperature is the most important environmental factor regulating development, survival and productivity. The following data illustrate the importance of effective temperature control for efficient rearing.

Table 6.2.1
Effect of temperature on survival during development

LIFE STAGE	% SURVIVAL AT INDICATED TEMPERATURE 15°C	20°C	25°C	30°C
EGG-LARVA	95.04 ±0.89 (5)[i]	91.78 ±3.43 (6)	92.63 ±1.84 (6)	93.72 ±1.04 (5)
EGG-PUPA	75.83 ±1.75 (12)	79.17 ±3.52 (12)	74.50 ±3.11 (12)	63.67 ±4.17 (12)
EGG-ADULT	*ii*	75.75 ±5.69 (8)	63.00 ±2.95 (8)	24.00 ±3.22 (8)

i - number of experiments. For egg-larva, each experiment represents 100 eggs; for egg-pupa and egg-adult, each experiment represents 50 eggs originally set up on onion food source.
ii - pupae enter diapause

Table 6.2.2
Effect of temperature[i] on OM development

LIFE STAGE	DAYS REQUIRED AT INDICATED TEMPERATURES 10°C	15°C	20°C	25°C	30°C
1. EGG	10.32 ±0.03 (304)[ii]	5.63 ±0.02 (514)	3.08 ±0.01 (500)	1.96 ±0.01 (589)	1.69 ±0.003 (564)
2. LARVA (3 instars)	40.40 ±0.36 (75)	25.59 ±0.18 (107)	15.49 ±0.13 (107)	11.17 ±0.05 (118)	11.12 ±0.30 (84)
3. PUPA	*iii*	*iii*	12.18 ±0.20 (123)	9.87 ±0.11 (181)	8.35 ±0.06 (466)
4. ADULT (preoviposition)	37.00 ±7.00 (2)[iv]	16.25 ±1.55 (4)	8.67 ±0.33 (6)	6.50 ±0.34 (6)	6.50 ±0.43 (6)
TOTAL 1 + 2 + 3 + 4	-	-	38.75-40.09	28.99-30.01	26.87-28.45

i - LD 16:8; 60±5%RH
ii - number of individuals tested
iii- pupae enter diapause
iv - number of experiments. Each experiment comprised 20 pairs of newly emerged adult flies.

Table 6.2.3
Effect of temperature on egg production

# EGGS/♀ AT INDICATED TEMPERATURE			
15°C	20°C	25°C	30°C
32.5	211.4	139.5	70.9

20 pairs of newly emerged flies were set up in rearing cages and eggs collected every other day for 48 days (20, 25, 30°C) or 68 days (15°C).

7. SAMPLE SHIPMENT (also applies to CM and SCM)

OM may be readily shipped as pupae. Packing pupae loosely in moist soil in a small thermos bottle allows shipment of 1000 or more pupae with few problems due to desiccation or anoxia. Samples, however, must arrive at their destination within 10 days, prior to adult emergence.

8. CULTURE AVAILABILITY

The following strains of OM are available from the London Research Centre on request:

1. insecticide susceptible
2. cyclodiene resistant
3. white-eye homozygous recessive strain, cyclodiene resistant
4. parathion resistant
5. parathion resistant cross-selected with carbofuran
6. parathion resistant cross-selected with fonofos
7. white-eye homozygous recessive strain, parathion resistant

9. REFERENCES

Allen, W.R. and Askew, W.L. 1970. A simple technique for mass rearing the onion maggot on an artificial diet. Can. Ent. 102: 1554-1558.
Loosjes, M. 1976. Ecology and genetic control of the onion fly *Delia antiqua* (Meigen). Agric. Research Report 857, Centre for Agr. Publ. and Docu. & Wageningen. 179 pp.
Niemczyk, H.D. 1964. Mass rearing of the onion maggot, *Hylemya antiqua*, under laboratory conditions. J. Econ. Ent. 57: 57-60.
Perron, J.P. and Lafrance, J. 1961. Notes on the life-history of the onion maggot, *Hylemya antiqua* (Meig.)(Diptera: Anthomyiidae) reared in field cages. Can. Ent. 93: 101-106.
Rawlins, W.A. 1953. A method for rearing the onion maggot in insectary cultures. J. Econ. Ent. 46: 1101.
Ticheler, J. 1971. Rearing of the onion fly, *Hylemya antiqua* (Meigen), with a view to release of sterilized insect. In: Sterility Principles for Insect Control or Eradication (Proc. Symp. Athens, 1970). IAEA Vienna, pp 341-366.

DELIA PLATURA

J.W. WHISTLECRAFT, J.H. TOLMAN, and C.R. HARRIS

Research Centre, Agriculture Canada, University Sub Post Office, London, Ontario, Canada. N6A 5B7

THE INSECT

Scientific Name: *Delia platura* (Meigen)
Common Name: Seedcorn maggot
Order: Diptera
Family: Anthomyiidae

1. INTRODUCTION

The seedcorn maggot (SCM), *Delia platura* (Meigen), is an economically important pest on a variety of crops across North America and Europe (Miller and McClanahan, 1960; Eckenrode *et al.*, 1973). Depending on local climatic conditions, up to 4 generations of SCM may occur annually; highest populations usually coincide with spring seeding or transplanting. Female flies are stimulated to oviposit by a variety of signals including moist soil, decaying vegetation and germinating seeds (McClanahan and Miller, 1958; Barlow, 1965; Harris *et al.*, 1966b). Eckenrode *et al.*, (1975) also found that seed-borne microorganisms produce metabolites that elicit oviposition responses in the female. Eggs are laid in soil close to sources of food; newly hatched larvae either feed on rotting vegetation or directly attack germinating seeds. During heavy infestations young roots or stems of transplants are also attacked and may be destroyed.

The variety of crops attacked and the severity of damage caused by SCM has resulted in extensive research on both biology and control (*e.g.*, Harris *et al.*, 1962; 1966a; Eckenrode *et al.*, 1973; 1974). Such research requires production of large numbers of insects for experimental purposes. Several workers have devised techniques for limited or large-scale rearing programs using a variety of food sources including artificial diets (McClanahan and Miller, 1958; Harris *et al.*, 1966b). The procedure described below, modified from the technique described by Harris *et al.*, (1966b) has proven satisfactory for rearing both insecticide susceptible and cyclodiene resistant SCM strains at this laboratory for many years.

2. FACILITIES AND EQUIPMENT REQUIRED

2.1 Facilities

Rearing is carried out in controlled environment cabinets or rooms. Adults and continuously reared larvae are maintained at 19±1°C, 60±5% RH with LD 16:8. Light at 250-350 foot candles is provided by fluorescent tubes positioned vertically on the walls of the rooms or cabinets. Larvae scheduled for diapause induction and extended cold storage are reared in the dark at 10±1°C, 60±5% RH. Diapause pupae are stored in the dark at 1±0.5°C.

2.2 Equipment and materials

2.2.1 Adult emergence
- 1000 SCM pupae (maximum)(volumetric measure, 1000 pupae = 15.5 ml)
- refer to OM section 2.2.1 (Vol. II, p. 49) for additional details

2.2.2 Adult rearing and egg collection
- with the exception of the oviposition attractant, refer for details to

Fig. 1 and to OM section 2.2.2 (Vol II, p. 50)

- 15-20 swollen white bean seeds, *Phaseolus vulgaris* (L.), previously soaked in water for 2 h provide an oviposition stimulus for female SCM (Fig. 2)

2.2.3 Larval rearing

- pesticide residue-free white bean seeds
- 17.5 x 10 cm plastic bulb pots for larval rearing containers
- "standard soil mixture" as a rearing substrate
- large plastic bag
- plastic 15 cm petri dish lid

2.2.4 Pupal collection

- flotation device for pupae recovery consisting of a plastic pan (40 x 30 x 15 cm) with an exit tube 2.5 cm below the rim, a water source and a kitchen sieve to collect floating pupae (refer to OM Fig. 3)

2.2.5 Pupal storage

- plastic 15 cm petri dishes
- "standard soil mixture"
- 0.1% potassium permanganate ($KMnO_4$) solution
- spray mister for application of $KMnO_4$

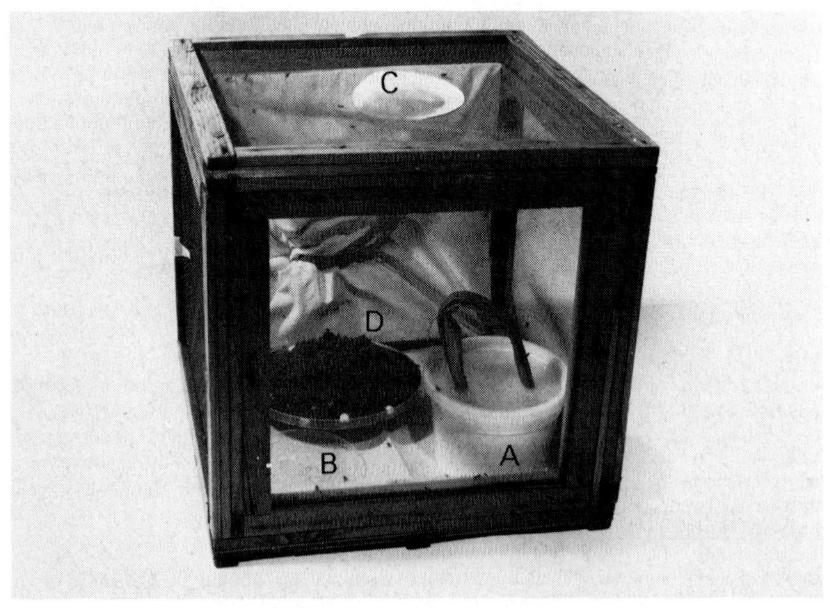

Fig. 1. Cage used for holding adults. A-water source; B-dry food; C-honey source; D-oviposition device.

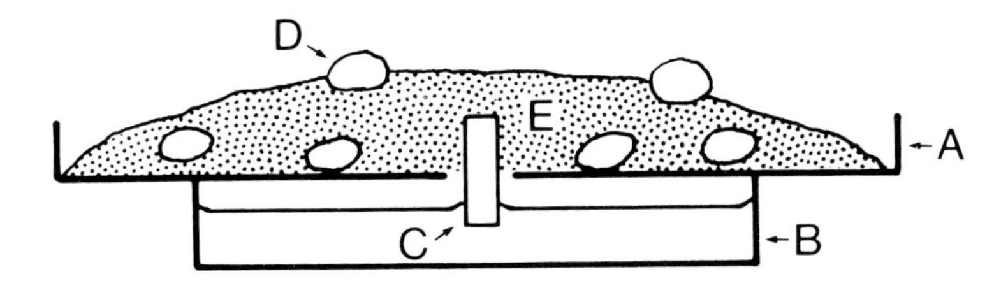

Fig. 2. Complete oviposition device. A-platform; B-reservoir; C-dental wick; D-swollen white bean seeds (attractant); E-"standard soil mixture".

3. ARTIFICIAL DIETS

3.1 Adult diet

3.1.1 Composition

- refer to OM Section 3.1 (Vol. II, p. 52)

3.1.2 Preparation

- refer to OM Section 3.2 (Vol. II, p. 52)

3.2 Larval diet

3.2.1 Composition

Ingredients	% by volume
soybean flour	19.3
whole wheat flour	19.3
brewers' yeast	9.65
potato or turnip	9.65
condensed milk	5.8
honey	3.86
corn oil	0.95
cholesterol	0.20
formic acid (90.5%)	0.10
distilled water	31.2

3.2.2 Preparation

- weigh desired quantity of soybean, wheat flour and brewers' yeast; mix thoroughly together in a large mixing bowl
- dissolve cholesterol in desired volume of corn oil
- blend potato or turnip, condensed milk, honey, formic acid and distilled water together in a large Waring blender until smooth

- add cholesterol: corn oil mixture and blend thoroughly
- add contents of blender to the dry ingredients in the mixing bowl and knead by hand into a dry dough
- the complete diet can be stored for up to 4 weeks in a refrigerated, sealed container

4. REARING PROCEDURE

The following technique is suitable for field collected or laboratory reared cultures. All equipment must be clean before use; although this technique is nonaxenic, cleanliness is a must.

4.1 Adult emergence
- mix 1000 pupae with 250-300 cc of moist "standard soil mixture"
- place pupae:soil mixture in 450 ml plastic tub, lay a plastic 10 cm petri dish on soil surface to prevent desiccation and insert completed preparation into adult rearing cage (Fig. 1)
- emergence begins within 8 days and continues for 4-5 days (Table 6.1)
- males generally emerge before females

4.2 Adult maintenance
- hold no more than 1000 flies (approximately 1:1 ratio ♂:♀) in each screened cage (Fig. 1)
- for additional details refer to OM Section 4.2 (Vol. II, p. 52)

PRECAUTION
Refer to OM Section 4.2 PRECAUTIONS (Vol. II, p. 52)

4.3 Egg collection
- at 19±1°C, oviposition begins 6-7 days after adult emergence, continuing for up to 3 weeks depending on adult density and the strain under culture
- divide eggs collected every 48 h from one cage during the peak oviposition period (*i.e.*, adults 10-20 days old) among 2 or 3 rearing pots
- refer to OM Section 4.3 for additional details

4.4 Continuous larval rearing and pupal collection
- mix 25-30 swollen white bean seeds with 450 cc of moist "standard soil mixture" and place in each bulb pot
- add either 1/3 or 1/2 of the soil:bean mixture containing 0-48 h old eggs from an oviposition device
- place a plastic 15 cm petri dish on the soil surface to reduce desiccation (Fig. 3)
- place the larval rearing container in a large plastic bag to maintain high humidity for the first 6 days
- add fresh swollen white bean seeds every other day and check the moisture level of the soil
- bury a 20 g portion of diet in the soil after 4-5 days, when the larvae have reached the second instar. Larvae are attracted to the diet, transferring readily from the white bean seeds
- additional diet may be required as the larvae grow
- recover pupae by flotation (OM Fig. 3) 2 weeks after the container is set with 0-48 h old eggs
- count collected pupae volumetrically and set up as in Section 4.1 to begin the cycle anew
- population increases by this method average 5x - 10x per generation

Fig. 3. Complete larval rearing container showing white bean seeds and plastic petri dish cover to minimize desiccation.

PRECAUTIONS

(i) Remove plastic bag after 6-7 days to prevent excessive moisture buildup.

(ii) If the larval population is too low the diet may become excessively mouldy.

(iii) Low populations can be successfully reared to pupation using white bean seeds as the sole food source.

(iv) Thoroughly stir the soil during pupal extraction to free small trapped pupae from large soil lumps and portions of diet.

(v) Do not wash out pupae during first 48 h after pupation as newly formed pupae do not float.

4.5 Insect quality

- pupal weight - at least 8.5 mg/pupa
- at 19±°C adult flies should live approximately 25 days, producing significant numbers of fertile eggs for 10-12 days

5. REARING SCM FOR DIAPAUSE INDUCTION AND LONG TERM STORAGE

By utilizing the tendency of larvae reared at low temperatures to enter pupal diapause (Harris *et al.*, 1966), labour costs can be significantly reduced. Diapausing pupae can be stored at least 1 year at 1±0.5°C with no effects on emerging flies; adults emerging from non-diapausing pupae stored more than 8 weeks at 1±0.5°C suffer reduced vigour and higher mortality.

5.1 Larval rearing conditions

- when diapausing pupae are desired, set up rearing containers and hold for 4 days as described in Section 4.4 to allow rapid egg hatch and establishment of larvae in the food source

- transfer rearing containers to diapause inducing conditions of 10±1°C, 60±5% RH and LD 0:24
- under these conditions the protective plastic bag may be removed and feeding frequencies may be reduced to once every 3-4 days
- recover pupae by flotation approximately 5 weeks after the rearing container is set with 0-48 h old eggs

PRECAUTIONS

In addition to those outlined in Section 4.4 PRECAUTIONS, discard old, consumed diet after removing any remaining larvae to reduce CO_2 buildup in rearing containers.

5.2 Pupal storage

- refer to OM Section 5.2.1 (Vol. II, p. 55)

PRECAUTIONS

Diapausing pupae must be stored at 1±0.5°C for at least 6 weeks to fulfill the "chilling" requirements of diapause.

6. LIFE CYCLE DATA

TABLE 6.1
Life cycle data under standard rearing conditions (19±1°C; 60±5% RH; LD 16:8)

LIFE STAGE	CONTINUOUS CONDITIONS Duration (days)	DIAPAUSE INDUCING CONDITIONS Duration (days)
ADULT		
preoviposition	6 - 7	6 - 7
oviposition	21	21
EGG	3	3
LARVA	10 - 14	4 30[1]
PUPA	8 - 13	in diapause[2]

[1] 10±1°C; 60±5% RH; LD 0:24
[2] pupae must be chilled at 1±0.5°C for at least 6 weeks to fulfill diapause requirements.

7. SAMPLE SHIPMENT

Refer to OM Section 7 (Vol. II, p. 57). Samples must arrive at their destination within 8 days, prior to adult emergence.

8. CULTURE AVAILABILITY

The following strains of SCM are available from the London Research Centre on request:

1. insecticide susceptible
2. cyclodiene resistant

9. REFERENCES

Barlow, C.A. 1965. Stimulation of oviposition in the seed-corn maggot fly, *Hylemya cilicrura* (Rond.)(Diptera: Anthomyiidae). Ent. exp. et appl. 8: 83-95.

Eckenrode, C.J., Gauthier, N.L., Danielson, D., and Webb, D.R. 1973. Seedcorn maggot: Seed treatments and granule furrow applications for protecting beans and sweet corn. J. Econ. Ent. 66: 1191-1194.

Eckenrode, C.J., Kuhr, R.J., and Khan, A.A. 1974. Treatment of seeds by solvent infusion for control of the seedcorn maggot. J. Econ. Ent. 67: 284-286.
Eckenrode, C.J., Harman, G.E., and Webb, D.R. 1975. Seed-borne microorganisms stimulate seedcorn maggot egg-laying. Nature 256: 487-488.
Harris, C.R., Svec, H.J., and Mazurek, J.H. 1962. Susceptibility of seed maggot flies, *Hylemya spp.*, to contact applications of aldrin, DDT, and diazinon. J. Econ. Ent. 56: 563-565.
Harris, C.R., Hitchon, J.L., and Manson, G.F. 1966a. Distribution of cyclodiene-insecticide resistance in the seed maggot complex in relation to cropping practices in southwestern Ontario. J. Econ. Ent. 59: 1483-1487.
Harris, C.R., Svec, H.J., and Begg, J.A. 1966b. Mass rearing of root maggots under controlled environmental conditions: Seed-corn maggot, *Hylemya cilicrura;* bean seed fly, *H. liturata; Euxesta notata;* and *Chaetopsis* sp. J. Econ. Ent. 59: 407-410.
McClanahan, R.J., and Miller, L.A. 1958. Laboratory rearing of the seed-corn maggot, *Hylemya cilicrura* (Rond.)(Diptera: Anthomyiidae). Can. Ent. 90: 372-374.
Miller, L.A., and McClanahan, R.J. 1960. Life-history of the seed-corn maggot, *Hylemya cilicrura* (Rond.) and of *H. liturata* (Mg.)(Diptera: Anthomyiidae) in southwestern Ontario. Can. Ent. 92: 210-221.

DELIA RADICUM

J.W. WHISTLECRAFT, J.H. TOLMAN, and C.R. HARRIS

Research Centre, Agriculture Canada, University Sub Post Office, London, Ontario, Canada. N6A 5B7

THE INSECT

Scientific Name: *Delia radicum* (L.)
Common Name: Cabbage maggot
Order: Diptera
Family: Anthomyiidae

1. INTRODUCTION

The cabbage maggot (CM), *Delia radicum* (L.), is a serious pest of a variety of cruciferous crops in temperate regions of North America, Europe and Asia. Depending on local climatic conditions, up to 3 generations of CM may occur each year (Soni, 1976). Adults emerge and mate in early spring and within 7 days females can lay eggs in the soil surrounding a suitable host plant. Newly hatched maggots invade the root tissues, feed, develop through 3 larval instars, and leave the plant to pupate in the soil. Aestivation may occur in newly formed pupae, at temperatures over 22°C (Harris and Svec, 1966), reducing numbers of flies present during extended periods of hot, dry weather. In late summer and early fall, soil temperatures below 15°C combine with short photoperiods to produce diapausing pupae able to withstand prolonged cold and/or freezing temperatures (McLeod and Driscoll, 1967).

Several techniques for rearing CM, generally using rutabaga as a food source, have been reported (Read, 1960; 1965; 1969; Harris and Svec, 1966; Finch and Coaker, 1969; Vereecke and Hertveldt, 1971; Keymeulen *et al.*, 1981). Dambre-Raes *et al.* (1976) also described a technique for rearing CM using an artificial diet. The technique described below, modified from that described by Harris and Svec (1966) has been used to maintain insecticide-susceptible and cyclodiene resistant CM strains at this laboratory for many years.

2. FACILITIES AND EQUIPMENT REQUIRED

2.1 Facilities

Rearing is carried out in controlled environment cabinets or rooms. Adults and continuously reared larvae are maintained at 19±1°C, 60±5% RH with LD 16:8. For additional details refer to onion maggot (OM) Section 2.1.(Vol. II, p. 49).

2.2 Equipment and materials

2.2.1 Adult emergence
- 2000 CM pupae (maximum)(volumetric measure, 1000 pupae = 32 ml)
- refer to OM Section 2.2.1 for additional details (Vol. II, p. 49)

2.2.2 Adult rearing and egg collection
- with the exception of the oviposition device, refer for details to Fig. 2.1 and to OM Section 2.2.2 (Vol. II, p. 50)
- oviposition device constructed as follows: (Fig. 2.2)
- 2, 17.5 cm diameter plastic bulb pots; the lower 17.5 x 14 cm deep pot serves as a water reservoir for the shallower 17.5 x 10 cm deep upper pot
- a 1.2 x 15 cm dental wick is suspended through a 1.5 cm hole in the bottom of the upper pot and is covered with a layer of paper towel to

Fig. 1. Cage used for holding adults. A-water source; B-dry food; C-honey source; D-oviposition device.

distribute the moisture
- 8 cm of moist soil is placed in the bottom of the upper pot
- a small 7-9 cm vertically halved rutabaga is pushed into the soil to serve as an oviposition stimulant
- each rutabaga half is separated by 2.5 cm of soil

2.2.3 Larval rearing
- pesticide residue-free supply of rutabagas (*e.g.*, cv. Laurentian)
- cold storage facilities to ensure adequate storage duration and quality of harvested rutabagas
- clean brick sand for use as a substrate for rutabaga halves and as a pupation medium
- plastic tubs (40 x 30 x 15 cm) having 4 (1 cm) drainage holes in the bottom for use as rearing tubs
- plastic sheeting (45 x 35 cm) to cover rearing tubs for 7-10 days to prevent desiccation of eggs and young larvae
- clay tile chips to loosely cover drainage holes in each tub, retaining the sand substrate while allowing good drainage of effluent produced by rutabaga breakdown during larval feeding
- metal trays to contain effluent drainage from rearing tubs

2.2.4 Pupal collection
- flotation device for pupae recovery consisting of plastic tub (40 x 30 x 15 cm) with an exit tube 2.5 cm below the tub rim; water source and kitchen sieve to collect the pupae that have floated from the sand and rutabaga mixture (refer to OM Fig. 3) (Vol. II, p. 51)

2.2.5 Pupal storage
- refer to OM Section 2.2.5 (Vol. II, p. 51)

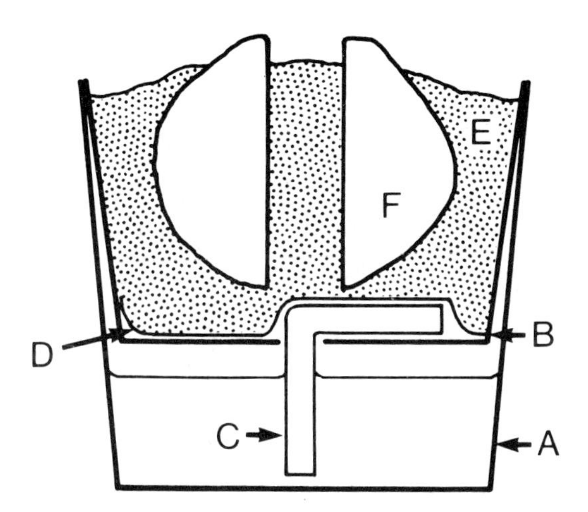

Fig. 2. Complete oviposition device. A-outer pot; B-inner pot; C-dental wick; D-paper towel; E-"soil mixture"; F-rutabaga.

3. ADULT DIET

3.1 Composition

- refer to OM Section 3.1 (Vol. II, p. 52)

3.2 Preparation

- refer to OM Section 3.2 (Vol. II, p. 52)

4. REARING PROCEDURE

The technique can be used for field collected or laboratory reared CM cultures. Although not an axenic technique, cleanliness is essential for continued productive rearing.

4.1 Adult emergence

- refer to OM Section 4.1 (Vol. II, p. 52)
- emergence begins within 12-14 days and continues for 10-11 days (Table 6.1)

4.2 Adult maintenance

- refer to OM Section 4.2 (Vol. II, p. 52)

PRECAUTIONS

refer to OM Section 4.2 PRECAUTIONS (Vol. II, p. 52)

4.3 Egg collection

- at 19±1°C, oviposition begins 6-7 days after adult emergence, continuing for up to 3 weeks, depending on adult density and the strain under culture
- divide eggs collected every 48 h from 1 cage during the peak oviposition period (*i.e.*, adults 12-21 days old) among 3 continuous rearing tubs or

Fig. 3. Complete rearing tub (minus plastic cover) showing rutabaga food source covered by soil containing eggs.

2 rearing tubs destined for diapause inducing conditions

- refer to OM Section 4.3 for additional details (Vol. II, p. 53)

4.4 Continuous larval rearing and pupal collection

- cover each of 4 drainage holes in the rearing tub with clay tile chips
- add clean, moist, brick sand to each tub to a depth of 6 cm
- tightly pack freshly cut rutabaga halves (cut lengthwise) root-end down, into the sand (Fig. 3)
- fill the spaces between the rutabaga pieces with more sand as most of the rutabaga should be buried to prevent desiccation
- scatter soil and eggs from each oviposition container over the sand surface and "water in" with approximately 250 ml of water
- cover each tub with a plastic sheet during the first 10 days
- remove the plastic cover once larvae have become established inside the rutabaga to allow drying of the substrate, slowing the rate of microbial breakdown of the food source
- recover pupae by flotation (OM Fig. 3) 4 weeks after the tub is set with 0-48 h old eggs
- count collected pupae volumetrically and set up as in Section 4.1 to begin the cycle anew
- population increase by this method averages 10x - 15x per generation

PRECAUTIONS

(i) maintain high humidity initially to ensure a good rate of egg hatch

(ii) provide at least 1 g of rutabaga for each egg set to ensure a uniform pupal size

(iii) maintain low humidity during latter stages of larval feeding to preserve rutabaga quality

(iv) do not wash out pupae during first 48 h after pupation as newly formed pupae do not float

4.5 Insect quality

- pupal weight - at least 15 mg/pupa
- at 19±1°C, adult flies live approximately 4 weeks, producing significant numbers of fertile eggs for at least 2 weeks

5. REARING CM FOR DIAPAUSE INDUCTION AND LONG-TERM STORAGE

The tendency of larvae exposed to low temperatures to enter pupal diapause can be used to advantage if one desires to accumulate large quantities of the insect for uses such as dispersal in a sterile insect release program or as a food source for rearing of predators and/or parasites. Alternatively, to minimize ongoing labour requirements, small cultures may also be placed in diapause. Diapausing pupae can be stored for at least 1 year at 1°C with little effect on emerging adults; adults emerging from non-diapausing pupae stored for more than 8 weeks suffer reduced vigour (Keymeulen *et al.*, 1981).

5.1 Larval rearing conditions

- when diapausing pupae are desired, set up rearing tubs and hold for 5 days as described in Section 4.4 to allow rapid egg hatch and establishment of larvae in the rutabaga food source
- transfer rearing tubs to diapause inducing conditions of 8 h light at 17±1°C followed by 16 h dark at 7±1°C (10.3°C average daily temperature)
- remove the plastic sheet after 10 days
- rutabaga breaks down quite slowly under diapause inducing conditions, allowing increased initial egg densities
- if extra food is required, remove the top layer of sand containing the maggot infested rutabaga halves, lay fresh rutabaga halves, cut-surface up, on the sand base and replace the maggot infested rutabagas and sand on top
- do not replace the plastic sheet after feeding
- recover pupae by flotation (OM Fig. 3) approximately 8-9 weeks after rearing tub is initially infested with 0-48 h old eggs

PRECAUTIONS

as in Section 4.4 PRECAUTIONS

5.2 Pupal storage

5.2.1 Small cultures

- refer to OM Section 5.2.1 (Vol. II, p. 55)

5.2.2 Mass cultures

- refer to OM Section 5.2.2 (Vol. II, p. 55)

PRECAUTIONS

diapausing pupae must be stored at 1±0.5°C for at least 16 weeks to fulfill "chilling" requirements of diapause

6. LIFE CYCLE DATA

TABLE 6.1
Life cycle data under standard rearing conditions (19±1°C; 60±5% RH; LD 16:8)

LIFE STAGE	CONTINUOUS CONDITIONS Duration (days)	DIAPAUSE INDUCING CONDITIONS Duration (days)
ADULT		
preoviposition	6 - 7	6 - 7
oviposition	21	21
EGG	3 - 4	3 - 4
LARVA	18 - 22	5
		45 - 50[1]
PUPA	12 - 24	in diapause[2]

[1] 8 h light at 17±1°C followed by 15 h dark at 7±1°C
[2] pupae must be chilled at 1±0.5°C for at least 16 weeks to fulfill diapause requirements

7. SAMPLE SHIPMENT

- refer to OM Section 7 (Vol. II, p. 57)

8. CULTURE AVAILABILITY

The following strains of CM are available from the London Research Centre, on request:

1. insecticide susceptible
2. cyclodiene resistant

9. REFERENCES

Dambre-Raes, H., Dambre, P., and Hertveldt, L. 1976. Nonaxenic rearing of the cabbage maggot, *Hylemya brassicae* (Bouché) on an artificial diet. Meded. Fac. Landbouw. Rijksuniv. Gent 41: 1575-1585.
Finch, S., and Coaker, T.H. 1969. A method for the continuous rearing of the cabbage root fly, *Erioischia brassicae* (Bch.) and some observations on its biology. Bull. Ent. Res. 59: 619-627.
Harris, C.R., and Svec, H.J. 1966. Mass rearing of the cabbage root maggot under controlled environmental conditions with observations on the biology of cyclodiene-susceptible and resistant strains. J. Econ. Ent. 59: 569-573.
Keymeulen, M. van, Hertveldt, L., and Pelerents, C. 1981. Methods for improving quantitative and qualitative aspects of rearing *Delia brassicae* for sterile release programs. Ent. Exp. & Appl. 30: 231-240.
McLeod, D.G.R., and Driscoll, G.R. 1967. Diapause in the cabbage maggot, *Hylemya brassicae* (Diptera: Anthomyiidae). Can. Ent. 99: 890-893.
Read, D.C. 1960. Mass rearing of the cabbage maggot, *Hylemya brassicae* (Bouché) (Diptera: Anthomyiidae) in the greenhouse. Can. Ent. 92: 574-576.
Read, D.C. 1965. Rearing root maggots, chiefly *Hylemya brassicae* (Bouché)(Diptera: Anthomyiidae) for bioassay. Can. Ent. 97: 136-141.
Read, D.C. 1969. Rearing the cabbage maggot with and without diapause. Can. Ent. 101: 725-737.
Soni, S.K. 1976. Effect of temperature and photoperiod on diapause induction in *Erioischia brassicae* (Bch.)(Diptera: Anthomyiidae) under controlled conditions. Bull. Ent. Res. 66: 125-131.

Vereecke, A., and Hertveldt, L. 1971. Laboratory rearing of the cabbage maggot. J. Econ. Ent. 64: 670-673.

DROSOPHILIDAE I: DROSOPHILA MELANOGASTER

JONG S. YOON

National *Drosophila* Species Resource Center, Department of Biological Sciences, Bowling Green State University, Bowling Green, Ohio 43403, U.S.A.

THE INSECT

Scientific Name: *Drosophila melanogaster* (Meigen)
Common Names: "Fruit fly", vinegar fly, pomace fly
Order: Diptera
Family: Drosophilidae

1.0 INTRODUCTION

The common "fruit fly," *Drosophila melanogaster*, was one of the first animals used for genetic studies and is still the most widely used species for this purpose. The fly is cultured easily, is highly prolific, and its generation time is only 10-14 days at room temperature. In addition, *Drosophila* has an abundance of genetic variability, and is a convenient and inexpensive organism to study.

Since the turn of the century, the "fruit fly" has been the subject of studies in genetics, cytology, physiology, and ethology. Recently molecular studies with the fly have been carried out intensively. In addition, the large polytene chromsomes found in salivary gland cells are very valuable for many types of chromosome studies, including phylogenetics. Culturing these flies, therefore, is an important part of the laboratory work.

2.0 FACILITIES AND EQUIPMENT NEEDED

2.1 Facilities

Rearing of flies is carried out in controlled environment rooms at 21-25°C, 45-55% RH, and an LD 12:12 photoperiod. Oviposition takes place under the same conditions. For small scale requirements, flies can be reared in an incubator with the above conditions.

2.2 Equipment and materials required for insect handling

2.2.1 Anesthesia

There are two commonly used methods for anesthetizing flies for examinations.

a. Etherizing procedure

Flies are etherized in order to keep them quiet while they are being examined and when they are transferred to culture vessels for matings. A few drops of ether will anesthetize a batch of flies, and they can be kept asleep for 1/2 hour or longer by re-etherizing at intervals. Several types of etherizers have been devised in various *Drosophila* laboratories (see Fig. 1). The simplest ones can be constructed with little effort from readily available materials. It is best to examine the flies on a white surface: either milk-glass, opaque-glass plate or plastic plate. Flies are moved around with a "pusher," which may be a small camel's hair brush (No. 0) or any pointed object.

If the flies should begin to wake up before examination is completed, they can be re-etherized without being returned to the etherizing bottle. The re-etherizer usually consists of a small petri dish with a strip of

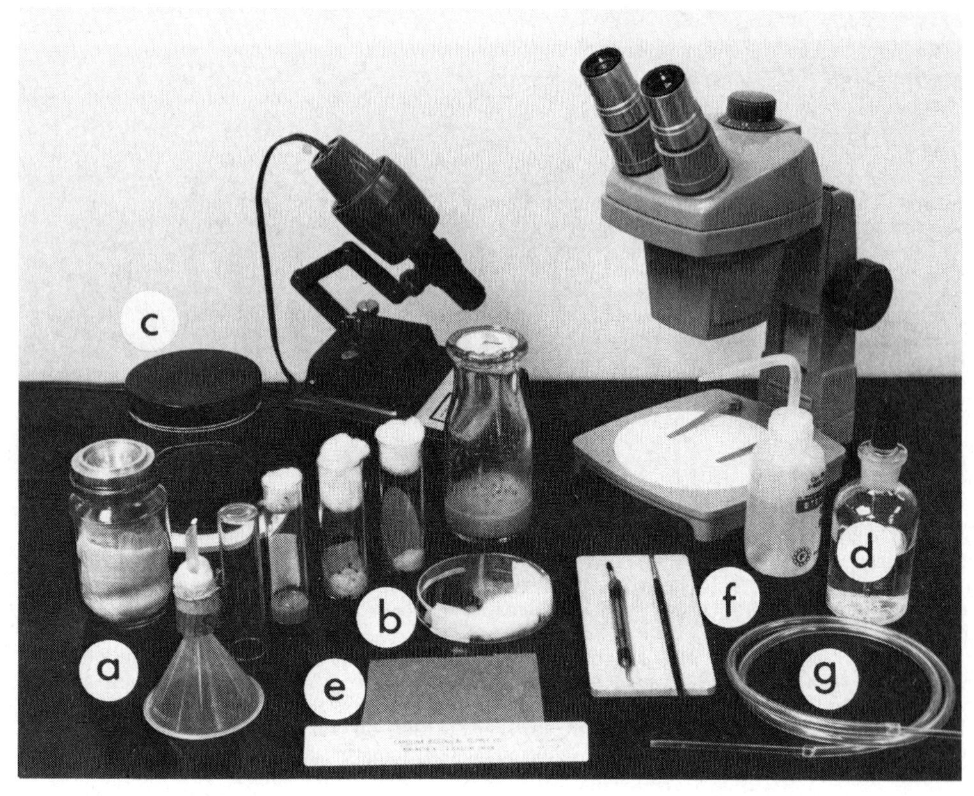

Fig. 1. Equipment and material required for handling Drosophila

a: etherizer
b: re-etherizer
c: morgue
d: ether
e: pad for bumping vials
f: white plate
g: mouth aspirator

absorbent paper or cotton fastened to the inside. A few drops of ether are placed on the paper or cotton, and the flies are covered with the re-etherizer until they go to sleep again.

Care should be taken not to overetherize flies, since that can prevent their further use or interfere with diagnosis. Flies killed by overetherizing have the wings extended at a 45° angle. Flies that have been examined and recorded, and are not required for further matings, can be discarded in a jar containing light mineral oil or alcohol (see "morgue" in Fig. 1).

b. CO_2 procedure

Another method of immobilizing flies is to use carbon dioxide anesthesia. Flies are exposed to a current of CO_2 to fall motionless, and can be kept in this state for many hours without any ill effects. The flies recover consciousness within 30 seconds after being transferred to a CO_2-free atmosphere.

2.2.2 The basic equipment for handling and examining *Drosophila*

-Stereo-dissecting microscope (preferably a range 10X-40X)
-Bright but cool light source
-Vials or bottles of culture medium, with plugs
-Etherizer or other anesthetic device (ex. CO_2)
-Re-etherizer
-Ether in dropping bottle
-White background plate for examining flies
-Fine camel's hair brush (No. 0) or teasing needles
-Fly "morgue" containing 75% ethyl alcohol or light mineral oil
-Pad for bumping vials

2.2.3 Insect mating and rearing

a. Individual (adult) mating and rearing
-Small (25mm x 95mm) or medium (30mm x 95mm) glass or plastic vials
-Non-absorbent cotton or sponge plugs (or absorbent rayon balls)
-Rubberbands for holding vials
-Tissues (cellucotton tissue strips) or paper tissues
-Pads for bumping vials

b. Mass mating or maintaining stocks
-Large (35mm x 100mm) vials or milk bottles (1/2 pint) (237 ml)
-Plugs (cotton or sponge), or cardboard milk-bottle caps
-Trays for holding vials/bottles

2.2.4 Larval handling

-Fine forceps
-Culture or concavity slides (45 x 75 x 5mm)
-Physiological saline solution (see Appendix for formula)
-Dissecting microscope, with black plate
-Light source (or illuminator)

2.3 Diet (medium)

2.3.1 Dry mix and preparation equipment

-Top loading balance (0.1 - 1000g)
-Mixing bowl
-Blender
-Beakers
-Measuring cylinders
-Measuring cups and scoops
-Stainless steel spatula
-Buckets
-Funnels
-Self sealing plastic bags
-Can opener

2.3.2 Medium cooking

-Asbestos gloves
-Towels
-Food wrapper
-Cooking equipment (steam kettle or range)
-Cooking utensil (basin or large pot)
-Muslin covers
-Food dispenser
-Refrigerator at 4°C

3.0 ARTIFICIAL DIET

3.1 Composition

3.1.1 Cooked culture media

a. Cornmeal medium (for 600 vials)

-Cornmeal (yellow, defatted)	700 g	5.88%
-Yeast (dry brewer's)	140 g	1.18%

-Agar (Carrageen Sea Kem Type 101)	77 g	.65%
-Soybean meal	20 g	.17%
-Soybean powder	10 g	.08%
-Water	10,000 ml	83.95%
-Light corn syrup	150 ml	1.26%
-Molasses (unsulfured)	500 ml	4.20%
-Tegosept®	12.5 g	.10%
-Benzyl benzoate	2.6 ml	.02%
-Ethanol (95%)	300 ml	2.52%
b. Banana medium (for 600 vials)		
-Banana (without skins)	1500 g	21.50%
-Yeast (dry brewer's)	112 g	1.61%
-Carragar	60 g	.86%
-Water	5,000 ml	71.69%
-Light corn syrup ("Karo")	133 ml	1.91%
-Malt	133 ml	1.91%
-Propionic acid	18 ml	.26%
-Ethanol (95%)	18 ml	.26%

3.1.2 Instant culture media

a. User-prepared instant medium

-Mashed potato mix	80 g	52.63%
-Yeast (dried brewer's)	20 g	13.16%
-Canned pineapple juice (or H_2O)	25 ml	16.45%
-Apple cider vinegar	25 ml	16.45%
-Moldex® powder	2 g	1.32%

b. Commercial instant medium

-Formula 4-24® (from Carolina Biological Supply Co.)

-H_2O

3.2 Brand names and sources of ingredients

Non-absorbent cotton (Chaston Medical & Surgical, P.O. Box 423, Lake Rd., Dayville, CT 06241); absorbent rayon balls (Chaston Medical & Surgical, P.O. Box 423, Lake Rd., Dayville, CT 06241); cellucotton (Chaston Medical & Surgical, P.O Box 423, Lake Rd., Dayville, CT 06241); cornmeal (Valley Farm Foods, 4223 South Ave., Toledo, OH 43615); agar (Marine Colloids, 2 Edison Place, Springfield, NJ 07081); soybean meal (Farm Feed Supplies, Mid-Wood Inc., Hub-Branch, 504 Ridge St., Bowling Green, OH 43402); soybean powder (DJ's Health Food Store, 115 W. Merry, Bowling Green, OH 43402); yeast (Local Grocery Store); corn syrup (Local Grocery Store); molasses (Stark & Co., 1104 N. Reynolds Rd., Toledo, OH 43615); Tegosept (Inolex-Goldschmidt Chemical, Div. of Wilson Pharmaceutical & Chemical Group, Jackson & Swanson Sts., Philadelphia, PA 19184); yeast (Yeast Products, Inc., 25 Styertowne Rd., Clifton, NJ 07012); malt (PMP Fermentation Products, 16840 Kercheval Ave., Grosse Pointe, MI 48230); Formula 4-24® (Carolina Biological Supply Co., Burlington, NC 27215); Difco Bacto Liver (Curtin Matheson Scientific, 4540 Willow Parkway, Cleveland, OH 44125).

3.3 Diet preparation procedure

3.3.1 Cooked culture media

a. Cornmeal medium

-Mix dry cornmeal, yeast, agar and soybean
-Add water little by little, stirring constantly
-Press mixture through sieve to remove lumps
-Bring to boil and cook 15 minutes
-While stirring add corn syrup, molasses and chemicals
-Heat 4 more min to boil again
-Using food dispenser or beaker, pour the food into vials or bottles
-Cover with muslin cover to avoid contamination
-Insert sterilized paper tissues into medium
-When the food is cool, spray or sprinkle with sterilized yeast

-Plug with sterilized cotton or sponge plugs
-Store for a day at cool room temperature before use (otherwise store at 4°C until use)

b. Banana medium

-Blend agar, yeast and chopped banana with most of the water sufficient to make a pourable mixture (save the rest of water for later)
-Pour banana mixture into cooking vessel, add remainder of water and stir well
-Bring to boil and cook 15 min, then add Karo syrup, malt and chemicals while stirring
-Cook 4 min more to boil
-Pour out medium into sterilized vials using food dispenser or beaker
-Cover with sterilized muslin cover to avoid contamination
-When the food is cool, insert sterilized paper tissue into the vials
-Plug the vials and store for a day at cool room temperature before use (otherwise store medium at 4°C until use)

3.3.2 Instant culture media

-Mix mashed potato mix with the inactivated brewer's yeast and Moldex® powder
-Mix the pineapple juice (or water) with vinegar
-Place some of the dry mix in vials to a depth of 1 cm, then add the liquid (pineapple 1:vinegar 1) to bring the volume up to 2.5 cm
-Allow a few minutes for the fluid medium to solidify
-Sprinkle a few granules of active yeast on the surface of the food, and plug the vials

PRECAUTION

All equipment, tools, plugs, cover and paper tissues must be sterilized for media preparation

4.0 REARING AND STOCK MAINTENANCE

4.1 General information (or rules)

Drosophila is a holometabolous insect, meaning that it has larval and adult stages separated by a distinct pupal stage during which metamorphosis occurs. Flies are reared in controlled environment rooms at 20-25°C, 45-55% RH, and an LD 12:12 photoperiod. Oviposition is carried out in the same conditions. Usually the rearing of these flies is carried out in the same vial or bottle throughout the entire life cycle. Special care should be given to maintain these cultures at the recommended temperature and humidity as higher temperature and humidity are conducive to the growth of bacteria, molds, and mites. The cultures should not be exposed to direct sunlight. Excess heat will sterilize the males.

4.2 Eggs

The eggs are usually fertilized at about the time of laying and early embryonic development takes place within the egg cases. The egg of *D. melanogaster* is about 0.5 mm long and has two filaments at the anteriodorsal surface. These keep the egg from sinking into soft food. Within two days after the eggs have been laid, very small larvae will have hatched out.

4.3 Larvae

The larva undergoes two molts, so that the larval period consists of three stages. In the final stage (or third instar), the larva is about 5 mm long if grown on food rich in yeast.

PRECAUTION

Special care is necessary in growing the larvae whose salivary glands are to be used. One of the major factors in obtaining large polytene chromosomes is the selection of well-fed larvae raised at a low temperature (18°C). Crowded culture bottles should be avoided. Salivary glands should be taken from full grown third instar larvae, shortly before pupation.

4.4 Pupae

Pupation occurs at about 6-7th day after hatching. The third instar larvae crawl up out of the culture medium and adhere to a relatively dry surface, such as the side of the vial or a piece of paper toweling inserted into the food. The soft larval skin dries and gradually develops a brown pigmentation. The metamorphosis from the larval to the adult form takes place within this pupa case. After about three days, the adult emerges by forcing its way through the anterior end of the pupal case.

4.5 Adults

The newly emerged adult fly is very long with unexpanded wings. Upon emergence, flies are relatively light in color, but they darken during the first few hours.

4.6 Sex determination

The sex ratio is approximately 1:1. It is very important to know how to identify the two sexes of *Drosophila* adults, and sometimes it is desirable to know the sex of the larvae and pupae.

4.6.1 Sexing larvae

- Remove the 3rd instar larvae onto a microscope slide in a little saline solution
- View with a good light and a black background
- The gonads are located about one-third of the body length from the posterior end
- Male has elliptical gonads (testes)
- Female has more round gonads (ovaries)
- The ovaries are one-fifth the size of the testes

4.6.2 Sexing pupae

- Select two or three mature, darkened pupae
- Examine the first pair of legs
- Male has dark "sex combs"
- Females does not have these "sex combs"

4.6.3 Sexing the adult fly

- The sex of *Drosophila* is most reliable through examination of the genital organs with magnification
- The male genitalia are surrounded by heavy dark bristles which do not occur on the female
- In older flies, the posterior part of the abdomen is quite dark in males and considerably lighter in females
- The tip of the abdomen is more rounded in males and elongated in females
- Only males have the "sex combs" on the front pair of legs

4.7 Collecting virgin females

A female *Drosophila* can store and utilize sperm from a single insemination for the major portion of her reproductive life. Therefore, females which have any chance of being nonvirgin should not be used for genetic crosses. The males need not be virgin. To insure virginity a procedure must be followed to collect females which are no more than 8 hours old.

Females tend to emerge somewhat sooner than males raised in the same vial. Hence, more virgin females can be obtained from a vial which has just begun producing adults than from an old culture. One should keep in mind that most of the flies will emerge during the early part of the day. One may get four times as many flies from a vial that was emptied at 8 a.m. and the flies collected at about 4 p.m. than if the vial was emptied at 10 a.m. and collection was made at 6 p.m.

- Shake out all adult flies from a culture bottle and make sure none are hiding in the paper (early in the morning)
- No longer than 8 hours later, shake out the newly hatched flies. The females from this latter group can be presumed to be virgins
- The virginity of the females may be tested by keeping them by themselves in a culture vial for 3-4 days before utilizing them for genetic crosses
- If larvae appear in the vial that contained only the females, then they were not all virgin and the crosses will not be meaningful.

NOTE: Virgin flies, like virgin chickens, will lay infertile eggs after several days without mating. You should not be disturbed if you find eggs in a vial of supposed virgins, unless you find larvae.

4.8 Insect quality

Insect quality must be checked continuously by keeping record of adult fecundity, egg viability, and adult hatchability. When stocks are kept by an inexperienced person, it is well to examine each fly carefully whenever new recommended cultures are prepared, so that any contamination (by "foreign" flies from other cultures outside) may be detected in time to prevent loss of the original stock.

In order to maintain healthy stocks, they should be changed every three or four weeks, depending on the temperature. At least two-three cultures of each stock should be kept, in case one should be unsuccessful. It is important not to keep cultures too long, and to clean old culture bottles as soon as they are discarded. Old vials or bottles may become infested with bacteria, mold, and mites, which can cause a great deal of trouble and interfere with viability of the flies, especially of some mutant types. Every stock bottle should have a label showing the stock number, the names of strains or mutant characters (in symbols) and the date of transfer.

4.9 Special problems

The principal difficulties in maintaining healthy cultures continue to be undesirable growth of bacteria, molds and mites. The most important working rules in the laboratory are:

1. All utensils and working area must be clean at all times.
2. Discarded culture must be sterilized and washed immediately.
3. The cultures should not be kept longer than one month.
4. Laboratory personnel need to observe good hygiene; frequent handwashing, washed lab coats, never leave media uncovered.

4.9.1 Control of molds

Drosophila cultures sometimes become infested with molds. Tegosept® (methyl-p-hydroxybenzoate) reduces the growth of molds. However, it is very important not to add more Tegosept than is required to control the mold, for it also hinders the growth of yeasts and of flies.

4.9.2 Control of mites

Mites are tiny, dot-sized members of the class Arachnida, having eight legs and a rounded body. Though there are many species of mites, the common laboratory varieties are white and feed on mold, *Drosophila* food, and dried pupal cases. The reproductive powers of mites are even greater than flies, and they may rapidly infect a laboratory and decimate the *Drosophila* population. Therefore, any infected culture should be removed immediately from the laboratory and sterilized.

The best prevention against an infection of mites is cleanliness. All utensils and working areas should be kept clean. Laboratory personnel should observe strict hygienic rules, including handwashing and cleaning of lab coats. A culture should not be kept longer than one month. If the stock is essential, some adult flies and larvae can be examined and transferred to a new culture. Within a few days, the same adults are rechecked and transferred: continue until the final culture is free of mites.

It has been known that an effective agent against mites is benzyl benzoate. A 2% solution of benzyl benzoate in ethanol (95%) can be spread on shelves containing cultures and population cages. (If the shelves are of metal the solution will last for at least a few weeks). Lindane-treated shelf paper will kill mites which walk across it. If culture vessels of *Drosophila* are set on such paper, mites cannot cross the paper and the flies in the culture vials are in no way injured (Flagg, 1973).

4.9.3 Control of bacteria

Other difficulties in maintaining healthy cultures may be due to the growth of bacteria and occasionally undesirable yeasts. Hanks *et al.* (1968) and Felix (1969) described newer methods for the control of bacterial infections of *Drosophila* media. Gonzales and Abrahamson (1968) described a method of controlling undesirable yeast growth on media.

5.0 LIFE CYCLE DATA

The life cycle and survival data during development of various stages under optimum rearing conditions of 25°C, 45-55% RH, and an LD 12:12 photoperiod are as follows:

5.1 Developmental data (*D. melanogaster* wild type)

Age (hours from oviposition)	Stage
0	Egg layed
0-22	Embryo
22	Hatching from egg (1st instar)
47	First molt (2nd instar)
70	Second molt (3rd instar)
118	Formation of puparium (= pupation)
215	Adult emerges from pupa case

5.2 Survival and other data

Egg viability	98 ± 2%
Preoviposition period	24 ± 2 h
Peak egg-laying period	3rd-5th day
Mean fecundity (maximum 500 eggs in 10 days)	55 ± 5 eggs/day/♀
Mean adult longevity	37 ± 7 days

6.0 PROCEDURES FOR SUPPLYING INSECT

6.1 Pre-requisites

- -Research and development of rearing procedures must be completed
- -The founder colony should be healthy and free from any contamination
- -The duration of developmental stages at various temperatures and humidity must be known
- -The best diet should be used
- -Standard rearing methods should be adopted
- -Adequate equipment and labor should be organized

6.2 Placing an insect order

A *Drosophila* order form must be prepared and mailed no less than 2 weeks in advance.

A Copy of

Drosophila Order Form

Requester: ______________________ Date: ______________

Address: ______________________ Phone Number: ( ) ________

____

____

Stock No. Species (or strains) Stages

____

____

____

____

____

When Required: ____

Nature of Use for These Stocks: ______________________________

____

Remarks:

6.3 Availability of stocks

All aspects of *Drosophila* culture are treated in the annual editions of *Drosophila Information Service* (D.I.S.), including lists of stocks in the various laboratories of the world. Two formal centers for the maintenance and distribution of *Drosophila* stocks are established in the United States and will supply specific stocks upon request:

Mid-America Drosophila Stock Center (for mutants of *D. melanogaster*)

National Drosophila Species Resource Center (for other *Drosophila* species and strains, and their mutants)

Both at Department of Biological Sciences, Bowling Green State University, Bowling Green, Ohio 43403 U.S.A.

7.0 FURTHER READING

No single publication can cover all aspects of *Drosophila* culture adequately, since techniques have been improved year after year. However, reading the Technical Notes section of *Drosophila Information Services* is one of the most satisfactory means of keeping aware of the changes. On the other hand, many basic techniques have not changed much. The useful references are: Demerec & Kaufman (1964), Flagg (1979), Spencer (1950), Strickberger (1962), and Wheeler (1976).

8.0 REFERENCES

Demerec, M. and Kaufmann, B.P. 1961. Drosophila Guide, 7th ed., Carnegie Institute of Washington. Washington, D.C. pp. 42.

Felix, R. 1969. Control of bacterial contamination in Drosophila food medium. D.I.S. 44: 131.

Flagg, R.O. 1973. Drosophila manual. Carolina Biological Supply Co. Burlington, NC. pp. 23.

Gonzales, F.W. and Abrahamson, S. 1968. Acti-dione, a yeast inhibitor facilitating egg counts. D.I.S. 43: 200.

Hanks, G.D., King, A.L. and Arp, A. 1968. Control of a gram negative bacterium in Drosophila cultures. D.I.S. 43: 180.

Spencer, W.P. 1950. Collection and laboratory culture. In Biology of Drosophila, ed. Demerec, M., Wiley, New York.

Strickberger, M.W. 1962. Experiments in genetics with Drosophila, Wiley, New York. pp. 143.

Wheeler, M.R. 1976. Fruitflies. In The UFAW Handbook on the Care and Management of Laboratory Animals. 5th ed., Livingstone, Edinburgh, P. 548-553.

DROSOPHILIDAE II: *DROSOPHILA* SPECIES OTHER THAN *D. MELANOGASTER*

JONG S. YOON

National *Drosophila* Species Resource Center, Department of Biological Sciences, Bowling Green State University, Bowling Green, Ohio 43403, U.S.A.

THE INSECT

Scientific Name: *Drosophila* spp.
Common Names: "Fruit fly", vinegar fly, pomace fly
Order: Diptera
Family: Drosophilidae

1.0 INTRODUCTION

The cosmopolitan species of *Drosophila* are readily cultured on rather simple media. For example, most species of the subgenus *Sophopora* (e.g., *D. melanogaster* species group, *D. obscura* species group, etc.) can be cultured on the cornmeal medium. A banana medium is widely used for some members of the subgenus *Drosophila* (e.g., *D. immigrans* group, *D. funebris* group, *D. virilis* group, etc.). The two basic media were described in the previous chapter.

However, hundreds of species require special diets, such as the high-protein medium or cactus-containing food. For example, the members of the *D. repleta* species group, one of the largest species groups in the genus, require unique plants, such as species of the Cactaceae, for food. Most species in this group are primarily from the New World deserts.

On the other hand, most species of Hawaiian *Drosophila* require a medium especially rich in proteins, vitamins, and minerals. After a number of unsuccessful attempts, Wheeler and Clayton (1965) devised a food medium, on which a large number of species can be maintained with greater success than on any other medium so far devised.

In this chapter, therefore, rearing techniques for these two unusual major groups of *Drosophila*, the New World desert species and the endemic Hawaiian species, are described. The *D. repleta* species group will represent the desert species.

2.0 FACILITIES AND EQUIPMENT NEEDED

2.1 Facilities

Hawaiian *Drosophila*: Rearing of the Hawaiian flies is carried out in controlled environment rooms at 17-19°C, 75-80% RH, and an LD 12:12 photoperiod.

The *repleta* group: Flies can be reared at 22-25°C, 40-45% RH, and an LD 12:12. For small scale requirements, flies can be maintained in an incubator with the above conditions.

2.2 Equipment and materials required for insect handling

2.2.1 Anesthesia
There are two commonly used methods for anesthetizing flies for examinations (See 2.2.1, Vol. II, p. 75)

2.2.2 Basic equipment for handling and examining flies
The equipment is the same as for *D. melanogaster* (see Section 2.2.2, Vol. II p. 77) with the following additions for the Hawaiian *Drosophila*:

-Collecting vacuum pump (Fig. 2)
-Mouth aspirator

2.2.3 Insect mating and rearing
-Wide-mouth Mason® jars with moistened sand (Fig. 1)
-Water sprayer (bottles)
-Humidifier

2.3 Diet (medium)

2.3.1 Dry mix and preparation equipment (see Vol. II, p. 77)

2.3.2 Medium cooking (Vol. II, p. 77)

3.0 ARTIFICIAL DIET

3.1 Composition

3.1.1 Cooked culture media

a. Modified Wheeler-Clayton medium (for 125 small vials)

-Water	920 ml	78.76%
-Agar (Carrageen® Sea Kem Type 101)	8 g	.68%
-Baby cereal (Gerber's Hi Protein®)	15 g	1.28%
-Wheat germ	8.8 g	.75%
-Kellogg's Special K® Cereal	15 g*	1.28%
-Cornmeal (yellow, defatted)	30 g	2.57%
-Yeast (dry brewer's)	20 g	1.71%
-Gerber® baby banana (1 jar)	128 g	10.96%
-Ethanol (95%)	12 ml	1.03%
-Propionic acid	12 ml	1.03%

*easier to store and measure if ground up in blender first

b. Banana-cactus medium (for 1 tray = 220 vials)

-Yeast (dry brewer's)	20 g	1.28%
-Agar (Carrageen® Sea Kem Type 101)	14.5 g	.93%
-Banana (without skins)	270 g	17.34%
-Water	800 ml	51.38%
-Malt	22.5 ml	1.44%
-Light corn syrup ("Karo®")	22.5 ml	1.44%
-Cactus (paste)	400 ml	25.69%
-Propionic Acid	4 ml	0.26%
-Ethanol (95%)	4 ml	0.26%

3.2 Brand names and sources of ingredients

Gerber's Hi Protein cereal, Gerber's Banana baby food, and Kellogg's Special K cereal (local grocery store); All other ingredients (see the previous chapter).

3.3 Diet preparation procedure

3.3.1 Cooked culture media

a. Modified Wheeler-Clayton medium
-Measure water and weigh out dry ingredients
-Pour about 50 ml of water in blender and add dry ingredients to water
-Add about half of remaining water (400 ml) and blend well
-Add Gerber's banana baby food
-Add remaining water and blend well
-Pour into pan and heat to boiling while stirring constantly
-Upon boiling reduce heat and slowly boil for 10 min.
-Remove from heat and add ethanol and propionic acid
-Stir well and pour into hot sterilized vials to a depth of 1 cm, using beaker
-Cover with sterilized muslin cover to avoid contamination
-When the food is cool, insert sterilized paper tissue into the vials
-Plug the vials and store for a day at cool room temperature before use

Fig. 1. Wide-mouth jars with moistened sand for rearing Hawaiian *Drosophila*

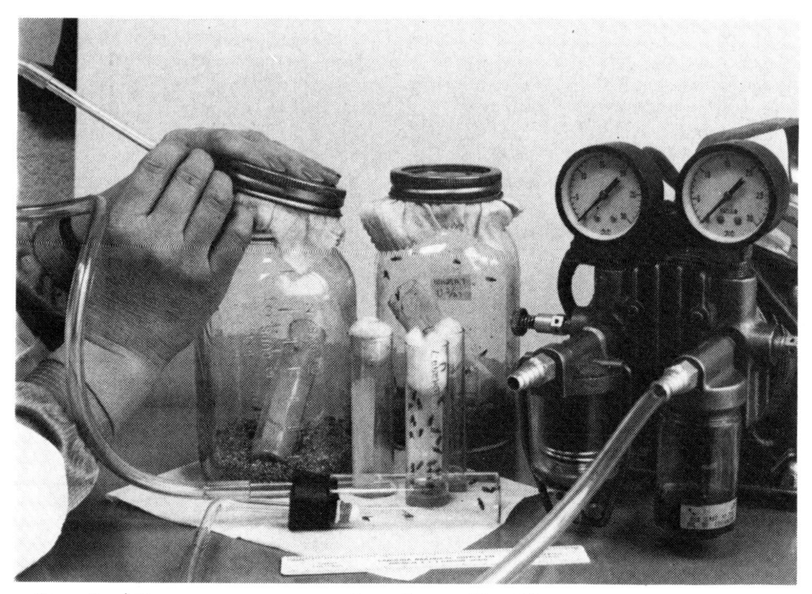

Fig. 2. Vacuum pump apparatus for collecting flies

b. Banana-cactus medium
- -Blend yeast, agar, and banana with most of the water sufficient to make a pourable mixture (save the rest of water for later)
- -Pour the mixture into cooking vessel, and add remainder of water and stir well
- -Add malt, Karo syrup, and cactus paste* to the mixture
- -Bring to boil and cook 12 min, then add ethanol and propionic acid while stirring
- -Cook 4 more min. to boil
- -Pour out medium into sterilized vials using food dispenser or beaker
- -Cover with sterilized muslin cover to avoid contamination
- -When the food is cool, insert sterilized paper tissue into the vials
- -Plug the vials

*The prickly pear cactus (Optunia lindheimeri) can be gathered throughout the year in the field or greenhouses. The cactus is diced into about 3-5 cm pieces, autoclaved for 45 min at 118°C and ground thoroughly in a food blender. In order to improve circulation during blending, about 200-250 ml of water is added to each batch (about 1000 ml) of cactus being ground. This slurry must be strained through 63.5 mm (=1/4 inch) wire mesh to remove the vascular tissue which was not cut by the blender knives to prevent clogging the food dispensing pump (Richardson and Kambysellis, 1968). This cactus paste must be used immediately or frozen for later use.

PRECAUTIONS

All equipment, tools, plugs, cover and paper tissues must be sterilized for media preparation. Do not use bottles for the modified Wheeler-Clayton media nor banana-cactus medium. These media are too soft to use in wide surface.

4.0 REARING AND STOCK MAINTENANCE

4.1 General information

(a) Hawaiian Drosophila: The eggs display interspecific and intergroup variety of considerable extent, especially involving size, sculpturing, and number of chorionic egg filaments (Throckmorton, 1966). The Hawaiian Drosophila provide a spectacular example of adaptive radiation, in which species use a wide variety of substrates, including the leaves, stems, flowers, or fruits of certain plants, slime flux, fungi, or even spider egg masses (Heed, 1968; Carson et al., 1970). Many of them are substrate specific and most cannot yet be cultured in the laboratory, although Wheeler and Clayton (1965) have devised a medium which is adequate for a number of species including other species difficult to rear on conventional media. Most Hawaiian species prefer to pupate in sand rather than on the sides of a vessel or on paper tissues.

The length of the life cycle is unusually long when compared to that of other Drosophila species. The newly emerged adults typically reach sexual maturity 10 to 12 days after emergence (the females reach their peak of receptivity 14 to 18 days after emergence). For the Hawaiian flies, the temperature requirements appear to be quite exact. Thus, in rearing these species in the laboratory, it is mandatory that the temperature be kept at lower than room temperature, and that the relative humidity must be high. At the National Drosophila Species Resource Center, these Hawaiian flies (eggs through adults) are maintained at 18±1°C with a RH of 80%. Under these conditions eggs hatch 5 to 7 days after oviposition and the period of time from oviposition to adult emergence varies from 20 to 48 days, depending on the species. A typical culture technique for the Hawaiian Drosophila (Yoon et al., 1972) is as follows: Newly eclosed adult flies are aspirated from the sand jars where they have been allowed to develop during their pupal period. They are put into vials containing a reduced amount of brewers yeast in the standard cornmeal medium. The vials are prepared with a moist paper tissue folded neatly and placed on one side of the vial to increase the humidity as well as to absorb seepage from the food. It is desirable to lay the vials on their sides rather than leaving them standing so that adults do not stick to the food surface. When the adults are sexually mature (12-19 days), they are introduced into a fresh vial of the yeast-sprayed cornmeal medium where they prefer to lay eggs on the food surface. Eggs hatch at about six days (5-7 days) after they have been oviposited. The larvae feed on the yeast for several days before burrowing into the medium itself. It is advisable, especially in cases where there are large number of larvae, to remove most of them to a fresh vial of the same food preparation. This is done by gently scraping them off the surface with a flat spatula. Since the developmental period of this stage is longer, it is recommended that the food supply not be too old by the time the larvae are ready to pupate. When larvae are ready to pupate, the unplugged vials are put into quart size jars containing about a cup of moist sterilized sand. The larvae pupate in the sand. Eclosion occurs about 10-15 days later (see the "LIFE CYCLE DATA" in this chapter).

(b) The desert flies: Most members of the species originated from the New World deserts have special nutritional requirements. They are not satisified by standard cornmeal medium or banana medium. For instance, a "banana-cactus" medium (Richardson and Kambysellisi, 1968) has been devised, which increases the viability of *D. aldrichi* and its related species of the *repleta* species group. With some modification of this banana-cactus medium, many other species from the New World desert can be cultured (see Yoon, 1984, for details)

4.2 Insect quality

See 4.8, Vol. II, p. 81.

5.0 LIFE CYCLE DATA

The life cycle and survival data during development of various stages under optimum rearing conditions are as follows:

5.1 Developmental data

The life cycles of all *Drosophila* species are longer when compared to that of *D. melanogaster*. For example, *D. hydei* of the *repleta* group has a single generation time of about 14 days at 24°C. In addition males of *D. hydei* require 8 to 9 days from the time of eclosion to become sexually matured. Developmental data of some representative species of the *repleta* and Hawaiian *Drosophila* are shown in the table.

Stage	Age (day)		
	D. hydei	*D. mimica*	*D. grimshawi*
Hatching from egg	2	5	7
Formation of puparium	5	6	14
Adult emergence from pupa case	6	7	13
1st oviposition	8	12	20

The life cycle data and optimal rearing conditions including the best media for more than 300 species of the genus *Drosophila* are available from the National *Drosophila* Species Resource Center.

6.0 PROCEDURES FOR SUPPLYING INSECT

6.1 Pre-requisites and placing an insect order

See Chapter .

6.2 Mailing or transport of flies

For mailing or transporting long distance, Spieth (1966) devised a novel sugar-vial method of feeding the flies en route without using the customary culture media.

The sugar-vial formula and procedure are as follows:

water	1,000 ml
agar	15 g
syrup (Karo®)	50 ml

- -Add agar to water, then heat to dissolve the agar completely
- -Add syrup to the agar solution
- -Pour into the filter paper-lined small vials to about 8 ml depth
- -Autoclave (sterilize) the vials; most liquid media should be absorbed into the paper lining and only a very thin film should remain on the bottom of the vial.

When the flies are placed in the vials, they feed readily upon the surface of the impregnated paper and can cling to it easily. For long distances, it is best to add to the insulated box or container a frozen can of Scotch Ice® or Magic Cold®, wrapping the vials in paper to avoid excessive chilling. In great extremes of temperature, flies have to be carried in polystyrene or polyfoam plastic boxes to protect them from excessive heat or cold.

6.3 Availability of stocks

Drosophila species and strains, and their mutants are available from the *National Drosophila Species Resource Center* at the Department of Biological Sciences, Bowling Green State University, Bowling Green, OH 43403, U.S.A. The Center's *Drosophila Species Stock List*, which includes stock numbers, original collection localities, and codes for optimal media for each species, will be supplied on request.

7.0 FURTHER READING

The useful references on *Drosophila* species other than *D. melanogaster* are: Ashburner and Thomson (1978), Sang (1978) and Yoon (1984).

8.0 REFERENCES

Ashburner, M. and Thompson, J.N. 1978. The laboratory culture of *Drosophila*. In *The Genetics and Biology of Drosophila*. ed. Ashburner, M. and Wright, T.R.F., Academic Press, London, P. 1-109.

Carson, H.L., Hardy, D.E., Spieth, H.T. and Stone, W.S. 1970. The evolutionary biology of the Hawaiian Drosophilidae. In *Essays in Evolution and Genetics in Honor of Th. Dobzhansky*, ed. Hecht, M.K. and Steere, W.C., Appleton-Century-Crofts, New York, p. 437-443.

Heed, W.B. 1968. Ecology of the Hawaiian Drosophilidae. Univ. Texas Publ. 6818: 387-419.

Richardson, R.H. and Kambysellis, M.P. 1968. A cactus-supplemented banana food for cultures of repleta group *Drosophila*. D.I.S. 43: 187.

Sang, J.H. 1978. The Nutritional Requirements of *Drosophila*. In *The Genetics and Biology of Drosophila*. ed. Ashburner, M. and Wright, T.R.F., Academic Press, London, p. 159-192.

Spieth, H.T. 1966. A method for transporting adult *Drosophila*. D.I.S. 41: 196-197.

Throckmorton, L.H. 1966. The relationships of the endemic Hawaiian Drosophilidae. Univ. Texas Publ. 6615: 335-396.

Wheeler, M.R. and Clayton, F. 1965. A new *Drosophila* culture technique. D.I.S. 40: 98.

Yoon, J.S. 1984. N.D.S.R.C. *Drosophila* species stock list-1984. National *Drosophila* Species Resource Center, Bowling Green, Ohio. pp. 16.

Yoon, J.S., Resch, K. and Wheeler, M.R. 1972. Cytogenetic relationships in Hawaiian species of _Drosophila_. I. The _D. hystricosa_ subgroup of the modified mouthparts species group. Univ. Texas Publ. 7213: 179-199.

APPENDIX

Some species* that can be successfully reared on these two media are:

(a) Modified Wheeler-Clayton	(b) Banana-cactus
D. aglaia	_D. aldrichi_
D. affidisjuncta	_D. ellisoni_
D. biseriata	_D. longicornis_
D. crucigera	_D. miranda_
D. eurypeza	_D. nigrospiracula_
D. grimshawi	_D. richardsoni_
D. hystricosa	_D. serido_
D. mimica	_D. spenceri_
D. silvarentis	_D. sticta_
D. uniseriata	_Scaptomyza adusta_
Samoaia leonensis	_S. anomala_

*See Yoon (1984) for a complete list of media for approximately 300 species representing seven genera in the family Dorsophilidae.

FANNIA CANICULARIS

BERNARD GREENBERG AND JEANETTE GEORGE

University of Illinois at Chicago, Department of Biological Sciences, Chicago, Illinois 60680, U.S.A.

THE INSECT

Scientific Name: Fannia canicularis (Linnaeus)
Common Name: Lesser house fly
Order: Diptera
Family: Muscidae

1. INTRODUCTION

Fannia canicularis is commonly associated with human activities and habitations. It is of Northern hemispheric origin and is now cosmopolitan in distribution. The larvae occur in decomposing organic material of many kinds, but they are primarily dung feeders, secondarily scavengers and sometimes cause myiasis. Large populations are found on poultry farms and in dairy barns with adults invading nearby houses, favoring the upper floors of buildings. With Musca domestica this fly is among the most abundant flies of human habitations. Both sexes prefer cool, shaded environments, the male fly being found more frequently in buildings and the female outdoors nearer potential breeding sites. Activity decreases with increasing summer temperatures. Although this fly visits excreta, human food, and the home, its importance as a vector of human pathogens is uncertain (Greenberg, 1971, 1973); there is evidence that it may transmit Newcastle disease virus (Rogoff et al., 1975). Larvae of the nematode, Thelazia californiensis, have been reared in F. canicularis; they have been found in the wild in Fannia benjamini (Burnett et al., 1957). Fannia is sometimes implicated in human myiasis. The adult female is attracted to human (and possibly other animals) genital discharge -- both normal and purulent. Larvae consume the matter and then may crawl into the urethra or rectum. In cases of urogenital myiasis they may even reach the bladder. Being unable to complete their life cycle within the body, they eventually are passed out. If they invade the distal part of the intestine, rectal myiasis results; they may even be able to complete their life cycle here under some conditions. It is also possible for intestinal myiasis to result if food containing living larvae is consumed; in this case, the larvae may be passed with the feces but they are not able to complete their life cycle higher in the intestinal tract. This condition is called pseudomyiasis (James, 1974; Zumpt, 1965). Fannia larvae have long, paired flagelliform dorsal and lateral spines. Debris adheres to the body quite easily and it may be necessary to scrub the body surface to obtain a clear view of the entire spine. Morphological details, adult and larval keys, and illustrations of diagnostic characters may be found in Zumpt (1965) and Greenberg (1971, 1973). See Section 6 for Fannia bionomics.

2. FACILITIES AND EQUIPMENT REQUIRED

These are the same as those required for rearing blow flies. See Section 2, Vol. II, p. 26.

3. DIET

3.1 Natural adult diet composition

- Water
- Sugar
- Milk (dry or skim)

3.1.1 Natural adult diet preparation procedure

- Place dry ingredients in separate containers (e.g. petri dish bottom)
- Place liquid ingredients in separate containers (500 ml erlenmeyer flask if water; 25-50 ml erlenmeyer flask if liquid milk) which have been fitted with wicks (see Sec. 3.1, Vol. II, p. 27 for wick construction).

PRECAUTIONS

(i) If the colony does not thrive on dry food components, a switch to their liquid versions (30% sugar syrup, skim milk) is recommended.

(ii) Replace liquid milk daily to avoid souring and coagulation.

(iii) Replace wicks when they become soiled.

3.2 Natural larval diet composition

- CSMA medium (33.3% wheat bran, 26.7% alfalfa meal, 40% brewer's grain)
- Water

3.2.1 Natural larval diet preparation procedure

- Mix 350 ml water with 400 ml (dry volume) CSMA medium
- Fill a container (wide-mouth jar, beaker, or equivalent) approximately 2/3 full with this moistened medium

3.3 Brand names and sources of ingredients

- Sugar and dry milk, local supermarket
- CSMA medium (Chemical Specialities Manufacturers' Association), Ralston-Purina, St. Louis, Missouri

4. REARING AND COLONY MAINTENANCE

4.1 Colony start-up

- Use flies, larvae, or pupae collected in the field. Fannia are attracted to poultry houses) or obtained from other stock colonies
- Set up a cage as described for blow flies, Section 4, Vol. II, p. 28
- Provide diet appropriate for the life stage of the specimens
- Add specimens
- Maintain the colony at 20-25°C and 50% RH. A LD 12:12 photoperiod is acceptable

PRECAUTIONS

(i) Make sure that specimens used to start a colony are free of disease and parasites. Adults may be visually checked; the larval or pupal stages may be allowed to develop to the adult stage before final selection.

(ii) Confirm specimen identification as *Fannia canicularis*.

4.2 Adult oviposition

- 3 days after adult emergence, place a container with larval diet (moistened CSMA medium) in the cage; females will oviposit after mating; 10-40 eggs per clutch may be produced at 2- to 4-day intervals.

4.3 Egg collection

- Using a fine brush, forceps, or probe, collect eggs from the surface of the larval diet medium; they may also be found on the surface of the wick used with liquid milk.

4.4 Egg sterilization (if required for sterile rearing or other subsequent purposes)
- Use the egg sterilization procedure as outlined for blow fly eggs in Section 6.2, Vol. II, p. 30

4.5 Diet inoculation
- Place previously collected eggs or larvae below the larval food surface

4.6 Pupal collection
- Examine the larval diet material daily after 7-8 days of development; pupariation occurs throughout the food medium, including the dried surface crust

4.7 Adult emergence and collection
- Place the desired number of pupae in a petri dish, test tube, or other appropriate container (it may be fitted with a nylon net or gauze top) and check for emergence daily starting about 16 days from egg stage or about 7-8 days from pupariation

4.8 Sex determination
- Check the distance at the top of the head between the eyes of adult flies; the eyes of males almost touch, the eyes of females are farther apart

4.9 Rearing schedule
- Maintain the colony at about 50% RH and between 20-25°C; a LD 12:12 photoperiod is suitable
- Replenish water and dry and liquid adult food as required
- Remove dead flies weekly or as needed
- Set out fresh oviposition bait/larval diet every 2-3 days, until the new generation of adults emerge; if rapid build-up of the colony is desired, continue at this frequency until the desired adult density is attained. Thereafter, set out fresh bait/diet at reduced frequency, enough to maintain that density; 7- to 10-day intervals are recommended. Female flies may be expected to oviposit approximately 10-40 eggs per clutch, every 2-4 days.
- If generational separation is desired, remove the container of larvae and/or pupae to a freshly prepared, empty cage before eclosion

NOTE: Add more larval diet if the medium dries out before the larvae reach prepupal stage.

4.10 Insect quality
- Check a representative sample of 3rd instar larvae, prepupae, or pupae for length and/or weight; undersized larvae will develop into undersized and less fecund adults; see Section 5, Vol. II, p. 29

5. OTHER SPECIES

This rearing method may be used for other species of *Fannia*, as well as *Musca domestica* and other Muscoid fly species with similar life styles (i.e., with garbage and/or fecal material used by female flies for oviposition).

6. BIONOMICS OF *FANNIA CANICULARIS*

Stage	Duration
Egg	32-35 h
Instar 1	36-60 h
Instar 2	96-120 h
Instar 3	120-348 h
Prepupa	228-252 h
Pupa	348-372 h
Total immature	420-444 h
Emergence	18 d; simultaneous both sexes, circadian rhythm
Time to first copulation	96-120 h F; 48-72 h M at 17.6-19.0 °C
Time to first oviposition	72-96 h at 21-25 °C; mate just prior
Time span of mating activity and oviposition	66 d; both sexes still fertile
Adult life	up to 12 weeks

h=hours; d=days; F=female; M=male
Surface temperature of medium, 24.5-28.5 °C
Cumulative times calculated from introduction of oviposition medium, zero hour.

(Largely after Tauber, 1968)

7. REFERENCES

Burnett, H. S.; Parneke, W. E.; Lee, R. D.; and Wagner, E. D., 1957. Observations on the life cycle of *Thelazia californiensis* (Price). Jour. Parasit. 43: 433.

Greenberg, B., 1971. Flies and disease, Vol. I. Ecology, classification, and biotic associations. Princeton Univ. Press, Princeton, New Jersey. 856 p.

Greenberg, B., 1973. Flies and disease, Vol. II. Biology and disease transmission. Princeton Univ. Press, Princeton, New Jersey. 447 p.

James, M. T. 1947. The flies that cause myiasis in man. USDA Misc. Publ. No. 631, U.S. Govt. Printing Office, Washington, D.C. 175 p.

Rogoff, W. M.; Carbrey, E. C.; Bram, R. A.; Clark, T. B.; and Gretz, G. H., 1975. Transmission of Newcastle disease virus by insects: detection in wild *Fannia* spp. (Diptera: Muscidae). Jour. Med. Entomol. 12: 225-227.

Tauber, M. J., 1968. Biology, behavior, and emergence rhythm of two species of *Fannia*. Univ. Calif. Publ. Entomol. 50: 1-86.

Zumpt, F., 1965. Myiasis in man and animals in the Old World. Butterworths, London. 267 p.

GLOSSINA SPP.

P. A. LANGLEY

Tsetse Research Laboratory, ODA/University of Bristol, Langford House, Langford, Bristol BS18 7DU, England

THE INSECT

Scientific Name: *Glossina* spp.
Common Name: tsetse flies
Order: Diptera
Family: Glossinidae

1. INTRODUCTION

Tsetse flies belong to the genus *Glossina*, the sole genus of the family Glossinidae within the order Diptera. Their natural distribution is confined to the African continent where they are the vectors of human and animal trypanosomiasis. Twenty-two separate species exist, but together with sub-species and races approximately 30 different types of tsetse fly are recognised.

Only 6 species of tsetse are implicated as vectors of trypanosomiasis which may be described as important in terms of human and animal health. However, it is likely that all species are potential vectors.

Tsetse are divided into 3 major groupings on taxonomic grounds and according to habitat: (i) the *morsitans* group which inhabits savannah woodlands, (ii) the *fusca* group which typically infests areas of tropical rain forest or forest outliers in savannah and (iii) the *palpalis* group which also infests the tropical rain forest but extends into the dry savannahs in riparian vegetation, except on river systems draining into the Indian Ocean. Of these, only some species of the first and third groups are of major economic importance in the transmission of trypanosomiasis and not surprisingly it is examples of these groups which have received the greatest attention in terms of the development of laboratory rearing methods.

The incentives for rearing tsetse in captivity have been the provision of a regular supply of research material for (a) insecticide testing, (b) development, (c) the study of tsetse/trypanosome relationships, and (d) study of the basic biology of the insect with a view to developing new control methods which are more environmentally acceptable than conventional methods.

In many respects tsetse flies present special problems of rearing in the laboratory. The sole source of food for the adult fly is vertebrate blood and both sexes must feed regularly. An essential requirement is that the fly must pierce a membranous surface in order to feed. In addition to this special feeding requirement, tsetse flies reproduce by adenotrophic viviparity. A single egg is ovulated and retained in a distensible oviduct where embryogenesis and the whole of the life of the larva is spent. The larva is nourished by a nutrient secretion of the highly modified accessory gland of the female and the birth-product is a fully grown third instar larva. At a maintenance temperature of 25°C the females of all tsetse studied ovulate their first egg between the 8th and 10th day of adult life. The period between ovulation and parturition (larviposition) is 9-10 days, and the newly deposited larva undergoes barrelling, hardening and darkening to form a puparium within 1-2 hours whether or not it is given the opportunity to burrow. There then follows a period of intra-puparial life without diapause before the adult emerges some 30 days later. Females develop faster and emerge a day or two earlier than males.

The paired ovaries of the female contain two polytrophic ovarioles each and oocytes develop in sequence, alternating between one ovary and the other. After larviposition the next mature egg is ovulated immediately, is fertilised by spermatozoa stored in the paired spermathecae and the reproductive "cycle" begins again. Hence, the reproductive rate of tsetse is low compared to most insects. However, both sexes live in captivity for relatively long periods and a fertilised female can theoretically produce up to 9 offspring in a lifetime of 100 days, following a single mating. Males can be expected to live half this long and can be used to inseminate females up to 6 times. In practice, males are not generally used more than twice and are not kept for more than 21 days in a breeding colony.

The disadvantages of a low rate of reproduction are that many more adults must be maintained to produce a useful disposable surplus of offspring than is necessary with an oviparous insect. However, there are also certain advantages in that the females only require to be mated once, and that no special techniques are required for larval feeding and maintenance. Thus, the number of labour-intensive operations in the handling of the insects is reduced to mating, separation of the sexes, adult feeding, and collection of puparia. The major difficulty for most researchers is the provision of food for the flies and the problems encountered differ according to whether live hosts or membrane feeding systems are used.

2. FACILITIES AND EQUIPMENT REQUIRED

2.1 Facilities

Adults are held in controlled climate rooms maintained at 25°C ± 1°C, 60-70% RH (*G. morsitans*), or 80% RH (*G. pallidipes*, *G. austeni* and *G. palpalis*) and a LD 12:12 photoperiod provided by low intensity but adjustable tungsten lighting. Puparia are maintained in the same environment as the adults.

A fly-handling room, separate from the holding room is recommended. This need not have climatic control provided a comfortable ambient temperature is maintained, and daylight can be permitted to enter.

A fly-feeding room, adjacent to the handling room should have climatic control and be maintained exactly as the holding room.

2.2 Equipment and materials for insect handling

2.2.1 Adults fed on live hosts

- Rectangular soldered wire cage frames, 25 x 12.5 x 5 cm or 15.5 x 8.5 x 5 cm
- Rectangular plastic end pieces 12.5 x 5 cm or 8.5 x 5 cm with circular hole to take 2.5 cm ø cork (ø= diameter)
- 250 denier black terylene netting
- Needle and nylon thread, or sewing machine
- Metal or plastic trays to support cages
- Wire mesh tray inserts to keep cages off tray surface
- Racks or trolleys for stacking trays of cages

2.2.2 Adults fed through membranes

- 5 cm lengths of 12.5 cm or 15 cm ø PVC drainpipe with a circular hole bored to take a 2.5 cm ø cork
- 250 denier black terylene netting
- PVC cement
- 2 cm wide paint brush
- Strong rubber bands
- All other requirements as for 2.2.1

2.2.3 Pupal collection and maintenance

- Dry, coarse sand
- 2 cm wide paint brush
- Small container for pupal collection
- Tray for pupal storage
- Perforated cover for pupal storage tray

- Sieve for separating pupae from sand
- 42 x 10 x 4 cm aluminium rectangular emergence tray
- Rubber bands
- Two 42 x 9.5 cm aluminium rectangular slides
- 42 x 42 x 10 cm rectangular soldered wire emergence cage frame
- 250 denier black terylene netting
- Needle and nylon thread or sewing machine

2.2.4 Emergent adults - collection and sexing
- Chest-type commercially available freezer
- Replacement thermostat, range -50 to +20°C
- Air circulation fan rated for continuous running at low temperatures
- 18 gauge aluminium sheeting
- Metal or plastic louvres
- Formica or other washable plastic surface
- Artists' paint brush
- Glass or plastic collection tubes with corks

2.3 Requirements for adult feeding: live hosts

2.3.1 Goats
- Suitable animal; trained, tractable and unaffected by fly bites
- Suitable restraint
- Elastic belting with teazel strip fastenings for joining ends
- Plain cloth cover
- Electric hair clippers

2.3.2 Rabbits
- Suitable animal is the half lop-eared variety with large ears
- Suitable restraining box with neck yoke
- Rectangular cushion of soft material
- Elastic belting with teazel strip fastenings

2.3.3 Guinea pigs
- No special requirements as to breed
- Suitable restraining box with strap covers for each animal

2.4 Requirements for adult feeding: membrane technique

2.4.1 Equipment sterilization
- Autoclave
- Metal trays for bottle sterilization
- Porous sterilization bags
- Autoclave tape
- Dust-free "sterile" store
- Oven set at 100°C

2.4.2 Blood collection from pigs at slaughter
- 'Ekstam' hollow stainless steel knife (made in own workshop)
- 2 m of 2.5 cm i.d. rubber tubing
- 2.5 liter autoclavable, wide-mouthed glass jar
- Metal screw-top lid with two circular holes: one central, 0.5 cm ø, for paddle shaft, one 2.5 cm ø for rubber tubing
- Stainless steel paddle
- Electric motor with variable, low gearing to operate stainless steel paddle
- Metal framework to hold electric motor above metal lid of wide-mouthed jar
- Aluminium cooking foil

2.4.3 Blood collection by jugular venepuncture of live cow
- 14 gauge hypodermic needle
- 2 m of ¼.5 cm ø autoclavable plastic tubing
- Other requirements as for 2.4.1 excluding 'Ekstam' hollow knife and 2.5 cm ø rubber tubing

2.4.4 Blood dispensing and storage
- Downflow sterile air cabinet (e.g. Microflow®)
- Magnetic stirrer
- 2 liter conical flask with glass and teflon tap
- Polypropylene filter funnel (20 cm ø)

- Gauze filter
- 50 ml nominal capacity screw-top bottles
- Refrigerator
- Disposable gloves and clean overalls

2.4.5 Bacteriological testing of blood
- Incubator set at 37°C
- Combined agar slope and broth culture bottles
- Sterile hypodermic syringe (5 ml)
- Hypodermic needles

2.4.6 Silicone rubber membrane
- 50 cm square x 1 cm thick sheet of perspex with 0.75 cm x 0.5 cm deep groove around perimeter and 10 cm in from edge
- Olive oil
- 50 cm square piece of terylene netting (mesh size and colour not important)
- Silicone rubber
- Hardener
- Electric mixer with dough hook attachment
- 50 ml syringe with 0.4 cm ø nozzle
- Sheet of 500 gauge polythene, 50 cm square
- Heavy rolling pin or photographic plate roller

2.4.7 Feeding plates
- 43 cm square plates of inert material (glass or polycarbonate) with parallel grooves 1-2 mm depth

2.4.8 Heating plates
- Aluminium sheet, 16 or 18 gauge
- 4 x 100 watt electric light bulbs
- Thermostatic sensor
- Silicone rubber
- Aluminium plate 1 cm thick

2.5 Brand names and sources of equipment

Cage netting, quality No. 6480, 250 denier black terylene, (J.C. Small & Tidmus, Perry Rd., Sherwood, Nottingham NG5 3AF, U.K.); PVC drainpipe and Bartol(®) PVC cement, (obtainable from any supplier of building or plumbing materials); modified freezer fitted with RS 509-226 mm, 240 v A.C. fan, (R.S. Components, P.O. Box 427, 13-17 Epworth St., London EC2P 2HA); modified freezer fitted with replacement thermostat -5° to +20°C, model KP71 60L 1113, (Dean & Wood (London) Ltd., 83-85 Mansell St., London E1 8AS); teazel strip fastenings for elastic belting are Velcro(®), (Selectus Ltd., Biddulph, Stoke-on-Trest, ST8 7RH, U.K.); glass jars for blood collection, (aluminium screw-topped sweet jars obtainable from retailers or wholesalers); stainless steel paddle driven by model SD15 (200 rpm) electric motor, (Parvalux Ltd., Wallisdown, Bournemouth, U.K.); PTFE and glass tap (autoclavable), Interflow(®) ¾ mm 1St/3/B stopcock, (Springham & Co. Ltd., South Rd., Temple Fields, Harlow, Essex, U.K.) fitted to 14/23 Quickfit(®) adaptor, (Corning Ltd., Stone, Staffs., U.K.) joined to 2 l conical flask; Rolon(®) disposable gloves, (Richardsons of Leicester Ltd., Evington Valley Rd., Leicester, LE5 5LJ, U.K.); Vacuneda(®) MW800 combined nutrient broth and agar medium, (Medical Wire & Equipment Co. (Bath) Ltd., Corsham, Wilts., U.K.); Silicone rubber, RTV M539 and hardener T35, (Wacker Chemie, Munich, W. Germany, supplied in U.K. by Micro Products, 22 The Green, West Drayton, Middlesex UB7 7PQ); Polycarbonate grooved sheets, Makrolon, pattern 281 R, (Aalco Bristol Ltd., Stafford St., Bristol BS3 4DA, U.K.).

3. DIET PREPARATION PROCEDURE

3.1 Blood collection from pigs

Fresh blood is collected aseptically from pork-weight pigs (50-60 kg live weight) at slaughter. They are electrically stunned, suspended head down from an overhead rail and a 15 cm incision is made through the skin of the neck from the manubrium to the intermandibular space with a sharp knife. The skin on either side of the incision is then parted, using retractors which can be made by bending the tynes of an ordinary table fork at right angles to the handle. A 22 mm external diameter, 1.5 mm wall, hollow stainless steel knife based on the Ekstam* design (Fig. 1) is plunged through the thoracic inlet parallel and

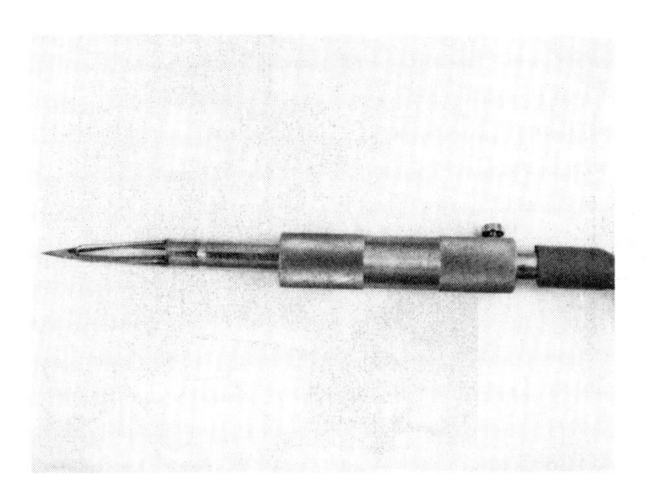

Fig. 1 Hollow, stainless steel knife used for collection of blood from pigs at slaughter.

immediately dorsal to the sternum until the anterior vena cava or aorta is severed. (The hollow knife can be made by any competent workshop technician and is not available commercially.) The blood flows for approximately 15 seconds through a 2.5 cm ø rubber tube attached to the hollow knife and into a 2.5 liter capacity glass (or polypropylene) jar where it is stirred continuously with an electrically driven stainless steel paddle until defibrination is complete in about 10 min. (Fig. 2). After filtering through gauze the blood is dispensed into 50 ml screw-cap bottles (Fig. 3) and stored at 4°C until used. About 1.5 liters of defibrinated blood is obtained from each pig.

3.2 Blood collection from cattle

Fresh cow blood is collected aseptically from the jugular vein of a live cow, using a 14 gauge needle. The blood flows via a 5 mm ø silicone rubber tube into a 2.5 liter jar, where it is defibrinated and dispensed in the same way as for pig blood. Up to 2 liters of blood can be removed from a cow once each week.

3.3 Diet quality control

A 5 ml sample of blood from every collection is incubated in a combined nutrient broth and agar medium for routine diagnostic use. In order to standardise the procedure, recently dispensed blood is held in 50 ml bottles for one week at 4°C before testing for bacterial contamination. This gives time for the normal bacteriostatic and bacteriocidal properties of blood to act, and for the white corpuscles to die. Incubation of such a blood sample at 37°C with the medium over the next seven days then provides an indication of infection which is not always reliable when freshly collected blood is incubated. Using this method, low levels of bacterial contamination are occasionally detected and these batches of blood are discarded. Hence, no blood is used for fly feeding until it has been stored 2 weeks at 4°C, and provided it is kept at 4°C it can be held for a further 2 or 3 weeks before use.

*Dr. M. Ekstam, KBS Abattoir, Kristianstad, Sweden.

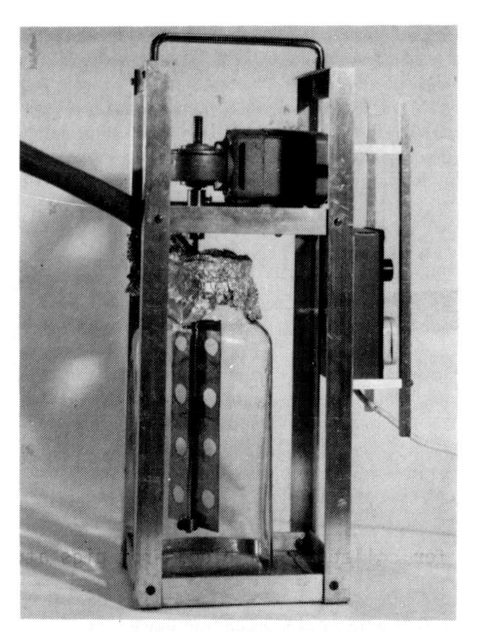

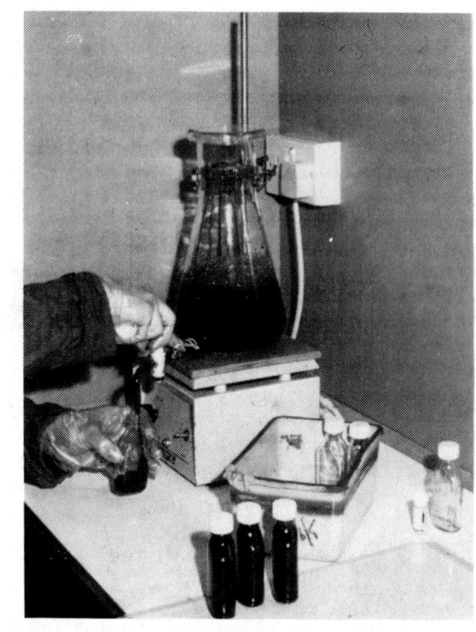

Fig. 2 Glass jar containing stainless steel paddle attached to electric motor mounted on carrier.

Fig. 3 Filtered, defibrinated pig blood is dispensed into 50 ml screw-cap bottles. The 2 liter capacity conical flask has an autoclaved PTFE and glass tap.

PRECAUTIONS

Apparatus associated with blood collection and storage is autoclaved for 30 min at 1.4 kg cm^{-2} before use. All blood is dispensed, using aseptic procedures, in a laminar downflow hood. All personnel involved in aseptic procedures wear sterile caps and gowns, face masks and disposable polythene gloves.

4. MANUFACTURE OF EQUIPMENT

4.1 Silicone rubber membranes

A 40 cm square membrane is recommended for use with 50 ml blood.

A mould consisting of a 50 cm square x 1 cm thick sheet of perspex has a 0.75 cm x 0.5 cm deep groove cut into it to provide a raised rim to the finished membrane (Fig. 4). The surface of the mould and the groove is covered with a thin layer of pure olive oil to act as a releasing agent. A piece of white terylene netting (mesh size is not important) is cut to cover the flat surface of the mould. It should extend beyond the groove and can be trimmed later.

150 g silicone rubber is mixed very thoroughly with 6 ml hardener, but not so violently as to include air bubbles in the mixture. An electric mixer run at slow speed with a dough hook attachment is ideal. The hardener may be obtained with a red dye included which facilitates observation of the mixing process. There is no reason to believe that the hardener has a limited shelf life. When thoroughly blended the mixture is less viscous and at this point some is placed into a 50 ml polypropylene syringe, the orifice of which is

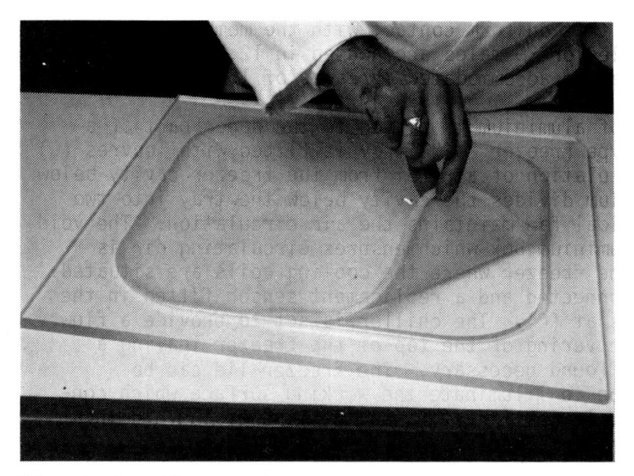

Fig. 4 Freshly cured silicone rubber membrane being removed from perspex mould.

enlarged to about 0.4 cm ø. With this the mixture is 'piped' into the groove in the mould until it is filled. Next, half the remaining mixture is placed in the centre of the mould with a spatula. The terylene netting is laid across the mould and the other half of the mixture placed on top in the centre of the mould. The mould and mixture is then covered with a sheet of 500 gauge polythene wiped with olive oil, and the mixture is rolled out in all directions from the centre with a firm pressure, using a heavy rolling pin or a photographic plate roller. Care must be taken to exclude air bubbles and the final thickness of the membrane is determined by the amount of pressure used and the thickness of the terylene netting. A thickness of 0.2 to 0.3 mm is ideal. The mould is now left for about 16 hours at room temperature when careful lifting of the polythene sheet will indicate whether curing is complete. After removal of the polythene sheet the membrane is released from the mould and left for a further 24 h to cure completely. The membrane is then rinsed in water, dried, and is ready to sterilize, which is most conveniently achieved by leaving in an oven at 100°C for 12 hours. Autoclaving of membranes is not recommended as it causes deterioration.

4.2 Fly cages

All cages are covered with black terylene netting since this permits the flies to be viewed easily.

4.2.1 Live hosts

For feeding on goats a rectangular cage is used. This consists of a frame of steel rods (preferably stainless steel which must be silver soldered) 25.0 x 12.5 x 5.0 cm covered tightly with a closed sleeve of netting. The open end of the cage is closed with a 12.5 x 5.0 cm rectangle of hard material held in place with sticky cloth tape. A hole is bored in the end piece to take a cork. Such a cage is used routinely to hold 25 adult female flies or up to 35 adult males.

Smaller cages of the same design, 15.5 x 8.5 x 5.0 cm, house 10 females or 15 to 20 males and are known as "Geigy 10" cages. These are used for feeding flies on guinea pigs or on the ears of rabbits (Fig. 9).

The manufacture of rectangular cages can be tedious and expensive. A simple alternative is to place a 5 cm length of circular PVC pipe in hot water (60°C) until it can be deformed to make an oblong instead of a circular shape. Upon cooling the shape is retained and netting is glued to it in the same way as for a circular cage.

4.2.2 Membrane feeding

Cages are made from 5 cm lengths of PVC drainpipe. A circular hole is bored through the side to accept a cork and each open end of a cage is covered with terylene netting which is glued in position with PVC cement, while the netting is held taut with a thick rubber band. The netting must be held taut while the cement sets in order that uniform contact with the membrane is made during feeding. Up to 15 females or 20 males are housed in 12.5 cm ø cages while 30 females or 40 males can be accommodated in cages of 15 cm ø.

4.3 Chiller

A tray insert (Fig. 5A) of aluminium is fitted to the upper part of a commercially available chest-type freezer. The tray is fitted with louvres (C) at opposite sides to allow circulation of air (B) from the freezer cavity below. A vertical partition of aluminium divides the cavity below the tray into two unequal portions and an electrical fan maintains the air circulation. The void (E) is enclosed by a hollow aluminium box which ensures circulating air is driven close to the sides of the freezer where the cooling coils are situated. The freezer thermostat is disconnected and a replacement sensor fitted in the air stream close to the louvres at (C). The chiller is set to provide a flow of air (B) at 4-5°C. Partial covering of the top of the freezer leaving a small working area open may be found necessary. The freezer lid can be equipped with a fluorescent tube to illuminate the working surface which consists of a rectangle of plastic laminate resting on the base of the aluminium tray (A).

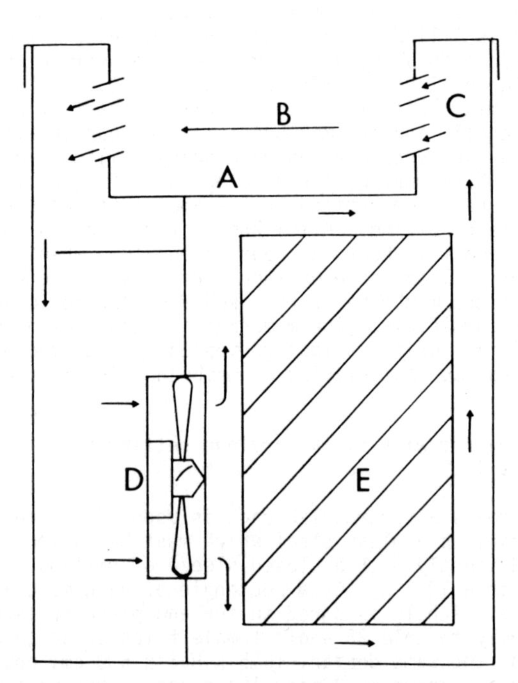

Fig. 5 Diagram of modified chest-type freezer used as a chiller.

4.4 Heating plate

A heating plate supports the feeding plate, raising its temperature to 37-39°C. An electrically heated plate can be made from any material of suitable heat capacity and conductivity, but for durability the following is suggested. Four x 100 watt electric light bulbs are wired in series/parallel with the centre points linked, such that the bulbs work at half voltage. A thermostatic sensor is embedded into the heating plate with silicone rubber, to control the operation of the bulbs which are positioned as shown in Fig. 6. The box is covered with an aluminium plate and a small hole is drilled in it above each light bulb to check that each is working. Such a heating plate has only a 1° or 2° temperature differential over its surface.

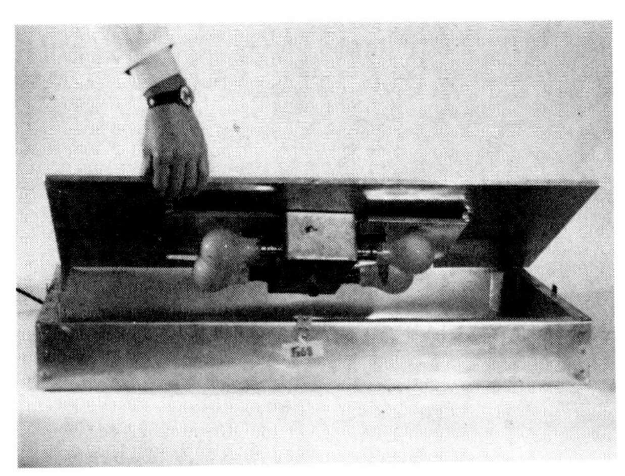

Fig. 6 Electrically heated plate used to support feeding plates and membranes. Top raised to show bulbs.

5. REARING AND COLONY MAINTENANCE

5.1 General

Cages of females are supported on square mesh ('weldmesh') wire frames in aluminium trays (7¼ x 43 x 6 cm). Each tray accommodates 200 to 300 females. Ten trays are stacked at a shallow angle on a metal framed trolley 175 cm high, 77 cm wide and 42 cm deep (Fig. 7). Cages of male flies are stacked on trays on a horizontal platform at the bottom of each trolley.

5.2 Puparial collection and storage

Larvae crawl through the cage netting and fall on to the angled tray where they gravitate to the lower edge and pupariate. Puparia are collected by sweeping them through a hole in the lower end of the tray which is sealed with a cork when not in use. They are collected in a shallow tray of sand over a period of 3 weeks and then transferred to a rectangular aluminium emergence tray (42 x 10 x 4 cm) during the fourth week. This is covered by an emergence cage (42 x 42 x 10 cm) held in position with elastic bands (Fig. 8).

5.3 Adult emergence and sexing

An emergence rate in excess of 95% is to be expected. The emergence cage is cleared daily on 5 or 6 days per week. A rectangular aluminium slide closes the emergence tray and a second slide seals the end of the emergence cage which is then removed from its tray by detaching the rubber bands at each end (Fig.8).

The emergence cage is laid flat inside the chiller. This immobilises all flies which are gently shaken out on to the working surface within the chiller. The sexes are separated, assigned to their respective cages and the cages removed where, at room temperature, the flies are allowed 15 min to recover before feeding. They should be chilled only for long enough to immobilise them for sorting.

Fig. 7 Trolley with sloped stack of aluminium trays containing fly cages.

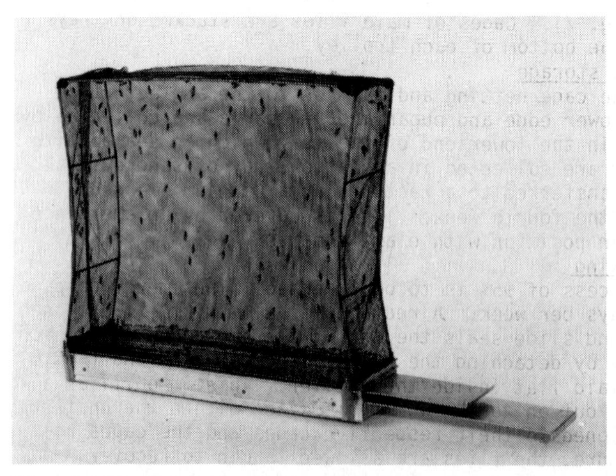

Fig. 8 Emergence cage in position on tray containing puparia. Aluminium slides permit cage to be separated from tray for removal of flies.

5.4 Feeding

5.4.1 General

All flies are offered the opportunity to feed in indirect tungsten lighting of low intensity.

Cages of flies are left in contact with feeding surfaces for 15 min, and in the initial stages of technique development it is best to offer food daily for 6 or 7 days per week. It is possible, with experience, that a 5-day per week feeding regime will be adequate and in some instances feeding on alternate days has been practised successfully. The daily feeding time may also be reduced to 5 or 10 min, but in general, fly survival, reproductive rate and offspring size will dictate the optimum technique to be adopted.

5.4.2 Live hosts

Limitations on the use of individual animals as hosts depend on the degree of skin reaction to the bites of the flies, the general immune response of the host to the saliva of the fly which can be deleterious to both host and fly, as well as on the tractability of the host animal. A large goat should not be exposed to the bites of more than 600-700 flies every three days, while figures for rabbits and guinea pigs are closer to 200 and 100 respectively. Flanks must be shaved regularly to maintain the shortest possible hair covering, but rabbits' ears do not require this treatment.

The degree of restraint of the host animal depends largely on the care of its handling and training. A well-trained goat requires no more than a neck collar restraint. Rabbits require to be boxed and a neck yoke employed. However, the rabbit must also be restrained from attempting to jump which can result in a broken neck. Guinea pigs are boxed and restrained but the sides of the box are open so that cages of flies may be placed in contact with the flanks of the animal. The most suitable type of box for guinea pigs is that in which several animals may be held, such that each cage of flies placed between them makes contact with the flanks of adjacent guinea pigs. A rectangular cushion of soft material is placed under the ear of a rabbit and a cage of flies placed on the upper ear surface. The whole assembly is held in position by elastic belting with teazel strip fastenings and rests on a horizontal platform which is made as an integral part of the rabbit box restraint. Two or 3 cages of flies can be applied to each flank of a goat and secured by an elastic belt, passing round the animal and secured with teazel fastenings. The different devices described are illustrated in Fig. 9a,b,c.

5.4.3 Membrane feeding

Use of aseptic techniques is emphasized because bacterial contamination of the diet is lethal to tsetse flies. Hence, feeding membranes and feeding plates must be heat-sterilized (oven at 100°C) and surfaces which are to make contact with the diet must only be exposed briefly. When making up large numbers of feeding units it is convenient to stack membranes in an aluminium box and to stack feeding plates face-to-face in pairs. The operator should be dressed as for carrying out aseptic procedures. Separation of a pair of sterile feeding plates is then followed rapidly by pouring blood into the centre of each and covering with a membrane. The blood may be spread under the membrane with a hand-held roller or with a dry sponge and, of course, aseptic procedures can cease from this moment. Feeding plates are maintained at 37-39°C by standing them on a heating plate. A 40 x 40 cm membrane covering 50 ml blood accommodates 8 x 12.5 cm diam. cages or 4 x 15 cm diam. cages simultaneously, and four changes of cages are possible before serious depletion of the blood under the membrane occurs. Cages are placed upside down on the membrane so that the cleaner netting is on contact with the feeding surface. Since bacterial infections acquired by flies can be transferred from fly to fly during feeding, the youngest flies are always fed first. In this way any infection can only be passed to older flies in the colony and cannot be passed up to younger flies. This acts as a fail-safe method to prevent extermination of a colony in the event that a group of flies accidentally acquires an infection. No such precautions are necessary when flies are fed on live hosts.

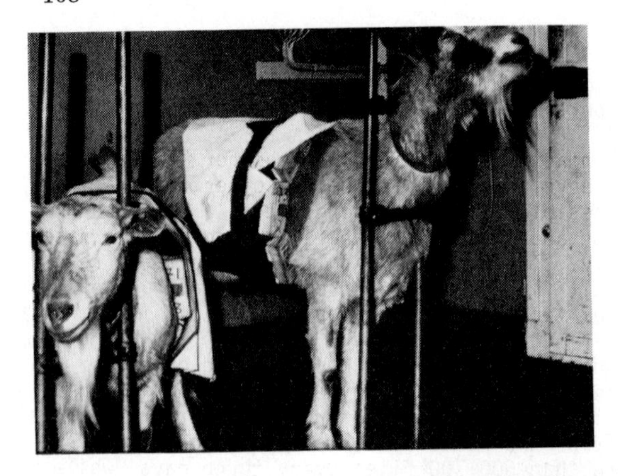

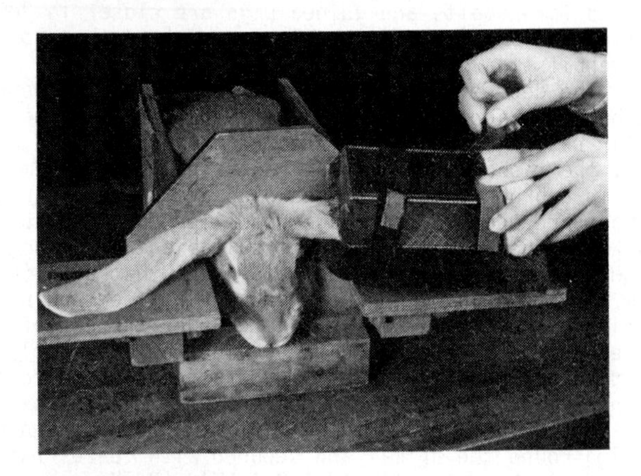

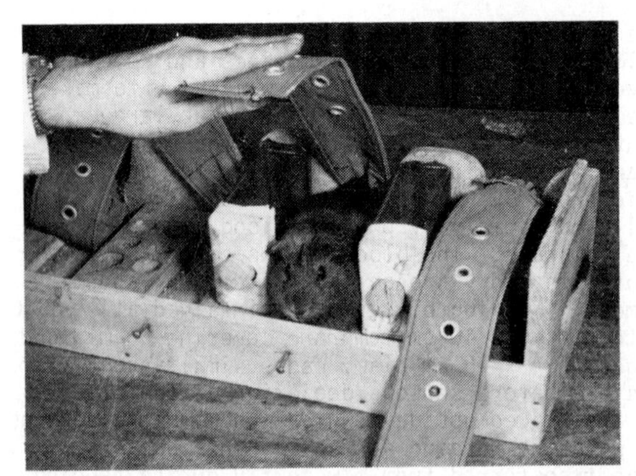

Fig. 9 Various types of restraint for animals used as hosts for tsetse flies, (a) goat, (b) rabbit, (c) guinea pig.

PRECAUTIONS

The tsetse responds to a complex set of stimuli provided by the host animal, but in the laboratory the most important is temperature. Hence, ensure that the animals used for feeding are warm. This is most important when using rabbits since they are able to withdraw much of the peripheral circulation from the ear when cold. Other stimuli may be described as negative in that they discourage the fly from feeding on a host. These are bright light and operator movement, which cause flies to cease attempting to feed and to congregate on the cage surface away from the host. Hence, the feeding area should be subjected to dim illumination with indirect tungsten, rather than fluorescent lighting, and daylight should be reduced to a minimum. Cages of feeding flies should be covered with a clean cloth as an added precaution against disturbance. The feeding area should be maintained as close to 25°C as possible and temperature fluctuations should be avoided. It appears unnecessary to regulate the relative humidity in the feeding area.

A large facility for membrane feeding probably requires air conditioning to maintain the air temperature close to 25°C.

Recently fed flies must not be chilled.

5.5 Mating

Mating techniques vary according to species. Optimal insemination rates in *G. m. morsitans* are obtained when an excess of males, more than 7 days old, is introduced into a cage containing 3-day old females. Flies should be fed 3 to 6 hours previously and pairing is generally immediate. The sexes are separated again, after chilling, 18 to 24 hours later, but before feeding. *G. austeni* females are paired on the day after emergence, following feeding, with an excess of males more than 7 days old, and left together for 48 hours. With this species pairing does not occur so rapidly as with *G. m. morsitans*. Techniques developed for *G. m. morsitans* are also inappropriate for *G. p. palpalis* in which species it is found expedient to house 5 males more than 7 days old with 15 newly emerged females and to leave the sexes together for life. Optimal insemination rates in *G. pallidipes* are achieved only when 7-10-day old females are caged with 12-day old males for 24 hours.

It must be emphasised that many strategies developed to optimise insemination rates in the laboratory have no bearing on the behaviour of flies in the field. Therefore as a guideline in the laboratory, females should be inseminated before they ovulate their first mature egg at 8-10 days at 25°C; female receptivity appears to be maximal during the first week of adult life and it subsequently declines with age. Males of all species undergo a period of maturation and are likely to inseminate best when 1-2 weeks old. If used more than once they should be rested for 48 hours between matings.

6. COLONY MANAGEMENT

6.1 General

Three parameters provide an indication of the success of a tsetse colony: adult longevity, female fecundity, and offspring size. The survival of adult males is not vital beyond the time necessary for them to be mature and to be used for mating. However, their pattern of survival is a useful guide to the success of maintenance techniques employed. Both sexes suffer up to 10% mortality during the first week of adult life but thereafter survival rates should be high until the onset of senescence at which mortality rates increase dramatically. This can be expected to occur in *G. morsitans* after 2 months in males and after 3 months in females. Any other pattern of mortality should be viewed with suspicion.

In building a tsetse colony to provide a surplus of flies for research, attempts should be made to maintain a stable age distribution. The rate at which cages of newly emerged flies is added to the colony should not exceed the rate at which cages of old flies are removed once the final colony size has been achieved. During the time of building up the colony it is important that additions of a constant size are made at regular intervals rather than additions of fluctuating size at irregular intervals which will upset the age distribution.

Female flies of approximately the same age (± 3 days) can be pooled to make groups of a given size (60 to 200 females per group is recommended) to aid record keeping. A constant group size also allows comparisons to be made so that the success of the techniques employed is constantly monitored in terms of fly performance.

Females have a tendency to produce offspring of a size characteristic of the individual, and each female is capable of producing several offspring in her lifetime. Hence, care must be taken in attaching significance to differences between mean puparial weights from different groups of flies. Therefore, size trends should be monitored and puparia can be weighed in groups rather than individually. Overall mean puparial weights should exceed 30 mg in G. morsitans and G. palpalis, while G. austeni produce offspring weighing 23-25 mg and G. pallidipes in excess of 40 mg. Since there is a rapid loss of water from puparia during the first few days, and thereafter nutritional reserves are being consumed, puparia should be weighed as soon as possible after collection.

6.2 Record keeping

The female tsetse produces an offspring every 9 or 10 days at 25°C and it is convenient to divide the life of the female into 9-day periods for the purposes of record keeping. A daily form is used to record the number of flies alive on each day, the number of offspring produced and the number of abortions in each group. Patterns of mortality and fecundity will vary according to techniques of maintenance used and thus there can be no specific ruling on the length of time that a group of females should be kept before discarding. Clearly, a group should be discarded when the number of flies in a cage and the number of offspring produced falls to a low level. In general this occurs after approximately 90 days.

Information gathered on a daily form is transferred to a master sheet where ten groups of flies may be compared and judgements concerning the performance of consecutive groups of flies can rapidly be made. Basically, the number of flies alive half-way through each 9-day period (the pivotal age) and the numbers of puparia produced during the whole of each 9-day period are transcribed to the master sheet. At the end of the useful life of the group of flies, the total number of puparia produced by the group is recorded in the box for that group on the right-hand side of the sheet. Also, the aggregate number of flies alive for each of the 9-day periods is totalled to give the number of reproductive cycles that occurred in the group. However, for this purpose the first 9-day period and two-thirds of the second 9-day period are ignored because no flies can produce an offspring during this time. Dividing the number of puparia produced by the total number of reproductive cycles recorded gives a value for the fecundity of the surviving females. Of course, the fecundity per initial female will be lower than this and can be calculated by dividing the total number of puparia by the maximum possible number of reproductive cycles if no mortality had occurred. Female survival is expressed as the total of the aggregate number of flies alive during each 9-day period divided by the number had no mortality occurred.

Although the above calculations provide absolute values for comparison within the colony, it is often useful to convert the information to the numbers of offspring produced per surviving female per week or per month.

In practice, the fecundity of a female fly rises to a maximum value in the third or fourth 9-day period, after which it falls. The rate at which it falls and the extent will dictate the age at which the group of flies should be removed from the colony. A fecundity value close to 1.0 will only be achieved for a short time and the average for a group over 90 days will be nearer 0.6 puparia per reproductive cycle per surviving female.

In order for a colony to be self-supporting, a female fly must produce in excess of three offspring in her lifetime. Over 90 days the maximum number of offspring a female can produce is 8.3 and therefore a fecundity of 0.6 means each surviving fly produces five offspring. However, remember that mortality among females will reduce the output and the number of offspring per initial female must exceed three or a fecundity of about 0.35 per initial female for a colony to be self-supporting.

6.3 Producing a surplus

Assume a sex ratio of 1:1 for most species; for G. palpalis fewer males than females are produced. Then the number of puparia which must be kept for colony maintenance is equal to twice the number of females required, plus twice the number of puparia required to compensate for early female mortality. The number of females required per day or week depends on the group size and the number of groups being assembled in a given time. Obviously any puparia left over after satisfaction of colony requirements can be disposed of for research.

6.4 Fulfilling an order

Surplus puparia are maintained in open plastic tubes in the insectary and each day's collection is maintained separately, tubes being appropriately labelled. Care should be taken not to stack puparia more than 3 or 4 high as this leads to excessive moisture retention and possible increases in mortality.

Puparia are despatched in their plastic tubes after adding foam rubber to prevent excessive movement.

Perforated lids are held in position with rubber bands and tubes are mailed in padded envelopes.

Adults for short-term experiments are also mailed in their cages, packed in padded envelopes. Flies are fed before mailing. It is not advised that such flies are used to establish colonies or for long-term experiments since transportation often adversely affects survival.

7. TRIATOMA SPP. AND RHODNIUS PROLIXUS

7.1 Introduction

The techniques developed for feeding Glossina have been adapted for the routine maintenance of a number of blood-sucking bugs including Triatoma infestans (Klug), T. brasiliensis Neiva, Panstrongylus megistus (Burn) and Rhodnius prolixus Stal.

A diet of defibrinated pig blood is recommended since cow blood is apparently nutritionally inadequate.

7.2 Cages

Up to one hundred eggs are placed in a 7.5 x 2.5 cm glass tube containing a vertical strip of filter paper and closed with a square of voile held in place by an elastic band.

A combined holding and feeding cage for all instars consists of a 10 cm length of 11 cm diameter PVC pipe, closed at one end with a piece of stretched and glued terylene voile. The other end has a voile covering held in place with an elastic band.

Each cage contains a perching site consisting of a fluted strip of filter paper 8 cm high, bent and stapled into a circular shape.

Up to 100 insects from 1st to 4th instar are housed in such a cage. Fifth instar nymphs are housed at the rate of 50 per cage and adults 30 per cage (15 males plus 15 females).

7.3 Feeding membranes

A 1 cm length of 12 cm diameter PVC pipe serves as a frame upon which is stretched a 7 cm x 7 cm piece of Parafilm M (American Can Co., Neenah, Wisconsin, U.S.A.). Silicon membranes are prepared in the same way as already described for Glossina.

7.4 Feeding and handling

The first 3 instars of Rhodnius and at least the first instars of the larger species are fed through Parafilm membranes. These are swabbed with 70% alcohol and dried with a sterile cloth before placing over a blood pool. All other instars and adults of all species are fed through sterile silicone rubber membranes.

The fixed voile cover of each cage is placed in contact with the feeding membrane surface and left for 30 minutes. After feeding the cages are inverted on to a filter paper lining on a holding tray. All excrement is therefore collected on the removable voile cover which can be changed frequently.

Feeding intervals will be dictated by maintenance temperature, species, stage of development and experimental protocol. Adults maintain optimum egg production when kept at 28°C and offered food twice a week.

RECOMMENDED READING

Bauer, B. and Wetzel, H., 1976. A new membrane for feeding Glossina morsitans Westw. (Diptera, Glossinidae). Bull. ent. Res., 65: 563-565.

Jordan, A.M. and Trewern, M.A., 1976. Sulphaquinoxaline in host diet as the cause of reproductive abnormalities in the tsetse fly (Glossina spp.). Ent. exp. appl., 19: 115-129.

Langley, P.A. and Pimley, R.W., 1978. Rearing triatomine bugs in the absence of a live host and some effects of diet on reproduction in Rhodnius prolixus Stål (Hemiptera: Reduviidae)., Bull. ent. Res., 68: 243-250.

Mews, A.R., Langley, P.A., Pimley, R.W. and Flood, M.E.T., 1977. Large-scale rearing of tsetse flies (Glossina spp.) in the absence of a living host. Bull. ent. Res., 67: 119-128.

Nash, T.A.M. and Jordan, A.M., 1970. Methods for rearing and maintaining Glossina in the laboratory, pp. 441-452 in "The African Trypanosomiases" ed. H.W. Mulligan, London, Allen and Unwin, 1970.

Van der Vloedt, A.M.V., 1982. Recent advances in tsetse mass rearing with particular reference to Glossina palpalis palpalis (Rob.-Desv.) fed in vivo on guinea pigs, pp. 223-253 in "Sterile Insect Technique and Radiation in Insect Control, Proceedings of a Symposium, Neuherberg, 1981, I.A.E.A., Vienna, 1982.

Wetzel, H. and Luger, D., 1978. In vitro feeding in the rearing of tsetse flies (Glossina m. morsitans and G. p. palpalis, Diptera: Glossinidae). Tropenmed. Parasit., 29: 239-251.

ACKNOWLEDGEMENTS

Thanks are due to Dr. A.M. Jordan for permission to publish this article. The involvement of the entire staff of the laboratory in the development of rearing methods for tsetse flies is gratefully acknowledged.

The laboratory is supported by the Overseas Development Administration of the Foreign and Commonwealth Office of the U.K. Government.

HAEMATOBIA IRRITANS

SIDNEY E. KUNZ and CHARLES D. SCHMIDT

U. S. Livestock Insects Laboratory, Agricultural Research Service, U. S. Department of Agriculture, P. O. Box 232, Kerrville, Texas 78028 U.S.A.

THE INSECT

Scientific Name: *Haematobia irritans* (Linnaeus)
Common Name: Horn fly
Order: Diptera
Family: Muscidae

1. INTRODUCTION

The horn fly is an obligate bloodsucking parasite of cattle throughout North America. It is endemic to Europe from where it was introduced into the United States in the 19th century. The pest is host specific and reproduces only in cattle feces. Adults will feed to a lesser extent on sheep, goats, horses, and wildlife. Populations vary throughout the distribution range, and it is not uncommon for up to 1000-1500 flies per animal to occur on cattle in the southern United States.

At 28°C eggs hatch in 18-24 hours and larval development is completed in 4-5 days. Adults start to emerge from pupae in 3-4 days and oviposit in 4 days. In nature diapause occurs in all but the southernmost parts of Texas and Florida. Diapause, however, does not exist in horn fly colonies maintained under laboratory conditions.

Colonies can be established from the field, but time required for laboratory adaptation can be lengthy. The recommended method of establishing a laboratory colony is to obtain pupae from the U.S. Livestock Insects Laboratory, Kerrville, Texas, or Veterinary Toxicology-Entomology Research Laboratory, College Station, Texas, or the Department of Entomology, New Mexico State University, Las Cruces, New Mexico. For colonies to be established outside North America, it would be best to collect locally.

2. FACILITIES, EQUIPMENT, AND PROCEDURES

Two methods of adult rearing are used:
(a) Rearing *in vitro* (off host)
(b) Rearing *in vivo* (on host)

2.1 Rearing *in vitro*

2.1.1 Facilities should include separate, insect-proof, adult and larval rearing rooms sufficient in size to supply the needs of the laboratory which the colony is expected to support. Rooms can range in size up to 4 x 4 meters, or refrigerator-size incubators have been used very successfully (Bay and Harris 1978). Adult rooms are maintained at 29°C, 60% RH with an L:D 8:16 regimen.

2.1.2 Adult cages

- Construct 25-cm^2 plexiglass-framed screenwire cages with a 12 cm opening in front. Melt 16-mesh (1.2 mm) screenwire into the plexiglass frame to make the cage sturdy and secure. (Adequate cages can also be constructed from several available materials such as 1- or 4-liter paper or plastic cartons, the size depending on the needs of the colony or the experiment.)

2.1.3 Egg collection

- Place the adult cage on a wire rack over a 35 x 25 x 5 cm deep plastic tray (or any size tray to accommodate cages as necessary) that contains absorbent cotton which is covered with cloth in water to a depth of ca 1 mm to prevent desiccation of the eggs over the 24-h collection period.
- Wash the eggs from the trays with a stream of water into a 50-mesh sieve, then into graduated 100-ml centrifuge tubes.
- Collect eggs daily; estimate the number of eggs on the basis of 8000 eggs/ml.
- Prepare aliquots for implantation into larval medium.
- Take a quality control measure of a 100-egg sample at least weekly.
- Count the empty egg cases 1-2 days later to estimate % hatch; expected hatch should be in excess of 85%.

2.1.4 Pupal collection

- Spray water on the medium 5 days after implantation of the eggs to moisten the surface and to separate the pupae from the medium.
- On the 6th day, place the medium in water, agitate the mixture, and collect pupae as they float to the top.
- For larger amounts of medium, use an agitator device to separate pupae from the medium. Agitate 2-3 min until the pre-wetted medium sinks to the bottom, then collect the free-floating pupae on the water surface.
- After collected, air-dry and separate the pupae from the remaining chaff as they are collected. Larger numbers of pupae can also be dried in specially prepared tumbler dryers which have a fan blowing on the pupae to facilitate the drying. Dry pupae just to the point to not adhere to one another.
- Take samples of at least 100 pupae once a week to determine the % adult emergence as a quality control standard; emergence from a healthy colony should be at least 85-90%.

PRECAUTION

Excessive drying causes increased mortality of the pupae and, in turn, lowers production of adults.

2.1.5 Handling and Emergence

- Put ca 75 ml (7500-9000) of the separated, air-dried pupae in small containers, such as paper sacks or cartons, on top of paper toweling and then into adult-holding cages. The paper toweling facilitates adult transfer from the pupal container to the adult cage.
- Adults should begin to emerge in ca 7-9 days after egg implantation; a majority of the flies will usually have emerged within 4 days after pupation. On the 4th day after emergence of the first adults, place the cages, which contain the adults, over egging trays and hold there ca 4 days for maximum egg production. Discard the cages at this time unless demands of the colony dictate otherwise.

2.1.6 Feces collection

- Stanchion male bovine animals on an elevated 1.2 x 1.8 m steel platform, which is ca 30 cm above the floor, with a 45-cm wide section of grating 30 cm from the rear edge. The grating covers a gutter positioned to collect and drain urine. (The raised platform is optional, but its use will facilitate prevention of urine contamination.)
- Flush gutter several times a day with sufficient water to control ammonia odors.
- Use rubber matting on platform to provide soft footing for animal.
- Place plastic-lined 60 x 90 cm fiberglass, plastic, or metal trays immediately behind stanchion to catch fresh feces from animals.
- Collect feces daily and refrigerate until used.

3. DIET PREPARATION

3.1 Larvae

The larval diet consists of a mixture of bovine feces and artificial diet (extender).

- Collect feces from animals being fed a combination of alfalfa-plus-dried grass hay diet or alfalfa alone.
- Dry mix or extender to be mixed with the feces contains by weight:
  - 40 parts of ground sugarcane bagasse
  - 8 parts whole-wheat flour
  - 4-6 parts fishmeal (65 percent protein)
  - 1 part sodium bicarbonate (baking soda)
- A ratio of 2 parts feces to 1 part dry mix produces maximum pupae.
- Also, to produce more pupae in the 2:1 mixture keep the moisture content at no more than 70 percent. However, if corrugated boxes are used, a mix of 10 parts feces, 3 parts of dry mix, and 10 parts of water, which produces a mix of ca 78% water is more productive.

PRECAUTION

It must be pointed out that feces contaminated with urine are unsatisfactory for horn fly production.

3.1.1 Larval pans

- Put 8-10 kg of medium in a pan ca 45 x 35 x 13 cm deep.
- Place ca 5 ml of eggs on top of the medium.
- Cover with a cloth and store in a room maintained at 27°C, 50-60% RH, for larvae to develop.
- Wash the pupae, which form after 3-4 days, from the medium 6 days after egg implantation.

PRECAUTIONS

(i) The eggs are placed on top of the medium and covered lightly with medium or a damp paper towel to prevent desiccation.

(ii) Eggs will suffocate if implanted too deeply (5 mm).

3.2 Adult diet

- Bovine blood
- Sodium citrate
- Cotton padding for feeding pads (sanitary napkins or similar absorbent cotton pads)
- Collect bovine blood at the abattoir. Mix immediately with sodium citrate solution (5 g sodium citrate in 5 ml water per liter of blood) to prevent coagulation.
- Feed adult horn flies the bovine blood twice a day at about 8:00 a.m. and 4:00 p.m.
- Place the blood in shallow pans and allow to absorb into feeding pads.
- Place the soaked pads on top of the cages and cover to delay drying. A routine feeding per cage requires about 100 ml of citrated bovine blood.

PRECAUTIONS

For storage and refrigeration at 40°C in the laboratory, a 25-ml water mixture containing 250,000 units of oral-grade nystatin and 0.5 g of kanamycin sulphate powder is mixed into 1 liter citrated blood. This required medication prevents the growth of hazardous microorganisms that reduce longevity and productivity of the flies. (It may be eliminated for extended periods of time if reduced production or increased adult mortality is not observed.)

4. REARING *IN VIVO*

Details of rearing adult horn flies on a host are provided by Berry et al. (1974) and Miller et al. (1979). Adult colonies can be started either from field-collected material or from already established colonies. Once the colony is established, it is necessary to start new larval containers daily at least 5 days of the week.

4.1 Rearing rooms

- A fly-free building is needed; a 4 x 4 m room will house one animal. This unit should be air-conditioned to maintain a 26-28°C temperature and 60-70% RH.

4.2 Larval diet preparation procedure

- Collect feces at least once a day and hold 24 h past voiding to allow a majority of the eggs to hatch before the feces are mixed with additional rearing medium.
- Add water to feces and mix to form a paste or slurry.
- Add this slurry, which contains the newly hatched 1st-instar larvae, to artificial media.
- The artificial media (extender) is a dry mix containing (parts by weight):
  - 40 ground bagasse
  - 8 whole-wheat flour
  - 6 fishmeal (65 % protein)
  - 1 sodium bicarbonate (baking soda)
  - 100 water
- Mix 2 parts feces to 1 part dry mix
- More pupae are produced in the 2:1 mixture when the moisture content is no more than 70 percent, but it is difficult to separate the pupae from the 70%-water medium when clean pupae are required because more and larger pieces of debris float with the pupae.

4.3 Adult diet preparation

- Is not necessary since flies are maintained on the host animal.

4.4 Pupal and adult collection

- For the production of adults for release, place the medium containing pupae on a rack in an emergence room until the adults begin to emerge over a 4-day period, usually 8-11 days after mixing. Adults can be attracted to a light port for collection (Miller et al. 1979). Pupae can be floated to the surface in water containers, or the medium can be placed into an agitating machine, such as a washing machine, to separate the medium from the pupae, and the floating pupae can be collected from the top of the water.

4.5 Quantity of flies

- It is possible to maintain a population of ca 6500 flies on a steer by returning ca 1000 pupae per day to the colony. A population of 10,000 flies can be maintained by returning ca 2000 pupae per day. Excess pupae are used for research.

4.6 Selection and care of animals

- Steers can be used successfully as long as the population of flies do not exceed 10,000 flies per head, after which fully grown bulls are more suitable. Steers become more irritable and "act up" which prevents feces production of desired quality. Bulls of British breed or British crosses, 400-500 kg in weight, have been successfully used.
- Groom the bulls periodically to remove fly feces, scabs, and scales of skin from the hair.
- Replace animals on a periodic rotation to maintain condition of the skin, freedom from soreness or swelling, and freedom from infection or other parasites.

PRECAUTIONS

(i) Arthropods, such as other feces-breeding flies or mites, can be very detrimental to good horn fly production making it necessary to have flyproof buildings.

(ii) Freshly collected feces can be frozen to eliminate infestation of other arthropods; however, frozen feces may not be as productive as feces that have not been frozen.

REFERENCES

Bay, D. E. and Harris, R. L., 1978. Small scale laboratory rearing of the horn fly. Southwest. Entomol. 3(4): 276-278.

Berry, I. L.; Harris, R. L.; Eschle, J. L. and Miller, J. A., 1974. Large-scale rearing of horn flies on cattle. U. S. Agric. Res. Serv. Bull. ARS-S-35, 6 pp.

Miller, J. A.; Schmidt, C. D. and Eschle, J. L., 1979. Mass rearing of horn flies on a host. U. S. Sci. Educ. Adm. AAT-S-8, 12 pp.

RECOMMENDED READING

Harris, R. L., 1962. Laboratory colonization of the horn fly, *Haematobia irritans* (L.). Nature 196: 191-192.

Harris, R. L.; Frazar, E. D. and Grossman, P. D., 1967. Artificial media for rearing horn flies. J. Econ. Entomol. 60: 891-892.

Schmidt, C. D.; Harris, R. L. and Hoffman, R. A., 1967. Mass rearing of the horn fly, *Haematobia irritans* (Diptera: Muscidae), in the laboratory. Ann. Ent. Soc. Am. 60: 508-510.

Schmidt, C. D.; Harris, R. L. and Hoffman, R. A., 1968. New techniques for rearing horn flies at Kerrville. Ann. Entomol. Soc. Am. 61: 1045-1046.

Schmidt, C. D.; Dreiss, J. M.; Eschle, J. L.; Harris, R. L. and Pickens, M. O., 1976. Horn fly: Modified laboratory rearing methods. Southwest. Entomol. 1(1): 50-51.

FOOTNOTES

Mention of a commercial (or proprietary) product in this paper does not constitute an endorsement of this product by the USDA-ARS.

LIXOPHAGA DIATRAEAE

E. G. KING and G. G. HARTLEY

Agricultural Research Service, USDA , P. O. Box 225,
Stoneville, Mississippi 38776

THE INSECT

Scientific Name: *Lixophaga diatraeae* (Townsend)
Common Name: Cuban fly
Order: Diptera
Family: Tachinidae

1. INTRODUCTION

The Cuban fly, *Lixophaga diatraeae* (Townsend), is a selective parasite of the sugarcane borer, *Diatraea saccharalis* (F.). This parasite is native to several Caribbean Sea islands but has been colonized on the sugarcane borer in other areas, including the states of Louisiana and Florida (Bennett 1969). Typically, *L. diatraeae* is attracted to its host initially by volatile substance(s) given off by larvae feeding on the sugarcane plant. Larviposition occurs in response to fore tarsi contact (by the parasite) on the host's frass (Roth et al. 1978, Roth et al. 1982). Roth et al. (1978) reported that *L. diatraeae* females most often deposit only 1 maggot at a time, then move away and groom themselves before resuming search activity. King et al. (1976) observed that regardless of the number of maggots entering 3rd to 4th stage sugarcane borer larvae, only one typically completed development. Laboratory studies indicated that the early 5th larval stage of the host sugarcane borer could produce more than a single parasite and was most suitable for rearing *L. diatraeae* (Miles and King 1975, King et al. 1976). In fact, the parasite most often parasitized 4th or later stage sugarcane borer larvae in the field (King et al. 1981).

During the period 1973 through 1976 we evaluated the feasibility of controlling the sugarcane borer in sugarcane in Louisiana and Florida by augmentative releases of *L. diatraeae* (Summers et al. 1976, King et al. 1981). Methods reported for rearing *L. diatraeae* were time consuming and cumbersome for large-scale rearing (see Bennett 1969). These were: (1) flies were held in cages for emergence, mating, and holding until maggot extraction; (2) uteri were dissected from flies and placed in a water droplet and, as maggots exited from egg shells (hatched), 1-3 were placed via a fine brush on each host larva; and (3) food and water were supplied to the flies daily. Thus, in 1973 and 1974 we developed a rearing program for *L. diatraeae* using sugarcane borer larvae as the host. And in 1975 and 1976, a mass rearing procedure was developed for *L. diatraeae* on the greater wax moth because they were easy and inexpensive to mass produce (Hartley et al. 1977, King et al. 1979). We report here only our procedures for large scale production of *L. diatraeae* on sugarcane borer larvae because they are also suitable for small scale rearing.

2. FACILITIES AND EQUIPMENT REQUIRED

2.1 Facilities

Adult flies are held in controlled environment rooms maintained at 26°C and 80% RH on a LD 14:10 schedule. Light intensity makes little difference to fly mating. Lights can be either incandescent, flourescent, or a combination. Fly maggots are reared inside their host larvae, the sugarcane borer, at 26-28°C and 80% RH in complete darkness.

2.2 Equipment and materials required for insect handling

2.2.1 Adult holding and mating
- Aluminum frame cages (25.4 x 22.9 x 15.2 cm)
- Tube gauze - size T-1 (used for cage cover)
- 400 ml water glass
- 6 mm Plexiglas
- Aluminum window screen
- Rubber band
- Narrow neck 240 ml plastic bottle

2.2.2 Adult collection and maggot extraction
- 100-200 ml glass collection jar
- 6 mm rubber hose (used as a fly aspirator or construct an aspiration device as described by Gantt et al. 1976). For small scale rearing fly maggots can be removed by hand from the fly uterus with dissecting forceps eliminating need for the following equipment: Photo Flo; sodium hypochlorite; stainless steel semi-micro blender container; rheostat autotransformer; and 24 mesh stainless steel screen

2.2.3 Host parasitization
- Maggot counting grid (constructed from graph paper containing 0.4 cm^2 cut in a circle containing 171 0.4 cm^2) (King et al. 1979).
- Stereoscope
- Maggot dispensing machine (Gantt et al. 1976). For small scale rearing, a 2.5 ml repeating dispenser (Hamilton Model PB 600-1) can be used. A fine camel hair brush can be used to place the fly maggot directly on its host.

2.2.4 Parasitized host holding
- 22.5 ml plastic cup
- Cell pak or other suitable container used for holding cups.

2.2.5 Puparia collection
- Forceps
- Paper towels
- Small plastic pan

3. DEVELOPMENT OF INSECT STAGES AT DIFFERENT TEMPERATURES

3.1 Maggot

Developmental time (days) of the (parasitic) maggot stage of *L. diatraeae* and percentage parasitism when held at different constant temperatures. (a)(b)

Temp. (C)	Developmental time: Male Mean	Male Min	Male Max	Female Mean	Female Min	Female Max	% parasitization(c)
16.0	25.2	20	30	25.8	20	33	94.0
18.4	14.4	11	21	16.1	13	23	79.2
22.0	10.3	9	13	10.6	9	13	87.7
26.0	7.3	7	9	8.4	8	11	91.5
28.0	6.9	6	11	7.1	6	10	94.2
30.0	6.2	6	9	6.8	6	9	97.2
32.0	6.5	6	9	6.6	6	8	96.8

a. Tests at 16 and 32°C conducted at different time than others due to malfunction of temperature chamber.

b. Maggots did not complete development at 34°C.

c. Host larva considered parasitized if maggot emerged or maggot present in host larva when dissected.

4. REARING AND COLONY MAINTENANCE

The original colony of Lixophaga diatraeae was collected from sugarcane fields in south Louisiana.

4.1 Puparia and adults

- Collect puparia from cups containing parasitized sugarcane borer larvae 11 days after inoculation with L. diatraeae maggots.
- Disinfect puparia in sodium hypochlorite solution (1%) active ingredient for 3 minutes followed by a 15 minute rinse in water.
- Place 200 L. diatraeae puparia in a 240 ml narrow neck plastic bottle containing a cardboard strip which runs from the bottom of the bottle to 5.0 cm above the top. This strip helps guide the flies from the bottle and provides a surface to which the flies can attach while expanding their wings.
- Prepare holding cages by stretching tube gauze over the aluminum frame cage.
- Prepare watering devices - fill 400 ml glass with water and place screen followed by Plexiglas over the mouth of the glass; a rubber band is placed from top to bottom around the glass to hold the device together. Screen and Plexiglas are cut about 2.5 cm more in diameter than the mouth of the glass. Invert glass of water over screen and Plexiglas.
- Cover bottom of aluminum cage with paper towels to absorb any excess moisture.
- Place watering device and plastic bottle containing puparia in cage (lay plastic bottle on its side).
- Close cage ends with wire twist tabs.
- As soon as flies begin to emerge place about 1 g of raw sugar on the top of the cage surface and sprinkle with water 3 times daily - add extra sugar as needed.

4.2 Fly collection and disinfection

- 12 days after emergence flies are aspirated from the holding cages and placed into a collection jar containing sodium hypochorite (1%) active ingredient plus 0.16% Photo Flo for 3 minutes. This task can be accomplished by using a hand to mouth aspirator or by constructing an aspiration device as described by King et al. (1979).

4.3 Maggot extraction and disinfection

- Rinse sodium hypochlorite from the flies.
- Place 200 to 800 flies with 50 ml of formalin solution (0.7%) active ingredient into a stainless steel semi-micro container that can be used with a Waring Blender, which is attached to a rheostat autotransformer so as to produce 8500 revolutions/minute.
- Blend flies three times for 3 seconds each time. This separates the fly abdomen from the thorax and ruptures the uterus thereby releasing the maggots into the formalin solution. The formalin solution effectively eliminates the bacterium, Serratia marcescens (Bizzio) which often contaminates the maggots.
- Wash maggots for 5 minutes in the formalin solution (wash periods exceeding 5 minutes may result in maggot mortality).
- Pour solution with maggots, fly parts, and formalin through a 24 mesh stainless steel screen into a beaker - the screen retains large fly particles but allows the maggots to flow through.
- Decant the formalin solution from the maggots and thoroughly rinse maggots in distilled water (maggots sink).

4.4 Determination of maggot number

- Suspend maggots in 0.15% agar solution and place in a petri dish positioned over a grid pattern. Count maggots in 11 randomly selected squares (0.4 cm^2 per square). Calculate the averge number of maggots per sample square and multiply by 171 (number of 0.4 cm^2 per dish) to obtain the total number of maggots per dish. Add additional agar solution to obtain the desired maggot density in a given solution.

4.5 Host parasitization

- The agar-water solution containing the maggots was metered into 30 ml cups containing early fifth-stage host larvae of the sugarcane borer at a density of 3 to 4 maggots per host larva.
- The machine used for dispensing *L. diatraeae* maggots consists of a positive-displacement piston pump (activated by a solenoid switch) attached to a discharge needle that energized the motor automatically by penetrating the lid of the cup which was used to rear the host larva.
- For smaller scale rearing a 2.5 ml repeating dispenser (Hamilton Model PB 600-1) was used (King et al. 1975).
- For extremely small scale rearing a fine camel hair brush could be used to transfer maggots directly to the host larva.

4.6 Puparia collection

- Parasite puparia were harvested on the 11th day after parasitization and were removed from the cups containing sugarcane borer larvae with forceps. Puparia retained for the reproductive colony were disinfected in sodium hypochlorite (1%) active ingredient for 3 minutes followed by a 15 minute rinse in water.

4.7 Adults

- The sex ratio is approximately 1:1.

4.7.1 Sex determination

- Flies were sexed after emergence from puparia by the following criteria: male pulvilli more elongate and lighter in coloration than female; also female larger than male with abdomen bulbous.

4.7.2 Mating

- Flies are held at 26°C and 80% RH with LD 14-10 photoperiod.
- Light type or intensity does not affect fly mating, but flies will not mate in complete darkness.

4.8 Insect quality

- Insect quality is monitored by recording puparia weights, percentage parasitization, number of maggots per female, and adult longevity. Quality should be as follows: puparia weights, male = 14 mg, female = 20 mg; percent parasitization, 90%; number maggots/female, 70; and maximum adult longevity, male = 29 days, female = 24 days.

4.9 Special problems

a. Diseases

Contamination of the fly maggots by the bacterium, *S. marcescens*, can cause septicemia in parasitized sugarcane borer larvae (King et al. 1975). Washing of maggots in formalin (0.7%) active ingredient for 5 minutes immediately before being used to parasitize host larvae effectively controlled the disease.

b. Human health

All work with formalin should be conducted under a fume hood.

5. LIFE CYCLE DATA

5.1 Developmental data - at 26° C; 80% RH

Stage	Average period
Maggot to puparia	♂ = 7.3 days; ♀ = 8.4 days
Puparia to adult	♂ = 10.9 days; ♀ = 11.5 days
% Fly Emergence	♂ = 95% ♀ = 95%

6. PROCEDURES FOR SHIPPING INSECTS

Puparia shipped to field stations are packaged between layers of cotton in 0.47 liter cylindrical containers (500 per container) and placed in styrofoam boxes that contain ice packs. Ice packs were prevented from coming in direct contact with containers of puparia by sandwiching them between sheets of plywood perforated with 2.5 cm diameter holes. A temperature of 18°C to 24°C is recommended for about 1 day during transit.

7. REFERENCES

Bennett, F. D. 1969. Tachinid flies as biological control agents for sugarcane moth borers. *In* J. R. Williams, J. R. Metcalfe, R. W. Montgomery, and R. Mathes (eds.), Pests of Sugarcane, pp. 117-148. Elsevier Publishing Co., New York.

Gantt, C. W.; King, E. G.; and Martin, D. F. 1976. New machines for use in a biological insect-control program. Trans. ASAE 19: 242-243.

Hartley, G. G.; Gantt, C. W.; King, E. G.; and Martin, D. F. 1977. Equipment for mass rearing of the greater wax moth and the parasite *Lixophaga diatraeae*. U.S. Agric. Res. Serv. (Rep.) ARS-S-164, 4 pp.

King, E. G.; Miles, L. R.; and Martin, D. F. 1976. Some effects of superparasitism by *Lixophaga diatraeae* of sugarcane borer in the laboratory. Entomol. Exp. Appl. 20: 261-269.

King, E. G.; Hartley, G. G.; Martin, D. F.; Smith, J. W.; Summers, T. E.; and Jackson, R. D. 1979. Production of the tachinid *Lixophaga diatraeae* on its natural host, the sugarcane borer, and on an unnatural host, the greater wax moth. USDA, SEA, Adv. Agric. Technol., Southern Series, No. 3, 16 pp.

King, E. G.; Sanford, J.; Smith, J. W.; and Martin, D. F. 1981. Augmentative releases of *Lixophaga diatraeae* (Dip.: Tachinidae) for suppression of early seson sugarcane borer populations in Louisiana. Entomophaga 26: 59-69.

Miles, L. R.; and King, E. G. 1975. Development of the tachinid parasite *Lixophaga diatraeae*, on various developmental stages of the sugarcane borer in the laboratory. Environ. Entomol. 4:811-814.

Roth, J. P.; King, E. G.; and Thompson, A. C. 1978. Host location behavior by the tachinid, *Lixophaga diatraeae*. Environ. Entomol. 7: 794-798.

Roth, J. P.; King, E. G.; and Hensley, S. D. 1982. Plant, host, and parasite interactions in the host selection sequence of the tachinid, *Lixophaga diatraeae*. Environ. Entomol. 11:

Summers, T. E.; King, E. G.; Martin, D. F.; and Jackson, R. D. 1976. Biological control of *Diatraea saccharalis* in Florida by periodic releases of *Lixophaga diatraeae*. Entomophaga 4: 359-66.

8. FOOTNOTE

Mention of a trademark, proprietary product or vendor does not constitute a guarantee or warranty of the product by the U.S. Department of Agriculture and does not imply its approval to the exclusion of other products or vendors that may also be suitable.

MUSCA AUTUMNALIS

FRED W. KNAPP

Department of Entomology, University of Kentucky, Lexington, Kentucky 40546-0091

THE INSECT

Scientific Name:	Musca autumnalis DeGeer
Common Name:	face fly
Order:	Diptera
Family:	Muscidae

1. INTRODUCTION

The face fly is indigenous to Europe, Russia, China, North Africa, Korea, Japan, and Iran. It was first discovered in North America in 1952 and is now found throughout southern Canada and all the continental United States excepting Arizona, Florida, New Mexico and Texas (Pickens and Miller, 1980). The face fly's habit of clustering around cattle and horses and feeding on their ocular and nasal secretions causes excessive annoyance. This disrupts grazing activities resulting in weight loss of ca. 114 g/day and milk production loss of 25% (Pickens and Miller 1980). Face flies are vectors and intermediate hosts of Thelazia spp. eyeworms and suspected vectors of pinkeye and IBR (infectious bovine rhinotraceitis) pathogens in cattle. Feeding by face flies can also cause lesions on the conjuctiva of the eyes of cattle.

Three larval stadia occur, taking 3 to 5 days in the laboratory at 27°C. The pupal period requires approximately 7 to 8 days, and mating occurs at 4 to 5 days post-emergence. Under laboratory conditions, 11 to 12 days are required for the cycle from egg to adult (Wang, 1964; Arends and Wright, 1981). In field conditions, an overwintering diapause is triggered in August or September by short day length and lower temperatures.

No artificial larval diet has ever been successfully utilized, thus a fresh source of bovine feces is essential for laboratory rearing.

2. FACILITIES AND EQUIPMENT REQUIRED

2.1 Facilities

Rearing is carried out in controlled environmental rooms maintained at ca. 27°C, 50-70% RH and a 24 hour continuous light, although a 16:8 LD cycle may be used. Lighting is provided by cool-white lights

2.2 Equipment and materials required for insect handling

2.2.1 Adult maintenance

- Bioquip® 1450 Series collapsible cages 61 cm (Bioquip Products, 1320 E. Franklin Avenue, El Segundo, CA 90245)
- Tubeguaz® size 78 (Hospital Product Div., Chicago IL 60610)
- 100 x 15 mm disposable petri dishes
- 230 ml light duty wax treated Dixie® no. 358 squat containers (American Can Co., Greenwich, CN 06830)
- Non-sterile absorbent cotton
- Durosorb® disposable polypropylene backing underpads for bottom of cages (Professional Medical Products, Inc., Greenwood, SC)

- Water
- Adult diet

2.2.2 Egg collection and storage
- 100 x 15 mm disposable petri dishes
- Reynolds 914 Film (Reynolds Metals Co., Richmond, VA 23261).
- Fresh bovine feces - best from pasture or hay diet

2.2.3 Egg hatch and larval rearing
- Fresh bovine feces
- Water
- 24 x 31 x 5 cm metal pans
- 45 x 35 cm lipped fiberglass trays (cafeteria serving trays)
- Sifted, clean sand
- Large stainless steel spatula
- Small stainless steel spatula
- 3.3 liter plastic bucket

2.2.4 Pupal collection and storage
- No. 10 soil sifter
- 92 ml Solo® no. P50 poly lined souffle cups (Solo Cup Co., Urbana, IL 61801).
- 7.6 cm x 12.7 cm index cards

2.3 Diet preparation equipment
- 3.8 liter glass or plastic container with lid

3. DIET

3.1 Larval diet

Fresh bovine feces is used for larval rearing. The best feces is from cattle feeding on a pasture. If this is not available, feces from cattle on a grass hay is used. A dry-lot dairy herd on a high-roughage diet with a daily supplement is a good source of winter feces. Feces may be held at room temperature up to 24 h before use.

3.2 Adult diet

3.2.1 Composition
- 50% (by volume) sucrose
- 25% (by volume) dried egg product
- 25% (by volume) nonfat dry milk crystals

3.2.2 Brand names and sources of ingredients

Granulated Pure Cane Sugar (Colonial Sugars, Inc., Mobile, AL Distributors, Inc., Stockon, VA 95204); Pasteurized Instant Nonfat Dry Milk Crystals (National Institutional Food Distributor Associates, Inc. Atlanta, GA 30325); Pasteurized School Packed Dried Egg Product (Nugget Distributors, Inc., Stockton, CA 95204).

3.2.3 Diet preparation procedure
- Place portions of sugar, dried egg product, nonfat dry milk crystals in a 2:1:1 ratio in a container and thoroughly mix.

4. INSECT HOLDING

4.1 Adults

Parent fly colonies should be maintained in 61 cm cages. If small numbers of adults are needed for specific tests, they are more easily obtained from a smaller container. Therefore, place the number of pupae required into a smaller container, such as a vented wide mouth 3.8 liter jar for emergence. With food and water, 300 to 400 adults may be held for 3 to 4 days.

4.2 Eggs

Eggs can be held for up to 22 hours within the oviposition media at 20°C.

4.3 Larvae

Larvae cannot be held, except in rearing media under normal rearing conditions.

4.4 Pupae

Pupae may be held for up to 12 days in an incubator set at 20°C.

5. REARING AND COLONY MAINTENANCE

Colonies may be established in any of three possible ways:

a. Pupae received from an established colony.
b. Collection of wild adult flies from cow feces.
c. Collection of larvae from cow feces. The larvae to pupate in sand. These pupae may be sifted out as in Section 5.3.

5.1 Egg collection

- Line a 100 mm x 15 mm petri dish with Reynolds® 914 Film, plastic wrap (Reynolds Metal Co., Richmond, VA 23261).
- Fill dish with fresh bovine feces, ca 9 g.
- Place dish into cage with adults.
- Leave dish in cage for ca. 1 h.

5.2 Inoculation of larval rearing medium with eggs

- Fill 24 x 31 x 5 cm metal pan with fresh bovine feces (ca 2.2 kg).
- Fill to rim and smooth out.
- Place metal pan on 45 x 35 cm fiberglass tray with approximately 300 g of sand under the pan to a height of 1-1.5 cm. Larvae migrate to the sand to pupate.
- Remove Reynolds Film with feces and eggs from petri dish.
- Cut out portion of seeded feces with ca 500 eggs from feces and place into depression formed in the dung in metal pan.
- Hold at ca. 27°C and RH 50-70%.

5.3 Pupal collection

- When gray-white pupae are seen on fiberglass trays, the larve have migrated out of the feces, pupates will be collected.
- Lift one end of metal pan up.
- Gently use a 7.6 x 12 cm index card and scrape pupae from bottom of pan onto the sand.
- Sift sand through No. 10 mesh soil sifter.
- Pupae may be transferred into ca 95 ml souffle cups.
- The souffle cups are then either placed into the Bioquip 1450 Series cage (colony maintenance) or into 3.8 liter olive jars (testing).

5.4 Adults

5.4.1 Emergence period

- Adults will emerge 7 to 8 days after pupation.

5.4.2 Oviposition

- Females begin mating 4-5 days after emergence.
- Females begin oviposition 2-4 days after first mating.
- Oviposition occurs throughout lifetime (3 weeks to 3 months) but optimum egg production is at ca 11-14 days of age.

5.4.3 Sex determination

- Abdomen of male orange-brown, female silvery.
- Thorax of male black, female gray with thoracic stripes.
- Inner margins of compound eyes almost touching in male, angled in female.

5.5 Rearing schedule

a. Alternate days

- Change water (supplied as water-soaked cotton in 230 ml treated cups).

b. Twice weekly

- Collect fresh feces.
- Place oviposition dishes in cage.
- Prepare larval media.
- Check larval trays for pupae.

c. Weekly

- Change Durasorb® pads at bottom of adult cages and remove dead flies

5.6 Insect quality

Insect quality is checked by keeping record of pupal weights. Normal pupae are nearly cylindrical, about 6.5 mm long and 2.6 mm in diameter with a weight of ca. 30 ± 2 mg.

6. LIFE CYCLE DATA

6.1 Developmental data

Stage	Time	% mortality
Egg	14-16 h	10-15
Larvae (3 instars)	4-5 days	15-20
Pupae	7-8 days	15-20
Total Development (egg to egg)	19-20 days	

7. REFERENCES

Arends, J. J. and Wright, R. E. 1981. Mass rearing of face flies. J. Econ. Entomol. 74:355-358.

Fales, J. H., Bodenstein, O. F., and Keller, J. C. 1961. Face fly laboratory rearing. Soap Chem. Spec. 37:81-83.

Pickens, L. G. and Miller, R. W. 1980. Biology and control of the face fly, *Musca autumnalis* (Diptera: Muscidae): Review article. J. Med. Entomol. 17(3):195-210.

Wang, C. M. 1964. Laboratory observations on the life history and habits of the face fly, *Musca autumnalis* (Diptera: Muscidae): Ann. Entomol. Soc. Amer. 57:563-569.

8. HISTORY OF THE COLONY

The following is a history of the first laboratory face fly colony in the United States as told to the author by Calvin M. Jones (USDA retired, Fort Payne, AL).

Face flies were colonized from a wild population netted in Maryland in June through August 1960 (see Fales et al. 1961). Calvin Jones established a second colony in Lincoln, Nebraska from Fales' colony in November 1961. The following summer he established a new colony from field flies and mixed them with Fales' colony. In 1962, he destroyed this colony and began a new Lincoln colony.

Pupae from this Lincoln colony were then sent to the following laboratories for starting colonies.

University of Kentucky - Feb. 1962 and July 1971
Clemson University - Sept. 1962
Wooster, Ohio - July 1963
Kansas State University - June 1964
Illinois Natural History Survey - Nov. 1966
University of California - Feb. 1969 and Feb. 1970
Guelph, Ontario - Jan. 1970
University of Massachusetts - Nov. 1971

Some laboratories, such as the University of Kentucky and the University of Massachusetts, also received pupae from the Beltsville laboratory. The University of Kentucky maintains this colony of face flies with both the Lincoln and Maryland gene pool. In addition, Kentucky has established a Kentucky strain of face flies. Other states have also established laboratory colonies from wild flies within their own areas.

MUSCA DOMESTICA

PHILIP B. MORGAN

Insects Affecting Man and Animals Research Laboratory, USDA, ARS ,
P. O. Box 14565, Gainesville, FL, 32604

THE INSECT

Scientific Name: *Musca domestica* L.
Common Name: House Fly
Order: Diptera
Family: Muscidae

1. INTRODUCTION

The common house fly, *Musca domestica* L. has been extensively utilized as a test organism to screen candidate insecticides, chemosterilants, and insect growth regulators by scientists, in public or private research institutions both here and abroad. In addition, all stadia have been utilized as a source of food for parasites and/or predators. Since the immature stadia can survive in various substrates, entomologists have normally used materials that were available locally (Spiller, 1964, 1966, Louw, 1964; Sawiciki, 1964; Keiding and Arevad, 1964; and Schoof, 1964). This has caused variation in the quality of flies that were produced, making it difficult to replicate test results. Here at the Insects Affecting Man and Animals Research Laboratory, USDA, ARS, we have been able to produce a consistent quality insect at an economical cost by rearing the house fly according to the methods described by Morgan (1980, 1981), Morgan and Patterson (1978) and Morgan et al. (1978).

2. FACILITIES AND EQUIPMENT NEEDED

2.1 Facilities

2.1.1 Maintain the adult colonies in a controlled environmental room at 25.5 ± 2°C, 75 ± 5% relative humidity and LD 12:12.
  a. Humidifier: Herrmidifier Model 707TW (3.1)

2.1.2 The larvae and pupae are maintained in a controlled environmental room at 28.3 ± 2°C and 70 ± 5% relative humidity.
  a. Humidifier: See 2.1.1a

2.2 Equipment and materials required for insect handling

2.2.1 Adults
  a. Adult fly cage constructed of aluminum framing (45 cm x 35 cm x 35 cm) covered with aluminum mesh window screen. Attach a 60 cm section of tubular orthopedic stockinette at one end by rubber beading for easy access to the cage (3.2).
  b. Cage label

Musca domestica
Pupae in cage ____________
Adult flies emerged __________
Oviposited ________________

c. Watering containers
- 1 liter waxed container
- 1.5 cm^2 pieces of styrofoam

d. Adult food
- Granulated sugar
- Powdered non-fat milk
- 250 ml plastic container

e. Adult oviposition
- Old larval medium
- Black cloth (unsized) 13 cm^2
- 250 ml plastic container

f. Egg collection and storage
- 500 ml glass beaker
- Incubator (26.1 ± 2°C) (70 ± 5% RH)

g. Egg measuring device
- 15 ml plastic centrifuge tube

h. Egg viability
- Black cloth (1 cm x 5 cm)
- Fresh larval medium (100 g)
- Paper container (400 ml)
- Unbleached muslin (12 cm^2)
- Brush No. 00

i. Hydrothermograph

2.2.2 Larvae

a. Larval medium
- Brewers dried grain (3.3)
- Soft wheat bran (3.4)
- Pelletized coastal Bermuda (3.5)

b. Larval rearing container (3.6): Plastic tray (50 cm x 40 cm x 10 cm)

c. Larval container label

Musca domestica
Trays seeded with eggs ________
Pupation occurred ____________
Pupae to be floated ___________

d. Racks: 7.5 cm slotted angle iron

e. Hydrothermograph and a soil temperature probe

2.2.3 Pupae

a. Flotation chambers (3): 15 liter capacity

b. Aluminum window screen: 7 cm x 10 cm

2.2.4 A source of high pressure steam for cleaning larval medium containers and adult cages: 137°C - 7 kg/cm^2

2.3 Diet preparation

2.3.1 Adult

a. Granulated sugar 6 parts (3.7); non-fat powdered dry milk 6 parts (3.8); powdered egg yolk 1 part (3.9)

b. Since this is a dry mix, it can be easily mixed by hand, mix and store in a sealed container at 25.5 ± 2°C

2.3.2 Larval

a. Brewers dried grain 40% or 1.04 kg (3.3)

b. Soft wheat bran 33% or 0.85 kg (3.4)

c. Pelletized coastal Bermuda 27% or 0.7 kg (3.5)

d. Add 4 to 5 liters of water

e. Mixture stirred to insure uniform distribution of water as well as the 3 ingredients

2.4 Preparation of oviposition containers

2.4.1 Wrap ca 50 g of used larval medium in the moistened black cloth and place in a 250 ml plastic container

3. MATERIALS AND SUPPLIES

3.1 Humidifier Model 707 TW
Herrmidifier Company, Inc.
1770 Hempstead Rd.
Lancaster, PA USA

3.2 Tubular Orthopedic Stockinette
Burlington Industries
P. O. Box 610
Asheboro, NC USA

3.3 Brewers Dried Grain
Howlands Feed Co.
Live Oak, FL USA

3.4 Soft Wheat Bran
Brownlee Feed and Seed, Co.
Gainesville, FL USA

3.5 Pelletizied Coastal Bermuda
Whistling Pines Ranch
Gainesville, FL USA

3.6 Plastic Rearing Containers
Panel Controls Corporation
P. O. Box 66
Mt. Lakes, New Jersey 07046

3.7 Granulated Sugar
Publix Stores, Inc.
Gainesville, FL USA

3.8 Non-fat Powdered Milk
Publix Stores, Inc.
Gainesville, FL USA

3.9 Powdered Egg Yolk
Tharp Brothers
Pangburn, Arkansas 72121

4. REARING AND COLONY MAINTENANCE

4.1 The colony was established from field collected adults in 1942, and has been reared on an artificial diet for ca 960 generations

4.2 Adult

4.2.1 Place 15,000 house fly pupae in the adult cage (see 4.4.1e to determine numbers). Approximately 80% (12,000 adults) will emerge. A daily survival rate of 0.8 will provide ca 5000 adults or 2500 five-day-old females. At 40 eggs/female should collect ca 100,000 eggs.

a. Remove empty puparia within 48 h following completion of adult eclosion
b. Place container of water in cage
   - Add styrofoam (1.5 cm^2) to act as landing platforms to reduce adult mortality due to drowning
   - Replace water container every 3rd day
c. Place 250 g of food in cage
   - Replace food daily

4.2.2 Adult oviposition

a. Adult colony room temperature increased to 27.7 ± 2°C and 90 ± 5% RH 1 h prior to introducing the oviposition containers

b. Place oviposition containers in adult fly cage
c. Two h later remove oviposition containers
d. Return adult colony room to 25.5 ± 2°C and 75 ± 5% RH
e. Cap and store containers of eggs in the incubator at 26.1 ± 2°C and 70 ± 5% RH until needed. Eggs cannot be held longer then 8-10 h
f. Gently remove eggs from all oviposition containers and place in a 500 ml beaker containing 250 ml of tap water
g. Break up egg masses by gentle stirring. Please note that some of the egg masses will float due to pockets of air within the masses. However, individual viable eggs will sink, while individual dead eggs will float. Remove all individual floating eggs
h. Three ml quantities of eggs are measured into individual 15 ml graduated plastic centrifuge tubes (ca 10,000 eggs/ml). The eggs are allowed to settle for 5 min to provide accuracy in measurment

PRECAUTION
Embryonic mortality occurs if the eggs are allowed to remain in water longer than 15 min

4.2.3 Egg viability
a. Place 100 freshly oviposited eggs on the piece of moistened black cloth
b. Transfer the cloth and eggs to the surface of 100 g of fresh larval medium in a 435 g paper carton
c. Cover the carton with the 12 cm^2 of unbleached muslin
d. Place carton containing medium and eggs in the incubator (28.3 ± 2°C and 75 ± 5% RH)
e. Twenty-four h later determine percentage egg hatch
f. Six days later determine percentage pupation
   - should obtain 80 pupae/100 eggs
g. Four days later determine percentage adult emergence
   - should obtain 95 to 100 adults per 100 pupae

4.3 Larvae

4.3.1 Larval medium
a. Place 2.9 kg of freshly prepared larval medium in larval rearing container
b. Place 3 ml of eggs on the surface of the larval medium
c. The medium will begin fermenting and will reach a maximum of 40°C in 36 h

4.4 Pupae

4.4.1 Larvae will pupate close to the surface of the medium
a. Gently transfer the medium containing the pupae to the 1st flotation container containing ca 10 liters of water (25.5 ± 2°C).
b. Gently stir the pupae and medium. The majority of the medium will absorb moisture and sink to the bottom of the flotation container
c. Gently transfer the pupae to the 2nd and 3rd flotation container
d. Allow the pupae to dry at room temperature
e. Determine number of pupae by weight (13 mg to 15 mg)
   - Count out 100 pupae
   - $\frac{100 \text{ pupae}}{x \text{ pupae}} = \frac{\text{wt of } 100 \text{ pupae g}}{\text{wt of } x \text{ pupae g}}$

PRECAUTION
Do not subject the pupae to stress by severe handling

5. LIFE CYCLE

5.1 Under laboratory conditions the duration of house fly generation averages 15 days and consists of ca 1 day for the egg stage, 1, 2, and 3 days for the 1st-, 2nd-, and 3rd-larval instars, respectively, 4 days for the pupal stage, and a 3-day preoviposition period for adults, with oviposition on the 4th day. The adult's life span averages 30 days and the total number of eggs laid by each female is ca 1080, with 120 eggs per gonotrophic cycle or 40 eggs per ovipositing female per day. The egg, larval and pupal daily survival rate is 0.735, while the adult daily survival rate is 0.8.

6. INSECT QUALITY

6.1 Insect quality is checked continously by recording adult fecundity, egg viability and pupal weight.

a. 120 eggs/gonotrophic cycle
b. 95% to 100% hatch
c. 85% to 90% pupation
d. 13 mg to 15 mg pupal weight

7. SPECIAL PROBLEMS

7.1 Parasites

There are three genera of pupal parasites of the order Hymenoptera - Family Pteromalidae, that can be a problem in house fly colonies. The genus *Spalangia* sp. has infested our fly colonies more than the other 2 genera, *Muscidifurax* sp. and *Pachycrepoideus* sp. The genus *Nasonia* sp has never been a problem. However, good rearing techniques and sanitation will easily control the parasites. With the adult fly colony rooms maintained at 25.5 ± 2°C, the fly pupal stage lasts 4 days. Since the parasitic wasps require a minimum of 13 days to develop, removing the pupae from the adult colony cages by the 6th day will eleminate the parasite problem.

7.2 Predators

Mites of the genus *Macrocheles* sp. have been a problem and have drastically affected house fly production. Generally they have been found in the larval trays, feeding on the house fly eggs. Again proper techniques and sanitation will control these predators. Holding all larval media at 0°C for a minimum of 10 days prior to use as well as steam cleaning all larval rearing trays and adult cages on a daily basis eliminated the mite problem.

7.3 Pathogens

Entomophthera sp. has also infested our adult house fly colonies. However, steam sterilization of the adult colony cages and thoroughly sponging the colony room ceiling, walls, shelves, floor, etc., with 0.13% aqueous solution of Zephiran® eliminated the problem.

8. EQUIPMENT MAINTENANCE

8.1 Steam clean adult cages between adult populations
8.2 Stockinette cleaned for reuse
8.3 Steam clean larval rearing containers between larval populations
8.4 Clean tray racks, shelving, surface of laboratory benches and floors weekly with 1.0% aquous solution of sodium hypochlorate
8.5 Change the hydrothermograph charts weekly
8.6 Clean the humidifiers quarterly

9. PROTECTIVE EQUIPMENT

9.1 Particle masks
9.2 Gloves
9.3 Laboratory coats
9.4 Protective eye glasses

10. REFERENCES

Keiding, J., and Arevad, K. 1964. Procedure and equipment for rearing a large number of housefly strains. Bull. Wld. Hlth. Org. 31: 527-528.

Louw, B. K. 1964. Physical aspects of the laboratory maintenance of muscoid fly colonies. Bull. Wld. Hlth. Org. 31: 529-533.

Morgan, P. B. 1980. Mass culturing three species of microhymenopteran pupal parasites, Spalangia endius Walker, Muscidifurax raptor Girault and Sanders, and Pachycrepoideus vindemiae (Rondani) (Hymenoptera: Pteromalidae). VIII Reunion Nacional de Control Biologico. Tecomon, Colima, Mexico. April 22-25, 1980.

Morgan, P. B. 1981. Mass production of Musca domestica L. In Status of Biological Control of Filth Flies. Proceedings of a Workshop, February 4-5, 1981. Insects Affecting Man and Animals Research Laboratory, USDA, ARS, SR, Florida-Antilles Area and University of Florida, Gainesville, Florida. 212 p.

Morgan, P. B., LaBrecque, G. C., and Patterson, R. S. 1978. Mass culturing the microhymenopteran parasite, Spalangia endius Walker. J. Med. Entomol. 14: 671-673.

Morgan, P. B., and Patterson, R. S. 1978. Culturing microhymenopteran pupal parasitoids of muscoid flies. In Facilities for Insect Research and Production. USDA Tech. Bull. 1576. 86 p.

Sawicki, R. M. 1964. Some general considerations of housefly rearing techniques. Bull. Wld. Hlth. Org. 31: 535-537.

Schoof, H. F. 1964. Laboratory culture of Musca, Fannia and Stomoxys. Bull. Wld. Hlth. Org. 31: 539-544.

Spiller, D. 1964. Nutrition and diet of muscoid flies. Bull. Wld. Hlth. Org. 31: 551-554.

Spiller, D. 1966. House Flies. In Insect Colonization and Mass Production. Ed. Carroll N. Smith. Academic Press, New York. 618 p.

FOOTNOTE

Mention of a commercial or proprietary product does not constitute an endorsement by the USDA.

RHAGOLETIS CERASI and CERATITIS CAPITATA

ERNST F. BOLLER

Swiss Federal Research Station for Fruit Growing, Viticulture and Horticulture, CH-8820 Waedenswil, Switzerland

THE INSECTS

Scientific Name: *Rhagoletis cerasi* (Linnaeus)
Common Name: European cherry fruit fly
Order: Diptera
Family: Tephritidae

Scientific Name: *Ceratitis capitata* (Wiedemann)
Common Name: Mediterranean fruit fly
Order: Diptera
Family: Tephritidae

1. INTRODUCTION

The rearing method described here was originally developed for the European cherry fruit fly, *Rhagoletis cerasi*, later it became evident that the liquid larval diet was also suitable for the medfly, *Ceratitis capitata*. Two other fruit fly species, the Caribbean fruit fly, *Anastrepha suspensa* (Loew), and the olive fly, *Dacus oleae* were also reared on a small scale for several generations.

The need for a standardized diet for *Ceratitis capitata* emerged during the development and application of the International RAPID Quality Control System (Boller et al. 1981). Comparative quality profiles established on medfly strains showing different rearing histories, different environmentally induced impacts caused by manipulation, shipment in unsuitable containers over long distances, emergence of adults in transit etc., revealed that the intrinsic or genetically determined quality of a given fruit fly strain could only be assessed after rearing the test material for one generation on a standardized and relaxed rearing system. The system should also allow high production of offspring in field-collected wild fruit fly strains.

Of special importance in this context is the optimal design of all aspects of the rearing procedures that interfere with essential components of adult behavior, especially mating (leading to high egg fertility) and oviposition (permitting a high percentage of the females to participate in the reproduction process and to realize a high individual fecundity). For further details concerning the rationale of quality control in fruit flies reference should be made to Boller (1972) and Boller and Chambers (1977).

The relaxed rearing system described here is not a substitute for the mass-rearing procedures developed and utilized in large-scale SIT programs. It is rather a standardized rearing program for research purposes or - by expanding the number of rearing units - for the production of larger numbers of flies of defined specifications that might be required for special field experiments such as those conducted in 1978 in Guatemala and 1979 in Spain (Chambers *et al*. 1983).

The development of the larval diet for *R. cerasi* started in 1967 when for the first time cherry fruit fly pupae were produced on a modification of a solid diet developed by Maeda *et al*. (1953). Reports about research in other laboratories on liquid diets (Monro 1968, Mittler and Tsitsipis 1973)

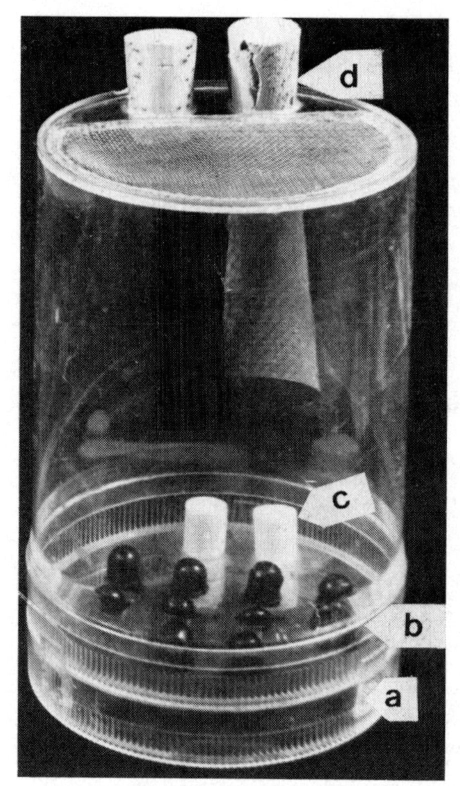

Fig. 1. Oviposition cage for R. cerasi with (a) water reservoir, (b) dome plate carrying 10-mm-diam ceresin oviposition domes, (c) cotton wicks as water supply, and (d) ventilation screen and servicing holes with food strips.

stimulated investigations on liquid larval diets in combination with suitable inert carriers such as textiles, foam rubber, styrofoam beads and finally (after many negative results) with hygrophilic cotton pads. The correct description of the Waedenswil diet would therefore be a 'liquid larval diet on a cotton carrier'. Its basic components have been published by Katsoyannos, Boller and Remund (1977).

2. FACILITIES AND EQUIPMENT REQUIRED

2.1 Facilities

Rearing of adults and pupae is carried out in controlled environment rooms maintained at 25 ± 1°C, 50-60% RH and an LD 18:6 photoperiod. The larval stages are reared at 25 ± 1°C, 80% RH and either in a dark room or at LD 18:6 photoperiod.

2.2 Equipment and materials required for insect handling

2.2.1 Oviposition

a. R. cerasi and D. oleae: Both species do not easily lay eggs into perforated spherical structures but require oviposition substrates that

Fig. 2. Standard RAPID Plexiglass® cage for *C. capitata* with water supply, dry food mix and perforated plastic oranges fitted over water containers serving as oviposition devices. Four food strips as additional food supply, resting, and mating sites.

can be penetrated by the ovipositor. The cylindrical oviposition cage (Fig. 1) consists of: (a) lid to serve as water reservoir; (b) lid to carry 10-20 ceresin wax domes of 10 mm diam, black, hollow, and 2 dental cotton wicks; (c) polystyrol container, 1-liter size, with hardware covered ventilation window; and (d) 2 corks holding the dry-food paper strips.

b. *C. capitata*: oviposition into perforated plastic tubes, funnels or spheres with high relative humidity inside the device. The oviposition devices shown in Fig. 2 consist of plastic oranges (or orange plastic balls), 7 cm diam, perforated with 1-mm holes, fitted over 5-cm-diam glass containers with tap water (providing the necessary humidity level inside the oviposition device).

2.2.2 Egg collection and transfer to diet
- Water squirt bottle
- Erlenmeyer flask (500 ml)
- Glass funnel
- Fine nylon gauze (15 x 15 cm)
- Fine, camel hair paint brushes (size 00)

2.2.3 Larval rearing and pupation containers
- Polystyrol petri dishes (13.5 cm diam, 2 cm high)
- Sterile cotton pads (10 x 10 cm, 0.5 cm thick)
- Plastic bags (24 x 36 cm)
- Masking tape
- Stackable plastic trays (40 x 60 x 6 cm, Fig. 3) with 8 lateral ventilation holes (1 cm diam) serving as floor or cover

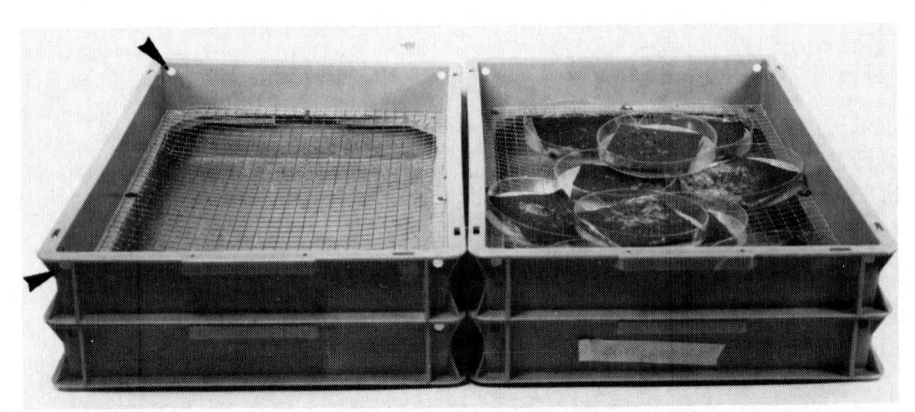

Fig. 3. Empty (left) and loaded pupation units (right). Arrows point to ventilation holes.

part, or with the bottom removed and replaced by a 2-cm-mesh wire screen as a holding unit for the larval diet containers
- Fine washed quartz sand

2.2.4 Diet preparation equipment
- Top loading balance (0.1-1000 g)
- 2-liter Waring® blender
- 3 glass beakers (500 ml)
- 2 glass beakers (100 ml)
- 1 Erlenmeyer flask (2000 ml)
- 1-liter measuring cylinder
- 50-ml measuring cylinder
- Medium size spatula or scoop
- 10-ml glass pipette
- 1 glass funnel
- 1 glass plate or petri dish (20 cm diam)
- Forceps

2.2.5 Specification and manufacturers of materials
- Soft ceresin wax: type MP 121 (penetration index DIN 51579 = 44-55 at 25°C), 25-kg blocks (E. Weisser & Co., Birsigstr. 50, CH-4000 Basel, Switzerland)
- Cotton pads: no. 7053, 0-B sponges, 2 ply (10 cm x 10 cm) (Kendall Company, Hospital Products, Boston, MA 02101, USA)
- Stackable plastic trays: RAKO 2 (Georg Utz AG, CH-5620 Bremgarten, Switzerland)
- Plastic oranges (your local store)

2.2.6 Ceresin dome preparation equipment and procedures

a. Equipment
- Hot plate and 1-liter steel pan
- Thermometer (0-150°C)
- 2 glass beakers (1 liter each), 1 with cold water and 1 with soap solution
- 1 dissecting needle
- 1 metal spoon
- Bunsen burner
- glass test tube (10 mm diam)
- 500 g ceresin, dyed by 1.5 g black candle-wax dye

b. Preparation according to Prokopy and Boller (1970) as follows:
- Heat ceresin wax up and maintain at 90°C for optimal dome characteristics
- Dip convex end of test tube first into soapy water and then for 1 sec or less about 1 cm deep into molten ceresin wax
- Immediately point the wax-covered end of the test tube upward to allow the small droplet of excess wax to run down the side of the tube before solidifying
- Dip test tube into cold water to harden wax; make a small hole into the apex with needle
- Strip wax dome from test tube and store in clean cold water until used (to avoid having several domes stick together)
- Slide ceresin dome into 10-mm hole of lid (domeplate b in Fig. 1) so that protruding dome forms perfect hemisphere
- Warm spoon on Bunsen burner and weld wax dome to carrier, whereby excess wax is simultaneously removed

2.2.7 Cages (adult emergence, holding, and oviposition cages)

Adult emergence, sexual maturation, and mating take place in the standardized plexiglas cages specified for the RAPID Quality Control System, (Boller *et al.* 1981). *C. capitata* (and other fruit fly species such as *Anastrepha* spp. and *D. dorsalis*, requiring perforated oviposition devices stay in this cage and are provided with suitable oviposition devices upon maturation. *R. cerasi* (and *D. oleae*) are transferred to their special oviposition cages described in section 2.2.1 (Fig. 1).

3. ARTIFICIAL DIET

3.1 Composition (1 liter of liquid larval diet)

Ingredients	Amount	% (Estimated)
Sugar (sucrose)	100 g	10.0
Brewer's yeast	100 g	10.0
Wheat germ	75 g	7.5
Citric acid	12 g	1.2
Propionic acid (conc.)	4 ml	0.4
Ascorbic acid	9 g	9.0
Water (tap)	700 ml	70.0

Acid content will adjust pH of diet at 3.9-4.0

3.2 Brand names and sources of ingredients

Sugar (local grocery store); brewer's yeast (item no. 103085 Torula yeast, ICN Life Sciences Group of Nutritional Biochemicals, 26201 Miles Road, Cleveland OH 44128, USA); wheat germ (ICN item no. 903288); citric acid, propionic acid, ascorbic acid (Fluka AG, CH-9470 Buchs, Switzerland)

3.3 Diet preparation procedures

3.3.1 Preparation of slurry
- Put 1/3 of water into blender and at high speed add sugar, citric acid, propionic acid
- Blend for 2 min and add another 1/3 of water
- Add the other ingredients, rinse the containers with the rest of the water and add it also while blender is still stirring at high speed
- Blend for about 10 min, check proper pH and adjust with citric acid if too high (pH 3.9-4.0)
- Pour liquid diet through funnel into Erlenmeyer flask, cover top with inverted youghourt (or plastic) beaker and store diet at least 2 days at 5°C or until required
- Prepared diet has a shelf life of approximately 6 weeks at 5°C

3.3.2 Preparation of larval diet pads
- Take liquid diet and shake it thoroughly in the flask
- Pour it about 1 cm deep into the large petri dish
- Remove 1 cotton pad (10 x 10 cm) with clean forceps and pull it through the liquid diet in such a way that only 1 side of the cotton pad is in

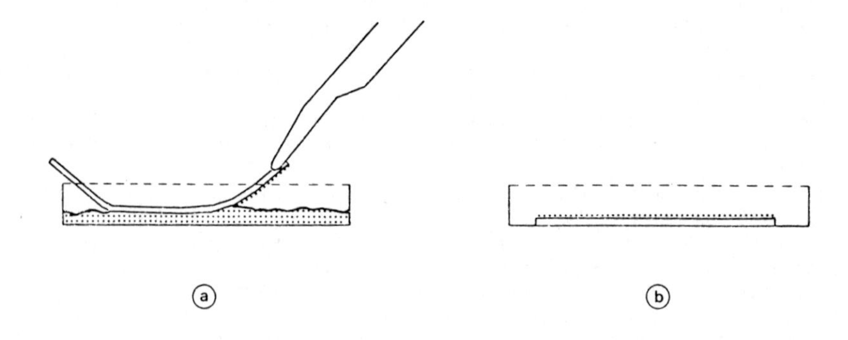

Fig. 4. (a) Cotton pads are pulled through the diet in such a way that only 1 side is in contact with the liquid, and (b) are placed wet side up into petri dishes.

contact with the liquid (Fig. 4). This is the most important step in the entire preparation. Complete soaking of the cotton carrier results in poor larval development probably due to excess water on the surface. A moisture gradient through the pad is an important prerequisite for optimal larval development
- Place impregnated cotton pad into polystyrol petri dish, close the dish
- Store the prepared dishes for 2 days at cool temperature before use

PRECAUTIONS
(i) All equipment must be clean, the cotton pads and petri dishes must be sterilized.
(ii) Propionic acid must be handled in a fume cupboard.

3.3.3 Preparation of adult diet

Adult diet consists of the standard mixture of sugar and yeast hydrolysate (4:1). It is fed in dry form (in small petri dishes) when flies are maintained in the larger Plexiglas cages. The food is administered as food strips in the small cylindrical oviposition cages that are suspended from the ceiling. The food strips are prepared as follows:
- Cut paper towels into pieces 3 cm wide
- Mix 4 parts of sugar, 1 part of yeast hydrolysate in a glass beaker and add 3 parts of water
- Stir the mixture for about 30 min
- Soak the paper strips in the slurry, remove excess food with a fork and hang the strips in a drying cabinet

4. INSECT HOLDING

4.1 Environmental conditions

4.1.1 Adults and pupae

Fruit fly adults are held at 25 $\pm$ 1°C or fluctuating temperatures (according to the rearing strategy), 50-60% RH (higher humidity liquifies the adult food) and an LD 18:6 photoperiod (an LD 14:10 photoperiod for *C. capitata* if activity patterns under lab and field conditions are to be compared). Illumination provided by fluorescent light tubes (Chromalight

FL 40, General Electric) producing a light intensity of 1500 lux at the level of the oviposition devices. Pupae are held under identical conditions to synchronize the diurnal rhythms to the photoperiod.
(Pupae of R. cerasi exhibit an obligatory pupal diapause that requires a chilling period of 3-5 months at 4°C before the post-diapause development can take place. For details see Vallo et al. 1976.

4.1.2 Larvae and eggs

The larvae are held in the diet containers with high RH at 25 ± 1°C unless special rearing conditions are desirable (e.g., influence of fluctuating or low temperatures on quality parameters). During pupation a high humidity of 80% is maintained.

Since eggs are transferred directly from the oviposition devices to the larval diet, they are exposed to the same conditions as the larvae.

5. REARING AND COLONY MAINTENANCE

The founder colony of R. cerasi was established from field-collected pupae and has been reared on artificial diets for more than 10 generations (1 generation per year). Several wild strains of C. capitata and 1 wild strain of D. oleae were reared continuously by this rearing procedure.

5.1 Egg collection (twice a week)

- Remove plastic oranges from oviposition cage and separate plastic part from water container
- Rinse eggs with water squirt bottle or with hand atomizer into glass beaker and let them sink to the bottom
- Note Ceratitis strain and collection date on a piece of masking tape and stick it to the beaker (this identification tag will move through the entire rearing sequence and end in the notebook or data sheet when the material has been used in experiments).
- Decant carefully top layer of water with dirt particles and empty egg shells and pour contents of beaker through a fine nylon gauze; rinse with plenty of water
- Briefly put gauze with eggs on a piece of foam rubber or towel to drain excess of water, no sterilization is necessary

Eggs from ceresin domes are collected likewise except that the entire oviposition cage is held over a larger funnel and the eggs rinsed from the lumen of the domes with a relatively sharp water jet (hand atomizer). Flies do not have to be removed from the cage during egg collection.

5.2 Diet innoculation with eggs and neonate larvae

- Condition diet by allowing to stand at rearing temperature overnight
- The number of eggs transferred to the diet by means of a fine-hair brush depends on the fruit fly species and the desired pupal size (decreasing with increasing larval density). The largest pupae are obtained by seeding the following number of eggs per neonate larvae per gram of liquid diet or per total of 12 g diet per dish, respectively:

  R. cerasi: 2-3 per g or 30-40 per dish

  C. capitata: 10-20 per g or 200 per dish

  Increased larval density of C. capitata will lead to rapid food depletion and smaller pupae. Even at excessive numbers (1000 eggs per dish), large numbers of extremely small pupae can be produced (1.5 mm diam).
- After egg transfer (distribute evenly over diet surface), close petri dish and put 10 dishes into plastic bag
- Leave bag open (to avoid humidity build up), put identification tag on the bag and store in larval rearing room
- Keep C. capitata in this condition for 5 days, R. cerasi for 10 days

5.3 Pupation

- Take petri dishes from the plastic bag and transfer the open dish to the screen of the pupation unit (Fig. 3). One unit consists of the upper holding tray, the lower pupation tray with little sand and either an empty tray as cover or the pupation tray of the next pupation unit. It

can receive 10 dishes of a given larval bag. The lower pupation tray receives the proper identification tag.
- Stack as many pupation units as possible on the same pile (to avoid contamination of the different strains reared simultaneously in the same pupation room, do not arrange 2 piles side by side.)
- Check holding and pupation trays every 2nd day for fungal contamination, humidity build up, and pupation
- Remove diet pads when larvae have left (*Ceratitis* will jump from the diet pads to the pupation tray, *Rhagoletis* will either crawl out of the diet dish or pupate under the cotton pad and have to be collected there). This exodus will last about 3 days in *Ceratitis* and 1 week in *Rhagoletis*.
- Collect pupae from the sand after 5 days. Transfer *Ceratitis* pupae to emergence cage (will emerge 9 days after pupation); transfer *R. cerasi* to 4°C.

PRECAUTIONS

(i) Exceeding optimal larval densities in *Ceratitis* will produce high amounts of respiratory humidity in the larval containers and will require adequate ventilation. This phenomenon will also occur when the pupation units are not adequately ventilated (provision of 4-8 lateral ventilation holes of 1 cm diam necessary; see arrows, Fig. 3).

(ii) Manipulation and strong shocks to pupae less than 4 days old will cause ruptures of the flight muscles ('droopy wing syndrome' as described by Little *et al.* 1981). Handle pupae very gently.

(iii) The larval diet is quite resistant against fungal and bacterial diseases if normal cleaning procedures are observed. However, periodical inspection of pupation units should help to detect early signs of moulds and to remove contaminated units at an early stage.

(iv) A small amount of fine and sterilized sand is preferred for 'naked' pupation (prolonged prepupation time due to increased larval movement) and to coarse pupation substrates (problem of separating pupae from substrate).

5.4 Adults

5.4.1 Emergence period (at 25°C)

C. capitata adults start to emerge 9 days after pupation and *R. cerasi* 17-18 days after transfer from optimal chilling treatment to laboratory conditions. In both species emergence takes place predominantly during the morning hours. Duration of emergence depends on the age variation of the pupae and is extended in field-collected material. In laboratory reared *C. capitata*, with little variation in age of pupae, emergence takes place within 2 days. *R. cerasi* collected from sweet cherries emerge within 1 week and peak between day 20 and 23 after incubation, wild pupae from honeysuckle (*Lonicera* spp.) will produce 1st flies as early as the cherry race but extend the emergence period over several weeks.

5.4.2 Fly density in cages

High mortality occurs among adult flies of *C. capitata* collected in the field through irritation and unnatural densities in laboratory cages (with severe consequences to inadvertent selection for laboratory ecotypes). Hence, densities in the first step of laboratory propagation should be as low as feasible. F_1 flies and laboratory adapted strains prepared for quality profiling should not exceed 50 individuals per 1-liter holding cage (20 cm^3 per fly). In *R. cerasi* optimal fly densities refer to the ratio female-oviposition domes, which should not exceed the ratio of 1:1 if the oviposition devices are changed only every 2 weeks.

5.4.3 Oviposition

- *C. capitata*: starts to lay eggs after 4 days (wild females later), discard after 4 weeks
- *R. cerasi*: starts oviposition after 4-5 days (peaks during 2nd and 3rd week), discard after 3 weeks. Change domeplates when total number of

eggs per dome exceeds 50. Females lay an average of 200 eggs under laboratory conditions.
- Oviposition occurs in both species in the afternoon

PRECAUTION

These fruit fly species produce oviposition deterring pheromones (ODP) that reduce the attractiveness of artificial oviposition devices and hence the egg production. Oviposition devices must be rinsed frequently with hot water (plastic devices) or cold water (ceresin domes). The oviposition punctures in ceresin domes can be closed by dipping the dome plate for 1 sec into water at 45°C.

5.4.4 Rearing schedules

a. *C. capitata*

Daily
- Check conditions in rearing and pupation room

Tuesday and Friday
- Collect eggs and replace oviposition devices
- Inoculate larval diets
- Collect pupae and prepare emergence cages
- Refill or replace food and water
- Discard old strains and clean cages

Once per week (or in longer intervals)
- Prepare diets

b. *R. cerasi*

Same schedule as for *C. capitata* with the following modifications
- Transfer 20 females + 5 males to fresh oviposition cages (twice a week when 5-7 days old)
- Change oviposition plates every 2nd week
- Incubate new batch of pupae every 3rd week (unless incubation at shorter intervals)

5.4.5 Insect quality

Quality control procedures for fruit flies and especially for *C. capitata* and *R. cerasi* have been developed and described in detail (Boller and Chambers 1977, Boller *et al.* 1981, Chambers *et al.* 1983). They include methods for monitoring the production and performance characteristics. Whereas large mass-rearing programs have quality control departments that measure quality traits on a daily, weekly, or monthly basis and conduct more elaborate and expensive field tests, quality control programs in research colonies have to be simpler in order to be applied regularly.

A combined production/performance monitoring is carried out on a monthly (or generation) basis in *C. capitata* which could incorporate the following tests:
- Female fecundity
- Pupal size (calibration machine for pupal diam)
- RAPID Ability Test (emergence rate, dead pupae, minimal flight performance)
- Longevity test
- Mating propensity

6. LIFE CYCLE DATA

Life cycle and survival data during development of various stages under optimum rearing conditions of 25°C are given.

6.1 Development data

Stage	Number of days R. cerasi	C. capitata
Eggs	4	2
Larvae	11-16	7-8
Pupae	17-26	9-10
Preoviposition period	4-6	4-6

6.2 Other data

Parameter	R. cerasi	C. capitata
Female fecundity	200	ca 500-1000
Egg fertility	80-90%	>90%
Survival on diet		
neonate larva/pupa	40-45%	>90%
Average female longevity	32 days	35 days
Mean pupal weight	4.0 mg	9 mg

7. REFERENCES

Boller, E., 1972. Behavioral aspects of mass-rearing of insects. Entomophaga 17: 9-25.

Boller, E. F. and Chambers, D. L. 1977/5. Quality Control - An Idea Book for Fruit Fly Workers. IOBC/WPRS Bull. 162 p.

Boller, E. F., Katsoyannos, B. I., Remund, U. and Chambers, D. L., 1981. Measuring, monitoring and improving the quality of mass-reared Mediterranean fruit flies, Ceratitis capitata. 1. The RAPID Quality Control System for early warning. Z. ang. Entomol. 92: 67-83.

Chambers, D. L., Calkins, C. O., Boller, E. F., Itô, Y. and Cunningham, R. T., 1983. Measuring, monitoring and improving the quality of mass-reared Mediterranean fruit flies, Ceratitis capitata. 2. Field tests for confirming and extending laboratory results. Z. ang. Entomol. 95: 285-303.

Katsoyannos, B. I., Boller, E. F. and Remund, U., 1977. Beitrag zur Entwicklung von Methoden zur Massenzucht der Kirschenfliege, Rhagoletis cerasi, auf künstlichen Substraten. Mitt. Schweiz. Entomol. Ges. 50: 25-33.

Little, H. F., Kobayashi, R. M., Ozaki, E. T. and Cunningham, R. T., 1981. Irreversible damage to flight muscles resulting from disturbance of pupae during rearing of the Mediterranean fruit fly, Ceratitis capitata. Ann. Entomol. Soc. Am. 74: 24-26.

Maeda, S., Hagen, K. S. and Finney, G. L., 1953. Artificial media and the control of microorganisms in the culture of tephritid larvae (Diptera: Tephritidae). Proc. Hawaiian Entomol. Soc. 15: 177-185.

Mittler, T. E. and Tsitsipis, J. A., 1973. Economical rearing of larvae of the olive fly, Dacus oleae, on a liquid diet offered on cotton towelling. Entomol. Exp. Appl. 16: 292-293.

Monro, J., 1968. Improvements in mass rearing the Mediterranean fruit fly, Ceratitis capitata Wied. Proc. IAEA/FAO Panel on 'Radiation, radioisotopes and rearing methods in the control of insect pests'. Tel-Aviv 1966. pp. 91-104.

Prokopy, R. J. and Boller, E. F., 1970. Artificial egging system for the European cherry fruit fly. J. Econ. Entomol. 63: 1413-1417.

Vallo, V., Remund, U. and Boller, E. F., 1976. Storage conditions of stockpiled diapausing pupae of Rhagoletis cerasi to obtain high emergence rates. Entomophaga 21: 251-256.

ACKNOWLEDGEMENT

I acknowledge with thanks the technical contribution of Dr. B. I. Katsoyannos who participated in the development of the liquid diet.

SIMULIIDS (MAINLY SIMULIUM DECORUM WALKER)

JOHN D. EDMAN and KENNETH R. SIMMONS

Department of Entomology, University of Massachusetts, Amherst, Massachusetts, USA 01003

THE INSECT

Scientific Name: *Simulium decorum* Walker
Common Name: black fly
Order: Diptera
Family: Simuliidae

1. INTRODUCTION

Black flies occur in both temperate and tropical regions that have fast-flowing, unpolluted streams. Dense blood-feeding populations cause severe annoyance and injury to man, domestic animals and wildlife. Simuliids also vector important nematode and hematozoan parasites of birds and mammals, including the filaria *Onchocerca volvulus*, which causes river blindness disease in 20-40 million people in Africa and parts of South and Central America.

Immature stages normally are lotic dwellers. There are both univoltine (with obligate egg diapause) and multivoltine species. Adults emerge from submerged pupal cases within an air bubble between mid-spring and summer in temperate regions but may occur year-around in the tropics. They mate soon after emergence in aerial swarms or on substrates. Host-seeking then follows in anautogenous species; eggs mature 3-5 days after blood-feeding. Autogenous species emerge with developing eggs that mature in 2-5 days without a blood-meal. Autogenous species may seek blood after ovipositing. Oviposition often occurs around sunset. Eggs are either dropped singly during flight and sink to the stream bottom or are deposited in masses on objects just beneath the water. Embryonation and hatch require 3-5 days in multivoltine species but several months in diapausing, univoltine species.

Larval instars (6-7) attach to substrates in stream current and feed mainly by filtering seston with their head fans. Early instars utilize small particles (< 50μm); i.e. later instars feed on particles up to about 150um in diameter. Larval development requires as little as 8-10 days in tropical species but up to 3-4 months (avg. about 25 days) in temperate ones. Diapause eggs laid during spring may hatch in autumn; larvae then develop throughout the winter. Others hatch between mid-winter and early spring. Pupae occur on larval substrates but on less turbulent surfaces. Duration of the pupal stage is 2-10 days but may be prolonged when temperatures are unseasonably cold.

Adult black flies are diurnally active; peak activity varies with environmental conditions but often occurs in mid-morning and late afternoon. Only females take blood but both sexes feed on plant sugars.

Rearing and colonization of black flies have been difficult (Colbo & Thompson 1978, Mokry et al. 1981, Muirhead-Thompson 1966). Recent development of efficient larval rearing procedures has contributed greatly to the successful colonization of several species. With larger numbers of healthy and synchronous lab-reared flies, it is possible to obtain a few individuals that will mate, blood feed, and oviposit under favorable laboratory conditions. Simulium decorum has been in colony in our laboratory (Simmons & Edman, 1981) for 27 generations. This species has also been maintained at Cornell (Brenner & Cupp, 1980) and Memorial University in Newfoundland (Colbo, pers. commun.). One sibling of the African Simulium damnosum complex has been maintained for 5 generations in our laboratory (Simmons & Edman, 1982) and in Ghana (Raybould & Boakye, in press). Simulium vittatum and Simulium pictipes recently have been colonized at Cornell (Cupp, pers. commun.). Boophthora erythrocephalum was colonized for a few generations in Germany during the early 1970's (Grunewald 1973). Recently, this species and Simulium lineatum were colonized in England (Ham & Bianco, pers. commun.). Cnephia dacotensis (Fredeen 1959a) and the parthenogenetic form of Stegopterna mutata (Mokry 1978) also were technically colonized but the obligate egg diapause in these species prevents continuous rearing.

Colonies are best initiated from material already in colony at another laboratory. Special equipment, technical expertise and considerable time are required to maintain a continuous colony. Biological characteristics that suit S. decorum to colonization are: 1) widely distributed throughout N. America, 2) easy to locate on dam structures (man-or beaver-made) at pond outlets, 3) multivoltine, non-diapausing, 4) eggs layed in masses on substrate, 5) eggs can be stored for extended period, 6) adults mate on substrate, 7) autogenous for first egg batch, anautogenous for subsequent batches, and 8) larvae are tolerant to a wide range of physical and hydrochemical conditions.

2. FACILITIES AND EQUIPMENT NEEDED

2.1 Facilities

Larvae and pupae are reared in an artificial stream system housed in a separate temperature controlled room (wet lab) with a cement floor and drain. Each artificial stream requires about 4 sq m of floor space. Electrical outlets are the outdoor type and are placed 1.5 m above the floor. Photoperiod is 15:8 LD with a 0.5 h crepuscular period between the day and night light changes. Lights are cool white fluorescent bulbs. Intensity is 1800 foot candles (FC). The crepuscular light is a 40 W incandescent lamp with an intensity of 10 FC. The adult oviposition chamber is kept in the wet lab.

Adults are held in cages in a separate room with the same photoperiod as the larvae. RH is maintained at 90+ % and temperature is 22-24°C.

2.2. Equipment and materials required for insect rearing and handling

2.2.1 Adult mating (two methods are used)

Adults are induced to mate under crowded conditions in mouth aspirators and in holding cages altered to phototactically attract flies to an aggregation site

a. Mouth aspirator is made of 1 x 40 cm Plexiglas® tubing with a 20° angle bend 10 cm from the tip, fine mesh wire screen for a stop and 1 x 50 cm latex tubing for a mouth piece

b. Adult holding cages are wood frame with bottom, back and one side of white formica, top of Plexiglas, front of muslin surgical stocking, and other side of nylon screen. To form a mating aggregation site the top of the cage is covered with aluminum foil. A 2 cm hole is cut in the foil near the side of the cage with the muslin stocking, which is pushed slightly into the front of the cage to form a ramp where the flies aggregate in the light from the hole in the foil.

2.2.2 Adult oviposition

The oviposition chamber is made of 6.4 mm Plexiglas. Dimensions are 60 cm long x 45 cm wide x 50 cm high, and it is divided into 3 compartments. The middle compartment is 15 cm wide and has a removable divider. Current is produced against the divider wall in the middle compartment with a Plexiglas bubbler (3 mm diam Plexiglas tubing with 0.5 mm holes every 1 cm) attached to an air pump. The entire chamber is covered with brown paper. Sunset conditions are created with a 40 watt incandescent lamp (with reflector) directed at the water line.

2.2.3 Egg collection and storage

- No. 5xxx grade corks
- Binocular microscope with fiber optic light
- No. 5 series forceps bent slightly inward on the point of one of the tynes
- 5.5 cm petri dishes lined with polyester fiber
- Insulated portable ice chest

2.2.4 Larval rearing system (Figure 1)

The rearing system consists of a reservoir tank (outer tank) inside of which is a second tank (inner tank) that holds the 16 troughs and gravel filter. The tank is filled with ca 325 liters of water; temperature is regulated with an Aqua-Chiller (Jewel Industries, Inc., Chicago, IL 60639).

a. Outer tank dimensions

Angled Plexiglas ribs (6.4 mm) are attached along the inner midlines of the bottom and each sidewall for structural support and to form a frame to hold the removable inner tank in position

b. Inner tank dimensions

The inner tank is made of 6.4 mm Plexiglas. It is 66 x 66 cm wide at the top and narrows to a 16.5 cm square opening at the bottom, to which is attached a 14 cm deep Plexiglas junction box. The junction box has 2 Plexiglas tubes (1.2 cm diam) which are connected by polyethylene tubing leading through one side of the outer tank to the exterior mounted pumps. Total height of the inner tank is 57 cm. Walls of the inner tank slope inward 64° to maximize the water volumne in the outer reservoir tank. The 4 troughs on each inside wall of the inner tank slope 5° in alternating directions. Trough lengths from top to bottom are 56, 45, 40 and 35 cm and their width is 5 cm. Gates opening to each of the 4 top troughs are 2.7 cm high, 5 cm wide, and are located on the wall opposite the troughs so that water flows straight into each trough. Water is recirculated with 1 pump (Model AC-3C-MD, March Mfg., Inc. Glenview, IL, 60025). Flow rate is regulated by a brass faucet valve attached in-line to each pump. A second pump can be added (see Fig. 1) if rearing a species with a faster current requirement is desired.

2.2.5 Adult collection

- Emergence cage (90 cm x 90 cm x 15 cm high) with nylon screen top which fits on the rim of the outer tank. Four holes, stoppered with corks are cut in the nylon to insert aspirators to remove emerged adults.
- Aspirator (see 2.2.1.a)

<u>2.3 Diet preparation equipment</u>

<u>2.3.1.</u> Larval diet equipment

- Large mortar and pestle
- 30.3 cm diam 150 micrometer mesh brass sieve
- 5 cm^2 38 micrometer mesh stainless steel screen
- Waring blender (20,500 rpm)
- 2 liter flask

2.3.2 Adult diet equipment

2.3.2.1. Sugar diet

- 30 ml serum vials
- Cotton wicks

2.3.2.2. Blood meals

- 28 g (1 oz) clear plastic cups (Premium Plastics, Inc., Chicago, IL) fitted with 1.2 mm mesh cloth (absorbant) netting over the open end. Hold netting in place with elastic bands. Drill 1 cm hole in bottom of cup and stopper with a cork

- 4 cm glass jacketed artificial membrane feeding apparatus
- Water bath
- Parafilm (for membrane)

3.0 DIET

3.1 Larval diet

- Tetra® - tropical conditioning food vegetable diet for fish (Tetra Werke, West Germany - available in most pet stores)

3.1.1 Larval diet preparation

- Grind Tetra to coarse powder with mortar and pestle
- Pass Tetra through brass sieve by grinding it over sieve with the stainless steel screen

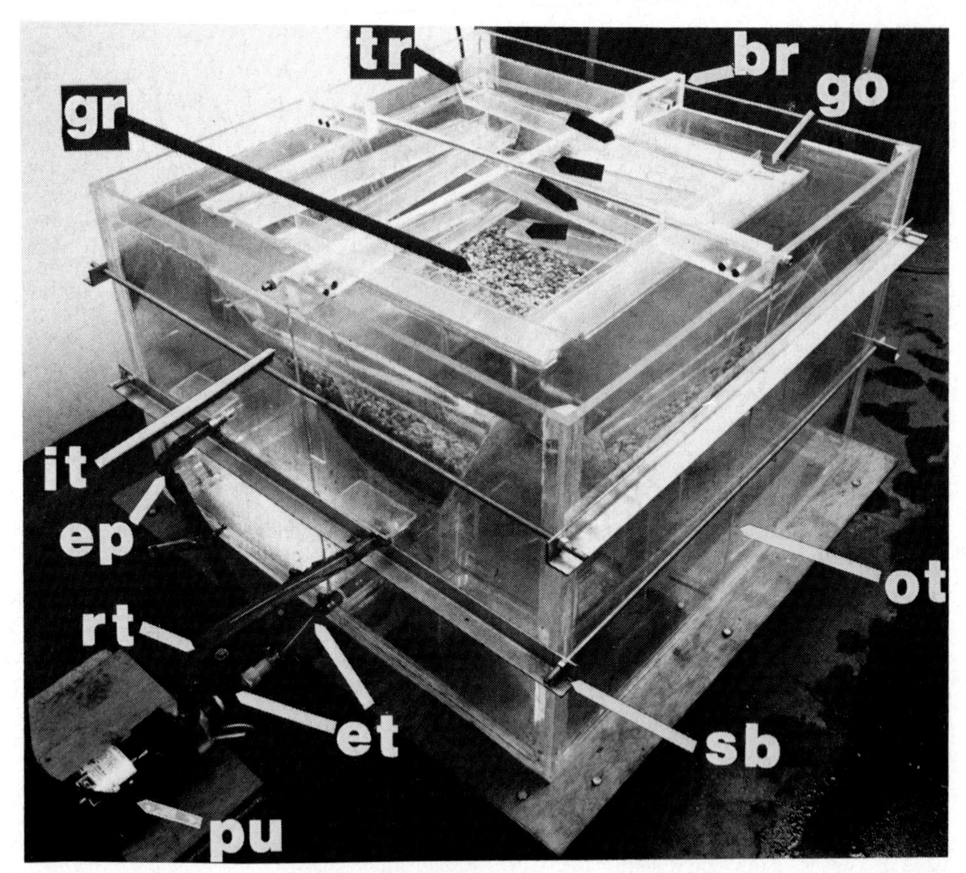

Fig. 1. Chamber for rearing S. decorum: IT, inner tank; GR, gravel filter bed in inner tank; BR, brace which holds inner tank in position; GO, gate opening to 1 of the 4 top troughs; OT, outer tank; SB, steel support brace; PU, pump; ET, exit tube and flow control valve; RT, return tube; EP, extra port for second pump; TR, trough - arrows indicate direction of flow down the 4 troughs. Newly emerged adults are retained in a screen cage which fits over the top of the outer tank. It is not illustrated. The Aqua Chiller is on the side of the tank which is not visible in this photograph.

- Mix pre-weighed ground Tetra with 150 ml of distilled water and blend at 20,500 rpm for 2-3 minutes
- Dilute to 2 liters with distilled water and let settle 2-3 h in refrigerator

3.2 Adult diet

3.2.1 Sugar diet
- 10% sucrose solution in distilled water
- Put solution in vials with cotton wick

3.2.2 Blood (2 methods are used)

a. Natural hosts
- Place a maximum of 25 females in plastic feeding cup and secure on hairless region of host (human bicep, shaved belly of restrained dog, horse or cattle)

b. Artificial feeding
- Use citrated (sodium citrate) or heparinized human (outdated blood available from hospital blood banks) or bovine blood (available at abattoirs)
- Place parafilm membrane on feeder
- Add blood to inner jacket of feeder
- Circulate 37°C water through the outer jacket to warm blood to 37°C
- Attach flies in feeding cup to the feeder

4. HOLDING SIMULIUM DECORUM AT LOWER TEMPERATURES

4.1 Eggs
- Store up to 60 days packed on wet polyester fiber in covered 14 mm petri dishes packed in crushed ice in ice chest (place chest in refrigerator to slow melting of ice)
- After placing egg mass on wet fiber prod with forceps to separate eggs

4.2 Larvae and pupae
- Place on wet filter paper in petri dish and store as above; keep well dampened
- Will survive 5-7 days if kept moist

4.3 Adults
- Place in unwaxed 1/2 (.275 liter) or 1 pint (.55 liter) ice cream containers and store as above or in refrigerator
- Place moist paper towel in container; keep moist
- Will survive up to 14 days

5.0 REARING AND COLONY MAINTENANCE

5.1 Egg collection and hatching

5.1.1 To initiate a colony from nature
- In eastern North America, egg masses of *S. decorum* can be found at outfalls to small man- or beaver-made dams from June through February
- The best way to assure the species identity of eggs is to observe females ovipositing at dusk; *S. decorum* females can readily be separated from other species with the naked eye due to their large size, silvery thorax and yellowish abdominal sternites
- Scrape eggs from natural substrates with a sharp forceps and place on moistened polyester filter fiber; return to the lab
- Place fiber with eggs 1) in the head current of the artificial stream rearing system or 2) in petri dishes on wet ice until needed
- Stream systems should be operated for at least 1 day prior to the introduction of eggs for hatching
- *S. decorum* pupae also can be readily collected in the field by removing them with forceps from their substrates, allowed to emerge in the lab, mated and held to produce eggs for initiating a colony

5.1.2 For maintenance of existing colonies

- Remove filter fiber containing excess numbers of eggs from cold storage and place in rearing system to hatch

5.2 Egg sterilization

Eggs from an existing colony do not need to be sterilized. When field eggs are mixed with colony eggs they should be sterilized by 5-10 min submersion in 5% sodium hydroxide followed by a thorough rinse in fresh water (Fredeen 1959b). Place eggs in a wire basket to fascilitate sterilization and washing.

5.3 Larval/pupal rearing

- Set water temperature at 18°C
- Institute daily feeding schedule as soon as 1st instars hatch from the eggs
- Add 1/2 the normal daily food ration during the first 6-8 days or until larvae are 3rd instars; provide full daily ration (ca 1 mg dry Tetra/larva) for the remainder of larval growth period
- Add prepared Tetra directly to the rearing water
- Divide daily food ration into 2 or more separate feedings (the more the better)
- When larvae reach the 3rd instar, thin the population to the desired density by scraping off the excess individuals
- 1000-2000 larvae per generation is ample for colony maintenance; larger numbers (5000-10,000) are desireable when first establishing a colony from wild material
- Terminate daily feeding when the majority of larvae have pupated.

5.4 Adults

- 2 days after the first pupae appear place the emergence cage on the rim of the outer tank
- Remove emerged adults in groups of ca 30 (15♂ & 15♀) from the rearing unit in late afternoon each day with a mouth aspirator (see 2.2.1a); stopper the aspirator tube and allow flies to remain crowded in the tube for about 1 h to promote mating; then transfer to a standard holding cage containing 10% sugar
- Keep each daily cohort in a separate, labeled holding cage
- During the first 2 days post-emergence, cover the Plexiglas top of the holding cages altered (see 2.2.1. b) to promote aggregation and mating.

5.5 Oviposition

- Transfer 5 day old females directly onto the corks in the oviposition cage after room lights go off; retain over night under sustained twilight conditions for egg laying
- The following day, remove eggs from the cork substrate and transfer to polyester filter fiber for storage in petri dishes on crushed ice; break egg masses up to separate the individual eggs.

5.6 Sex determination

- Male black flies are smaller, have thinner elongate abdomens, and have larger eyes (holoptic) than females (dioptic).

5.7 Rearing schedule

- Rear one complete generation (1000-2000 flies) every other month for routine colony maintenance
- Store eggs during interum
- Rear experimental flies during off months or in separate rearing systems
- An 8-10 day period of continuous daily attention is required during each generation when adult flies are emerging, mating and ovipositing
- Daily feeding of larvae requires only a few minutes and can be fully automated to further save time and enhance efficiency.

5.8 Insect quality

Physically small flies resulting from stressed rearing conditions do not mate as well or survive as long as larger flies. Adequate nutrition, water quality and flow rate should be maintained to minimize stress.

Wild females can be periodically added to the colony to maintain genetic diversity. These should be brought in from the field as pupae, allowed to emerge and then held with colony males for mating. Combining wild eggs with colony eggs is not recommended due to the possible introduction of disease organisms.

5.9 Special problems and considerations

a. Disease
Disease problems often occur in larvae, pupae and adults from eggs brought directly in from the field since wild black flies commonly harbor a variety of pathogens. This mainly presents problems, though not insurmountable ones, during initial colonization of field material, and can be minimized by collecting pupae instead of eggs

b. Sanitation
To further minimize problems with disease and contamination, the rearing system should be drained, cleaned, sterilized with 2% chlorox, (ca 8 liters) and thoroughly rinsed between each generation

c. Egg storage
Storage of eggs beyond 60 days is not recommended unless a large excess of eggs is available. Storage for up to 30 days is recommended.

6. STANDARD LIFE CYCLE UNDER LABORATORY CONDITIONS

6.1 Duration of S. decorum life stages in the laboratory

Stage		Days to-Completion min	mean	max
Egg		3	4	6
Larvae/pupae				
	♂♂	14	20	32
	♀♀	14	23	32
Age to mating		<1	<1	5
Age to oviposition		4	5	6
Total	♂♂	18	25	43
	♀♀	21	32	44

6.2 Survivorship of colonized S. decorum

(a) Eggs: minimum of 90% if stored less than 1 month.
(b) Larvae/pupae: estimated at 80-90%.
(c) Adults: ♂ 1 day = 92%, 4 days = 83%
♀ 1 day = 98%, 5 days = 86%
maximum = 20 days

6.3 Fecundity of colonized S. decorum

	min	mean	max
♀♀ (no. mature oocytes)	397	562	987
♂♂ (no. sperm/spermatophore)	1621	4224	6892

7. PROCEDURES FOR OBTAINING INSECTS

It is recommended that those investigators wishing to establish a colony of black flies contact one of the laboratories with recent experience in maintaining a colony in order to explore the current availability of material and to arrange for special shipping. Unembryonated eggs can be shipped on damp substrates in sealed containers provided they are kept cold (cold packs work well) and transition does not exceed 6-8 days. Hatch eggs immediately upon receipt.

8. OTHER SPECIES WHICH CAN BE REARED WITH THE METHODS DESCRIBED
   - *Simulium damnosum* complex "Beffa" form

9. REFERENCES

Brenner, R.J. and Cupp, E.W., 1980. Rearing black flies (Diptera: Simuliidae) in a closed system of water circulation. Tropenmed. Parasit. 31:247-258.

Colbo, M.H., and Thompson, B.H., 1978. An efficient technique for laboratory rearing of *Simulium verecundum* S. & J. (Diptera: Simuliidae). Can. J. Zool. 56:507-510.

Fredeen, F.J.H., 1959a. Rearing black flies in the laboratory (Diptera: Simuliidae). Can. Entomol. 91:73-83.

Fredeen, F.J.H., 1959b. Collection, extraction, sterilization and low-temperature storage of black-fly eggs (Diptera:Simuliidae). Can. Entomol. 91:450-453.

Grunewald, J., 1973., Die hydrochemischen Lebensbedingungen der praimaginalen Stadien von *Boophthora erythrocephala* DeGeer (Diptera, Simuliidae). 2. Die Entwicklung einer Zucht unter expeimentellen Bedingungen. Z. Tropenmed. Parasit. 24:232-249.

Mokry, J.E., 1978. Progress towards the colonization of *Cnephia mutata* (Diptera: Simuliidae). Bull. W.H.O. 56:455-456.

Mokry, J.E., Colbo, M.H. and Thompson, B.H., 1981. Laboratory Colonization of Blackflies, p. 299-307. *In* Blackflies. The Future of Biological Methods in Integrated Control. M. Laird (ed.). Academic Press.

Muirhead-Thompson, R.C., 1966. Blackflies, p. 127-144. *In* Insect Colonization and Mass Production. C.N. Smith. (ed.). Academic Press.

Raybould, J.N., 1981. Present Progress Towards the Laboratory Colonization of Members of the *Simulium damnosum* Complex, p. 307-318. *In* Blackflies. The Future for Biological Methods in Integrated Control. M. Laird (ed.). Academic Press.

Simmons, K.R. and Edman, J.D., 1981. Sustained colonization of the black fly *Simulium decorum* Walker (Diptera:Simuliidae). Can. J. Zool. 59:1-7.

Simmons, K.R. and Edman, J.D., 1982. Laboratory colonization of the human ochocerciasis vector *Simulium damnosum* complex (Diptera:Simuliidae), using an enclosed, gravity-trough rearing system. J. Med. Entomol. 19:117-126.

STOMOXYS CALCITRANS

SIDNEY E. KUNZ and CHARLES D. SCHMIDT

U. S. Livestock Insects Laboratory, Agricultural Research Service, U. S. Department of Agriculture, P. O. Box 232, Kerrville, Texas 78028 U.S.A.

THE INSECT

Scientific Name:	*Stomoxys calcitrans* (Linnaeus)
Common Name:	Stable fly
Order:	Diptera
Family:	Muscidae

1. INTRODUCTION

Stable flies are cosmopolitan in distribution. In the United States, they are of primary importance around confined animal operations and, also, as human pests, especially on the beaches of the southeastern United States. Colonization attempts are recorded as early as the mid-1920's, but most developments have been made since the 1950's. The colonization of stable flies became necessary as investigators began to search for ways to control this pest with chemicals and also for biological investigations.

The life cycle of the stable fly consists of eggs, 3 larval instars, pupae, and adults. The most efficient rearing temperature for the immature form is at 29°C when the average duration of egg hatch is 30 h; 1st instar, 31 h; 2nd instar 40 h; 3rd instar, 118 h; and pupation, 143 h. Time from eclosion to oviposition is 6 days. Stable flies overwinter as 3rd-instar larvae with pupation and adult activity initiated with the onset of moderating temperatures.

Colonies can be established from field collections, but the time required for laboratory adaptation can be lengthy. Newly colonized flies are reluctant to feed or mate; it may take 5-6 generations before significant production can be achieved. Unless field-collected flies are required for a specific purpose, material from an established source greatly facilitates establishment, and the time saved before tests are begun is considerable. The most certain method to establish a laboratory colony is to obtain pupae from an existing colony. Stable flies are currently colonized at several locations including the U. S. Livestock Insects Laboratory, Kerrville, Texas, and the Veterinary Toxicology-Entomology Research Laboratory, College Station, Texas, and the Insects Affecting Man and Animals Laboratory, Gainesville, Florida.

2. REARING

2.1 Facilities, Equipment, and Procedures

2.1.1 Facilities should include air-conditioned, fly-secure rooms adequate in size to meet the demands of the colony. Adult-fly rearing rooms should be maintained at ca 26-28°C and 50-60% RH with a LD 8:16 regimen.

2.1.2 Adult cages

- Fasten 16-mesh (1.2 mm) screenwire onto a sheet metal frame with a solid back and bottom.
- Secure a stockinette sleeve to the front to provide entry. A cage, 30 x 30 x 40 cm, will accommodate ca 3000 flies.
- Place paper in the bottom of the cage to facilitate cleanup.

- Plastic or paper containers (4-liter) may be modified for use, but, if so, replace the ends and a majority of the sides with screenwire to provide adequate air flow throughout the cage. A modified rigid plastic container could provide an adequate frame for a cage.
- Place the pupae, contained in paper bags, into the cage. Once this is done, the cage is not opened again until postproduction cleanup, except to withdraw flies as test organisms.

2.1.3 Egg collection

- Place water-soaked (not dripping) pads wrapped or covered with black cloth on top of the cages for oviposition before feeding to eliminate the need to insert oviposition devices into the cages.
- Remove the egging pads after 1-2 h, and wash the eggs with water into a sieve and onto a wet cloth.
- Count the eggs that settle to the bottom of a water-filled container by measuring with a graduated centrifuge tube; 1 ml is ca 11,000 eggs. New eggs, (< 1 h old) can be held in water for less than 1 h without reduction of egg hatch.

2.1.4 Pupal collection

When larval development is complete, the older 3rd-instar larvae migrate to the top of the media and begin to cluster and pupate in localized areas.

- To save time in removal and cleanup, collect only the clustered pupae if 80-90% pupal recovery is adequate. Place the medium in a container of water, 4-5X the volume of media if all the pupae need to be harvested; the medium sinks to the bottom, and pupae float to the top for easy collection.
- Lightly wet the surface of the medium 12-16 h before washing to facilitate cleanup since smaller quantities of the debris will float to the surface with the pupae.
- Separate the pupae from the media with agitation devices, such as an agitating-type washing machine or containers with air bubbled from the base. Stop agitation after 2-4 minutes, allow debris to sink to the bottom; collect or float the pupae off into a screen-catch basin.
- Separate large amounts of media with a series of containers or small barrels; connect containers in a series, with each container at a lower level than the previous one. As water rises in the first container, pupae float out into the next. When this procedure continues for 2 or 3 levels, the transfer of less and less debris from one to the other successively cleans the pupae.
- Dry the pupae until they do not adhere to each other after collection.
- Estimate numbers of pupae by weighing 100 pupae and recording their weight or place 100 pupae into a graduated volumetric cylinder and record the volume; the pupae can then be weighed or measured by volume to determine the number of pupae in a given batch. These values are useful for determining production, for subdividing into smaller lots for testing, or for placement into colony cages.

PRECAUTIONS

(i) Avoid desiccation of pupae by maintaining at least 85% RH.

(ii) Routine cleaning of rearing pans and related equipment with sodium hypochlorite solutions will prevent Pseudomonas infestations in larval medium which can reduce larval and pupal product ion.

(iii) If predacious mites, Macrocheles muscaedomesticae(Scopoli) (Machrochelidae) and Tyrophagus spp. (Acaridae) become problems, the addition of a miticide, Tedion® (Williams et al. 1981), is used. Also, be sure to observe proper washing and/or steam cleaning procedures of all trays, cages, and other rearing equipment.

3. DIET

3.1 Larvae

- The larval medium consists of a mixture of CSMA (Chemical Specialties Manufacturers Association) and bagasse:
  - 11.35 kg of CSMA
  - 4.1 kg of coarse-ground sugarcane bagasse

  (CSMA fly medium contains alfalfa meal, brewers' dried grain, and wheat bran, Ralston Purina Co., St. Louis, MO)
- Mix 3500 ml (700 g) of this dry medium with 2000 ml water (this consistency allows a few drops of water to be squeezed from a handful) to fill a 30 x 30 cm container to a depth of 6 cm.
- Grind bagasse through a 13-mm screen and add to the medium for necessary bulk. Optional bulk materials are sawdust, wood shavings, vermiculite, and coarse-ground grass hay.
- Optional rearing media used successfully are:
  - 2000 g bagasse (coarse-ground Bermuda grass hay will substitute)
  - 200 g whole wheat flour
  - 100 g fish meal
  - 2000 ml water

  or
  - 1 part wheatbran
  - 1 part sawdust
  - 2 parts water

  or
  - 1 part (by volume) pelletized sugar cane bagasse
  - 3 parts wheatbran
  - 6 parts water

PRECAUTIONS

(i) Resins from certain woods cause mortality; therefore, the quality and type of wood shavings and sawdust to mix with media are problems.

(ii) Vermiculite and similar materials float and make it difficult to separate pupae by flotation.

3.1.1 Larval-medium pans

- Place medium in plastic or enameled pan; cover with a cloth held tightly in place with a rubber band or similar device to prevent other insects from infesting the medium.
- Measure eggs volumetrically and seed into larval media at the rate of 0.4 ml/2700 g media. A moist paper towel spread on top of the egg mass or eggs lightly mixed into the top layer of the medium will prevent desiccation.
- CSMA-bagasse medium undergoes considerable fermentation with the addition of water, and temperatures up to 50°C are recorded in the center of the media pans.
- When specifically aged material is needed (e.g., larvicide testing against 2nd-instar larvae), larval medium should be preconditioned 4-5 days to achieve the appropriate temperature for determination of development rates (Kunz et al. 1977).

PRECAUTIONS

(i) Do not mix eggs too deeply into the medium, this temperature will be fatal or interfere with development calculations.

(ii) Watch for mold which will oftentimes form in newly mixed media within 48-72 h; breakup of media by stirring will usually eliminate this problem. This may also be an indication of inadequate moisture.

(iii) Be sure to protect medium from other insects; improper security can result in mixed colonies. House fly females will readily oviposit on freshly prepared stable fly medium, and the production of stable flies can be greatly reduced.

3.2 Adults

- Bovine blood
- Sodium citrate
- Feed adult flies once daily, usually in the morning, on citrated bovine blood collected from local abattoirs.
- Place cotton pad or sanitary napkins containing ca 100 ml of blood on top of the cages. Remove feeding pads after 2-4 h to discourage oviposition, especially on cages used for egg collections.
- Stir sodium citrate (5 g dissolved in 5 ml water) into 1 liter of blood immediately following collection from the animal. Store blood in a refrigerator at 1-4°C. Replenish the blood supply weekly; up to 2-weeks use from a single collection often times is satisfactory. It is not necessary to add bacteriostats for stable fly feeding.

PRECAUTIONS

The collection of blood contaminated with stomach contents from the donor animal must be avoided because putrefaction of the blood will result.

REFERENCES

*Bailey, D. L., Whitfield, T. L. and LaBrecque, G. C. 1975. Laboratory biology and techniques for mass producing the stable fly, Stomoxys calcitrans (L.) (Diptera: Muscidae). J. Med. Entomol. 12(2): 189-193.
*Campau, E. J., Baker, G. J. and Morrison, F. D. 1953. Rearing the stable fly for laboratory tests. J. Econ. Entomol. 46: 524.
*Doty, A. E. 1937. Convenient method of rearing stable flies. Ibid. 30: 367-369.
*Glaser, R. W. 1924. Rearing flies for experimental purposes with biological notes. Ibid. 17: 486-496.
*Kunz, S. E., Berry, I. L. and Foerster, K. W. 1977. The development of the immature forms of Stomoxys calcitrans. Ann. Ent. Soc. Am. 70(2): 169-172.
*McGregor, W. S. and Dreiss, J. M. 1955. Rearing stable flies in the laboratory. J. Econ. Entomol. 48: 327-328.
Williams, D. F., Patterson, R. S., LaBrecque, G. C. and Weidhaas, D. E. 1981. Control of the stable fly, Stomoxys calcitrans (Diptera: Muscidae), on St. Croix, U. S. Virgin Islands, using Integrated Pest Management measures. II. Mass rearing and sterilization. J. Med. Entomol. 18(3): 197-202.

FOOTNOTES

*Not cited in chapter but recommended for reading

TOXORHYNCHITES AMBOINENSIS

LEONARD E. MUNSTERMANN and LORRAINE B. LEISER

Vector Biology Laboratory, Department of Biology
University of Notre Dame, Notre Dame IN 46556, U.S.A.

THE INSECT

Scientific Name: *Toxorhynchites amboinensis* (Doleschall)
Common Name: None
Order: Diptera
Family: Culicidae

1. INTRODUCTION

Toxorhynchites amboinensis is representative of a mosquito genus whose 69 member species are found chiefly in tropical and subtropical regions throughout the world. The larvae of the genus are nearly always associated with phytotelmata such as leaf axils, cut or bored bamboo, seed husks, flower bracts or tree holes; however, effluvia of human civilization in the form of junkyards and discarded tires provide alternate habitats (Steffan and Evenhuis, 1981). *Tx. amboinensis* is a tropical species, native to Indonesia, Thailand and The Philippines.

Recently *Toxorhynchites* has come under intensive scrutiny for several reasons. First, the larvae are predators on other aquatic container inhabitants, especially other mosquito larvae. Since these habitats are particularly refractory to normal mosquito control measures such as insecticides or "source reduction" (swamp drainage, and so forth), *Toxorhynchites* species have been field-tested as biological control agents (Focks et al., 1980; Gerberg and Visser, 1978). Secondly, not only are these mosquitoes large (4 to 6 times the size of *Aedes aegypti*) and quite beautiful (white scales contrasting with irridescent blues and greens), but also the females are not bloodfeeders and are completely autogenous. Thirdly, its role as a living Petri dish for growth of several types of deadly arboviruses has recently become widely exploited (Rosen, 1981). Because of its size and non-biting habit, an infected *Toxorhynchites* can produce much larger quantities of virus with much less human risk compared with a normal mosquito host.

Because of increased interest in the genus *Toxorhynchites*, several free-mating colonies have been established. Among these are *Tx. amboinensis* (Steffan et al.1980b), *Tx. brevipalpis* (Trpis and Gerberg, 1973), *Tx. rutilus* (Focks and Hall, 1977), *Tx. splendens* (Furumizo et al., 1977) and *Tx. theobaldi*. Further information on these and other *Toxorhynchites* species can be found in an annotated bibliography of the genus by Steffan et al. (1980a) and Manning et al. (1983).

Proper diet is the most important factor in successful rearing of *Toxorhynchites* larvae. Although several non-living diets have been tried (Focks et al., 1978; Trpis, 1979), these slow development. Living prey in the form of mosquito larvae such as *Ae. aegypti*, *Ae. albopictus* or other species, promote the most rapid growth and successful emergence as adults. Methods of rearing and diet described here for *Tx. amboinensis* can be successfully employed for each of the 5 species listed in Section 8.

2. FACILITIES AND EQUIPMENT REQUIRED

2.1 Facilities

The entire life cycle is carried out in controlled environment rooms maintained at 27 $\pm 1^{\circ}$ C, 80% RH, and photoperiod at 16:8 LD (light intensity not critical).

2.2 Equipment and materials needed for insect handling

2.2.1 Adult oviposition

- 450 ml (1 pint) wide mouth Ball® jar (Fig. 2A)
- Black rust-retardant paint, e.g. Rust-oleum®
- Tap water

2.2.2 Larval rearing

a. Group rearing

- Enamel or plastic pans capable of holding 1 liter of water
- Square pieces of Plexiglass® to cover above pans
- Plastic sieve (Fig. 1B)
- Small mesh tea strainer
- Waterproof glue
- 1.5 mm diam wire
- Pantihose
- Prey (*Aedes aegypti* or other prey species)
- Liver powder
- Weighing balance
- 1000 ml Erlenmeyer dispensing flask with 10 ml volumetric head
- Tap water

b. Individual rearing

- Plastic ice cube trays
- Disposable 14.5 cm glass Pasteur pipet
- Pipet bulb
- Prey

2.2.3 Pupae

- 450 ml plastic coated paper cups (Nestyle)
- Neptune straight-sided 450 ml plastic coated paper cups
- Black nylon tulle netting

2.2.4 Adult maintenance

2.2.4.1 Plastic bucket cage

- 15.5 liter (34 lb) white plastic buckets and lids
- Black tulle netting
- Razor knife
- Plier stapler
- Stockinette
- Paper towel

2.2.4.2 Cage arrangement

- Cage measuring 60 x 60 x 60 cm or plastic bucket cage (see precaution below)
- 8 x 8 cm piece paper towel
- Paper cups (Nestyle)
- Cotton balls
- Mechanical aspirator (Fig. 2D)
- Plastic sheeting
- 12 x 6 cm piece of wire window screen (Fig. 2B)
- Paper clips

PRECAUTIONS

(i) If *Toxorhynchites* are obtained from an established lab colony, use the same size colony cage that is used by the source lab

(ii) If *Toxorhynchites* strains can mate in a cage smaller than 60 x 60 x 60 cm, use plastic bucket cage

2.3 Diet

2.3.1 Larval diet

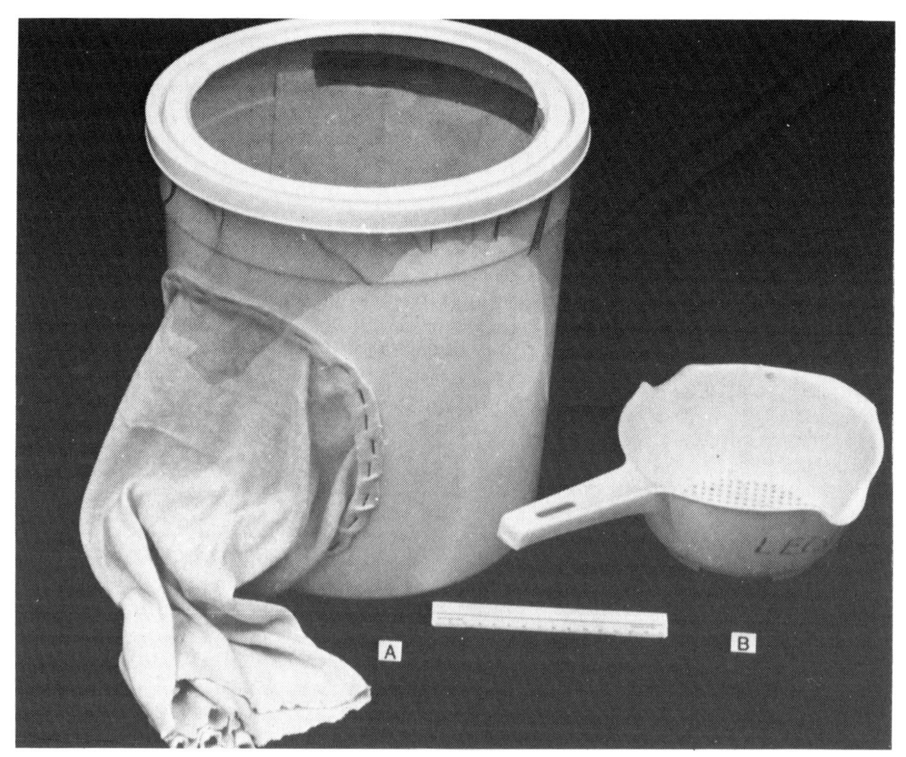

Fig. 1. Accessory equipment for Toxorhynchites rearing.
A. Plastic bucket cage. B. Plastic sieve.

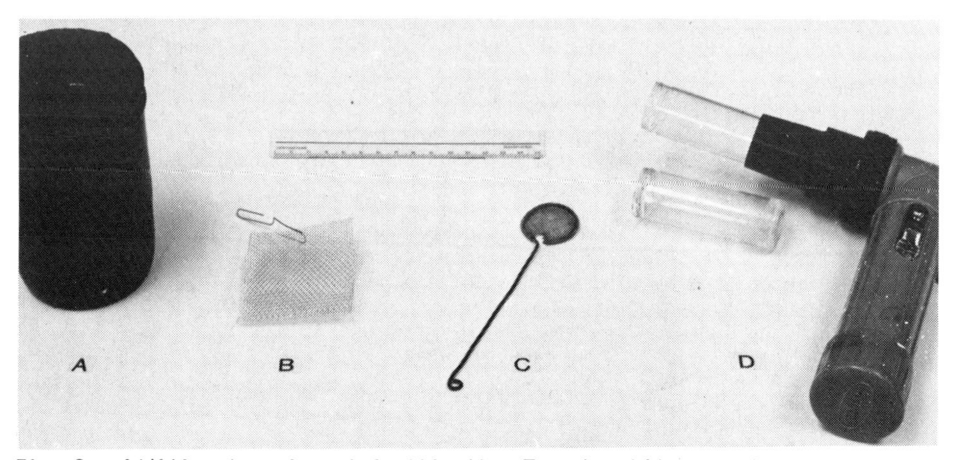

Fig. 2. Additional equipment facilitating Toxorhynchites rearing.
A. Oviposition jar. B. Holder for honey pads (wire screen).
C. Larva and pupa transferring net. D. Mechanical aspirator for adults.

a. 1st stage larvae
- 1st instar *Aedes aegypti* prey

b. 2nd stage larvae
- 2nd instar prey

c. 3rd to 4th stage larvae
- 3rd and 4th instar prey

2.3.2 Adult diet
- Honey pads
- Moisture source

2.4 Brand names and sources of supplies

Honey (commercially processed, locally retailed); plastic sieve and tea strainer (locally purchased); Nestyle 16 oz (450 ml) tall container without lid (Seairight Co., Inc., Fulton, NY 13069); enamel dish pans (No.32 basin, General Housewares Corp., P.O. Box 4066, Terre Haute, IN 47803); "chimney" (Neptune heavy duty pint container with rolled rim, Neptune Paper Products, Inc., Jamaica, Long Island, NY); plastic bucket (Imperial plastics, Inc., 101 Oakley St., P.O. Box 959, Evansville, IN 47706-0959, 34 lb [15.5 liter] size plastic container with lid); mechanical aspirator (Hausherr's Machine Works Toms River NJ 08753); flat black rust retardant paint (Rust-oleum Corp., Evanston, IL 60204); Dispo Pipets, 14.5 cm long, P5200-1 (Scientific Products, Div. of Am. Hospital Supply Corp. McGaw Park, IL 60085); pipet bulb, R5002-2 (Scientific Products, Div. of Am. Hospital Supply Corp. McGaw Park, IL 60085); absorbent cotton or rayon balls (purchased locally from physicans supply store); 1 pint [450 ml] wide mouth Ball® jar (locally purchased); plastic ice cube trays (locally purchased); liver powder (ICN Nutritional Biochemicals, P.O. Box 28050, Cleveland, OH 44128); stockinette (Tomac® tubular stockinette, American Hospital Supply, McGaw Park, IL 60084); tulle netting (purchased from local fabric stores); absorbent wadding (Curity®, Kendall Hospital Products Division, Boston, MA 02110)

3. PROCEDURES

3.1 Diet preparation

3.1.1 Larval diet preparation
- Follow general procedure for rearing prey as given in Section 4 of *Aedes aegypti* chapter (Munstermann and Wasmuth, this volume) with the following differences:
- Counting larvae for prey pans is unnecessary
- Use large 60 x 60 x 60 cm cage as colony cages for prey egg production
- Use restrained guinea pigs or rabbit to feed prey colony cages

3.1.2 Adult diet preparation
- Shred cellucotton (50 x 50 cm) into strips 5-10 cm wide and place in a large mixing bowl
- Moisten with 200 ml tap water
- Add 1 kg honey (approx. 2 lb) and knead until a uniform consistency is obtained
- Store at 5° C
- Discard if signs of fermentation (odor) appear
- A raisin, apple slice, or 15% sucrose solution absorbed in a cotton ball can be used as a substitute for the honey mixture

3.2 Liver powder suspension preparation
- Weigh 60 g liver powder into 1000 ml flask
- Mix the powder with 250 ml warm tap water
- Bring volume to 1000 ml with cool tap water
- Let stand until foam subsides (note: the resulting suspension must be agitated before each use)
- Store at 5° C; discard after 2 weeks

3.3 Larva and pupa transferring net preparation
- Bend heavy guage wire into a 2.5 cm diam loop with a 12 cm handle (Fig. 2C)
- Glue pantihose to loop with waterproof glue. For best results, stretch

pantihose slightly as it is applied.
- Wrap hose edges smoothly around the wire and glue

3.4 Oviposition jar
- Paint jar inside and outside with black paint
- Let dry for 4 days
- Fill 2/3 full with tap water (Fig. 2A)

3.5 Pipet for egg transfer
- Attach bulb to Pasteur pipet
- Break pipet tip so that the opening is 3 mm diam

3.6 Chimney preparation
- Punch out the top and bottom from the straight-sided cups (Neptune)
- Place a square of black nylon tulle on top of the cup
- Secure it with the rim

3.7 Plastic bucket cage preparation (Fig. 1A)
- Using the razor knife, remove the center of the bucket lid leaving 2-3 cm of the rim intact
- Cut a 14 cm diam hole in the side of the bucket
- Stretch a 40 cm long piece of stockinette around the hole
- Secure the stockinette with a continuous row of staples
- Tape a paper towel to the inside of the cage to provide a resting place for the adults
- Cut a piece of nylon tulle several cm larger than the lid
- Place tulle over bucket and secure with lid (rim)

3.8 60 x 60 x 60 cm cage and plastic bucket cage arrangement
- Lay one 8 x 8 cm piece paper towel on top of 6 cotton balls in a paper cup
- Add 100 ml water
- Place 4 of these cups inside the cage as an extra humidity source. If the plastic bucket cage is used, only one of these is necessary.
- Place honey pads on bottom of overturned cup or hang pads from the top of the cage inside folded screen (Fig. 1B). Use bent paperclips to hang screen from the top of the cage.
- Cover the cage with sheets of plastic to maintain high humidity

4. REARING AND COLONY MAINTAINENCE

4.1.1 Egg collection and hatch (group rearing procedure)
- Place oviposition cup in colony cage for 24 hours
- Remove cup and check for oviposition. Eggs are relatively large (0.61 x 0.42 mm), white, and float in a cluster on the water's surface.
- Separate 30-40 eggs into clean rearing pan filled with water
- Add prey egg paper containing 300-400 eggs
- Add 10 ml liver powder solution to water
- Cover pan with Plexiglass® lid

4.1.2 Egg collection and hatch (individual rearing procedure)
- Place oviposition cup in colony cage for 24 hours
- Remove cup and use pipet to place eggs singly in plastic ice cube trays
- Add approximately 25 1st instar prey or eggs to each tray section

PRECAUTION

Tx. amboinensis eggs are hydrophobic and may bounce out of the oviposition container if it is bumped

4.2.1 Larval rearing

4.2.1.1 Group rearing procedure
- Beginning on the day the *Toxorhynchites* eggs are hatched, place an additional prey egg paper containing approximately 400 eggs in a 2nd rearing pan and add 10 ml liver powder suspension. Do this every day until *Toxorhynchites* pupation is nearly completed. If 2 pans of

Toxorhynchites eggs are hatched, then 2 additional pans of prey should be hatched, and so forth. If several pans of *Toxorhynchites* are being reared at once, hatch 1 heavy egg paper (>10,000 eggs) per day instead of several smaller egg papers; separate the prey into the same number of pans as there are predator larvae.

- On the 3rd day after the *Toxorhynchites* eggs are hatched, pour the "oldest" pan of prey through the tea strainer, and add prey to the pan of *Toxorhynchites* larvae
- Continue every day until all the *Toxorhynchites* have pupated
- Use large plastic sieve to separate 4th instar larve and pupae from prey
- Discard uneaten prey or rear to adults and add to colony cage for prey egg production

PRECAUTIONS

(i) Since *Toxorhynchites* are cannibalistic, care must be taken that prey are always present. Too little food will result in slower growth and stunted larvae.

(ii) Prey numbers must be regulated so that prey do not pupate and emerge in the *Toxorhynchites* pans.

(iii) Killing but not consuming prey may be observed, especially in the 4th instar. This is normal behavior.

4.2.1.2 Individual rearing procedure

- Hatch prey larvae daily as above
- Add prey larvae to ice cube trays containing *Toxorhynchites* larvae as needed

4.3 Pupae

- Collect pupae every day with the small net and place in a chimney-covered cup 1/2 filled with water

4.4 Adults -sex ratio is approximately 1:1

4.4.1 Collection

- Place the entire emergence cup in the cage, remove the chimney to free the adults
- Use the mechanical aspirator to transfer adults from cage to cage

4.4.2 Oviposition

- The pre-oviposition period is from 5-12 days
- Oviposition is diurnal with a peak between 17 and 19 hours. It occurs continuously during the female's lifetime.
- Oviposition behavior: The female flies in a vertically oriented elliptical loop. As the loop path becomes smaller and close to the water, she ejects a white, oval egg downward onto the water's surface.

4.4.3 Sex determination

- Male pupae are distinguishable by the enlarged external genitalia at the base of the swimming paddles
- Male adults are distinguished by conspicuous plumose antennae with dense setae, and clasping genitalia

5. LIFE CYCLE PARAMETERS

- Hours to hatch-42
- Days as 1st stage larva-1.
- Days as 2nd stage larva-3.9
- Days as 3rd stage larva-2.2
- Days as 4th stage larva-6.5
- Days as a pupa-6.3
- Days egg to adult-21.7
- Days egg to egg-26.7-33.7

6. REARING SCHEDULE

a. Daily

- Feed larvae
- Set up prey pans
- Release newly emerged adults into colony cage
- Collect pupae

b. Alternate days
- Feed prey pans

c. Every week
- Change honey and humidity cups in adult cages

d. Every 2-3 weeks, or as needed
- Collect eggs and hatch

7. SHIPMENT OF *Toxorhynchites*

7.1 Eggs
- Place eggs between 2 pieces of very damp filter paper
- Fold paper twice and place in plastic petri dish or small 20 dram bottle with snap on lid
- Secure container lid with tape. The eggs must remain damp during shipping.
- Pack well and mail by fastest method

7.2 Fourth stage larvae
- Place early fourth stage larvae in individual 4 dram vials filled 2/3 full with water
- Seal vial with tape
- Pack in an insulated container and mail by fastest method

8. *Toxorhynchites* SPECIES PRESENT AT THE UNIVERSITY OF NOTRE DAME VECTOR BIOLOGY LABORATORY

- The following species are maintained at the Vector Biology Lab. They may be obtained by writing the Director, PROF. GEORGE B. CRAIG, JR. at the following address:

VECTOR BIOLOGY LABORATORY
DEPARTMENT OF BIOLOGY
UNIVERSTITY OF NOTRE DAME
NOTRE DAME IN 46556, U.S.A

- *Toxorhynchites rutilus rutilus* (Coquillet)
- *Toxorhynchites theobaldi* (Dyar and Knab)
- *Toxorhynchites amboinensis* (Doleschall)--wild type and yellow larva strains
- *Toxorhynchites brevipalpis* Theobald
- *Toxorhynchites splendens* (Wiedemann)

9. REFERENCES

Focks, D.A., Dame, D.A., Cameron, A.L., Boston, M.D., 1980. Predator-prey interactions between insular populations of *Toxorhynchites rutilus rutilus* and *Aedes aegypti*. Environm. Entomol. 9: 37-42.

Focks, D.A., Hall, D.W., 1977. Laboratory colonization and biological observation of *Toxorhynchites rutilus rutilus*. Mosq. News 37: 751-755.

Focks, D.A., Seawright, J.A., Hall, D.W., 1978. Laboratory rearing of *Toxorhynchites rutilus rutilus* (Coquillett) on a non-living diet. Mosq. News 38: 325-328.

Furumizo, R.T., Cheong, W.H., Rudnick, A., 1977. Laboratory studies of *Toxorhynchites splendens*. Part I. Colonization and laboratory maintenance. Mosq. News 37: 664-667.

Gerberg, E.J., Visser, W.M., 1978. Preliminary field trial for the biological control of *Aedes aegypti* by means of *Toxorhynchites brevipalpis*, a predatory mosquito larva. Mosq. News 38: 197-200.

Manning, D.L., Evenhuis, N., Steffan, W.A., 1983. Annotated bibliography of *Toxorhynchites* (Diptera: Culicidae). Supplement I. J. Med. Entomol. 19: 429-486.

Rosen, L., 1981. The use of *Toxorhynchites* mosquitoes to detect and propagate dengue and other arboviruses. Am. J. Trop. Med. Hyg. 30: 177-183.

Steffan, W.A., Evenhuis, N.L., 1981. Biology of *Toxorhynchites*. Ann. Rev. Entomol. 26: 159-181.

Steffan, W.A., Evenhuis, N.L., Manning, D.L., 1980a. Annotated bibliography of *Toxorhynchites* (Diptera: Culicidae). J. Med. Entomol. Suppl. 3: 1-140.

Steffan, W.A., Stoaks, R.D., Evenhuis, N.L., 1980b. Biological observations of *Toxorhynchites amboinensis* (Diptera: Culicidae) in the laboratory. J. Med. Entomol. 17: 515-518.

Trpis, M., 1979. Development of the predatory larvae *Toxorhynchites brevipalpis* (Diptera: Culicidae) on nonprey diet. J. Med. Entomol. 16: 26-28.

Trpis, M., Gerberg, E.J., 1973. Laboratory colonization of *Toxorhynchites brevipalpis*. Bull. W.H.O. 48: 637-638.

ACKNOWLEDGEMENTS

This work was supported by NIH Grant No. AI-02753 to Prof. G. B. Craig, Jr. Many thanks are due P. Hodges for her assistance and helpful suggestions and to S. Durso for use of unpublished data and manuscript review.

ADOXOPHYES ORANA

G.W. ANKERSMIT

Vakgroep Entomologie, Binnenhaven 7, 6709 PD Wageningen, The Netherlands.

THE INSECT

Scientific Name: *Adoxophyes orana* Fischer von Röslerstamm
Common Name: Summer fruit tortrix
Order: Lepidoptera
Family: Tortricidae

1. INTRODUCTION

Adoxophyes orana is the most important insect pest in dutch apple orchards. It has two complete generations a year. The young larvae overwinter on the trees in cracks and crevices, become active in April and feed on the leaves. The moths emerge and deposit their eggs on the leaves in June and the beginning of July. The first generation or summer larvae feed again on the leaves but also on apples and cause the most injury. The second flight occurs in August or September. On hatching the small larvae feed on the foliage and fruits to a small extent. They hibernate as second or third instar larvae.

The area of distribution extends from Western Europe through the USSR and China into Japan (Barel, 1973). The species is polyphagous and a list of hostplants is given by Barel (1973). The diet used was originally described by Ankersmit (1968) and later modified Ankersmit et al. (1977a) who developed a mass rearing method for sterile male release trials. The laboratory colony was established in summer 1965 and reared on *Vicia faba*.

2. FACILITIES AND EQUIPMENT REQUIRED

2.1 Facilities

Eggs, larvae, pupae and moths are kept in controlled environment rooms maintained at 20 ± 1°C, 70% RH and LD 16.5:7.5 photoperiod (longer photoperiods are also possible) (light source, Philips fluorescent tubes 40 W No 33).

2.2 Equipment and materials required for insect handling

2.2.1 Oviposition

- Plastic, transparant 2-liter jars (Fig. 1) (22 cm high, upper diameter, 11.7 cm)
- Tightly fitting cover (green)
- Polythene sheet (0.03 mm) 16 x 16 cm (sheets cut from strips 16 cm wide and kept in water for 1 week to dissolve toxicants)
- Glass tube 50 x 15 mm
- Cotton wool (fitting in tube)
- Adhesive tape
- Filter paper disc (Schleicher and Schüll diameter 9 cm)

2.2.2 Egg collection and storage

a. Individual rearing

- Glass tube 70 x 20 mm diameter
- Cotton wool
- Felt tip pen
- Scissors
- Refrigerator 5°C (optional)

b. Mass rearing
- Refrigerator box 16 x 16 cm at bottom
- Tray (30 x 30 x 5 cm) with 0.1% commercial detergent in water
- Scissors
- Felt tip pen
- Refrigerator 5°C (optional)

Fig. 1. Oviposition jar.

2.2.3 Egg hatch and larval inoculation of diet

a. Individual rearing
- Glass vial 70 mm high 20 mm diameter
- Cotton wool
- Fine paint brush (size 00)
- Bottle with 1% Clorix® (commercial sodiumhypochlorite 10 g active Cl/100 ml).
- Towel
- Glass plate 50 x 30 cm (optional)

- White filter paper fitting under glass plate
- Sterile room or hood

b. Mass rearing
- Paper towels
- Cotton wool
- Polythene garbage bag 80 x 60 cm
- Scissor
- Rearing containers (Fig. 2)
- Paper clips
- Spatula
- 3.7% AI formaldehyde
- Bottle with 1% Clorix®
- Sterile room or hood

Fig. 2. Disposable containers. Open container shows plastic strips.

2.2.4 Larval rearing containers

a. Individual rearing
- Glass tubes (50 x 15 mm)
- Cotton wool
- Container or tube rack

b. Mass rearing on agar diet
- Disposable containers (Fig. 2 rim size 14.5 x 20.5 cm)
- Same container but with hole, diam 15 mm on either side, as cover

and 4 cm high
- Flexible plastic strips 61 cm x 3 cm
- Paper towels
- Paper clips
- Sprayer with 0.37% AI formaldehyde

c. on lucerne meal diet
- As above but cover without holes

2.2.5 Emergence of moths

a. Individual rearing
- Emergence cage with sleeves
- Aspirator (connected with vacuum cleaner, or central vacuum installation)

b. Mass rearing
- Wooden boxes (Fig 3) size 41 x 42 x 50 cm
- With 3 holes (diam 9 cm) and lid on opposite side
- Plastic funnels fitting on these holes, at narrow end 2.8 cm diameter
- Polythene bags without unneccessary folds made from a tube, size 28 x 20 cm
- Support for plastic bag
- Sealing apparatus (to make bags)
- Rubber bands
- Adhesive tape
- Lamp

2.2.6 Holding of moths
- Cold room 2-4°C
- Plastic transparant jars (2-liter Fig. 1)
- Paper towels

2.3 Diet preparation equipment

a. Individual rearing on agar diet
- Balance 0.05-2000 g
- Blender (14000 min^{-1} in water)
- Paper (weighing)
- 2 liter measuring cylinder
- 10 ml measuring cylinder
- Stirring spoon
- Scoop
- Thermometer 0-100°C
- Squeeze bottle
- Towel or gloves
- Pan 3 liter
- Heating apparatus
- Refrigerator at 5°C
- Paper bags, plastic bags size fitting to tube rack

b. Mass rearing on agar diet
- Similar as for individual rearing
- Stainless steel spatula
- Container 38 x 22 x 5 cm

c. Mass rearing on lucerne meal based diet
- Balance 0.05-2000 g
- Handblender (Household type)
- Bowl for blender
- 1 liter measuring cylinder
- 10 ml measuring cylinder
- Scoop

Fig. 3. Wooden emergence boxes

3. ARTIFICIAL DIET

3.1 Composition, to prepare 1017 g diet

a. Based on Agar

Ingredients	Amount (g)	% (weight)
Agar	20	2.0
Carboxymethylcellulose	5	0.5
Casein	35	3.4
Wheat germ	35	3.4
Sugar	30	3.0
Brewers yeast	30	3.0
Linseed oil	4.7 (5ml)	0.5
Choline chloride	4.0	0.4
Sorbic acid	1.5	0.15
Methyl 4-hydroxy benzoate	1.5	0.15
Streptomycine sulphate	0.2	0.02
Water	850	83.6
Total	1016.9	

b. Based on lucerne meal, to prepare 992 g diet

Lucerne meal	250	25.2
Casein	35	3.5
Wheat germ	35	3.5
Sugar	30	3.0
Brewers yeast	30	3.0
Linseed oil	4.7 (5 ml)	0.5
Choline chloride	4.0	0.4
Sorbic acid	1.5	0.15
Methyl-4-hydroxy benzoate	1.5	0.15
Streptomycinesulphate	0.2	0.02
0.1 M citric acid	600	60.5
	991.9	98.87

3.2 Brand names and sources of ingredients.

Agar, (agar (Ag 172) Brocacef Maarsen): Carboxymethylcellulose, (Aku CMC HZ858, AKZO Arnhem);
Casein, (Spray Casein 90 mesh, Melkindustrie Veghel, Netherlands);
Sugar, (household product); Brewers yeast; (Saccharomyces cerevisiae siccum, Brocacef B.V. Maarsen, Netherlands);
Linseed oil, (Oleum lini, Lamers en Indemans, Chemicals B.V. 's-Hertogenbosch, Netherlands); Choline Chloride, (Baker Chemicals B.V. Deventer, Netherlands);
Sorbic acid, (Merck NF XIV E200. Darmstadt DBR);
Methyl 4-hydroxybenzoate, (BDH Chemicals Ltd. Poole, England);
Streptomycine sulphate no 5-6501, (Sigma Chemical Company Saint Louis, Missouri, USA).

3.3 Diet preparation procedure

3.3.1 Agar diet

a. Individual rearing

- Dissolve agar at 95°C in tapwater
- Place all other ingredients in blender (except the linseed oil)
- Add the agar and linseed oil
- Blend for 1 minute at high speed (14.000 min^{-1} in water)
- Pour the hot diet in squeeze bottles
- Pour quickly about 1-1½ cm layer of diet in each rearing tube
- Cool tubes in racks in paper bag at room conditions over night or for 6 hours
- Prepared diet in plastic bag can be stored at least 1 week in a refrigerator at about 5°C

b. Mass rearing on agar diet

- Pour the hot diet in containers (38 x 22 x 5 cm) in layers of about 1 cm

- Cool overnight for 6 h at room conditions in a paper bag. Diet can be stored in a plastic bag in a refrigerator of ca 5°C for at least 1 week

3.3.2 Lucerne meal diet, mass rearing
- Place all ingredients in a bowl, first dry ingredients
- Blend with a hand blender
- Add a layer of 1-2 cm in the rearing container for immediate use

PRECAUTIONS

(i) All equipment must be sterilized; plastic containers are sterilized by keeping them overnight in garbage bag with pads drenched in 3.7% AI formaldehyde.

(ii) Keep diet preparation room and rearing room separate.

4. HOLDING INSECTS AT LOWER TEMPERATURES

4.1 Adults

Moths can be stored in groups of ca 500 at 8°C for 1 day in plastic 2-liter jars with paper towels to provide resting places. Moths survive longer storage but egg production at 20°C then declines, and reduces hatchability of eggs. Storage at 4°C proved more detrimental.

4.2 Eggs

Shortly before hatching these can be held at 5°C for not more than 4 days. Newly laid eggs are killed at 5°C. Almost no eggs hatch when kept at 13°C but at 15°C hatchability is about 70%.

4.3 Larvae

Individually larvae can be reared at 15°C. In mass rearing molds develop at low temperatures.

5. REARING AND COLONY MAINTENANCE

The colony of *A. orana* was established in 1965 and reared since 1966 on artificial diet. The diet was modified in course of time slightly. The number of generations was not exactly followed but with ca 7 generations annually at 20°C ca 120 generations have been reared.

5.1 Egg collection

- Uncover the oviposition jar
- Take a new polythene sheet
- Place the new sheet on top of the old sheet
- Remove the old sheet with eggs carefully by gently pulling and tapping to remove moths underneath the new sheet. Most eggs are laid on this sheet
- Collect sheets, keep them at desired temperature, in container placed in tray with water + detergent to prevent escape of young larvae; no sterilisation method is appled to eggs
- For individual rearing, pieces of egg sheets are collected into glass tube (70 x 20 mm) closed with a cotton plug

5.2 Diet inoculation

a. Individual rearing in tubes
- Tubes with diet are placed overnight in a sterile room or clean place
- Clean working table with a 1% sodium hypochlorite solution
- Remove rack with tubes from the bag
- Open tube with neonate larvae
- Transfer one larva carefully into the tube using a fine brush
- Close the tube tightly with a cotton plug
- Place the tube, cotton plug end upright in rack and transfer filled racks to rearing cabinet

b. Mass rearing on agar diet
- Clean working table with 1% sodium hypochlorite solution
- Spray bottom of containers with a little 0.37% AI formaldehyde
- Cut diet in container in cubes of ca 1 cm and place one layer of cubes to the rearing container (Fig. 2)

Fig. 4. Mass rearing in rearing cabinet

- Fit plastic strip into container making as many curves with it as possible and press it to the bottom (Fig. 2)
- Place one folded paper towel on top of the rearing container without covering the whole open upperside of the container
- Place egg sheets cut from oviposition sheet with about 300 eggs in total on the paper towel
- Close the open upperside of the rearing containers with 2 partly unfolded paper towels making 4 layers of paper in total
- Place the cover with holes over container with diet and close cover and container with 10 paper clips on the rim
- Transfer to rearing cabinet (Fig. 4)

c. Mass rearing on Lucerne meal diet

- As above but use covers without holes

5.3 <u>Emergence period</u>

In both individual and mass rearing development from oviposition to emergence of first moth takes 45 days (35 days for larval and pupal development). Emergence is nearly complete in mass rearing 57 days after oviposition. Some stragglers may still emerge but it is not economical to keep the emergence box in use.

5.4 Adults

The sex ratio is approximately 1:1; males emerge earlier than females.

Moth collection:

a. Individual rearing
- Remove cotton plug from tube
- Place rack with open tubes in a cage with sleeves
- Collect daily emerged moths with an aspirator through the sleeve

b. Mass rearing
- Uncover the containers
- Upper paper towel can be removed (probably no pupae)
- The two paper towels underneath (containing many pupae) are unfolded and placed in emergence box (Fig. 3)
- Rearing containers with diet, larvae and pupae are stacked crosswise in other emergence box
- Funnels are fitted with their wide mouth over the holes in the box with tape
- Fit the plastic bags, blown up, with rubber bands to the funnels (Fig.3)
- Place light in front of the funnels to attract the moths
- Bags with moths can be collected daily

5.5 Oviposition

At LD 16:5:7:5 photoperiod and 20°C, preoviposition period is one day and peak egg laying occurs on the 2nd and 3rd day. After 5 days the moths are discarded though still some eggs are laid.
- Clean oviposition jars (we used plastic 2-liter pots)
- Filter paper at the bottom
- Tube with water and cotton fixed to the wall of the pot by tape
- Plastic sheet 16 x 16 cm on top of the pot
- About 40 moths, not more, (half of them females) are entered into the jar by emptying a plastic bag from moth collecting box (lower numbers of moths will produce more eggs per female)
- Fix plastic sheet with cover of jar to the jar

5.6 Sex determination
- Last instar male larvae are distinguished by the 2 testes being visible through integument as bean-like bodies
- Female pupae and adults are usually larger than the males
- The female pupa has on the ventral side of the 8th abdominal segment a visible opening for the bursa copulatrix
- Male moths have a plume of hairs at abdominal tip

5.7 Daily rearing schedule
- Collect eggs
- Collect moths
- Start new oviposition jar (optional, depends upon size of rearing, once every 5 days is possible)
- Prepare new rearing containers and inoculate with eggs (optional)
- Place containers with pupae due to become moth in moth emergence box (optional)
- Discard contents of old moth emergence boxes (optional)
- Sterilise the moth emergence boxes

Rearing schedule in course of time at 20°C

Day	Activity
1	Oviposition
2	Collecting of eggs
9	Preparation of diet
10	Start of larval culture
45	Opening of rearing containers
	Transfer of containers to emergence boxes
57	End of culture, discard old rearing containers
	Clean boxes

5.8 Insect quality

This was determined by regular weighing and determination of sex ratio from samples. Weight of males ranged from 8.7 mg at the end of the collecting period to 10.5 mg at the beginning. Females weighed from 19.2 to 24.2 mg in the course of this period. In sterile male release trials, mating competitiveness was compared with native moths (Ankersmit et al., 1977b)

5.9 Special problems

Microbial control of diet. Strict sanitary measures, work in clean environment. Keep diet preparation and inoculation separate from moth emergence room. A constant temperature reduces chances for moisture condensation on innersides of containers and diet.

6. LIFE CYCLE DATA

6.1 Development in individual rearing on agar diet

Stage	Number of days 20° C	25° C
Egg	8.9	6.5
Larva	24.0	18.5
Pupa	8.6	5.6

In mass rearing development is somewhat slower.

6.2 Daily moth production of 41 containers started at one day

Days after inoculation	Number ♀	♂
36	84	294
37	294	511
38	497	620
39	479	493
40	244	286
41	356	271
42	307	225
43	151	124
44	146	89
45	123	79
46	121	66
47	81	54
48	67	30
	2950	3142

6.3 Survival data

a. Individual rearing (°C) 13 14 15 20 25
Egg viability (%) 3 19 72 87 93

Individual rearing (°C)	13	14	15	20	25
Egg viability (%)	3	19	72	87	93

Neonate larvae to adult usually more than 85%.

b. Mass rearing

Total survival from egg to adult about 50% on agar diet but on lucerne meal diet between 30-50%.

In mass rearing multiplication rate between generations was 15 times.

6.4 General information

Preoviposition period 1-2 days at 20°C
Peak egg laying period 2-3 days after emergence
Fecundity between 200-300 eggs
Weight of moths in mass rearing (mg), ♀ 23, ♂ 10; varying between 19.2 and 24.2 for ♀♀ and 8.7 and 10.5 for ♂♂.

7. REFERENCES

Ankersmit, G.W. 1968. The photoperiod as a control agent against Adoxophyes reticulana (Lepidoptera, Tortricidae). Entomologia exp. appl. 11: 231-240.

Ankersmit, G.W., Rabbinge, R. and Dijkman, H., 1977a. Studies on the sterile male technique as a means of control of Adoxophyes orana (Lepidoptera, Tortricidae). 4. Technical and economical aspects of mass rearing. Neth. J.Pl. Path. 83: 27-39.

Ankersmit, G. W., Barel, C. J. A., Mobach, J. D., Schout-Parren, J. and Wassenberg-de Vries, G., 1977b. Studies on the sterile male technique as a means of control of Adoxophyes orana (Lepidoptera: Tortricidae). 5. Release trials. Neth. J. Pl. Path. 83: 73-83.

Barel, C. J. A., 1973. Studies on dispersal of Adoxophyes orana F. v. R., in relation to the population sterilisation technique. Meded. Landb. Hogesch. Wageningen 73 (7): 1-107.

AGROTIS IPSILON

ROBERT G. BLENK, R. J. GOUGER, T. S. GALLO, L. K. JORDAN, E. HOWELL

ICI Americas Inc., P. O. Box 208, Goldsboro, NC 27530

THE INSECT

Scientific Name: *Agrotis ipsilon* (Hufnagel)
Common Name: Black cutworm (BLCU)
Order: Lepidoptera
Family: Noctuidae

1. INTRODUCTION

The black cutworm is a world-wide pest of herbaceous and woody plants and is often associated with moist conditions and weedy fields. It is found in all 50 of the United States, on all other continents and in many island communities (Rings et. al, 1976). At least 53 crop hosts had been listed by Rings et al. (1974) with additional references to such broad host categories as: cereal grains, pasture and turf grasses, greenhouse crops, vegetables and weeds.

A. ipsilon attacks seedlings and young plants, usually at night, cutting the entire plant off near the soil line or otherwise inflicting severe damage by heavy feeding in larger plants. It has two generations a year as far north as Canada and up to four generations a year in the southern United States.

Eggs are laid singly or in small groups on plants or in cracks in the ground and hatch after five days. All larval stages live in the soil but emerge, usually at night, to feed on plants. After at least two weeks, the larvae pupate in the soil and diapause can often occur at this stage. Moths normally emerge from their pupal cases in one to eight weeks.

The original colony at ICI Americas was obtained in 1973 from a laboratory strain at the University of Maryland. It was interbred at a 1:1 ratio in 1980 with a field-collected strain from Florence, South Carolina, after both strains were found equally susceptible to a broad range of insecticides.

2. FACILITIES AND EQUIPMENT

2.1 Facilities

Larval rearing and pupation is done in an environmental chamber and adult emergence and egg collection is accomplished in a multi-purpose insectary. Both rooms are kept at 27±1°C and 65-75% RH with a day/night regime of 14/10 hours under white fluorescent lighting. For test purposes egg, late instar larval and pupal development are frequently slowed in separate rooms at 15 or 22°C. A small room is utilized solely for diet prepartion and another for larval transfer to diet.

2.2 Equipment and materials for maintenance of colony

2.2.1 Adult oviposition

- Metal cage (60 x 30 x 15 cm)
- Expanded mica (Vermiculite, "Terra-Lite", W. R. Grace, Grade 3)
- Waxed feeding cup (45 ml) containing absorbent cotton
- Screen underlid (6 mm mesh) "Hardware cloth".
- Masking tape (5 cm wide)

- Paper hand toweling (Viva) green or yellow
- Metal cage lid with 1.5 cm frame and standard mesh window screen insert
- Sucrose (10% solution) squirt bottle
- Tap water squirt bottle
- Industrial oven (0-500°F, TA-500, Grieve Corp. Round Lake, IL)

Fig. 1 Adult Emergence Cage Showing Crumpled Paper Toweling for Oviposition

2.2.2 Egg collection and sterilization

- Porcelain trays, 36 x 20 x 5 cm
- Scissors
- Fume hood
- Formalin dilution (0.37% AI in tap water)
- Mechanical timer, GE, (0-120 min)
- Metal frame and cord "washline"

2.2.3 Egg storage
- Glass beaker, 4 liter
- Plastic film wrap (Borden "Sealwrap")
- Rubber band (10 x 0.5 cm)
- Incubation chamber/room

Fig. 2 Egg-Collection Beakers Showing Eggs in "Blackhead" Stage and Larvae Beginning to Emerge

2.2.4 Egg hatch and diet infestation
- Contaminant-free transfer room or area
- Small camel hair artist brush
- Cup lids, unwaxed cardboard, 3.7 cm diam (Standard Cap & Seal, Chamblee, GA)
- Microwave oven, (Litton "Minute Master")

2.2.5 Larval Rearing and Holding
- Plastic jelly cups, 35 ml, clear (Filrite, Newark, NJ)

- Pressed fiber tray (36 unit, 25 cm square, Cell-Pac, Diamond International Corp., New York, NY)

2.2.6 Pupal collection/storage

- Forceps, hard and soft tip
- Flat tray (20 x 14 cm, Mobilfoam, Mobil Chem. Co., Covington, Georgia)
- Plastic petri dish (15 cm, Fisher Scientific, Raleigh, NC)
- Plastic autoclaving bag (61 x 76 cm, Fisher)
- Holding chamber/room

2.3 Diet preparation equipment (in contaminant-free room)

2.3.1 For individual ingredients:

- Double pan balance (Ohaus "Harvard", 0-2 kg)
- Top loading balance (Mettler, P163, 0-160 g)
- Plastic weighing boats (Fisher, assorted sizes)
- Graduated cylinders (glass, assorted sizes)
- Glass flasks (assorted sizes)
- Rubber spatula (5 x 8 cm paddle)
- Plastic container (3 liter)
- Microwave oven (33 x 22 cm opening)
- Refrigerators, two (under-counter)
- Glass test tubes (25 ml) and rack
- Metal spatula
- Magnetic stirrer/heater (Fisher "Thermix")
- Waring blender ("Commercial", 3-speed 4 liter)
- Ear protection (David Clark Co. "Straightaway", E-310)

2.3.2 For finished diet

- Thermometer, probe (Bailey Instrument Co., BAT-4)
- CO_2 tank (20 lb)
- Pressure regulator, two-stage (Air Products, Atlanta, GA)
- Spray tank, stainless steel, 15 liter (Firestone Steel Products Co., Ann Arbor, Michigan)
- Trigger-type kitchen spray nozzle and hose
- Shelving or mobile rack
- Plastic autoclaving bags
- Incubator, (Fisher "Isotemp", 0-60°C)

3. ARTIFICIAL DIET

The ICI artificial diet is a modification of wheat germ diets (Vanderzant, 1962).

3.1 Composition to make 3.8 kg finished diet

	Ingredient	Amount	% of Total
a.	Agar mixture		
	Agar	50 g	1.30
	Distilled water	2200 ml	57.60
b.	Blender mix		
	Distilled water	1000 ml	26.16
	KOH solution	30 ml	0.78
	Formalin (37% AI)	3.5 ml	0.08
	Methyl paraben	7 g	0.18
	Vanderzant wheat germ diet	400 g	10.46
	Wheat germ coarse	60 g	1.57
	Sorbic acid	7 g	0.18
	Ascorbic acid	15 g	0.39
	Choline chloride solution	36 ml	0.94
	α-Tocopherol suspension	6 ml	0.15
	Inositol	1 g	0.03
	Bacitracin	0.2 g	-
	Vitamin mixture	6 ml	0.15

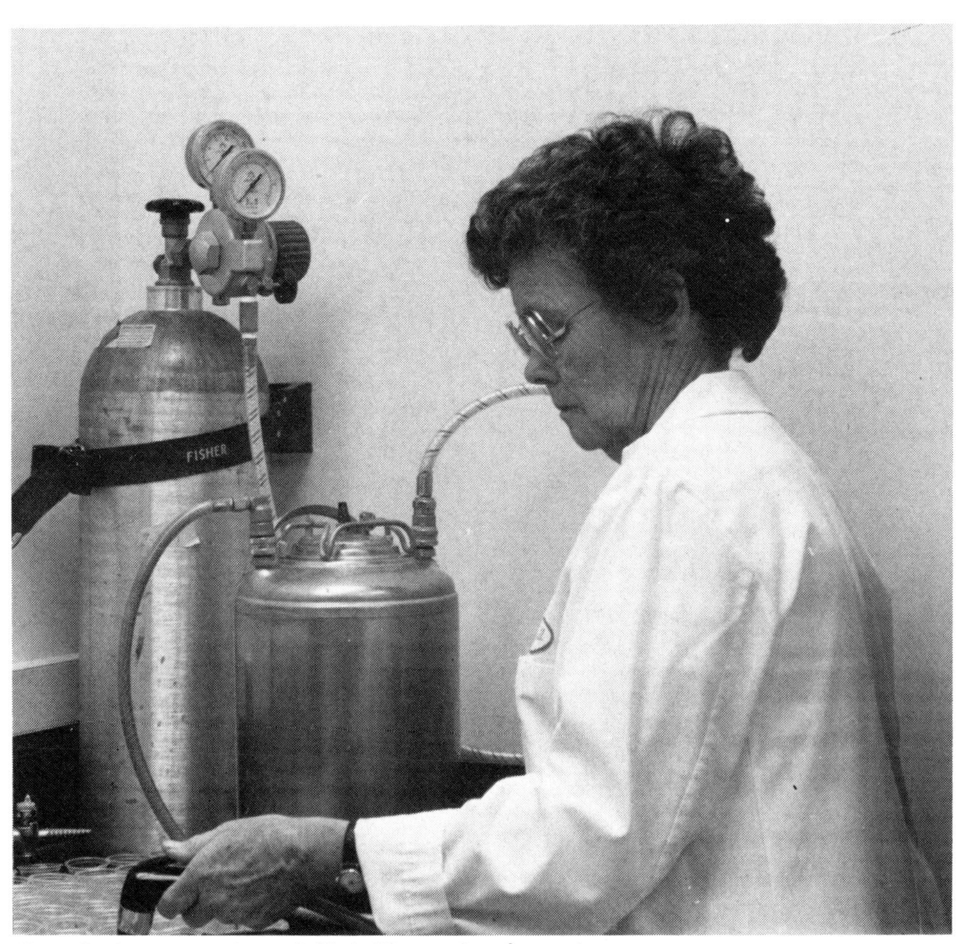

Fig. 3 Demonstration of Diet Dispensing Apparatus

c. Preparation of stock solutions/suspensions
   i. KOH = 2.5% w/v in distilled water
   ii. Choline chloride = 10% w/v in distilled water
   iii. α-Tocopherol = 1% v/v in linseed oil

d. Preparation of vitamin mixture

Ingredient	Amount
Distilled water	100.0 ml
Nicotinic acid	0.6 g
Calcium pantothenate	0.6 g
Riboflavin	0.3 g
Thiamine hydrochloride	0.15 g
Folic acid	0.15 g
Pyridoxine hydrochloride	0.15 g
Vitamin B_{12}	0.001 g
Biotin	0.01 g

3.2 Brand names and sources of ingredients

All diet ingredients (Section 3.1) are purchased from ICN Life Sciences, Cleveland, Ohio or from Fisher Scientific, Raleigh, North Carolina.

3.3 Diet preparation procedure

a. Agar mixture
- Measure water into plastic container
- Add agar, stir with rubber spatula
- Place in microwave oven set for 16 minutes
- Open oven and stir @ 5 and 10 minutes
- Bring to temperature of 80-85°C

(Above can be done while blending the following)

b. Blender mixture
- Measure water into blender
- Add all dry and liquid ingredients (Section 3.1b)
- Cover blender, start at low speed, move through medium to high speed, blend 2 1/2 minutes, move through medium to slow speed, shut off.

c. Final blend/dispensing
- Cool agar mixture (a) to 75°C
- Add agar mixture (a) to mixture (b) in blender
- Blend 1 1/2 minutes, moving through speeds as in (b)
- Pour immediately into stainless steel spray tank, seal tank top, close pressure relief valve
- Pressurize spray tank from CO_2 cylinder immediately: open main valve, set tank valve at 18 psi
- Dispense diet into individual cups in trays as quickly as possible, using trigger nozzle
- Shut off both CO_2 valves, open pressure relief valve on spray tank, remove lid, flush with warm water
- Set all filled trays on shelves to "age" 2 hours for infestation or to store in plastic cook bags in 15°C chamber

PRECAUTIONS

(i) Wear ear protection when using large blender

(ii) Post microwave warning signs on all entrances to diet prep room to warn those wearing pacemakers

(iii) Anchor CO_2 cylinder to wall and insure proper order of use of valves to pressurize and to release pressure

(iv) Clean all equipment immediately with warm water to prevent caking of diet

4. HOLDING INSECTS AT LOWER TEMPERATURES

4.1 Adults

The adults are usually maintained at or near 27°C.

4.2 Eggs

Egg collection towels in sealed beakers (see Section 5.1) are frequently held at 15°C to regulate hatch for transfer to diet. Holding time at this temperature seldom exceeds three days, since the older (5-6 day) eggs begin hatching.

4.3 Larvae

a. Neonate larvae are held at 15°C for up to 24 hours prior to transfer to diet.

b. Once transferred to diet larvae are rarely held below 27°C through third instar.

c. Later instars are frequently held at 15 or 22°C for one to three days to insure test insects of a standard weight. Untreated insects so exposed are ordinarily returned to the main colony.

4.4 Pupae

When pupae are fully colored (dark brown) they may be held at 15 or 22°C one to several days to regulate emergence dates. Most oviposition cages are set up with the majority of the pupae reared at 27°C.

5. REARING AND COLONY MAINTENANCE

5.1 Egg collection and sterilization (Mon.,Wed.,Fri.)

- Remove paper toweling from oviposition cages
- Place that day's collection of towels in porcelain pan. Maximum of 6 towels/pan
- Flood with dilute formaldehyde solution (2.2.2)
- Set mechanical timer for 15 minutes
- Wash several times with tap water
- Hang towels individually on washline until near dry, about one hour
- Place either in 15 or 27° chamber

PRECAUTIONS

(i) If towels reach dryness on "washline", remoisten with 1-2 ml distilled water from a squirt bottle prior to sealing with wrap.
(ii) Do not allow towels to hang more than 3 hours beyond dryness.
(iii) Store concentrate of formaldehyde in prescribed safety area for hazardous chemicals.
(iv) Store dilute formaldehyde in fume hood or in metal cabinet.
(v) Use nontoxic particle mask to avoid inhalation of moth scales in all handling of oviposition cages.
(vi) Yellow or green towels have consistantly induced greater oviposition than towels of other colors, including white.

5.2 Egg storage

- Place all towels for one day in 4 liter glass beaker or a maximum of 4 towels/beaker for large scale rearing
- Mark beaker for species, if more than one is reared on site
- Cover with two layers of plastic wrap and pull taut to allow no air pockets at lip of beaker. (Seals in moisture, prevents larval escape)
- Secure overlap of wrap with wide rubber band
- Date plastic wrap top with ball point pen

PRECAUTION When more than one species or strain is reared as at ICI, each towel strip and each beaker top (wrap) must be coded to insure their identity.

5.3 Diet infestation

- Allow fresh diet to age minimum of 2 hours after gel or incubate cold-held diet at 50°C for 20-30 minutes
- Transfer one larva per cup using camel hair brush, disinfect brush with 70% isopropyl alcohol solution between use for different beakers
- Cap each cup
- Reseal beaker with wrap for further use
- Date and code (see 5.2 precaution) caps of two diagonally opposite corner cups in each 36-unit fiber tray (2.2.5)
- Place in 27°C chamber
- Return beaker(s) to 15 or 27°C chamber

PRECAUTIONS

(i) For small scale rearing utilize all available neonate larvae each generation to protect against loss of the colony
(ii) For continuous rearing do not continue to transfer from a beaker in which cannibalism has occurred if larvae are available from a newer beaker

(iii) Kill unwanted larvae by pouring hot water over and submerging egg strips in beaker, allow to stand in sink until cool.

(iv) Although not as cannibalistic as _Heliothis zea_, the black cutworm is the largest larva reared at ICI, and must be reared singly in diet cups to insure normal development.

5.4 _Pupal collection_ (when pupae are 2-8 days old)

- Remove caps from cups, one tray at a time, using hard tip forceps
- Carefully remove pupae from cells in diet using soft tip forceps. Place on flat tray (Mobilfoam) exposed to air for 2 hours
- For use that day in oviposition cage (see 5.5.1) collect 100-150 pupae (sex ratio is very nearly 1:1)
- Place all discarded diet cups in plastic cook bag (see Section 5.9)

5.5 _Adults_ (physical handlng of adults is not necessary)

5.5.1 Oviposition cage

- Place pupae on 2-3 cm layer of moistened vermiculite in metal cage and set two feeding cups in cage
- Place two crumpled paper towels on vermiculite in base of cage for egg collection
- Tape course screen to top of cage (once in place this under lid prevents escape of emerging adults and is removed only to change egg-collection towels)
- Place strip of towels to cover underlid, hold in place with window screen top lid
- Moisten top toweling, using water squirt bottle

5.5.2 Adult maintenance

- Remove window screen top lid and top paper toweling
- Maintain food in cups using sucrose solution squirt bottle
- Remove coarse screen to replace crumpled egg-collection towels with new ones
- Re-moisten vermicultite and new towels using water squirt bottle
- Replace paper top toweling and window screen top lid
- Re-moisten upper toweling, using water squirt bottle

PRECAUTIONS

(i) Use nontoxic particle mask for all handling of adult cages to prevent inhalation of moth scales.

(ii) Maintain moisture on paper towel egg collection strips on a daily basis and RH in rearing room at 55% or above.

(iii) _A. ipsilon_ will not oviposit on the upper toweling as do other species reared at ICI, they must be supplied oviposition sites (toweling) in the base of the cage.

5.6 _Sex determination_

- Female and male pupae can be readily segregated under a low-power microscope (Butt and Cantu, 1962).

5.7 _Rearing schedule_

5.7.1 Daily (7 day/week)

- Maintain moisture in adult cages and on oviposition toweling and humidity in rearing room
- Alternate egg beakers between 15 and 27°C rooms to slow/hasten hatch
- Cook all discarded diet material and unwanted insects in oven (see 2.2.1, 5.9.3).

5.7.2 Mon.-Wed.-Fri.

- Replenish sucrose solution in adult feeding cups
- Collect, surface sterilize and incubate eggs
- Prepare diet
- Transfer neonate larvae

5.7.3 Weekly/bi-weekly
- Collect pupae
- Set up oviposition cage(s)
- Discard, clean oldest egg beaker(s) and oviposition cage(s)
- Make fresh diet stocks, sucrose solutions, formaldehyde dilutions

5.8 Insect quality

In rearing large numbers for testing, quality control is built into the daily operation in terms of rate of larval growth to desired test size, weight of pupae (female = 691 ± 83 mg, male = 603 ± 55 mg), and egg hatch.

5.9 Special problems

5.9.1 Diseases

a. Surface sterilization of eggs prevents transfer of bacterial, fungal contaminants from adult cages to diet. Clean transfer, diet prep rooms are essential.
b. Individual diet cups allow easy discard of accidentally introduced contaminants in large scale rearing.
c. For very small colonies individually contaminated cups may be retained, surviving pupae surface sterilized in a 1.5% sodium hypochlorite solution and returned to the breeding stock.

5.9.2 Identity of colonies

Rearing of large numbers of several species and strains at one facility necessitates care in the labeling of all egg beakers, larval cup trays and oviposition cages.

5.9.3 Waste disposal

All waste materials from diet cups and oviposition cages are placed in plastic autoclaving bags and cooked in a large oven at 218°C for 30-45 minutes before discard to prevent undue infestation of the surrounding agricultural area. All unused eggs, neonate larvae are killed with hot water.

5.9.4 Human health

a. Fume hoods and protective masks should be used where indicated in rearing procedures.
b. Ear protection is necessary for use of the large blender in a small diet room.

6. LIFE CYCLE DATA (FOR 27°C, 65-75% RH, 14/10 PHOTO PERIOD)

6.1 Development time

Stage	Mean No. Days
Egg	3 to 4
Larva	19
Pupa	12
Egg to adult	35
Egg to egg	42

6.2 Survival

Stage/Period	%
Egg viability	90+
Neonate larva to pupa	95+
Pupa to adult	90+

6.3 Other

Peak egg laying (days after emergence)	7
No. egg collections/cage	5-6
No. larval instars in lab	5

7. PROCEDURES FOR SUPPLYING INSECTS

7.1 Timing (for *A. ipsilon* or any of the several species on hand at ICI Americas)

7.1.1 Starter colonies: contact ICI personnel at any time (919-731-5200). Small quantities of pupae (the preferred shipping stage) are usually readily available.

7.1.2 Large quantities or several shipments: Make contact at least 6 weeks before date of first desired shipment.

7.2 Legal shipment

Ascertain requirements for introduction of the desired species into your state/country. For the US your state department of agriculture will determine the need for a shipping permit. These are available from the USDA in Hyattsville, Maryland. Allow at least one month to obtain the permit(s).

Fig. 4 Egg Collection Method for S. frugiperda, the Fall Armyworm

8. OTHER SPECIES

diet has been used successfully in our laboratory to rear Heliothis virescens, H. zea, Spodoptera exigua, S. frugiperda, Agrotis ipsilon, Trichoplusia ni and Ostrinia nubilalis continuously for several years. In addition, many field-collected strains of these species and a recently-acquired colony of Pseudoplusia includens have adapted readily to it. The primary difference in the rearing method is the manner of egg collection. Most of the above species oviposit on the underside of toweling covering the adult cages.

9. REFERENCES

Butt, B. A. and Cantu, E. 1962. Sex determination of lepidopterous pupae. USDA, ARS-33-75, 7 pp.

Rings, R. W., Arnold, F. J., Keaster, A. J., Musick, G. J. 1974. A worldwide annotated bibliography of the black cutworm. Ohio Agri. Res. and Dev. Center, Res. Circ. 198, 106 pp.

Rings, R. W., Musick, G. J., Keaster, A. J., Luckman, W. H. 1976. Geographical distribution and economic importance of the black, glassy and bronzed cutworms. USDA Coop. Plant Pest Rpt. 1:16, p. 178.

Vanderzant, E. S. 1962. Rearing the Bollworm on Artificial Diet. J. Econ. Entomol. 55:1, p. 140.

ANTICARSIA GEMMATALIS

N. C. LEPPLA

Insect Attractants, Behavior, and Basic Biology Research Laboratory, ARS-USDA, Gainesville, Florida 32604 USA.

THE INSECT

Scientific Name:	*Anticarsia gemmatalis* Hübner
Common Name:	Velvetbean caterpillar
Order:	Lepidoptera
Family:	Noctuidae

1. INTRODUCTION

The velvetbean caterpillar is a New World species that occurs throughout most of North, Central, and South America and the Caribbean. However, its distribution is limited to tropical and semitropical areas during the winter. In North America it apparently overwinters in Florida, and migrates north to Canada and Wisconsin by the end of summer (Ford et al. 1975).

The moths can be distinguished from other underwings (Catocalinae) by an oblique band of tan scales bordered by black extending from the posterior margin of the hindwing to the junction of the anterior and lateral margins of the forewing. Also, the males have prominent tufts of scales on the tibiae of the mesothoracic legs.

The females deposit single eggs on the leaves and stems of a variety of legumenous host plants. Larvae undergo 6 to 7 instars and sometimes seriously defoliate soybean, peanut, and alfalfa crops. They pupate beneath the surface of the soil. Adults emerge about a week later and begin to oviposit within 3 days.

The velvetbean caterpillar has been reared continuously for almost a decade using artificial diet (Greene et al. 1976). To rear this species successfully, it is essential to establish and maintain a colony that is free of pathogens, such as the fungus, *Nomuraea rileyi*. Adult holding cages must be relatively large and well ventilated for a noctuid. Also, for optimal levels of mating and oviposition, the moths require relative humidities in excess of 80% and a source of liquid food. The eggs are attached strongly to the oviposition substrate. Larvae are very active and feeding is disrupted when they are reared in high densities.

The procedure of King and Hartley (this volume) for the tobacco budworm also can be used for the velvetbean caterpillar with the use of a different oviposition cage. The oviposition cage is 30 x 30 x 45-cm and is constructed of 3-mm-thick plastic. The top is removable and equipped with two 3-mm-wide, 30-cm-long slots for insertion of wet strength paper towels to serve as an oviposition substrate. The back is fitted with a 2.5-cm-diam hole covered with a piece of rubber slit for connection to a scale collector. Pupae are sexed and 150 pairs are placed in each cage.

2. FACILITIES AND EQUIPMENT REQUIRED

2.1 Facilities (Leppla et al., 1978; Leppla et al., 1982)

Larvae are reared in holding rooms maintained at 27°C, 50% RH, and an LD 14:10 photoperiod (300-750 nm spectrum, 180-310 lux) with continuously recirculating air. Moths are held at 28°C and 80% RH with a 14-h photophase (same light quality as larvae plus a 0.25-W night light). An air filtration system used to control scales is composed of 2.0 x 0.7 x 1.4 m high cabinets with fiberglass furnace filters connected to a self contained dust arrestor (Arrestall® size 40, American Air Filter Co., Inc., 215 Central Ave., Louisville, KY 40277). The rooms are equipped with high and low temperature safety controls to protect the insects from temperature extremes caused by malfunctioning heating and cooling systems (Leppla and Freeman 1984).

2.2 Equipment and materials required for insect handling

2.2.1 Adult oviposition

- Holding cage, 41 x 41 x 31-cm-high plexiglass box with the bottom and hinged lid cut out and covered with hardware cloth. A large inspection port is located on the front and the inside corners are fitted with vertical wires to support the cloth oviposition substrate.
- Feeder cups, 59-ml paper soufflé cups, no. 200, Sweetheart®.
- Absorbent cotton balls, non-sterile (Surgical Supply, 305 S. W. 7th Terrace, Gainesville, FL 32604).
- Sugar (or beer).
- Honey, unprocessed.
- Deionized water.
- Organdy cloth, white, 35.6 x 76.2-cm or paper toweling of sufficient quality to withstand repeated agitation in 5% commercial bleach (5.25% sodium hypochlorite) solution for 5 min per wash.

2.2.2 Egg collection and storage

- Manual operation.
- Face mask, 8500 toxic particle mask (3M center, OH and SP Division, St. Paul, MO 55101).

2.2.3 Egg sterilization

- Portable washing machine (Rival 15-liter, Sears Wash-O-Matic®, No. 3400) modified by adding a deionized water supply port to the bottom, an internal strainer basket, and a vertical U-shaped 3-cm diam O.D. glass siphon tube to the outlet and by reducing the size of the agitator blades (Leppla et al. 1978).
- Commercial bleach.
- Egg collection net fabricated by attaching fine nylon organdy to a ringstand support ring.
- Beaker, 250 to 500-ml.
- Medicine dropper, 7.6-cm glass pipet with 2.5-cm suction cap (Bio-Quip Products, P. O. Box 61, Santa Monica. CA 90406).
- Rubber gloves.
- Sodium thiosulfate solution, laboratory grade (100 g/liter deionized water) (Fisher Scientific Co., 711 Forbes Ave., Pittsburg, PA 15219).

2.2.4 Egg hatch and larval inoculation of diet

- Paper toweling for surface-sterilized eggs.
- Casein glue.
- Drying table, 1.9 x 0.9 x 1.5-m high laminar-flow hood, Pure Aire 720B (Pure Air Corp., 8441 Canoga Ave, Canoga Park, CA 91304).

2.2.5 Larval rearing containers

- Plastic container, 30.5 x 30.5 x 12.7-cm deep I.D. Tupperware® #G-40, 6-liter.
- Plastic lid, 30.5 x 30.5-cm with 30% removed and replaced with a 28 x 28 cm sheet of 3-mm thick polypropylene 120-micron pore filter (Porex Materials Corp., P. O. Box 5481, Fairburn, GA 30213).
- Plastic insert, 28 x 28 x 1.10cm thick translucent white acrylic 1.2-cm cube louver no. TA 225 (Scientific Lighting Products, 10545 Baur Blvd., St. Louis, MO 63132).

2.2.6 Pupal collection, sexing, and storage
- Soft-nose #4750 forceps, spring steel (Bio-Quip Products, P. O. Box 61, Santa Monica, CA 90406.)
- Collander, 3-liter aluminum (Wear-ever® No. 3123).
- Beaker, 500-ml.
- Deionized water.
- Commercial bleach.
- Portable washing machine, Rival 15-liter, Sears Wash-O-Matic®, No. 3400.
- Paper toweling.
- Paper cups, 0.47-liter, unwaxed.
- Analytical balance, 0- to 30-g, No. A-30 (Mettler Instrument Corp., Hightstown, NJ 08520).
- Stereozoom microscope, 10X-70X (Bausch and Lomb, Rochester, NY 14602).

2.2.7 Insect holding
- Controlled-temperature room, 27°C, 80% RH, and L:D 14:10 photoperiod (see Facilities 2.1).

2.3 Diet preparation equipment

2.3.1 Dry mix
- Clean tunnel, high efficiency particulate air-filtration system, No. FM-48 (Pure Air Corp., 8441 Canoga Ave., Canoga Park, CA 91304).
- Freezer, double-door, 1.2 m^3, No. T1-4-AD (Foster Refrigerator Corp., Thermodynamics, Inc., Parsons, TN 39363).
- Freezer, walk-in, 16 m^3, No. CPE-72, modular aluminum box with add-on coils (Larkin Coils, Inc., Atlanta, GA 30371).
- Top loading analytical balance, 0- to 8200-g, No. PC-8200 (Mettler Instrument Corp., P. O. Box 71, Hightstown, NJ 08520).
- Weighing pans, stainless steel pie pans.
- Graduated cylinder, 100-ml.
- Spatulas.
- Magnetic stirrer, No. PC-353 (Corning Glass Works, Corning, NY 14830).
- Fume hood.
- Diet dispensing table.

2.3.2 Finished diet
- Cooker, steam-jacketed kettle, 18.9-liter, No. TDB/4-20 (Groen Division, Dover Corp., 1900 Pratt Blvd., Elk Grove, IL 60007).
- Blender, 3.8-liter, No. CB-6 (Waring Products Division, Dynamics Corp. of America, New Hartford, CT 06057).
- Rheostat, Powerstat® variable autotransformer, No. BP57515, 120-V input, 0- to 140-V output (The Superior Electric Co., Bristol, CT).
- Pitchers, 2-liter plastic
- Sponges
- Kraft paper, 91.4-cm-wide, 30-lb. test (local supply).

3. ARTIFICIAL DIET

3.1 Composition

a. Dry mix to prepare 15.2 liters finished diet.

Ingredients	Amount	% (Estimated)
Gelcarin HWG	184 g	5.4
Casein	400 g	11.8
Pinto beans, ground	1100 g	32.5
Torula yeast	500 g	14.8
Wheatgerm	800 g	23.7
Soy protein	400 g	11.8
Total	3384	100%

b. Finished diet

Dry mix	3384 g
Deionized water	10,800 ml
Vitamin mixture	120 ml
Ascorbic acid	52 g
Methyl paraben (MPH)	32 g
Sorbic acid	16 g
Formalin	60 ml
Tetracycline	4 (250 mg capsules)

c. Preparation of vitamin mixture (400 g Hoffman-LaRoche insect vitamin mix/1000 ml H_2O).

Vitamin	Amount/1000 gram (QS with dextrose)
Vitamin A	20,246,000 IU
Vitamin E	7,361.4 IU
Vitamin B_{12}	1.8 g
Riboflavin	459.0 mg
Niacinamide	921.1 mg
Calcium d-Pantothenate	921.1 mg
Choline chloride	46.0 mg
Folic acid	230.8 mg
Pyridoxine HCL	230.8 mg
Thiamine HCL	230.8 mg
d-Biotin	18.4 mg
Inositol	18.4 g

3.2 Brand names and sources of ingredients

Gelcarin HWG (Marine Colloids, Inc., 2 Edison Place, Springfield, NJ 07081); casein (National Casein Co., 601 W. 80th St., Chicago, IL 60620); pinto beans (local source, Dixie Lilly Co., Williston, FL 32696); torula yeast (St. Regis Paper Co., Lake States Division, 603 W. Davenport St., Rhinelander, WI 54501); wheatgerm (Earthwonder, 1735 E. Trafficway, Springfield, MO 65802); soy protein, Supro 610 (Raulston Purina Co., Checkerboard Square, St. Louis, MO 63188); vitamin mixture, ascorbic acid (Roche Chemical Division, Hoffman-LaRoche, Inc., Nutley, NJ 07110); methyl paraben (Tenneco Chemical, Turner Place, P. O. Box 365, Piscataway, NJ 08854); sorbic acid (American Hoechst Corp., Chemicals and Plastics Division, Rt. 202-206, Somerville, NJ 08876); formalin (Fisher Scientific Co., 711 Forbes Ave., Pittsburgh, PA 15219); tetracycline (Rugby Laboratories, Inc., Rockville Center, LI, NY 11570).

3.3 Diet preparation procedure

a. Preliminary
- Place 1-liter bottle of vitamin mixture on magnetic stirrer in clean tunnel.
- Cover diet dispensing table with kraft paper and arrange rearing containers to receive diet.
- Add deionized water to cooker.
- Weigh dietary ingredients and thoroughly mix the Gelcarin HWG, casein, pinto beans, torula yeast, wheatgerm, methyl paraben, and sorbic acid.
- Place beaters on cooker, carefully lower into position, set speed at 6 for a 15-liter batch (slower for a smaller batch) and actuate.
- Set thermostat on cooker at 6 1/2 and switch on.

b. Preparation
- Slowly sprinkle dry ingredients into cooker.
- Cook until red thermostat light goes off, ca 15 min.
- Add the formalin and cook 5 min, turn off heat for 2 min.
- Sprinkle in the ascorbic acid and stir without heat for 3 min.
- Turn off motor, remove beaters from cooker, and place in hot water to soak.

c. Dispensing

- Transfer first 4 liters of diet from cooker into a pan.
- Dispense second 4 liters into blender.
- Return first 4 liters to cooker.
- Add 30 ml vitamin mix and 1 tetracycline capsule to blender.
- Blend for 30 sec with rheostat set at 100.
- Pour 1000 g of finished diet from blender into each rearing container.
- Repeat second through sixth steps for each 4 liters of diet.
- Cool diet in rearing containers at room temperature for at least 1 h before use.

d. Clean up

- Immediately use 4 liters of water to clean the empty cooker.
- Wash all utensils and clean room.
- Return all ingredients to freezers, restock supplies, and record diet output.

4. HANDLING INSECTS AT LOWER TEMPERATURES

Rearing is continuous so no storage is required.

5. REARING AND COLONY MAINTENANCE

The velvetbean caterpillar colony was established originally in 1973 with field-collected moths from north central Florida and has been maintained continuously on artificial diet. Eggs from ca 50 wild-type females are added to this colony each year (see special problems 5.9b).

5.1 Egg collection

- Manually remove paper toweling or cloth substrate containing eggs from adult oviposition cages.

5.2 Egg sterilization

- Fill washer with 15 liters of deionized water, add 70 ml commercial bleach and mix.
- Place egg sheets into washer and wash for 5 min.
- After washing, rinse sheets repeatedly to dislodge any remaining eggs.
- Squeeze excess bleach solution from sheets and dry them.
- Drain bleach solution containing eggs through siphon into net.
- Rinse washer and add remaining eggs to net.
- Invert net and rinse eggs into 500 ml beaker with ca 300 ml deionized water.
- Neutralize with ca 150 ml sodium thiosulfate solution.
- Soak eggs in solution for 2 min and decant upper layer containing unwanted eggs, larvae, and debris floating on surface.
- Pour mixture containing usable eggs into net, rinse with deionized water for ca 30 sec.
- Place eggs into empty beaker with enough deionized water (ca 60 ml) to stir them freely.

5.3 Placement of eggs in rearing containers

- Apply eggs to 4 x 4-cm pieces of paper toweling with a medicine dropper (100 eggs per spot, 6 spots per larval rearing container).
- Attach a square of toweling with dry egg spots to the inside center of each container lid by using a few drops of casein glue.

5.4 Collection of pupae

- Transfer containers with 5- to 6-day-old pupae from development room to harvest area, remove lid, and carefully extract the insert containing most of the pupae and shake them into the colander.
- Use soft-nose forceps to pick remaining pupae from diet and frass in bottom of container.
- Fill pupal washer with 15 liters of deionized water and add 400 ml of commercial bleach.

- Place pupae into washer, wash 10 min, and drain into the ca 3-liter colander.
- Rinse pupae 2 min, spread to dry on paper toweling.
- After drying, remove all debris with forceps, count pupae, and weigh a random sample of 25 (pupal count = total weight of harvested pupae per average weight of sample pupae).

5.5 Adults

5.5.1 Emergence period

Females begin to emerge 19-20 days after larval rearing containers are established (5-6 day pupal stage). Males begin to appear 1-2 days later. The emergence period lasts about 6 days with a peak on the third for females and the fourth for males. The sex ratio is ca 1:1.

5.5.2 Collection

Pupae are placed into oviposition cages before emergence of adults (75 ♀ and 50 ♂).

5.5.3 Oviposition

Moths are maintained in an oviposition room at 28°C, 80% RH, and a 14 h photophase (see 2.1, Facilities). The preoviposition period is 2 to 3 days, peak egg production occurs after the moths are 4 to 5 days old, and they are discarded after ca 7 days.

Procedures

- Place 4 feeder cups containing cotton saturated with 5% honey and sugar solution (54 g sucrose + 35 ml unprocessed honey per liter of deionized water or beer) in the center of the cage.
- Place pupae in a plastic cup or petri dish and transfer to the cage.
- Wrap internal sides of oviposition cage with one layer of cloth substrate held in place with the vertical wires.
- After oviposition begins, remove egg sheets daily and replace with new substrate.

5.6 Sex determination

- Male pupae and adults usually are larger than females.
- Male pupae have paired elevated pads on the ventral side of the ninth abdominal segment whereas females lack these structures but have an orifice to the bursa copulatrix on the eighth (Butt and Cantu, 1962).
- Male moths have paired yellow tufts of setae that extend over several segments on the dorsal side of the thorax and also have the usual lepidopteran claspers.
- Females have an eversible ovipositor.

5.7 Rearing schedule (5-day work week)

a. Daily

- Prepare diet.
- Collect and surface-sterilize eggs.
- Place eggs on lids of larval rearing containers.
- Harvest pupae.
- Discard one old cage and set up one new oviposition cage every Tuesday and Thursday.
- Check temperature, RH, and photoperiod in adult holding and larval rearing rooms.

b. Alternate days

- Clean and sanitize used oviposition cages with 5% bleach solution.
- Replace old feeder cups with new ones in oviposition cages.

5.8 Insect quality

Records are kept on inclusion of ingredients in larval diet and use of new batches of ingredients. Rates of oviposition, egg fertility, and adult longevity are noted. Average yields and weights of pupae are determined. All rearing environments are monitored continuously.

5.9 Special problems

a. Diseases

Nomuraea rileyi fungus infection is prevented by visually screening insects used to establish and maintain the colony and by surface-sterilizing eggs. Bacteria are controlled with tetracycline and fungi are impeded by dietary "inhibitors," cleanliness, and consistent environmental control.

b. Laboratory adaptation

Each year the eggs from ca 50 wild-type and lab females are combined to establish a new colony. Efforts are made to prevent bottlenecks, particularly reductions in the proportions of ovipositing females.

c. Human health

Body parts of insects and microorganisms associated with rearing can be hazardous, so contact is avoided and protective masks and clothing are worn. Adequate ventilation is essential.

6. LIFE CYCLE DATA

Life cycle and survival data are given for development of eggs, adults and pupae at 28°C and 80% RH, and of larvae at 27°C and 50% RH.

6.1 Developmental data (Leppla et al. 1977)

Stage		Number of days		
		Min	Mean	Max
Eggs		2	3	4
Larvae	♂ and ♀	14	16	19
Pupae	♂	6	7	9
	♀	5	6	8
Adult emergence	♂	22	26	32
(from establishment)	♀	21	25	31
Total development (egg to egg, inclusive preoviposition)		24	28	34

6.2 Survival data (%)

- Egg viability ca 80.
- Neonate larva to pupa >75.
- Pupa to adult >95.

6.3 Other data

- Preoviposition period 3-4 days.
- Peak egg laying period 4-5 days after emergence.
- Mean fecundity ca 400 eggs per female.
- Mean adult longevity (discarded after ca 7 days).
- No. larval instars 5.
- Mean pupal weight 275 mg.

REFERENCES

Butt, B. A., and Cantu, E. 1962. Sex determination of lepidopterous pupae. U. S. Dept. Agric., Agric. Res. Serv. ARS-33-75. 7 pp.

Ford, B. J., Strayer, J. R., Reid, J., and Godfrey, G. L. 1975. The literature of arthropods associated with soybeans. IV. A bibliography of the velvet- bean caterpillar, Anticarsia gemmatalis Hübner (Lepidoptera: Noctuidae). Illinois Natural History Survey, Biological Notes No. 92.

Greene, G. L., Leppla, N. C., and Dickerson, W. A. 1976. Velvetbean caterpillar: A rearing procedure and artificial medium. J. Econ. Entomol. 69: 487-488.

Leppla, N. C., Carlyle, S. L., Green, C. W., and Pons, W. J. 1978. A custom insect rearing facility. In: N. C. Leppla and T. R. Ashley (eds.), Facilities for Insect Research and Production. U. S. Dept. Agric. Tech. Bull. No. 1576, 86 pp.

Leppla, N. C., Fisher, W. R., Rye, J. R., and Green, C. W. 1982. Lepidopteran mass rearing: An inside view, pp. 123-133. In: Sterile Insect Technique and Radiation in Insect Control. International Atomic Energy Agency IAEASM-255/14, Vienna, Austria.

Leppla, N. C., Ashley, T. R., Guy, R. H., and Butler, G. D. 1977. Laboratory life history of the velvetbean caterpillar. Ann. Entomol. Soc. Am. 70: 217-220.

Leppla, N. C., and Freeman, J. M. 1984. A failsafe device for controlled environment chambers. J. Econ. Entomol. (in press).

Mention of a commercial or proprietary product does not constitute an endorsement by the USDA.

ARGYROTAENIA VELUTINANA

EDWARD H. GLASS and WENDELL L. ROELOFS

Department of Entomology, New York State Agricultural Experiment Station, Cornell University, Geneva, New York 14456

THE INSECT

Scientific Name: *Argyrotaenia velutinana* (Walker)
Common Name: redbanded leafroller (RBLR)
Order: Lepidoptera
Family: Tortricidae

1. INTRODUCTION

The redbanded leafroller is native to eastern North America, occurring primarily in the Atlantic provinces of Canada and over most of the area in the United States east of the 100th meridian. It feeds on a large number of unrelated woody and herbaceous plants including common deciduous fruits, vegetables, weeds, flowers, and forest or ornamental trees and shrubs. Its primary native hosts are thought to be certain Rosaceae (Chapman and Lienk, 1971).

A. velutinana caused extensive losses in apple orchards from 1947 through 1960. Before then outbreaks were sporadic and since 1960 the use of insecticides has prevented most losses. This species also attacks plum, prune, peach, cherry, grape, raspberry, and strawberry. Early instar larvae usually feed on the undersurfaces of leaves under webbing. Later they feed as leaftiers between leaves, between leaves and fruits, and between fruits. They eat irregular shallow cavities which may extend over considerable areas and cause premature dropping of fruit. Losses on grape are by direct fruit feeding and potential contamination of the grape products.

The redbanded leafroller is multivoltine with facultative diapause in the pupal stage. Scotophases of 11 hours or more induce diapause at temperatures as high as 29°C (Glass, 1963). There are two full generations annually in northern United States and Canada. There are three in its more southerly range. Adults emerge in the spring at the time apple buds are showing 1/2 inch green tissue. The spring flight of moths lay eggs in masses on the bark, whereas the summer flight lay primarily on upper leaf surfaces. Pupation of the first generation occurs wherever the last instar larvae have been feeding. Second and third generation larvae developing in late summer are predisposed to diapause by shorter day lengths and spend 10 to 28 days longer than normal in the last (5th) instar. During this time larvae continue to feed but are restless and finally drop to the ground where they spin a hibernation shelter in rolled leaves before pupating (Glass and Chapman, 1952; Glass, 1963).

Prior to 1958 several investigators reared the redbanded leafroller during the spring and summer on apple or grape foliage. Hikichi and Wagner (1958) were the first to rear this species continuously, using Scarlet runner bean plants as food and continuous fluorescent illumination to prevent diapause. Glass and Hervey (1962) developed a technique for continuous rearing on fava bean plants in the greenhouse that has been used for maintaining a colony in our laboratory for over 20 years. Several artificial diets have been used with more or less success (Redfern, 1963; Rock et al., 1964; Roelofs and Feng, 1967; Roelofs, 1967; Karpel and Hagmann, 1968). More recently Roelofs and associates have found a pinto bean medium (Shorey and Hale, 1965) modified for the leafroller to be the most practical artificial diet. The Rock et al., 1964 diet was developed for nutritional research and is classified as a meridic casein diet in which all ingredients but one are chemically defined.

Thus there are satisfactory techniques available to rear the redbanded leafroller on growing plants, on a non-chemically defined artificial diet, and on a meridic diet. The fava bean plant method is very useful for maintaining colonies on a natural diet and producing substantial numbers of eggs and newly hatched larvae with a minimum amount of labor and expense. However, it does require greenhouse space and is less satisfactory for producing large numbers of mature larvae, pupae, and adults. We have found the pinto bean diet to be the best and least expensive available method for producing late instar larvae, pupae, and adults for laboratory experiments. The meridic diet is expensive but useful for nutritional studies.

We have evolved in our laboratory a system using the fava bean plant for routine maintenance of the redbanded colony and a daily supply of eggs and neonate larvae, and the modified pinto bean diet to produce pupae and moths for pheromone and toxicological research. These two methods are described in this chapter.

2. FACILITIES AND EQUIPMENT REQUIRED

2.1 Facilities

Larvae and pupae are reared on plants in greenhouses maintained at 21° to 24°C, and continuous fluorescent lighting at 200 f.c. or more in the cages at plant height. All other operations including rearing on artificial media are done in controlled environment rooms.

2.2 Equipment and materials required for insect handling

2.2.1 Adult oviposition

a. For routine production of eggs and neonate larvae
- -Wood box 250(W) X 150(D) X 150(H) mm open at front and rear. Front and inside surfaces covered with white cheese cloth, rear fitted with black cloth sleeve (Fig. 1a).
- -Petri dish or small cup
- -Absorbent cotton
- -Wax paper

b. For individual or separate colony matings
- -Transparent polyethylene bags, 250 X 440 mm or smaller
- -Absorbent cotton dental roll
- -Wire twist ties

2.2.2 Egg collection and storage
- -Shell vials 20 X 90 mm (Fig. 1b)
- -Cheese cloth
- -Absorbent cotton
- -Vapor tight plastic box 140 X 190 X 111 m
- -Plastic cup, small enough to fit in one end of above
- -Black cloth

2.2.3 Egg sterilization (needed only when rearing on artificial media)
- -0.26% AI sodium hypochlorite
- -Distilled water squirt bottle
- -Shell vials
- -Sterile cotton plugs
- -Absorbent paper towel

2.2.4 Larval transfer
- -Fine camel hair paint brush (size 000)

2.2.5 Larval rearing

a. On fava bean plants
- -Clay pots 75 mm
- -Galvanized pans 500 X 900 X 40 mm
- -Cages 700(W) X 1000(D) X 900(H) mm (Figs. 1c and d)
- -Fluorescent lights
- -Fava bean seed, *Vicia faba* (L.)

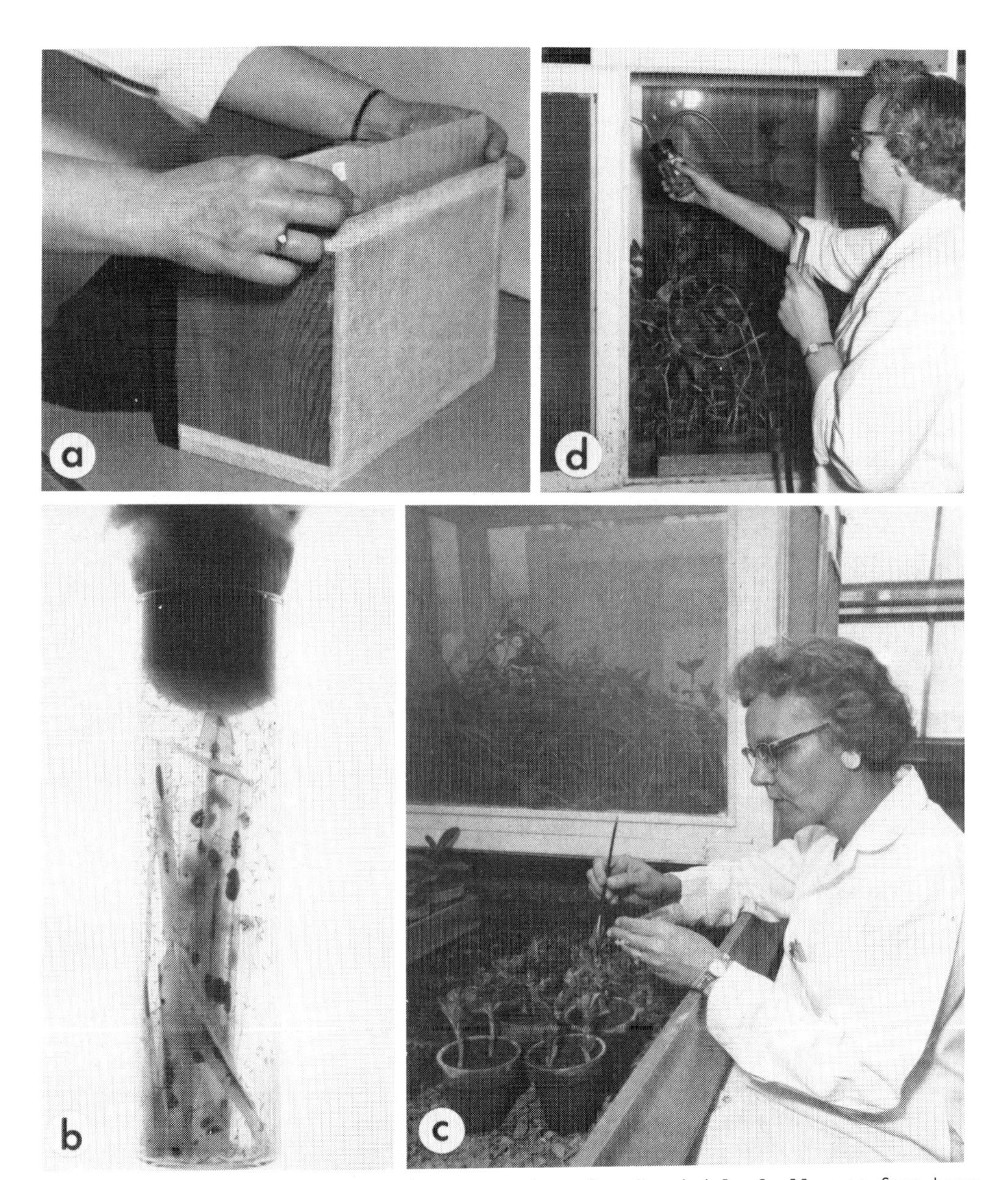

Fig. 1. Techniques for continuous rearing of redbanded leafroller on fava bean plants: a. wooden oviposition box; b. shell vial with egg papers and neonate larvae; c. transferring neonate larvae to young bean plants; d. collecting moths with an aspirator.

b. On artificial media
   - Cold drink waxed paper cup 400 ml with plastic lid
   - Media
   - Spatula

2.2.6 Pupal collection and storage equipment

a. Pupae are not collected or handled in any way in the fava bean plant rearing technique. They remain on the plants until adults emerge.

b. From artificial media
   - Petri dish
   - Paper towel
   - Plastic box with cover 140 X 190 X 110 mm
   - Plastic cup 30 ml
   - Absorbent cotton
   - Soft-nosed forceps

2.2.7 Adult collection and storage (Fig. 1d)

- Vacuum pump
- Plastic tubing
- Aspirator with wide-mouth collecting container
- Plastic box with cover 140 X 190 X 110 mm
- Plastic cup 30 ml
- Absorbent cotton
- Controlled environment rooms or cabinets

3. DIETS

3.1 Natural

3.1.1 Plant species

- *Vicia faba* (L.), fava or broad bean (Variety 96, Joseph Harris Co., Inc., Moreton Farm, Rochester, NY 14624, USA)

3.1.2 Plant maintenance

- Plant three fava bean seeds per clay pot in good greenhouse soil
- If aphids are a problem, pour 20 ml of a 1 to 2000 water mixture of demeton (2 lb./gallon formulation) onto the soil in each pot at planting time
- Maintain in greenhouse at 24° to $27^{o}C$ until plants are about 50 mm high

3.2 Artificial diet

3.2.1 Preparation equipment

- Balances
- Liquid measuring containers
- Stainless steel mixing bowl
- Hotplate
- Waring blender
- Miscellaneous spatulas and plastic spoons

3.2.2 Diet composition

Ingredients	g or ml/liter
Pinto beans	347.0 g
Brewers yeast	52.0 g
Ascorbic acid	4.9 g
Methyl p-hydroxybenzoate	3.3 g
Sorbic acid	1.6 g
Myacin	0.8 g
Propionic acid	1.2 g
Vitamin solution[a/]	40.6 ml
Formaldehyde	3.2 ml
Agar	21.1 g
Water	1000.0 ml

[a/] Composition and preparation of vitamin solution

Ingredients	g
Inositol	20.0

Calcium pantothenate	1.0
Riboflavin	0.5
Pyrodoxine HCl	0.25
Thiamine HCl	0.25
Folic acid	0.25
Biotin	0.02
B_{12} (0.1% in manitol)	1.0
Water (distilled)	1000.0

Weigh solid ingredients and add water.

3.2.3 Source of ingredients - Propionic acid (sodium salt, 99%) is from Aldrich Chemical Co., Inc., Milwaukee, WI 53233. Formaldehyde (certified ACS F-79, 37% W/W solution) is from Fisher Scientific Co., Chemical Manufacturing Division, Fair Lawn, NJ 07410. All other ingredients are obtained from Bio-Serv Inc., P. O. Box B.S., Frenchtown, NJ 08825.

3.2.4 Diet preparation procedure

- -Soak beans overnight in water, drain and rinse, cover with water, and bring to boil. Rinse beans and add 615 ml of cold water. Chop briefly (2 seconds) in blender.
- -Weigh dry ingredients, add to bean mix, and blend very briefly (1 second) twice.
- -Heat agar and 615 ml of water until becomes thickened.
- -Add agar to beans and mix well.
- -Add vitamin solution and formaldehyde to the blender, stir carefully with a spoon, and blend well.
- -Spoon into cups and smear over bottom and sides with a spatula.
- -Allow to cool and dry for 3 hours or longer until free water has evaporated from surface before adding larvae and capping.

4. INSECT HOLDING

4.1 Adults

Adults can be maintained for several days in plastic boxes provided they are supplied with a 5% sugar solution on absorbent cotton. Individual moths are best handled by placing a pupa in a 20 X 90 mm shell vial capped with aluminum foil. The foil is punctured to hold a wad of cotton which is kept moist with sugar solution after adult emergence to provide food and water. Reducing the temperature to 15°C reduces metabolic activity but does not ensure normal individuals after 2 or 3 days.

4.2 Eggs

Newly hatched egg masses can be stored for at least 7 days at 5° to 7°C, without observable ill effects, provided 100% RH is maintained in the storage container. Redbanded leafroller eggs on wax paper or plastic are very sensitive to low humidities and must be incubated at or close to 100% RH.

4.3 Larvae

Newly hatched larvae perish within a few hours unless maintained at high humidity and die within 24 hours if not provided with food. It is advisable to adjust the incubation temperatures to ensure hatching just prior to the time the larvae will be transferred to diets or used in experiments.

4.4 Pupae

These can be maintained for up to a week beyond normal eclosion by storing at 15°C. For prolonged storage, it is necessary to induce diapause by rearing the larvae on a LD 11:13 photoperiod. Diapausing pupae can be maintained for at least six months at 5°C.

5. REARING AND COLONY MAINTENANCE

Labortatory colonies can be readily established from field collected material or by obtaining eggs or pupae from an established colony. Eggs, larvae, and pupae of related species are easily confused with the redbanded leafroller, and emerging adults must be used to confirm the species (Chapman and Lienk, 1971).

5.1 Egg collection

5.1.1 From oviposition boxes

- Remove wax oviposition paper and replace with fresh creased paper daily (Fig. 1a).
- Cut out individual egg masses with scissors and place in a shell vial. Stopper vial with a cheesecloth-covered cotton wool plug (Fig. 1b).
- Incubate in plastic box in which 100% RH maintained and at 24°C or a temperature that results in hatching at a suitable time of day.

5.1.2 From polyethylene bags

- Open bag and transfer moths to a fresh bag if old moths are to be saved.
- Cut out egg masses and place in vials as above.
- Incubate as above.

5.2 Egg sterilization

Eggs that will be used to produce larvae to be reared on artificial media should be sterilized.

- Before cutting out egg masses, spray egg paper or opened plastic bag with 0.26% AI sodium hypochlorite.
- Rinse immediately with sterilized water.
- Gently dry with absorbent paper towel.
- Cut out egg masses and place in sterile shell vials.

PRECAUTION

This method is satisfactory for use with the pinto bean medium but not for axenic rearing (Rock et al., 1964).

5.3 Transfer of neonate larvae to diets

Newly hatched larvae are positively phototropic. By working near a single light source, it is possible to keep changing the orientation of the vials containing egg masses and larvae so that only a few larvae are near the open end where they are picked up with a camel hair paint brush. When touched, larvae wiggle rapidly and spin a web. Often it is possible to pick up a larva by its web.

5.3.1 To fava bean plants

- Using the fine brush, transfer 5 larvae to the plants in each pot (Fig. 1c).
- Place pots in pans in rearing cages in a greenhouse and rear at 24° to 27°C.

5.3.2 To artificial diet in paper cups

- Using the fine brush transfer 30 larvae to the pinto bean medium in each cup. Be certain there is no free moisture on diet surface.
- Place the lid on the cup.
- Place cups upside-down in the rearing chamber maintained at an appropriate temperature such as 27°C and a LD 16:8 photoperiod.

5.4 Pupal collection

5.4.1 From fava bean plants

- Do not collect or handle pupae from plants except for special needs. (It is easier and cheaper to rear on media for pupal collections.)

5.4.2 From artificial diets in paper cups

- Remove pupae using forceps after carefully dislodging them from the webbing.
- Place in uncovered petri dishes.
- Place petri dishes in plastic emergence boxes with paper towels on the bottom and 30 ml cups containing cotton wick kept moist with 5% sugar solution.

5.5 Adult collection

5.5.1 From fava bean rearing cages

- Collect moths daily using an aspirator (Fig. 1d).
- Place the moths in oviposition boxes.

5.5.2 From plastic emergence boxes

-Collect moths individually or in groups as may be required for their use in experiments or in rearing.

5.6 Oviposition

5.6.1 Oviposition boxes

-Roll strips of wax paper 175 mm (H) X 250 mm (W) on a pencil, remove the pencil and flatten the papers to make creases.
-Insert a paper into each box through the slot in the top of the box so it covers the cheesecloth on the inside (Fig. 1a).
-Remove and replace the wax papers daily.
-Maintain in each box an open petri dish containing absorbent cotton kept moistened with 5% sugar solution.
-Add the daily collections of moths to a box for a week.
-Start a new oviposition box each week and maintain old boxes until oviposition drops to a useless level.
-Hold boxes in cabinets or rooms maintained at 24° to 27°C with LD 16:8 photoperiods.

5.6.2 Plastic oviposition bags

-Place a 30 mm length of absorbent dental roll soaked in 5% sugar solution in each plastic bag.
-Inflate the bag and add moths from collecting bottles or vials.
-Tie bags shut with wire twist ties.
-Hold bags in cabinets or rooms maintained at 24° to 27°C with LD 16:8 photoperiods.

5.7 Sex determination of pupae and moths

Pupae - Females are larger than males and have three ventral abdominal segments, whereas males have four. Moths - Females are larger, with broader abdomens than males and have a large ovipository pore.

5.8 Rearing schedule

5.8.1 Natural diet rearing

a. Daily
-Collect newly emerged moths and place in oviposition box.
-Check soil moisture in pots and add water to pan only as needed.
-Moisten absorbent cotton in oviposition boxes.
-Remove wax egg papers, cut out egg masses, and place in vials in incubation boxes.
-Remove vials with newly hatched larvae (Fig. 1b) from incubation boxes for use in research or for setting up new rearing cages or cups.

b. Weekly
-Plant about 40 pots of fava bean for each rearing cage to be set up later.
-Add 5 newly hatched larvae to the plants in each of 32 pots and place in the pan in a clear rearing cage. Plants should be 50 to 100 mm high (Fig. 1c).
-Remove all pots, plant material, and debris from exhausted rearing cages and brush the inside of cage to remove any insect remains. It is desirable to have extra cages so there can be one week between clearing and reuse.

5.8.2 Pinto bean diet rearing

a. Daily
-Remove pupal dishes from emergence boxes.
-Set up new emergence boxes.
-Prepare media and place in rearing cups.
-Transfer newly hatched larvae to rearing cups.

b. Alternate days
-Collect, sex, and place pupae in an open petri dish emergence box.

c. Twice weekly
-Spot-check rearing cups for mold.
-Two weeks after setting up rearing cups, spot-check for pupae.

5.9 Insect quality

The quality of the insects should be monitored occasionally by checking pupal weights, adult fecundity, and egg viability. See the table in Section 6 for expected values. Maintaining our stock colony on fava bean plants helps to ensure "normal" insects.

5.10 Special problems

5.10.1 Diseases

A granulosis virus (Glass, 1958) can become a serious problem in laboratory colonies reared under crowded conditions. By sterilizing all reusable containers, by using new cups and bags, and by using eggs from the disease-free fava bean colony for artificial media rearing, we have eliminated the granulosis problem.

5.10.2 Microbial control in the artificial diets

Bacterial and mold growth is minimized by the procedures described; however, the diet should not be exposed longer than necessary and all operations should be done under hygienic conditions.

6. LIFE CYCLE IN THE LABORATORY AND GREENHOUSE

Development and quality data for *A. velutinana* reared on fava bean plants in the greenhouse maintained at 24° to 27°C. Values for individuals reared on the artificial diet do not differ significantly.

Average number eggs per mass	45
Incubation period (days)	6
Percent hatch	85
Minimum length of larval stage (days)	18
Minimum length of pupal stage (days)	7
Minimum length from egg deposition to moth (days)	32
Days from first to last moth	16
Adult longevity (days)	15
Preoviposition period (days)	2
Percent survival of neonate larvae placed on plants to moths	80
Average weight of male pupae (mg)	16.4
Average weight of female pupae (mg)	23.4
Average length of male moths (mm)	5.9
Average length of female moths (mm)	6.1

7. RELATED TORTRICID SPECIES REARED ON FAVA BEAN PLANTS AND PINTO BEAN MEDIUM

We have successfully reared several related Tortricidae on fava bean plants and the modified pinto bean medium using essentially the same oviposition and incubation techniques. These species are:

Pandemis limitata (Robinson)	threelined leafroller
Sparganothis sulfureana (Clemens)	Sparganothis fruitworm
Platynota flavedana Clemens	(no common name)
Platynota idaeusalis (Walker)	tufted apple bud moth
Platynota stultana Walsingham	(no common name)
Argyrotaenia citrana (Fernald)	(no common name)
Choristoneura rosaceana (Harris)	obliquebanded leafroller

Argyrotaenia citrana did well on the pinto diet but developed poorly on fava bean plants.

8. REFERENCES

Chapman, P. J. and Lienk, S. E., 1971. Tortricid fauna of apple in New York. Special Public. NY State Agr. Expt. Sta., 122 pp.

Glass, E. H., 1958. Laboratory and field tests with the granulosis of the red-banded leaf roller. J. Econ. Ent. 51: 454-457.

Glass, E. H., 1963. A pre-diapause arrested development period in the red-banded leaf roller, *Argyrotaenia velutinana*. J. Econ. Ent. 56: 634-635.

Glass, E. H., and Chapman, P. J., 1952. The red-banded leaf roller and its control. NY State Agric. Exp. Sta. Bull. 755. 42 pp.

Glass, E. H., and Hervey, G. E. R., 1962. Continuous rearing of the red-banded leaf roller, *Argyrotaenia velutinana*. Econ. Ent. 55: 336-340.

Hikichi, A., and Wagner, H., 1958. A technique for rearing the red-banded leaf roller, *Argyrotaenia velutinana* (Wlkr.) (Lepidoptera: Tortricidae), during the winter. Canadian Ent. 90: 732.

Karpel, M. A., and Hagmann, L. E., 1968. Medium and techniques for mass rearing the red-banded leaf roller. J. Econ. Ent. 61: 1452-1454.

Redfern, R. E., 1963. Concentrate media for rearing red-banded leaf roller. J. Econ. Ent. 56: 240-241.

Rock, G. C., Glass, E. H., and Patton, R. L., 1964. Axenic rearing of the red-banded leaf roller, *Argyrotaenia velutinana*, on a meridic diet. Ann. Ent. Soc. Amer. 57: 617-621.

Roelofs, W. L., 1967. Agarless medium for mass rearing the red-banded leaf roller. J. Econ. Ent. 1477-1478.

Roelofs, W. L., and Feng, K. G., 1967. Isolation and bioassay of the sex pheromone of the red-banded leaf roller, *Argyrotaenia velutinana* (Lepidoptera: Tortricidae). Ann. Ent. Soc. Amer. 60: 1199-1203.

Shorey, H. H., and Hale, R. L., 1965. Mass-rearing of the larvae of nine Noctuid species on a simplified medium. J. Econ. Ent. 58: 522-524.

CADRA CAUTELLA

ROBERT DAVIS and ROY E. BRY

Stored-Product Insects Research and Development Laboratory, ARS/USDA,
P. O. Box 22909, Savannah, GA 31403

THE INSECT

Scientific Name: Cadra cautella (Walker)
Common Name: Almond moth or Tropical warehouse moth
Order: Lepidoptera
Family: Pyralidae, Phycitinae or Phycitidae

1. INTRODUCTION

The almond moth is of tropical origin, but today is a common cosmopolitan pest in most of the temperate world as well as in colder areas where it is often maintained by the heated environment of warehouses and other stores. The almond moth is one of the most common insect pests of stored products. It attacks dried fruits and nuts, grain and their processed products, and oilseeds and their processed products. In the colder area, the almond moth will overwinter as larvae; but in warmer areas, it will remain quite active throughout the year, feeding and breeding continually; the life cycle is an average of 30 to 40 days with six or more generations per year.

The adult is a dull, greyish-brown moth with a wing spread of 14-20 mm. The forewings are narrow, especially at the base, greyish to yellowish with dark markings which may appear as obscure bands across the wings. The hind wings are a dirty whitish color. When at rest, the wings are folded closely together along the center of the body with the moth ranging in length from about 8 to 14 mm. The larva is whitish and when fully grown is about 12-15 mm long. It may be distinguished from the Indianmeal moth larva, often found in association, by the rows of black dots on each side of its body giving it a striated appearance. The larva spins a silken dragline which in heavy infestations will appear as masses of webbing intermingled with food and excrement.

In our laboratory, the almond moth has been maintained for more than 20 years. The laboratory reared insects are used for post-harvest entomological research, primarily for studies involving ecology, genetics, radiation biology, insecticide evaluation, insecticide resistance, pheromone evaluation and maintenance of and evaluations of parasite and predators. A laboratory colony can be started from either field collected individuals or from an established laboratory colony. The latter is recommended as wild strains often are infested with disease and also will not establish readily on standardized laboratory diets. This diet works equally well for the following phycitid moths, Anagasta kuehniella (Zeller), the Mediterranean flour moth, Cadra figulilella (Gregson), the raisin moth, Ephestia elutella (Hubner), the tobacco moth, and Plodia interpunctella (Hubner), the Indianmeal moth. A recent and comprehensive bibliography of the almond moth has been compiled by Abreu et al. (1982).

2. FACILITIES AND EQUIPMENT REQUIRED

2.1 Facilities

Rearing of larvae and pupae is carried out in controlled environment rooms maintained at 27+1°C and 60+5% RH. Lighting is with fluorescent lights (no less than 30-40 lux) with a 12:12 h L:D cycle. While RH of the room is important, actual rearings are accomplished inside gallon (3.785 liter) jars where the media and the insect activity maintain a much higher RH. Oviposition is in gallon (3.785 liter) jars at ambient room temperature and RH.

2.2 Equipment and materials required for insect handling

2.2.1 Adult oviposition

- 1-gal (3.785 liter) wide-mouth jars with 40-mesh screen lids
- Black construction paper (21.6 x 27.9 cm)

2.2.2 Rearing containers

- 1-gal (3.785 liter) wide-mouth jars with metal lids fitted with 8 cm holes and lined with a No. 1 filter paper disc 12.5-cm in diameter

2.3 Diet preparation equipment

- Hobart D-300(R) Mixer w/"B" beater, and a 30 qt mixing bowl w/splash cover
- Three stainless steel liter measuring containers

3. DIET

The Savannah Laboratory has used a modification of Haydak's mixture (Haydak, 1936) which has been widely used to rear many stored-product moths.

3.1 Composition

Ingredients	Liters	%
Cornmeal	4	27.6
Whole wheat flour	4	27.6
Rodent laboratory chow (ground)	2	13.8
Dried yeast (Brewer's)	1	7.0
Honey	1	7.0
Glycerine	1	7.0
Wheat germ	1/2	3.4
Oatmeal (rolled)	1	7.0

3.2 Brand names and sources of ingredients

- Stivers' Best Plain White Cornmeal, Southeastern Mills, Inc., Rome, GA 30161
- Stivers' Best Whole Wheat Flour, Southeastern Mills, Inc., Rome, GA 30161
- Quaker Old Fashioned Rolled Oats, The Quaker Oats Company, P. O. Box 484, Elizabeth, NJ 07207
- Rodent Laboratory Chow, Ralston Purina Company, Checkerboard Square, St. Louis, MO 63188
- Vita-Food Red Label Brewer's Yeast, Vitamin Food Co., Inc., Newark, NJ 07104
- Wheat Germ, Regular, Toasted, International Multifoods, Inc., Minneapolis, MN 55440
- Honey, Sioux Honey Association, Waycross, GA 31501
- Glycerine, Ashland Chemical Co., Savannah, GA 31401

3.3 Diet preparation

- This recipe will make enough medium for 20-22 cultures.
- Ingredients should be refrigerated (5°C) until needed to prevent infestation with insects and mites.
- Grind required amount of rolled oats and rodent laboratory chow together until they will pass through a U.S. Standard No. 8 sieve.

- Mix ingredients together in a large blender until thoroughly mixed. (This laboratory uses a Hobart D-300 mixer with a "B" beater).
- Mix fresh medium as needed. Place medium loosely in jars (do not pack, but avoid large air pockets) immediately after preparation to avoid having to break up the clumps that form as the medium hardens.

4. REARING AND COLONY MAINTENANCE

4.1 Finished diet

- Place approximately 400-500 g of diet loosely in each 1-gal (3.785 liter) rearing chamber. The rearing of *Cadra figulilella* will be more successful with about 40-50 g of raisins sprinkled on top of the medium.

4.2 Cultures started with adults

- Cultures may be started with adults by anesthetizing with CO_2 1 to 3-day old adults from several cultures to minimize inbreeding and randomly placing 20-25 randomly selected individuals (1:1 male:female ratio desirable) in each rearing container. Adults live ca. a week and it is not necessary to remove them from the cultures.
- Adults will start emerging in 3-4 weeks. This procedure works well to maintain the colonies or to supply test insects if large numbers are not needed.

4.3 Cultures started with eggs

- Clean gallon jars are set up as oviposition chambers. Each jar contains a 21.6 x 27.9-cm piece of construction paper folded (pleated) several times to provide surface area for resting. The top of this chamber is closed with a 40-mesh screen.
- Select 1 to 3-day old adults from several cultures to minimize in breeding and transfer all the adults to the oviposition chambers.
- The eggs will pass through the 40-mesh screen; therefore, for ease of egg collection, the chamber may be inverted over a clean container.
- Approximately 0.1 ml of eggs will equal 1000 eggs which is sufficient to seed one 1-gal rearing chamber. To facilitate seeding an appropriate volumetric scoop is recommended.
- If percent hatch date is desired, eggs can be placed in a planchet on the surface of the media (Boles and Marzke 1966).
- Yields will differ but at least 70% adult progeny should be expected from a given number of eggs.

PRECAUTIONS

Cultures should be examined daily. On the first appearance of diseases or contaminants such as mites, fungi or other insects the culture should be destroyed and the rearing chamber thoroughly cleaned and/or sterilized.

5. LIFE CYCLE DATA

At 27±1°C, 60±5% RH and a 12:12 h L:D cycle, the normal time for a generation of almond moth or Indianmeal moth is ca. 3 to 4 wk; for the raisin moth or the tobacco moth, ca 5 to 6 wk; and ca 6 to 7 wk for the Mediterranean flour moth.

6. PROCEDURES FOR SUPPLYING INSECTS

All life stages may be shipped in the rearing container which is prepared for shipping by placing a 40-mesh screen lid over the filter paper disc in the retaining ring. The culture jars are placed in corrugated cardboard boxes and are cushioned with a suitable packing material such as expanded polyethylene, polystyrene, strofoam or shredded newspaper to prevent breakage.

7. REFERENCES

Abreu, J. M. de, Williams, R. N., and Rude, P. A. 1982. Revised bibliography of the almond moth (Tropical warehouse moth) *Ephestia cautella* (Walker) (Lepidoptera: Phycitidae). *Trop. Stored Prod. Info.* 44: 15-36.

Boles, H. P. and Marzke, F. O. 1966. Lepidoptera infesting stored-products. pp. 259-270. *In* Insect colonization and mass production. [ed.] C. N. Smith. Academic Press, New York and London.

Haydak, M. H. 1936. A food for rearing laboratory insects. J. Econ. Entomol. 29, 1026.

CHILO PARTELLUS

PRAKASH SARUP, K. H. SIDDIQUI, and K. K. MARWAHA

Division of Entomology, Indian Agricultural Research Institute,
New Delhi-110012, India

THE INSECT

Scientific Name: *Chilo partellus* (Swinhoe)
Common Name: Maize stalk borer
Order: Lepidoptera
Family: Pyralidae

1. INTRODUCTION

The maize stalk borer (MSB), *Chilo partellus* (Swinhoe), abounds in nearly all the maize and sorghum growing tracts of the Indian Union and Pakistan. Its distribution extends from Afghanistan to Indonesia and Taiwan in the east and Sri Lanka in the south. Also, the MSB has been reported from East Africa. In India, it occurs as a major pest of maize and sorghum and is known to infest other cultivated plants like sugarcane, millets and paddy (Pradhan, 1969). Other host plants comprise Sudan grass, *Sorghum vulgare* var. *sudanensis* (Piper) Kitche; sarkanda, *Saccharum bengalense* Retz. (=*S. munja* Roxb.); and baru grass, *Sorghum halepense* (Linn.). However, sorghum is the preferred host.

The freshly hatched larvae scrape the epidermis and wander from 15 min to 8 h on leaf surface before entering the plant whorl. The larvae feed through the central whorl which on unfolding presents a severely fenestrated condition. A series of pinholes in horizontal plane appear on the leaf. Besides this characteristic symptom, the leaves reveal the presence of shot holes arranged in a row and long streaks and slits due to extensive damage caused by late larval stages. Even the mid-rib of leaf is often bored. In younger plants, the larvae attack the growing point and base of the central whorl, leading to the formation of typical "dead-heart". Hence, the severity of damage is more pronounced in earlier stages of the crop. Extensive larval feeding on pith of stem results in tunnelling. Also, tassel stalk, tassel and cob with unripe grains are damaged. Thus, every part of the maize plant except the roots are prone to larval attack. The estimated loss primarily due to this pest varied from 26.7 to 80.4% in different agroclimatic zones of India (Sarup 1980).

In northern India, there are 6-7 generations in a year. The MSB activity commences from early March and continues up to October. Thereafter, the larva enters hibernation and thus passes the winter season. In southern India, this unusual prolonging of the larval stage does not occur. The larval period lasts from 18-23 days. The pupation takes place inside the stem and pupal period ranges from 7-9 days. Adults are short lived, the life span being 2-5 days. Male and female moths copulate 1 day after emergence. Males mate only once in life. The creamy, yellowish white eggs are usually laid on the underside of the leaf tips. They turn black-head stage in 4-5 days at ambient temperature.

In India, MSB has been reared on several artificial diets (Pant *et al.* 1960, Chatterji *et al.* 1968, Dang *et al.* 1970, Lakshminarayana and Soto 1971, Siddiqui and Chatterji 1972, Siddiqui *et al.* 1977, Sharma and Sarup 1978, Seshu Reddy and Davies 1978, and Sarup *et al.* 1983). At the Indian Agricultural Research Institute, New Delhi, the MSB is reared on artificial diets based on (i) mixture of green gram, *Vigna radiata* (L.) Wilczek and dew gram, *Vigna* aconitifolia (Jacq.) Marichat, or (ii) mixture of dew gram and cowpea, Vigna unguiculata (L.) Walp. These 2 diets have been in use at the various maize research stations throughout the Indian Union for its mass rearing. The laboratory-reared MSB provide the base for insect-host-plant studies, particularly in the identification of sources of resistance among a large number of world maize germ plasms. Also, a regular supply of egg masses or larvae facilitates determination of economic threshold injury level and development of sound pest management. The various facets of insect toxicological and biological control investigations become feasible with the availability of insects in abundance.

2. FACILITIES AND EQUIPMENT

2.1 Facilities

Rearing of larvae of MSB is carried out in controlled rooms maintained at 27°C ± 2°C and 70 to 90% RH. These rooms are fitted with 3 m x 3 m racks made of slotted iron angles and wooden planks for keeping the glass rearing jars.

2.2 Equipment and other materials used

2.2.1 Adult oviposition

- Oviposition glass jars (15 cm x 10 cm)
- White tissue paper to cover the inner surface, bottom as well as the top of the jar
- Adhesive to paste the paper used to cover the inner surface
- River sand at the bottom (2 cm) of the jar (that was passed through a 40-mesh sieve and sterilized)
- Water to moisten the sand for adequate humidity

2.2.2 Egg collection

- Cut white tissue paper containing eggs and keep for incubation at 27°C ± 2°C
- Scissors for cutting the portions of tissue papers having egg masses
- Petri dishes (15 cm, 10 cm, 5 cm diam) for keeping egg masses
- Moisten absorbent cotton for maintaining required humidity for the proper development of eggs
- Forceps
- Camel-hair brush

2.2.3 Larval rearing jars

- Glass jars (15 cm x 10 cm); each rearing jar contains 170 g of diet (up to 2 cm height) - sufficient for about 40-45 developing larvae
- Coarse muslin cloth to cover the jar
- Rubber bands
- Black paper

2.2.4 Holding for eggs and pupae

- Incubators at 15°C

3. DIET

3.1 Natural

a. Sorghum plant - susceptible varieties, Swarna, CSH 1
b. Maize plant - susceptible varieties, Basi local and Vijay composite
c. Collect from the field tender pieces of sorghum/maize stem (apical portion of plants) for feeding of early instar larvae of MSB. For later instars, provide stem portions (9-12 cm long) as food.

3.2 Artificial

3.2.1 Equipment and other material

- Waring® blenders with 500-ml, 1-liter and 5-liter capacity stainless steel jars
- Hot plate
- Hot air oven (50°C to 250°C)
- Round stainless steel bowl (23 cm diam x 12 cm high)
- Stainless steel spatula
- Weighing balance 0.1 g to 100 mg

- Refrigerator
- Humidifier
- Hygro-thermograph
- Two room air conditioners (1 to 1 1/2 ton capacity)
- Thermometer
- 1, 10, 100 and 1000-ml capacity measuring cylinders
- Coarse muslin cloth
- Rubber bands
- Black paper
- Surgical gloves

3.2.2 Composition (in g or ml per 500 g diet)

Fraction A	Quantity	Percentage
Green gram:dew gram (1:1)	75.00 g	15.00
or	or	
dew gram:cowpea (1:1)	75.00 g	
Yeast powder	5.00 g	1.00
Wheat flour	20.00 g	4.00
Ascorbic acid	1.70 g	0.34
Sorbic acid	0.40 g	0.08
Methyl-p-hydroxybenzoate (methyl paraben)	0.80 g	0.16
Vitamin E (2 capsules of viteolin 100 mg each)	0.20 g	0.04
Formalin 40%	1.00 ml	0.20
Distilled water	260.00 ml	52.00
Fraction B		
Agar	6.00 g	1.20
Distilled water	130.00 ml	26.00
	500.10	100.02

Brand names and sources of ingredients:
Agar [M/s Cellulose Product of India Ltd., Ahmedabad/British Drug House/Glaxo Laboratories, Bombay (India)]; yeast powder (local pack); ascorbic acid [M/s Sarabhai M. Chemicals, Baroda (India)]; sorbic acid and methyl paraben (M/s E. Merck, Darmstadt, Germany); vitamin E - viteolin capsule [M/s Allenbury's Pharmaceutical Division and Division of Glaxo Laboratories Ltd., Bombay (India)]; formalin [M/s Indian Drugs and Pharmaceuticals Ltd., Hyderabad (India)]; pulses and wheat (local purchase)

3.3 Diet preparation procedure

All ingredients of fraction A are accurately weighed and transferred into the stainless steel jar (500 ml capacity) of a Waring blender and blended in 26 ml of distilled water for 3-4 min. The weighed quantity of agar of fraction B is then taken in a beaker with 130 ml of distilled water and heated to boiling for 5-6 min. The hot agar solution is added to the contents of the jar and blended for 2-3 min. The diet mixture is poured into glass jars (15 cm x 10 cm) to a height of 2 cm. The jars are covered with sterilized paper and kept at ambient temperature for about 12-24 h (depending upon the atmospheric conditions) for setting of the medium.

PRECAUTIONS

(i) Twelve to 24 h after the preparation of diet, the jars should be kept in a controlled temperature and humidity culture room to ensure its conditioning at rearing temperature.

(ii) All the glass wares used in the preparation of artificial diet must be thoroughly washed with water and then sterilized by keeping them in a hot-air oven at 120°C for about 30 min.

(iii) Coarse muslin cloth and black paper required to cover the jars should be heat sterilized.

4. INSECT HOLDING AT DIFFERENT TEMPERATURES

4.1 Adults

Moths can be stored individually up to 1 or 2 days at ambient temperature but no more as fecundity may drop.

4.2 Eggs

These can also be held at 15°C. For good results, store only brown stage eggs (2- to 3-days old) in petri dishes (15, 10, or 5 cm diam) or specimen tubes (7 cm x 2.5 cm) with some lightly moistened cotton wool to prevent the eggs from drying out. The black-head stage eggs can also be stored for about 24 h. Some lightly moistened cotton must be kept in the chamber.

PRECAUTIONS

(i) The containers must be checked daily for excess moisture.

(ii) Remove excess moisture as too much will encourage mould development.

4.3 Larvae

Freshly hatched larvae remain active for 10 h at 27°C ± 2°C. When released on artificial diet, they migrate and settle at the bottom.

PRECAUTIONS

(i) The freshly hatched larvae need careful handling with a fine camel-hair brush (No. 0).

(ii) Inoculation of larvae may encourage bacterial and fungal development due to injury in handling.

4.4 Pupae

The pupae can be stored at 15°C in glass or plastic containers, preferably petri dishes, for about a week. Slightly moistened cotton wool is to be kept in the container to prevent desiccation of pupae.

PRECAUTIONS

(i) Excess moisture should be removed to prevent mould development.

(ii) Storage for more than 7 days results in high pupal mortality.

5. REARING AND COLONY MAINTENANCE

The initial colony of MSB was established from moths obtained during March from the hibernating larvae resting in the stalks and stubbles for over 4 months. The moths were allowed to lay eggs on tissue papers hung inside field cages and provided the base for original culture.

5.1 Adult collection and oviposition

- Collect newly emerged moths daily from the larval rearing jars
- Keep about 12 pairs of moths in oviposition chambers for egg laying
- Prepare fresh oviposition chambers daily; line the glass jar with tissue paper (sides and bottom), cover the bottom with 2 cm of sand; keep paper in place with adhesive
- Keep egg laying chambers at 21°C ± 1°C
- Examine egg laying cages 2 days after caging
- Cut portions of white tissue papers containing eggs and keep for incubation at 27°C ± 2°C
- Transfer the moths which have already laid eggs into fresh oviposition jars for maximum egg laying
- Continue the process till the moths die

5.2 Egg collection

- Transfer moths from larval rearing jars to oviposition jars
- Remove tissue paper containing egg masses from the oviposition jars
- With scissors, cut pieces of tissue papers containing egg masses
- Place moist cotton wool in the container having tissue papers with egg masses to prevent desiccation

- Keep black-head egg mass in specimen tubes and tightly covered with moistened cotton wool to prevent desiccation and also larval escape, if eggs hatch

5.3 Release of eggs or larvae into the diet

- Condition the jars containing freshly prepared diet at ambient temperature for about 12-24 h so that excess moisture adhering to side-walls of the jars is evaporated before releasing the eggs or larvae into the diet.
- Release about 50 eggs (black-head stage) or neonate larvae per jar
- Cover the top of the jar with sterilized coarse muslin cloth. Cover the side-walls as well as the top with black paper leaving the lower portion of jar uncovered for the early and better establishment of the larvae in the diet.

PRECAUTIONS

(i) Remove the tissue paper containing egg shells from the jars 2-3 days after diet inoculation with egg masses to avoid fungal and bacterial growth in the diet jars.
(ii) Sterilize camel-hair brush by dipping in 70% alcohol and dry before transfer of neonate larvae.
(iii) Sterilize coarse muslin cloth and the black papers.

5.4 Pupation

Pupation and emergence of moths takes place inside the diet jar. A few larvae pupate on the coarse muslin cloth covering the jar.

5.5 Adults

The sex-ratio is approximately 1:1

5.6 Insect quality

The intensity of damage caused by the larvae of the MSB reared either 1 or 2 generations on artificial diets should be comparable with those obtained from natural food comprising maize plants. Depending upon the purpose, biological parameters like adult longevity, fecundity, egg viability and pupal weight should be comparable with natural populations.

PRECAUTIONS

(i) The culture room must be disinfected with phenyl or formaldehyde. Sterile conditions prevent bacterial infection of the diet. The diet should not be exposed to air for long periods.
(ii) All the glasswares, papers, and coarse muslin cloth should be sterilized in hot air ovens.
(iii) After the release of eggs or larvae, the diet jars must be checked at regular intervals for bacterial and fungal growth.
(iv) Keep plant material away from the culture room to prevent mite infestation.

6. LIFE CYCLE

6.1 Developmental data

Life cycle and suvival data during development of various stages under optimum rearing conditions Of 27°C $\pm$ 2°C and 70-90% RH are given in Table 1.

Table 1. Development data

Developmental stage	Artificial diet			Natural food		
	Min	Max	Mean	Min	Max	Mean
Developmental period from larva to adult (days)	29.0	30.0	29.5	24.0	26.0	25.0
Weight of female pupa (mg)	127.5	172.2	149.9	49.0	113.3	81.2
Weight of male pupa (mg)	64.0	87.0	75.5	26.3	64.9	45.6
Percentage of insects completing development from larva to adult	62.0	67.0	64.5	39.0	43.0	41.0
Sex ratio (male as unity)	1:1.15	1:2.78	1:1.97	1:0.97	1:1.08	1:1.03
Longevity of male moth (days)	5.3	5.9	5.6	4.2	4.9	4.6
Longevity of female moth (days)	4.0	5.2	4.6	4.9	5.0	5.0
Number of eggs per female which laid viable eggs	278.0	293.0	286.0	257.0	341.0	299.0
Incubation period (days)	4.2	5.3	4.8	4.3	5.0	4.7
Percentage viability	66.0	79.0	73.0	72.0	91.0	82.0

6.2 Other developmental data

Average pre-oviposition period is 2 days (at 21°C ± 1°C) after emergence. Average peak oviposition period is 2-3 days after emergence.

REFERENCES

Chatterji, S. M., Siddiqui, K. H., Panwar, V. P. S., Sharma, G. C. and Young, W. R., 1968. Rearing of the maize stem borer, Chilo zonellus Swinhoe, on artificial diets. Indian J. Ent., 30(1): 8-12.

Dang, K., Anand, M. and Jotwani, M. G., 1970. A simple improved diet for mass rearing of sorghum stem borer, Chilo zonellus Swinhoe. Indian J. Ent., 32(2): 130-133.

Lakshminarayana, K. and Soto, P. E., 1971. A technique for mass rearing of sorghum stem borer, Chilo zonellus (=partellus). Sorghum Newsl., 14: 41-42.

Pant, N. C., Gupta, P. and Nayar, J. K., 1960. Physiological studies on Chilo zonellus Swinhoe, a pest of maize crop. I. Growth on artificial diet. Proc. Natn. Inst. Sci. India, (B) 26: 379-383.

Pradhan, S., 1969. Insect pests of crops. National Book Trust, India, New Delhi, 208 p.

Sarup, P., 1980. Significant results of investigations carried out under maize improvement project in recent years. In: Joginder Singh (Editor), Breeding, Production and Protection Methodologies of Maize in India. Indian Agricultural Research Institute, New Delhi, pp. 190-192.

Sarup, P., Siddiqui, K. H., and Marwaha, K. K., 1983. Compounding artificial diets for mass rearing of the maize stalk borer, Chilo partellus (Swinhoe). J. Ent. Res. 7(1): 68-74.

Seshu Reddy, K. V. and Davies, J. C., 1978. A new medium for mass rearing of the sorghum stem borer, Chilo partellus (Swinhoe) (Lepidoptera: Pyralidae) and its use in resistance screening. Indian J. Pl. Prot., 6(1): 48-55.

Sharma, V. K. and Sarup, P., 1978. Formulation of suitable artificial diets for rearing the maize stalk borer, *Chilo partellus* (Swinhoe) in the laboratory. J. Ent. Res., 2(1): 43-48.
Siddiqui, K. H. and Chatterji, S. M., 1972. Laboratory rearing of the maize stem borer, *Chilo zonellus* Swinhoe (Crambidae: Lepidoptera) on a semi-synthetic diet using indigenous ingredients. Indian J. Ent., 34(2): 183-185.
Siddiqui, K. H., Sarup, P., Panwar, V. P. S. and Marwaha, K. K., 1977. Evolution of base-ingredients to formulate artificial diets for the mass rearing of *Chilo partellus* (Swinhoe). J. Ent. Res., 1(2): 117-131.

CHILO SUPPRESSALIS

SEIYA KAMANO and YASUO SATO

Entomology Division, National Institute of Agro-Environmental Sciences, Tsukuba, Ibaraki 305, Japan
Research Laboratories, Agricultural Chemicals Division, Takeda Chemical Industries, Ltd., Kyoto 606, Japan

THE INSECT

Scientific Name: Chilo suppressalis (Walker)
Common Name: Rice stem borer, asiatic rice borer, striped rice borer
Order: Lepidoptera
Family Pyralidae

1. INTRODUCTION

The rice stem borer, Chilo suppressalis, is one of the most important insect pests of rice crop. It is found throughout Asia, Hawaii, the Middle East, and the Mediterranean region. Rice, Oryza sativa, and water oat, Oryza latifolia are major hosts; sugarcane, wheat, Italian millet, and Indian corn are occasionally attacked.

Eggs are oviposited on leaves and hatch in about 7 days. The hatched larvae first bore into the leaf sheath and feed on the tissues for about a week then enter into the stem as 1st instars. The larvae live gregariously during the first three instars but disperse later. They become fully grown within 35 to 50 days after hatching and usually pupate in the stem or straw. The full-grown larvae, before pupating, cut exit holes in the internodes through which the moths emerge.

In temperate areas, such as Japan and China, the borer has one to three generations a year. In tropical areas where two or more rice crops are grown, the borer remains active throughout the year and can have up to 5 generations per year.

The initial injury by the larvae in the leaf sheath causes whitish discoloring at infestation sites. After this, the larvae bore into the stem and feed on internal tissues. If damage occurs during the vegetative growth, the central leaf whorl does not unfold, turns brownish, and dries out. This is known as a "dead heart". After panicle initiation, internal feeding severs the developing panicle at the node, so the panicle dries out and either does not emerge at all or does not produce grains. The emergent panicles become very conspicuous in the rice paddy, since they remain straight and are whitish. These are called "white heads".

C. suppressalis has been reared on several artificial diets by Ishii (1952) and Kamano (1971) and on rice seedling by Sato (1964). On artificial diet, it has been occasionally found that vitality of C. suppressalis decreased gradually because of inbreeding. To avoid the loss of vitality, Sato (1964) developed a mass-rearing method using rice seedlings. By this method, the vitality of C. suppressalis has been maintained for 70 generations over 10 years. These laboratory reared insects are used for sex pheromone studies, screening of potential insecticides, and tissue culture. A laboratory colony can either be established from field collected moths or by obtaining insects from another laboratory colony.

2. FACILITIES AND EQUIPMENT REQUIRED

2.1 Facilities

Rearing is carried out in an insectary or large incubator maintained at 28+2°C, 50-70% RH and 16:8 LD period provided by fluorescent lights (100-500 lux).

2.2 Equipment and materials required for insect handling

2.2.1 Adult oviposition

a. Field collected moths (Light trap)

- Transparent polyethylene bag (500 X 1000 mm)
- Rice plants (30 days after transplantation)
- Aspirator
- 20 watt blue fluorescent lamp or 60 watt incandescent lamp
- Large white sheet (2000 X 3000 mm)
- Electric cord
- Sweep net (300 mm diameter)

b. Successive rearing of moths

- Oviposition box (450 X 900 X 1300 mm) (Fig. 1)
- Potted rice plants (30 days after transplantation)
- Plastic trays (150 X 200 X 50 mm)
- Absorbent sponges (100 X 150 X 50 mm)

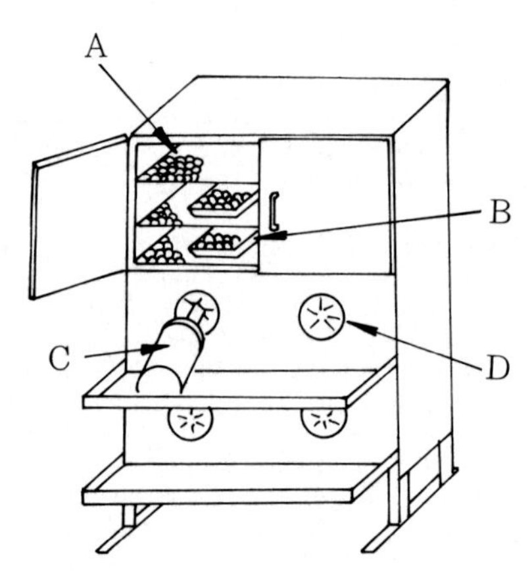

Fig. 1. Oviposition box for mass rearing of *C. suppressalis*. A: wire mesh screen; B: pupae in cotton stoppers; C: potted rice plant; D: plastic sponge.

2.2.2 Egg collection and storage

- Petri dishes (150 mm diameter)
- Whatman No.1 filter paper (150 mm diameter)
- Soft-nosed stainless forceps (100 mm long)
- Distilled water squirt bottle
- Refrigerator at 5°C

2.2.3 Egg sterilization

- 100 ml beaker
- Soft-nosed stainless forceps
- Scissors
- Whatman No.1 filter paper
- 70% ethanol
- 0.1% mercuric chloride
- 100 ml graduated cylinder
- Distilled water squirt bottle

2.2.4 Egg-mass inoculation and larval transfer to diet
- Soft-nosed stainless forceps
- Writing brushes or hair paint brushes
- Plastic trays
- 50 ml graduated cylinder
- 500 ml glass beaker
- Large flat stainless steel spatula
- Polyethylene bag (500 X 300 mm)

2.2.5 Larval rearing containers

a. Aseptic
- Erlenmeyer flasks (200 ml)
- Cotton wool
- Long stainless forceps

b. Septic
- Glass jar (110 mm diameter, 90 mm deep) (seedling jar) with screw metal cap (20 mm hole in center) (Fig. 2)
- Cotton wool

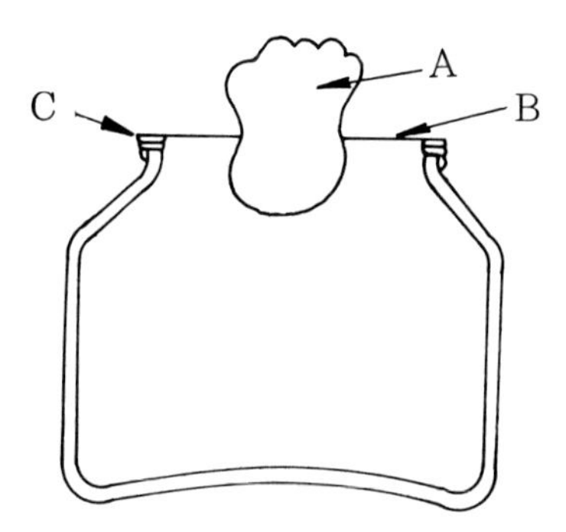

Fig. 2. Glass seedling jar used for rearing *C. suppressalis* larvae on rice seedlings. A: cotton stopper; B: screw metal cap; C: rubber ring.

2.2.6 Pupal collection and storage

a. Aseptic
- Pupation cage (metal cylinder, 90 mm diameter, 60 mm long) with wire-screened at bottom and top
- Cellophane tubes (3 X 40 mm)

b. Septic
- Soft-nosed stainless forceps
- Plastic trays (150 X 200 X 50 mm)
- Refrigerator at 5°C

2.2.7 Insect holding
- Room or incubator set at 28°C

2.3 Diet preparation equipment

2.3.1 Preparation of rice seed and storage
- Purchased rice seeds (300 kg/year)
- 50 liter plastic buckets
- Salt
- Hydrometer
- Top loading balance (0.1-10 kg)
- 1 liter graduated cylinder
- Metal or bamboo basket (300 mm diameter)
- Polyethylene or paper sheets (1 X 5 m)
- Rice stock stocker at 5°C

2.3.2 Preparation of rice seedlings
- 10 liter plastic bucket
- Metal or bamboo basket (300 mm diameter)
- 0.1-100 g balance
- 50 ml plastic cup
- 15 ml graduated cylinder

2.3.3 Preparation of artificial diet
- 200 ml Erlenmeyer flasks
- Cotton wool
- Mortar (150 mm diameter)
- Top loading balance (0.1-100 g)
- Drying oven
- Autoclave
- 1 liter glass beaker
- 100 ml graduated cylinder
- 10 ml pipette
- Stainless steel spatula
- Paraffin paper (120 X 120 mm)
- Non-toxic particle mask

3. DIET

3.1 Seed preparation procedure
- Immerse dry rice seeds in salt water (specific gravity 1.14)
- Discard floating immature seeds
- Collect seeds sunk in salt water
- Rinse seeds with running water
- Dry seeds on polyethylene or paper sheets in room for about 3 days
- Transfer seeds into large metal cans
- Keep metal cans in a stocker room (5-10°C) until used for the rearing

PRECAUTION

Rice seeds must be kept at 5-10°C to preserve germination ability.

3.2 Seedling preparation procedure
- Immerse seeds in a bucket of water over-night at 20-30°C
- Thoroughly rinse the seeds with water the next day
- Transfer washed seeds to baskets
- Put wet seeds into a rearing jar (60 g/jar)
- Add 15 ml water into a jar
- Seal with the metal cap
- Transfer jars to shelves in the insectary or incubator (28±2°C)
- Keep jars under illumination (100-500 lux) for 5-7 days

PRECAUTIONS

(i) Excess water will encourage growth of mold.

(ii) Excess water will depress the root growth of seeds.

3.3 Artificial diet preparation procedure

3.3.1 Composition of diet (see table in the next page)

3.3.2 Brand names and sources of ingredients

Agar powder, cellulose powder, and chemicals (Wako Pure Chemical Industries Ltd., 10 Doshomachi, 3-chome Higashi-ku, Osaka, Japan); wheat germ (Samitt-Seiyu Co., Ltd., 38 Shinminato, Chiba, Japan).

3.3.3 Diet preparation procedure
- Plug top of flask with cotton wool and sterilize in drying oven at 150°C for 15 min.
- Place agar, casein, cellulose, sucrose, starch, wheat germ, salt mixture, cholesterol into mortar
- Mix all of the above together with spatula
- Transfer dry mixture powder (8 g) into a flask

- Dissolve choline chloride, sodium ascorbate and L-cysteine HCl in 490 ml of distilled water and add 10 ml of vitamin solution
- Add 50 ml of mixture solution into a flask
- Mix fully using large spatula
- Cover cotton wool with paraffin paper
- Place flasks in autoclave for 30 min. at 112°C
- Remove flasks from autoclave
- Cool autoclaved diet to 50°C in room
- Shake until solution gels to make homogenous mixture
- Store flasks at 5°C until required for use

PRECAUTION:

A non-toxic particle mask should be worn when mixing the dry ingredients.

Ingredient	Amount/10 flasks	%
Agar powder	10 g	1.70
Cellulose powder	10 g	1.70
Sucrose	15 g	2.55
Starch	20 g	3.40
Casein	15 g	2.55
Wheat germ	10 g	1.70
Salt mixture[1)]	4 g	0.68
Cholesterol	0.2 g	0.03
Choline chloride	0.5 g	0.08
Sodium ascorbate	3 g	0.51
L-Cysteine HCl	0.3 g	0.05
Vitamin solution[2)]	10 ml	1.70
Water	490 ml	83.33

[1)] Salt mixture: K_2HPO_4, 49.75 g; KH_2PO_4, 18.00 g; $MgSO_4$, 16.00 g; $Ca(H_2PO_4)_2 \cdot H_2O$, 8.00 g; NaCl, 5.00 g; $Fe_2(SO_4)_3 \cdot 6H_2O$, 2.00 g; $MnSO_4 \cdot H_2O$, 0.50 g; $CuSO_4 \cdot 5H_2O$, 0.25 g; $Zn(C_2O_2H_3)_2 \cdot 2H_2O$, 0.05 g.

[2)] Vitamin solution: Thiamine hydrochloride, 200 mg; Riboflavine, 100 mg; Nicotinic acid, 200 mg; Pyridoxine hydrochloride, 100 mg; Calcium pantothenate, 200 mg; Folic acid, 20 mg; Biotin, 20 mg; p-Aminobenzoic acid, 200 mg; Inositol, 2000 mg; Water, 1000 ml.

4. HOLDING INSECTS AT LOW TEMPERATURES

4.1 Adults

Moths cannot be held at under 20°C as mating and flight activity cease.

4.2 Eggs

Egg-masses (all ages) attached to the leaves of rice plant can be kept for 1-2 weeks at 5°C safely in a petri dish containing moistened filter paper.

PRECAUTION

Too much moisture on the filter paper will decrease egg hatch because of oxygen deficiency.

4.3 Larvae

Larvae (all stage) feeding on seedlings in jars can be kept at 5°C safely for 1-2 weeks until new seedlings are prepared.

4.4 Pupae

Pupae can be kept at 5°C safely for 1-2 weeks. It is possible to regulate the emergence of moths.

5. REARING AND COLONY MAINTENANCE

The founder colony of *C. suppressalis* was established from field collected moths, and has been reared continuously on rice seedlings for over 70 generations.

5.1 Oviposition and egg-mass collection

a. From field collected moths
- Set up lamp and sheet in paddy field 1 hour before sunset
- Catch moths attracted by lamp using aspirator
- Transfer 50 moths into polyethylene bag with rice plants
- Place bags in darkened-room (20-30°C) over-night
- The next day remove rice plants with egg-masses from bags
- Wash down the scales of moths from leaves of plants
- Cut off individual egg-mass attached on the leaves (about 20 mm long) carefully using scissors
- Place egg-masses in petri dishes with moistened filter paper
- Incubate at 28°C or keep at 5°C if necessary

b. From laboratory moths
- Prepare oviposition box (Fig. 1) in the insectary
- Transfer cotton stopper with pupae to shelves of box. Moths emerge and mate in box.
- Place plastic trays with moistened sponge on shelves
- Set up potted rice plants
- Exchange potted rice plants every other day
- Spray water on rice plants with egg-masses, wash down moths scales from plants
- Cut off egg-masses as (a)
- Place egg-masses in petri dishes as (a)
- Incubate at 28°C

PRECAUTION

Keep humidity at about 70% in oviposition box to prevent desiccation of pupae and to increase the longevity of moths.

5.2 Egg sterilization

- Strip leaves of egg-masses (black head stage) using forceps
- Immerse egg-masses in 70% ethanol for a few seconds using forceps in aseptic room or clean bench
- Immerse egg-masses in 0.1% aqueous solution of mercuric chloride for 4 minutes
- Wash fully with 70% ethanol
- Place sterilized egg-mass (30-40 eggs) on the inside of glass wall of flasks containing diet with long stainless forceps

Note: This should always be done when rearing on artificial diets and is also recommended for seedling rearing.

5.3 Rearing of larvae

5.3.1 Inoculation with egg-masses

- Take off the metal caps on the seedling jars 5 days after seeds are put into jars
- Inoculate about 7 egg-masses (black head stage) per jar on the seedlings using stainless steel forceps
- Add 15 ml water in a jar
- Re-seal with the metal caps
- Keep about 10 days in insectary

5.3.2 Transfer of larvae

a. First transfer
- Take off the metal caps of the seedling jars 10 days after inoculation with egg-masses
- Remove the old seedlings consumed by the larvae, transfer the 3rd instar larvae from the rearing jars using a long spatula to a tray
- Transfer 100 larvae into a new seedling jar using a writing brush and

forceps
- Add 15 ml water to the seedling jar
- Re-seal with the metal caps
- Keep seedling jars for 7 days in the insectary

b. Second transfer
- Remove old seedling and 4th to 5th instar larvae from the rearing jars using long forceps and place in a plastic tray
- Transfer 50 larvae into a new seedling jar
- Add 15 ml water to the seedling jar
- Re-seal with the metal caps
- Keep rearing jars for 7-10 days in the insectary

5.3.3 Inoculation of sterilized egg-masses and aseptic rearing of larvae
- Place sterilized egg-mass on the glass wall in the inside of the flasks that contain artificial diet using long stainless forceps and sterile technique
- Plug flasks with cotton wool
- Keep about 25 days in insectary or incubator
- Remove cotton wool stopper when pupation starts
- Remove the mature larvae from the flasks and transfer to pupation cages with forceps (see below)

5.4 Pupal collection

5.4.1 Seedling rearing method
- Most of mature larvae bore into the cotton stopper in the metal caps and pupate
- Take off the metal caps from rearing jars
- Carefully remove the cotton stoppers from the metal caps
- Collect cotton stoppers in plastic trays
- Transfer the plastic trays to shelves in the oviposition box or in a refrigerator if necessary
- Re-seal rearing jars with new metal caps

PRECAUTIONS
(i) Pupae are easily crushed, so the cotton stoppers must be carefully removed from the metal caps.
(ii) Check the humidity in the oviposition box to prevent desiccation of pupae.

5.4.2 Artificial diet rearing
- Transfer mature larvae to pupation cage containing cellophane tubes
- Keep pupae cage in insectary at 25°C
- Pupation will occur inside the cellophane tubes

5.5 Adults

The sex ratio is approximately 1:1. The first males and females will emerge about 25 days after hatching. The males emerge before the females. Emergence is complete in 50 days with a peak occurring on 35-40 days. Adult survival from hatching larvae is about 60%.

5.6 Sex determination
- Wings of male moths are darker than wings of female moths
- Female and male moths are distinguished by morphological differences (ovipositor and clasper) in the abdominal tip

5.7 Rearing schedule

a. Daily
- Check moisture in petri dishes containing egg-masses
- Check the growth of seedling in jars and larval growth in rearing jars

b. Alternate days
- Exchange potted rice plants and cut off egg-masses from plants during oviposition period
- Collect cotton stoppers with pupae from rearing jars in the pupal stage, and replace metal caps

c. Weekly
- Put rice seeds into jars
- Inoculate egg-masses in seedling jars

- Transfer larvae from rearing jars to new seedling jars
- Wash rearing jars and metal caps
- Plug cotton wool in the hole of the metal caps

d. Monthly

- Wash oviposition box
- Wash plastic trays
- Wash floor of insectary with water and 0.1% formalin

5.8 Insect quality

Insect quality is checked continuously by recording final instar larval weight, number of eggs and percent hatching of egg-masses, and the susceptibility of larvae to insecticides. Accepted LD_{50} values are as follow:

Insecticides	LD_{50}* (μg/g)
Cartap	2.34
Fenitrothion	2.15
Phenthoate	1.57
Diazinon	3.82

*Topical application

5.9 Special problems

5.9.1 Diseases

If muscardine diseases occurs in larval stages all rearing jars must be sterilized or destroyed. Glass jars, metal cap and forceps must be disinfected by autoclaving.

5.9.2 Human health

Exposure to moth scales can be hazardous. A suitable protective mask should be worn when scales are evident.

6. LIFE CYCLE DATA

Life cycle and survival data of various stage under the optimum rearing conditions of 28°C, 50-70% RH and LD 16:8 photoperiod are given.

6.1 Developmental data

Stage	Number of days
Egg	5-6
Larvae	25-35
Pupae	6-7
Total development (egg to egg)	37-49

6.2 Survival data (%)

- Egg hatchability 80-90
- Neonate larva to pupa 65-80
- Neonate larva to adult 50-60

6.3 Other data

- Preoviposition period, 1 day
- Peak egg laying period 2-3 days after emergence
- Mean fecundity 235 eggs/female (min. 40, max. 700)
- Mean moth longevity 5 days (min. 4, max. 10)
- No. of larval instars, 6
- Mean pupal weight (mg), male 45, female 55

7. REFERENCES

Ishii, S. 1952. Some problems on the rearing method of rice stem borer by synthetic media under aseptic condition. Oyo-Kontyu 6: 93-98.

Kamano, S. 1971. Studies on artificial diets of the rice stem borer, Chilo suppressalis Walker. Bull. Nat. Inst. Agr. Sci., Tokyo, Ser. C 25: 1-45.

Sato, Y. 1964. A simple technique for mass rearing of the rice stem borer on rice seedlings. Jap. J. appl. Ent. Zool., 8: 6-10.

CHORISTONEURA OCCIDENTALIS and CHORISTONEURA FUMIFERANA

JACQUELINE L. ROBERTSON

USDA Forest Service, Pacific Southwest Forest and Range Experiment Station, P. O. Box 245, Berkeley, CA 94701, USA

THE INSECT

Scientific name: *Choristoneura occidentalis* Freeman
Common name: Western spruce budworm
Scientific name: *Choristoneura fumiferana* (Clemens)
Common name: Spruce budworm
Order: Lepidoptera
Family: Tortricidae

1. INTRODUCTION

The western spruce budworm is both the principal representative of the Abietoidae-feeding *Choristoneura* and the most destructive defoliator of forests in western North America (Furniss and Carolin 1977, Powell 1980). Distributed throughout the Rocky Mountain and Pacific Coast States and British Columbia, the western spruce budworm feeds primarily on Douglas fir and white fir. Other hosts include grand fir, subalpine fir, western larch, Engleman spruce, white spruce, and blue spruce. The spruce budworm occurs from Virginia to Laborador and west to the Yukon Territory (Baker 1972). Balsam fir is the preferred host, although white spruce, red spruce, black spruce, larch, hemlock, and pines may also be defoliated.

Larvae feed primarily in buds and on new foliage that develops each year. Sustained heavy feeding may cause virtually complete defoliation in 4-5 years. Among the effects of epidemics are decreased growth, deformation of trees, top killing, and death of trees over an extensive area. Western spruce budworm also causes cone and seed damage in western larch and Douglas fir in the northern Rocky Mountains.

There is 1 generation a year. Eggs, laid in July and August, hatch in about 10 days. First instars spin hibernacula under bark scales and lichens without feeding; they molt to the second instar before entering diapause. When development resumes the following spring, 2nd instars first mine into old needles, then into the new buds where they feed on the expanding needles. Third to 6th instars feed on needles as shoots elongate and the new growth matures. Adults emerge from puparia after about 10 days; they live for 5-7 days and do not feed.

Both species have been reared continuously on artificial diet both with and without diapause (Lyon *et al.* 1972, Robertson 1979). In our laboratory, *C. fumiferana* was reared without diapause for 30 generations over 4 years. *C. occidentalis* has been reared for 105 nondiapause generations over the last 15 years. Even after this long, diapause can be induced with proper rearing conditions. These insects are used for research on the genetic basis for response to insecticides, and for basic and applied toxicological investigations. Laboratory colonies can be started from field collections or by obtaining insects from an established laboratory colony.

2. FACILITIES AND EQUIPMENT REQUIRED

2.1 Facilities

Rearing of larvae is done in an environmentally controlled room maintained at 23-26°C and 30-50% RH with a LD 16:8 photoperiod. Fluorescent light intensity is 100-120 footcandles in the photophase and 10 footcandles in the scotophase. Mating and oviposition occur in a cabinet or container which excludes light.

2.2 Equipment and materials required for insect handling

2.2.1 Adult oviposition

Three methods may be used:

a. Individual mating
- Dixie® cup (237 ml) with clear or translucent plastic lid
- 2 waxed paper strips (approx. 2 x 20 cm)

b. Small group mating (10-15 pairs)
- No. 6 kraft paper bag, unwaxed
- 10 waxed paper strips (approx. 2 x 20 cm)

c. Large group mating (16-80 pairs)
- No. 46 kraft paper bag, unwaxed
- 25 waxed paper strips (approx. 2 x 20 cm)

2.2.2 Egg collection and storage
- Scissors
- Sheets of newsprint
- Dissecting forceps, curved, fine points
- Plastic container

2.2.3 Egg sterilization
- Plastic container (250 ml)
- Mechanical stirrer
- Fine nylon gauze or cheesecloth
- Rubber band
- No. 1 filter paper (15 cm diam), sterilized
- Sterile, disposable gloves

2.2.4 Larval rearing

a. With diapause
- White glue
- Parafilm®
- Surgical gauze (3 x 3-cm square)
- Sterile, plastic petri dishes (100 x 15 or 100 x 20 mm)
- Single-edged razor blade
- Cellophane tape
- Grinding tool or wheel
- Black paper bag (14 x 21 cm)
- Sterile, disposable gloves
- Sheets of newsprint
- Scissors
- Light table (fluorescent tubes under clear or opaque glass)
- Yellow wax pencil
- Refrigerator set at 4-5°C
- Refrigerator set at 0°C
- No. 3 camel-hair brushes

b. Without diapause
- Clear, plastic specimen container (200 ml) with unwaxed paper lid
- Scissors
- Aluminum foil (15 x 15 cm)
- Dissecting needle
- Dissecting forceps, curved, fine points
- Sterile, plastic petri dishes (100 x 15 or 100 x 20 mm)
- Sterile, disposable gloves
- Soft-nosed forceps
- Sterile spatula
- Sheets of newsprint

2.2.5 Pupal collection and storage
- Soft-nosed forceps
- Sheets of newsprint
- Sterile, plastic petri dishes (100 x 15 or 100 x 20 cm)
- Refrigerator set at 10°C

2.2.6 Insect holding
- Refrigerator set at 10°C

2.2.7 Anesthesia
- CO_2

2.3 Diet preparation equipment

2.3.1 Small batch (3750 g)
- Top-loading balance (0.1-1000 g)
- 8-liter stainless steel pot
- 3-speed blender with 3.785-liter container
- Metal tray (2.5 x 35 x 48 cm)
- Heavy duty aluminum foil
- Miscellaneous weighing containers
- Small, sterile, stainless steel spatula
- 2-ml (graduated 1/10) pipette
- Pipette bulb
- 25-ml volumetric cylinder
- 500-ml volumetric cylinder
- Autoclave
- Nontoxic particle mask
- Sterile, disposable gloves
- Thermometer (0-100°C)
- Refrigerator set at 10°C

2.3.2 Large batch (3750-15,000 g)
- Top-loading balance (0.1-1000 g)
- Two 8-liter stainless steel pots
- Industrial mixer with 18.9-liter mixing bowl, flat beater attachment and variable, timed speed controls
- 3-speed blender with 3.785-liter container
- 4 metal trays (2.5 x 35 x 48 cm)
- Heavy duty aluminum foil
- Large stainless steel spatula with wooden handle
- Small, sterile, stainless steel spatula
- Miscellaneous weighing containers
- 10-ml (graduated 1/10) pipette
- Pipette bulb
- 1-liter volumetric cylinder
- 100-ml volumetric cylinder
- Refrigerator set at 10°C
- Autoclave
- Nontoxic particle mask
- Sterile, disposable gloves
- Thermometer (0-100°C)

3. ARTIFICIAL DIET

3.1 Composition

3.1.1 Ingredients to prepare 15 kg finished diet

Ingredients	Amount	%
Total water		86.0
Group I		
Finely ground agar (USP food grade)	405 g	2.7
Distilled water	9000 ml	60.5
Group II		
Distilled water	2500 ml	16.8
4 M KOH	90 ml	0.6
Wheat germ	495 g	3.3

Linolenic acid	30 ml	0.2
Group III		
Vitamin mixture	180 g	1.2
Ascorbic acid	60 g	0.4
Group IV		
Methylparaben	25.5 g	0.17
Sorbic acid (K-salt, 99%)	18.0 g	0.12
Aureomycin (5.5%)	17.9 g	0.12
Formalin (37%)	9.0 ml	0.06
Benomyl (59%)	1.9 g	0.01
Group V		
Distilled water	1300 ml	8.8
Vitamin free casein	58.5 g	0.4
Alphacel	90.0 g	0.6
Sucrose	435 g	2.9
Salt mixture W	150 g	1.0

3.1.2 Prepare 4 M KOH stock solution (336 g KOH in 1500 ml distilled water)

3.2 Brand names and sources of ingredients

Agar (Moorhead & Co., Van Nuys, CA, 91401); aureomycin (Agricultural Division, American Cyanamid Co., Princeton, NJ 08540); benomyl (Benlate WP) (Cook Laboratory Products, Pico Rivera, CA 90660); alphacel, ascorbic acid, methylparaben, salt mixture W, vitamin free casein, vitamin mixture (ICN Pharmaceuticals Inc., Life Sci. Group, Cleveland, OH 44128); linolenic acid (Eastman Kodak Co., Rochester, NJ 14650); sorbic acid (Matheson, Coleman & Bell, Gibbstown, NJ 08027); wheat germ (Kretschmer regular). (All sources in USA)

3.3 Diet preparation procedure

- Molten agar
- Divide Group I ingredients evenly between 2 stainless steel pots (place all ingredients in 1 pot for small batch)
- Autoclave for 40 min at slow exhaust (liquid) setting
- Leave in autoclave for 15 min with the door shut, then remove
- Test temperature every 15 min
- Let cool to 75°C with frequent stirring
- Blend ingredients in Group II for 2 min at high speed
- Add Group III to blended Group II mixture and blend for 1 min at high speed
- Add Group IV to blended mixture of Groups II and III, blend for 1 min at high speed
- Place Group V ingredients in mixer with Groups II, III, and IV poured from blender and mix while Group I cools (for small batch, add Group V ingredients to Groups II-IV and blend at low speed while group I cools)
- Pour molten agar into mixer with Groups II-IV and mix complete diet for 10 min (for small batch, pour agar into the blender and blend at low speed for 3 min)
- Pour diet into aluminum foil lined pans to a depth of approx. 1 cm
- Cover with aluminum foil to prevent contamination of diet during cooling
- Allow to cool for 1 h before cutting with spatula for use or storage
- For storage, wrap in aluminum foil
- Prepared diet has a storage life of about 1 month at 5°C

PRECAUTIONS

(i) A nontoxic particle mask should be worn when mixing diet ingredients.
(ii) Sterile, disposable gloves should be worn while handling finished diet.
(iii) All equipment must be sterilized or thoroughly cleaned.
(iv) Exercise extreme caution when handling molten agar.

4. HOLDING INSECTS AT LOWER TEMPERATURES

4.1 Adults

- Should not be stored

4.2 Eggs

Sterile or unsterilized eggs can be stored at 10°C for 7 days in a plastic container. The top of the container should be closed with a paper lid or covered with aluminum foil. To avoid hatch, do not store eggs in which black headcapsules are visible for more than 3 days.

4.3 Larvae

- Larvae can be stored in petri dishes without food for 5 days at 10°C

4.4 Pupae

- Pupae can be stored for 1-2 weeks in petri dishes at 10°C

5. REARING AND COLONY MAINTENANCE

The nondiapausing *C. occidentalis* colony was established with field-collected larvae and has been reared continuously for 105 generations.

5.1 Egg collection

- Anesthetize adults in mating container with CO_2
- Remove waxed paper strips and place on newsprint on work table
- Carefully cut around individual egg masses with scissors, trimming as closely as possible to the eggs without damaging them
- Place egg masses into storage or washing container with dissecting forceps

PRECAUTION

Do not remove eggs from waxed paper, since eggs placed directly on diet do not hatch as well as those still attached to the paper.

5.2 Egg sterilization

- Place eggs at the bottom of a plastic or glass container (250 ml or larger)
- Pour 200 ml of 10% formaldehyde (final concentration active ingredients) over the eggs
- Add one drop of a wetting agent such as Tween 20®
- Stir gently for 15 min with a mechanical stirrer
- Remove stirrer, cover top with nylon screen or cheesecloth, affix to container with rubber band, and decant formaldehyde
- Fill container with 200 ml distilled water
- Stir gently for 10 min with mechanical stirrer
- Decant water
- Refill container with 200 ml distilled water
- Stir gently for 10 min with mechanical stirrer
- Decant water
- Remove egg masses from container with dissecting forceps and then singly onto sterile filter paper placed on sheet of newsprint
- Cover eggs with another piece of sterile filter paper
- Air dry for 1-3 h
- Store or use when dry

5.3 Larval rearing

5.3.1 With diapause

- Grind lugs from petri dish top to eliminate vents
- With white glue, affix the paper under a large disinfected egg mass to the bottom of the petri dish
- Using a rounded surface, such as the sides of a beaker, press a 3 x 3-cm gauze onto a square piece of Parafilm
- Stretch the Parafilm across the petri dish bottom so that the gauze is between the Parafilm and the egg mass
- Place the petri dish top over the Parafilm and firmly close the dish
- Seal the dish with 4 symmetrically placed pieces of cellophane tape
- Place the dish within a small, black paper bag, fold the unfilled portion of the bag over toward the bottom of the dish, and seal the flap to the rest of the bag with cellophane tape
- With a single-edge razor blade, cut a window out of the bag, just

behind the gauze and of about the same size
- Place the bag on a light table so that the light enters the dish through the gauze
- Date the bag with a wax pencil
- After 3 weeks, remove the bag and dish from the light table and place in a refrigerator set at 4-5°C
- After 1 week, transfer the bag and dish to a refrigerator set at 0°C and leave it for at least 16 weeks (200 days is optimal)
- Remove the bag and dish from cold storage
- Remove the bag, open the dish, and place a small cube of diet in the bottom
- Close the dish, reseal it with cellophane tape and place it back in the black bag with light entering through the bottom of the dish behind the diet
- Inspect the dish daily for second instars leaving the gauze and being attracted to the diet (peak emergence usually occurs about 3 weeks after a dish is placed at room temperature)
- Transfer 10-15 second instars with a sterile no. 3 camel hair brush to a petri dish (with lugs removed from top by grinding) containing 2 sections (8 x 1 x 1 cm) of diet
- Tape petri dish closed
- Refeed larvae in 10-14 days and as necessary thereafter

5.3.2 Without diapause

a. Main colony
- Fill each 200-ml specimen container 1/3-1/2 full with cubes of artificial diet 1 cm in each dimension (or, fill each cup 1/6-1/4 full and add fresh diet when larvae reach instars 3-4)
- With dissecting forceps, place a single, large egg mass or 2-3 smaller ones totalling about 150 eggs on the diet near the bottom of each cup with the waxed paper between the diet and the eggs
- To close the container, firmly press an unwaxed paper lid into the top against the inner rim (side tabs must be removed or folded so that the lid edge is flush with the cup)
- Cover each cup with aluminium foil so that light is excluded to the top of the diet (if light is not excluded, most first instars migrate to the cup-lid interface, spin hibernacula, and diapause)

b. Propagation colony
- With soft-nosed forceps, transfer ten 5th and 6th instars from rearing cups to each 100 x 15 or 100 x 20-mm sterile, plastic petri dish containing 1-2 pieces (9 x 1 x 1-cm) of artificial diet
- Use cellophane tape (single dishes) or rubber bands (stacks of 4 dishes) to close the dishes; store on shelves

5.4 Pupae
- Harvest pupae from petri dishes with soft-nosed forceps and place up to 50 individuals of the same sex in each sterile petri dish (males generally pupate 1 day before females of the same age)
- Store or set up individual or group matings

PRECAUTION

Allow pupae to fully tan before collecting.

5.5 Adults

The sex ratio is approx. 1:1.

5.5.1 Emergence period

a. Single-pair matings
- Matings are most successful when pairs mate within 24 h of their eclosion
- Avoid the use of any adult which eclosed more than 2 days before

b. Group matings
- Refrigeration of pupae tends to synchronize adult eclosion so that matings are highly successful

5.5.2 Oviposition

Optimum condition is 23-26°C in the dark. The preoviposition period is about 24 h. Peak egg laying occurs within 4 h of pairing. Adults are dead after 5 days.

a. Individual matings
- Place 1 male and 1 female pupa in the bottom of a 237-ml paper cup
- Place 1 or 2 waxed paper strips (approx. 2 x 20 cm) into the cup
- Close the cup with a clear or translucent snap lid
- Place in cabinet with door closed for 7 days
- Harvest eggs after adults are anesthetized with CO_2

b. Group matings
- Place 16-80 pupae of each sex into the bottom of a no. 46 unwaxed, kraft paper bag (or 2-15 pupae of each sex into the bottom of a no. 6 unwaxed, kraft bag)
- Place 25 waxed paper strips (approx. 2 x 20 cm) into the bottom of the bag (10 strips for the no. 6 bag)
- Rumple the strips so they are randomly dispersed
- Fold top of bag over and staple shut
- Place bag in cabinet with door closed for 10-14 days
- Harvest eggs after adults are anesthetized with CO_2

5.6 Sex determination
- Dark, paired testes are visible in male larvae as early as the 2nd instar and become increasingly distinct in later instars
- The female pupa has 4 ventral abdominal segments, the male has 5
- Female pupae and adults are generally larger than males
- Female adults have a large ovipository pore, male adults have a terminal abdominal brush

5.7 Rearing schedule

5.7.1 Without diapause

a. Large colony

Monday
- Harvest eggs from individual or group matings set up 12 days earlier
- Count and record numbers of dead pupae in mating containers for quality control records
- Sterilize eggs and store at 10°C
- Remove 10 egg masses at random and place in 60 x 15-mm snap-lid petri dishes without food for egg hatch estimate in quality control records
- Pick pupae from propagation dishes and store them at 10°C
- Select 20 male and 20 female pupae at random, then weigh and record their individual weights for quality control records

Tuesday
- Set up rearing containers using current week's eggs and previous week's refrigerated eggs (1:1) (refer to 5.8.2)
- Randomly select 10 rearing containers for larval survival quality control estimation ca 3 weeks later

Wednesday
- Set up propagation colony of 6th instars in petri dishes
- Set up mating containers and count numbers of male and female pupae in each

Thursday
- Dispose of excess rearing containers holding older larvae and pupae

Friday
- Do counts for egg hatch and larval survival aspects of quality control records

b. Small colony

Day 1
- Set up propagation colony of 6th instars in petri dishes

Day 8
- Harvest pupae from propagation dishes
- Randomly select 20 male and 20 female pupae and weigh them individually for quality control records

- Dispose of excess rearing containers
- Set up mating containers and count numbers of males and females in each

Day 20

- Harvest eggs
- Count and record numbers of male and female pupae in each
- Sterilize eggs
- Select 10 egg masses at random and place each in a 60 x 15-mm snap-lid petri dish without food for egg hatch estimate in quality control records
- Set up rearing containers with remaining egg masses
- Randomly select 10 rearing containers for larval survival quality control estimation to be done when next propagation colony is established

5.7.2 With diapause

a. Day 1
- Harvest and sterilize eggs
- Count number of dead pupae in mating container and determine percent of adult emergence for quality control records
- Randomly select 10 egg masses for estimation of percent hatch 14 days later
- Place individual egg masses in petri dishes for diapause rearing

b. Day 22
- Move petri dishes to refrigerator set at 4-5°C

c. Day 29
- Move petri dishes to refrigerator set at 0°C

d. Day 229
- Remove petri dishes from refrigeration
- Put artificial diet in each dish, then place back in bag with light source behind dish

e. Days 250-260
- Transfer emerging 2nd instars to petri dishes

f. Days 264-274
- Refeed larvae

g. Days 278-288
- Pick pupae from rearing containers
- Refrigerate pupae until enough are available for mass rearing, or mate individually

h. Days 290-300
- Harvest eggs and begin diapaused rearing procedure again

5.8 Quality control

5.8.1 Insect quality is checked continuously by keeping records of egg hatch, larval survival, pupal weights and adult emergence. Typical values of these parameters are: egg hatch-84%, larval survival-52%, female pupal weight-200 mg, male pupal weight-110 mg, and adult emergence-89%. Starch gel electrophoresis is useful in monitoring genetic quality (Stock and Robertson 1980). A recent investigation has shown that, while overall genetic identity of a large colony did not differ significantly from a wild population descended from one of the founder groups, reduced heterozygosity had occurred over the 10-year period since wild stock was last introduced. To prevent excessive homozygosity in a laboratory colony, wild stock collected in the same area as the founder group should be periodically introduced into a colony at 2- to 3-year intervals.

5.8.2 To prevent the formation of inbreeding lines within an existing colony, eggs and pupae from 1 rearing group should be refrigerated and mixed with those of the next group. For example, in a large nondiapausing colony, 50% of each week's pupae and eggs should be held at 10°C until the next week when they are combined with that week's pupae and eggs. It is important to mix the refrigerated pupae proportionately with the unrefrigerated animals to avoid continuing isolation of a line.

5.9 Special problems

a. Diseases

- Virus infection of the larval stages is prevented by egg sterilization; Nosema fumiferanae, a microsporidian protozoan to which the western spruce budworm is particularly susceptible, is suppressed by benomyl in the diet

b. Microbial control

- Bacterial and fungal infections will occur on the diet if proper sanitary conditions are not maintained. Escherichia coli, Aspergillus, and Penicillium may be particularly troublesome. Diet should not be left unwrapped or unrefrigerated for extensive periods; it should be handled only with sterile gloves and spatulas. All equipment used to handle insects, including brushes and forceps, should be sterilized and stored in sterile, closed containers such as enamel pans with lids. When in use, forceps should be wiped frequently with a surface sterilant.

c. Human health

- All work with formalin and propylene oxide should be done in a fume hood

6. LIFE CYCLE DATA

6.1 Developmental data

a. Nondiapause method

Developmental stage	Weight (mg) Mean ± S.D.	Weight (mg) Range	Duration (days) Mean ± S.D.	Duration (days) Range
Instar 1	0.2 ± 0.04[a]	0.1- 0.3[a]	8.9 ± 2.6	7-16
Instar 2	0.5 ± 0.3[a]	0.2- 1.0[a]	4.9 ± 1.6	3-8
Instar 3	2.4 ± 1.4[a]	1.0- 6.1[a]	6.6 ± 10.3	2-7
Instar 4	11.0 ± 5.0[a]	4.4- 22.0[a]	3.7 ± 5.1	1-4
Instar 5	31.0 ± 14.0[a]	10.0- 61.0[a]	4.5 ± 5.3	1-6
Instar 6	132.0 ± 71.0[a]	44.0-290.0[a]	7.0 ± 5.7	3-9
Pupa (male)	119.0 ± 29.0	48.0-186.0	7.3 ± 0.8	6-9
Pupa (female)	214.0 ± 41.0	148.0-279.0	7.4 ± 1.3	6-11
Adult (male)	58.0 ± 20.0	26.0-108.0	-----	--
Adult (female)	144.0 ± 33.0	102.0-218.0	-----	--

[a] Weights for first 4 days of the instar.

b. Diapause method

Development stage	Weight (mg) Mean ± S.D.	Weight (mg) Range	Duration (days) Mean ± S.D.	Duration (days) Range
Instar 1	------	--	-----	--
Instar 2	0.3 ± 0.2[a]	0.09- 0.6[a]	7.4 ± 2.7	5-11
Instar 3	2.3 ± 1.2[a]	0.34- 4.4[a]	3.8 ± 4.3	1-4
Instar 4	8.9 ± 4.5[a]	1.4 - 16.0[a]	5.2 ± 8.6	2-4
Instar 5	38.0 ± 22.0[a]	7.9 - 78.0[a]	3.1 ± 0.5	2-4
Instar 6	124.0 ± 72.0[a]	21.0 - 281.0[a]	6.6 ± 1.6	5-10
Pupa (male)	133.0 ± 24.0[a]	109.0 - 196.0	7.2 ± 0.8	5-8
Pupa (female)	214.0 ± 40.0	141.0 - 308.0	8.0 ± 4.8	5-9
Adult (male)	59.0 ± 13.0	42.0 - 85.0	-----	--
Adult (female)	139.0 ± 29.0	98.0 - 204.0	-----	--

[a] Weights for first 4 days of the instar.

6.2 Survival data

- Eggs (84%)
- Larva to pupa (52%)
- Pupa to adult (89%)

6.3 Other data

- Preoviposition period (ca 8 h)
- Peak egg laying period (ca 2 days after eclosion)
- Mean fecundity [136 eggs/female (minimum 70, maximum 210)]
- Mean adult longevity (5 days)
- No. of larval instars (6)

7. REFERENCES

Baker, W. L., 1972. Eastern Forest Insects. USDA Forest Service, Miscell. Pub. 1175. pp. 378-381.

Furniss, R. L. and Carolin, V. M., 1977. Western Forest Insects. USDA, Forest Service, Miscell. Publ. 1339. pp. 168-172.

Lyon, R. L., Richmond, C. E., Robertson, J. L. and Lucas, B. A., 1972. Rearing diapause and diapause-free western spruce budworm (*Choristoneura occidentalis*) (Lepidoptera: Tortricidae) on artificial diet. Can. Entomol. 104: 417-427.

Powell, J. A., 1980. Nomenclature of the Nearctic conifer-feeding *Choristoneura* (Lepidoptera: Tortricidae): historical review and present status. USDA, Forest Service, General Tech. Rep. PNW-100. 18 p.

Robertson, J. L., 1979. Rearing the western spruce budworm. USDA Miscell. Publ. Canada/United States Spruce Budworms Program. 18 p.

Stock, M. W. and Robertson, J. L., 1980. Inter- and intraspecific variation in selected *Choristoneura* species (Lepidoptera: Tortricidae): a toxicological and genetic survey. Can. Entomol. 112: 1019-1027.

CYDIA POMONELLA

MICHAEL D. ASHBY, PRITAM SINGH and GRAEME K. CLARE

Entomology Division, Department of Scientific and Industrial Research,
Private Bag, Auckland, New Zealand

THE INSECT

Scientific name: *Cydia (=Laspeyresia) pomonella* (Linnaeus)
Common name: Codling moth (CM)
Order: Lepidoptera
Family: Olethreutidae

1. INTRODUCTION

The codling moth is native to south-eastern Europe but its worldwide distribution makes it a key pest of international importance. It attacks apples, pears, walnuts, quinces, crab apples, hawthorn, passionfruit, oranges, plums, peaches, nectarines, cherries and apricots.

The life history is similar on all hosts with 1-3 generations per year depending upon climatic conditions. The egg-laying period extends through the late spring and summer. Female moths lay about 50, single, pancake-shaped eggs on or near fruit. Upon hatching, first instar larvae search for a suitable fruit entry site, usually the cheek or calyx. Damage is caused from late spring through until early autumn. Partial entry by neonates causes "stings" or shallow blemishes on the fruit surface and fruit quality is lowered as a result. Upon complete entry into the fruit, larvae usually penetrate to the seed cavity then tunnel their way back out. Some infested fruits drop from the tree and the larvae complete development on the ground. Mature fifth instars usually leave the fruit to seek a pupation site, or overwinter as diapause larvae in cocoons under loose bark or in the ground at the base of the host tree. With diapause termination in late winter or early spring, the larvae pupate and adults begin to emerge.

The codling moth has been extensively researched (Butt, 1975) and many of the artificial diets reported have been reviewed by Singh (1977). A laboratory colony can be established from field-collected specimens or by obtaining insects from an already established laboratory colony. Codling moth have been reared in our laboratory for over 140 generations for post-harvest disinfestation research, pheromone studies and virus and parasite production. Two diets are used in our laboratory. The diet of Singh (1983) is used for individual rearing and the production of diapause larvae. The preparation of this diet and methods for the individual rearing of larvae and pupae are exactly the same as described for the chapter on *Epiphyas postvittana*. The conditions required for the production of diapause larvae are outlined in the present chapter. A modified diet of Brinton *et al.* (1969) is used for group rearing and the description of this method follows.

2. FACILITIES AND EQUIPMENT REQUIRED

2.1 Facilities

Eggs, larvae, pupae and adults are reared in constant environment rooms with continuous air circulation at 25 ± 1°C, 60 ± 5% RH under an 18 h photoperiod. Light is supplied by four fluorescent tubes mounted in the ceiling and covered by transparent, multi-faceted perspex. Mating and oviposition are controlled in the same rooms under constant darkness.

2.2 Equipment and materials for insect handling

2.2.1 Adult collection

a. Individual collection
- 75 x 12 mm polystyrene test tubes
- Cotton wool

b. Group collection by phototaxic method
- Opaque polyvinylchloride (PVC) box (510 x 362 x 240 mm) open at bottom with handle on top (see Fig. 1). A 45 mm hole in one end has its centre located 145 mm from the bottom edge. The horizontal centre lies equidistant from the two sides.
- Perspex tube (44 mm external diam x 150 mm long)
- Adult oviposition container with lid with 44 mm hole in centre (see Fig. 2)
- Light source

Fig. 1 PVC rearing container cover and attachment of mating and oviposition container

2.2.2 Adult mating and oviposition
- Cylindrical cardboard container (185 mm long x 135 mm diam) with lid at one end, 10 meshes/cm gauze for ventilation at other end and 25 mm hole in side (see Fig. 2)
- Sheet of waxed lunch paper (450 x 300 mm) (Handee super waxed lunch paper, Caxton Printing Works Ltd, New Zealand)
- Cotton wool
- Open shelved cupboard 100 cm high x 92 cm wide x 45 cm deep with 3-4 adjustable shelves inside. Cupboard darkened with black linen cover which has two vertical zips on either side at front
- Scissors
- Stapler

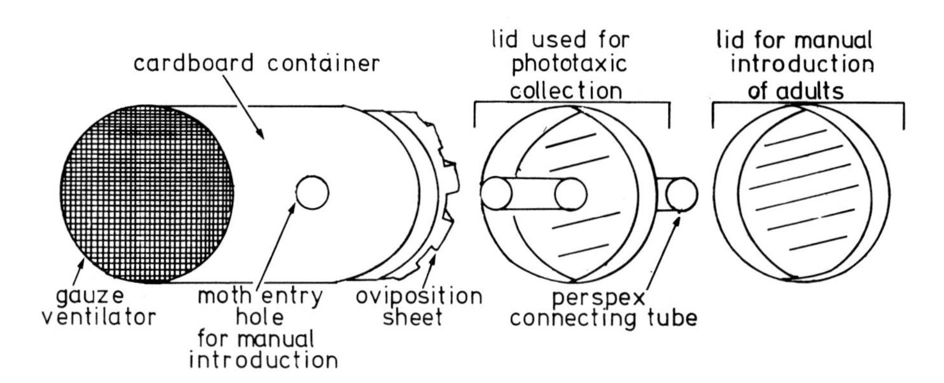

Fig. 2 Construction of oviposition container

2.2.3 Adult disposal
- Chloroform
- Cotton wool
- Waste disposal unit or garbage bag

2.2.4 Egg collection and storage
- Clear plastic No. 490 Savoy® containers (190 mm high x 210 mm diam) with lids (Savoy Housewares, Brisbane, Australia)

2.2.5 Egg surface-sterilization
- 2 plastic trays (600 x 350 x 75 mm)
- 3 litre container of 5% formalin (2% formaldehyde)
- Water supply
- 4 litre container of sterile water
- Thick soft paper towelling
- Long metal forceps or crucible tongs
- Fume cupboard or fume hood
- Plastic gloves

2.2.6 Larval rearing containers
- Containers with lower stainless steel tray (500 x 350 x 30 mm) and upper perspex box (484 x 352 x 210 mm) with 8 meshes/cm stainless steel top. Upper perspex section with hatch with 44 mm hole in centre [Ashby *et al.* (1982); see Fig. 3]
- Rubber bung (30 x 40 x 50 mm)

2.2.7 Holding insects at lower temperatures
- Incubators at 10, 15 and 20°C

2.3 <u>Diet preparation equipment</u>

2.3.1 Stock vitamin mixture
- Spoon
- Electric or manual micro-quantity dispensing spatula
- Disposable weighing trays
- 500 ml beaker
- 5 litre beaker
- 1 litre measuring cylinder
- Top loading balance (0.1 - 1000 g)
- Pan balance (0.01 - 100 g)
- Magnetic stirrer
- 4 cm stirring bar
- Plastic storage bottle
- Refrigerator at 4°C

2.3.2 Finished diet
- 10 quart Hobart® mixer (Model A200) with whisk attachment

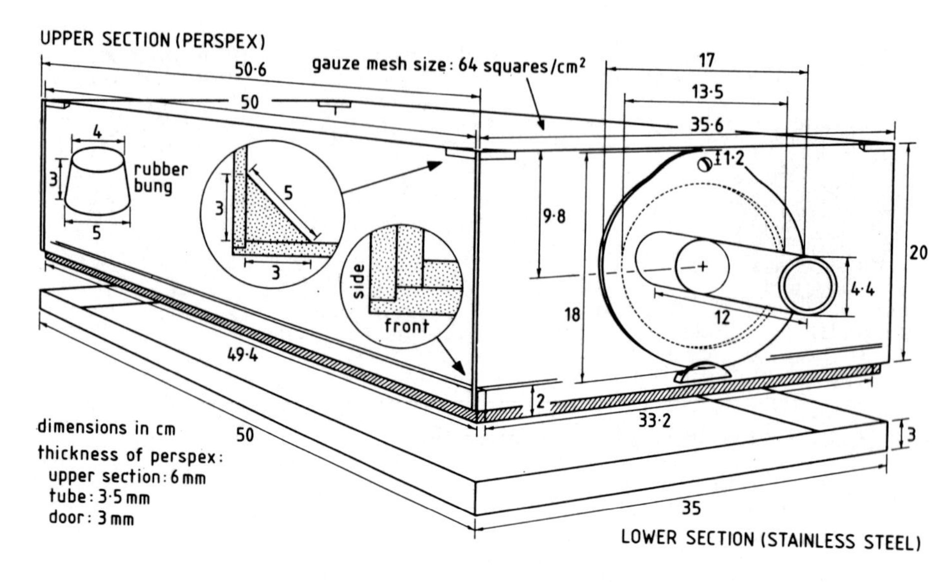

Fig. 3 Rearing container construction and dimensions

- Hotplate
- 5 litre cooking pot
- Plastic scoop (150 - 200 mm long)
- Top loading balance (0.1-1000 g)
- 10 ml measuring cylinder
- 100 ml measuring cylinder
- 1 litre measuring cylinder
- Oven or drying cupboard (up to 140°C)
- Stainless steel or enamel container (500 x 200 x200 mm)
- Transparent polythene bags (250 x 375 mm, 35 gauge)
- 0 - 100°C thermometer
- 3 stainless steel rearing container trays (see 2.2.6)
- Chiller at 4 ± 2°C
- 75% ethanol
- Cotton wool

2.3.3 Protective wax coating

- Gas ring with asbestos gauze or equivalent
- Paraffin wax with 60°C congealing point
- 1 litre stainless steel beaker
- 30-40 mm wide paint brush
- Polythene bag (250 x 375, 35 gauge)

3. ARTIFICIAL DIET

3.1 Composition

a. Ingredients to prepare 1 litre of stock vitamin mixture

Ingredient	Amount	Approx. %
Choline chloride	100.00 g	9.965
Nicotinic acid	1.00 g	0.100
Ca-pantothenate	1.00 g	0.100
Riboflavin	0.50 g	0.050
Thiamine HCl	0.25 g	0.025

Pyridoxine HCl	0.25 g	0.025
Folic acid	0.25 g	0.025
Cyanocobalamin	0.20 g	0.020
d-Biotin	0.02 g	0.002
Ethanol (95%)	700.00 ml	69.758
Water	200.00 ml	19.930
Total mixture approx.	1000.00 ml	

b. Ingredients to prepare 13.0 kg of finished diet (sufficient for 3 rearing containers)

Ingredient	Amount	Approx. %
*Untreated sawdust	1084.0 g	0.326
Bran	288.0 g	2.212
Sugar	432.0 g	3.318
Casein	432.0 g	3.318
Cellulose	181.2 g	1.391
Citric acid	144.0 g	1.106
Whole wheat flour	1584.0 g	12.167
Natural raw wheatgerm	144.0 g	1.106
Wesson's salt	100.0 g	0.768
Sorbic acid	12.0 g	0.092
Methyl *p*-hydroxybenzoate	6.0 g	0.046
Cholesterol	2.8 g	0.021
Ascorbic acid	57.6 g	0.442
Linoleic acid	7.2 ml	0.055
Vitamin mixture	144.0 ml	1.110
Water	8400.0 ml	64.522
Total diet approx.	13.0 kg	

* Particles (25 mm^2) stored for at least 2 months.

3.2 Brand names and sources of ingredients

Untreated sawdust (*Pinus radiata* from local sawmill), natural raw wheatgerm, whole wheat flour, bran (Sanitarium Health Food Co., Auckland, NZ), Wesson's salt mixture, cellulose powder (Bio-Serv Inc., PO Box BS, Frenchtown, New Jersey 08825, USA), sorbic acid, linoleic acid (BDH Chemicals Ltd, Poole, UK), methyl *p*-hydroxybenzoate, Ca-pantothenate, riboflavin, thiamine hydrochloride (Koch-Light Laboratories Ltd, Colnbrook, Bucks, UK), cholesterol, choline chloride, nicotinic acid, pyridoxine hydrochloride, cyanocobalamin, *d*-biotin (Sigma Chemical Co., PO Box 14508, St Louis, MO 63178, USA), sugar (Chelsea Refinery, NZ Sugar Co. Ltd, Auckland, NZ), casein (NZ Co-op Dairy Co., Casein Milling Station), citric acid (Labsupply Pierce (NZ) Ltd), ascorbic acid (Pharmaceutical Sales and Marketing, Auckland, NZ), folic acid (E. Merck, D61 Darmstadt, Germany).

3.3 Diet preparation procedure

a. Vitamin mixture

- Weigh choline chloride into 5 litre beaker and use spoon to break up any large lumps
- Weigh other dry ingredients into disposable weighing trays using spoon or microquantity dispensing spatula as required and add ingredients to choline chloride
- Measure out ethanol and water using 1 litre measuring cylinder and add to dry ingredients
- Mix suspension thoroughly using magnetic stirrer and 4 cm stirring bar

- Decant into plastic storage bottle and store in refrigerator at 4°C

b. Finished diet
   - Weigh sawdust into stainless steel or enamel container and place in drier or oven to sterilize at 140°C for 30 minutes. Mix
   - Measure out water, pour into cooking pot and heat until almost boiling
   - Swab Hobart mixing bowl and whisk using cotton wool soaked with 75% ethanol
   - Weigh sugar, casein, cellulose powder and citric acid and place in mixing bowl
   - Weigh whole wheat flour, bran, wheatgerm, Wesson's salt mixture, sorbic acid, methyl *p*-hydroxybenzoate and cholesterol into a polythene bag
   - Add hot water to mixing bowl
   - Whisk at high speed until froth rises to top of bowl
   - Sterilize stainless steel scoop with ethanol
   - Change to medium speed and gradually add flour mixture from polythene bag
   - When all of flour mixture added, increase to high speed and mix thoroughly until any remaining lumps have been completely broken up
   - Remove sawdust from oven
   - Sterilize stainless steel scoop with ethanol
   - Gradually add sawdust with whisk at low speed
   - When all added, increase to medium speed
   - Weigh ascorbic acid, measure out linoleic acid and shake vitamin mixture until uniform
   - Leave mixer on medium speed until diet cools to 55-60°C
   - Add ascorbic and linoleic acids
   - Reshake vitamin mixture, measure out and add to diet
   - Mix thoroughly for 10 minutes
   - Still at medium speed, wind down mixing bowl as low as possible to centrifugally remove diet stuck to whisk
   - Swab stainless steel trays with 75% ethanol and dispense diet

c. Protective wax coating
   - Melt 300 g paraffin wax in stainless steel beaker and heat until almost boiling
   - Coat diet with hot wax using paintbrush
   - Allow wax to set and use diet immediately or place in polythene bag and store in chiller at 4°C

PRECAUTIONS

(i) Ensure that sawdust is untreated and completely dry.

(ii) Sawdust quality may vary. It is always advisable to add 10% less water than is specified in the composition until quality is checked. Sterile water can be added if the finished diet is too dry.

(iii) Do not lay the paraffin coating on too heavily since neonate larvae may not be able to penetrate it. A better spread is obtained if the wax is very hot. Use even dabbing strokes, the edges first and then the centre.

(iv) Diet stored in a chiller must be conditioned for 24 h before use.

4. REARING AND COLONY MAINTENANCE

4.1 Adults

4.1.1 Emergence

The first adult will emerge 29-30 days after colonization of the diet by the neonate larvae. Emergence is complete by day 39 with a peak occurring on days 31 and 32. Adult survival from neonate larvae is 60-70% and a mean of 770 moths can be produced from one rearing container.

4.1.2 Collection and maintenance

Adults are collected individually or by photopositive response. Equal numbers of males and females are produced.

a. Individual collection
   - Remove rubber bung from hatch of rearing container

- Open hatch and collect one moth in each test tube
- Bung end of tube with cotton wool
- Store at 15°C until use

b. Phototaxic collection
- Prepare mating and oviposition container (see 4.1.3 and Fig. 2)
- Remove rubber bung from hatch
- Place opaque cover over rearing container
- Insert perspex tube through hole in cover and hole in hatch
- Use mating and oviposition container lid with 44 mm hole in centre
- Fit container over perspex tube with gauze end near light source
- Replace container twice per day or more often during peak emergence using rubber bung or cotton wool to plug hole

General
Adults may be fed a 10% honey solution through entry hole in side of container. Although this will increase lifespan it will not significantly increase fecundity.

4.1.3 Adult oviposition

Adults are introduced into a mating and oviposition container which is placed in the mating and oviposition cupboard. Allow 1 day for the pre-mating period and 1 day for the pre-oviposition period. Peak egg laying occurs 2-3 days after mating.

a. Preparation of mating and oviposition container (see Fig. 2)
- Cut out 450 x 300 mm sheet of waxed lunch paper
- Pleat sheet with sharp creases about 1.5 cm apart
- Make a series of 4.5 cm long cuts about 2.0 cm apart along the length of both sides of the sheet
- Fold the cut strips on both sides toward the centre of the sheet
- Draw the sheet in at both ends to fit the oviposition container and staple the ends together
- Place sheet inside container and cut hole for moth entry if necessary
- Replace lid

b. Introduction of moths into mating and oviposition container
- Place 30 pairs of moths into container either individually through hole in side or by phototaxic method
- Plug hole in side of container or lid with cotton wool and place container in oviposition cupboard

PRECAUTION
If too many moths (>35 pairs) are allowed to enter the container, oviposition will be inhibited.

4.1.4 Adult disposal
- Soak a little cotton wool in chloroform and place in sink with waste disposal unit
- Remove lid from mating and oviposition container
- Shake moths into sink
- Dispose of dead moths

4.2 Eggs

4.2.1 Egg collection
- Remove oviposition sheet and place in Savoy container at 25°C

General
Eggs may be collected every day and moths transferred to a newly prepared container. Incubation of eggs can then occur at different temperatures to allow hatch of eggs to be synchronized. The transfer of moths is made easier by chilling at 4-5°C for a few minutes.

4.2.2 Surface sterilization of eggs

Egg sterilization should be carried out if mould or other contaminants have developed on eggs. Under normal conditions surface-sterilization is not necessary.

- Use fume cupboard or fume hood with sink and wear plastic gloves
- Decant formalin into plastic tray
- Place up to 3 eggsheets in formalin and agitate gently for 10-15 minutes using forceps or crucible tongs
- Use second tray to rinse sheets in gently running tap water for 10 minutes
- Replace tap water with sterile water and agitate gently for a further 5 minutes
- Remove sheets, place at 25°C, and dry by laying flat on soft tissue paper and covering with tissue paper

PRECAUTIONS

(i) Do not surface-sterilize eggs until 4-6 days old.
(ii) Do not place more than 3 eggsheets in either the formalin or water baths since large numbers of eggs may be lost due to friction.
(iii) Ensure that trays, tissue paper and tongs are sterile before use.

4.3 Larvae

4.3.1 Diet inoculation with neonate larvae from eggsheets

- Condition diet by allowing to stand at rearing temperature for 24 h
- Remove eggsheets from Savoy containers
- When eggs are at blackhead stage cut sheets into strips ca 4 cm wide
- Lay strips with eggs facing down onto diet
- Fit perspex top of rearing container to bottom tray of diet
- Place rubber bung in hatch hole
- Remove and discard strips 3-4 days after first egg hatch

General

The paraffin wax coating on the diet should prevent growth of mould and bacteria. However, should these occur they can be removed by cutting around and removing the infected area. Swab the exposed edges of the remaining diet in the tray with mould inhibitor solution (see section 3.1 c. of the chapter on *Epiphyas postvittana* for mould inhibitor preparation, Vol. II, p. 275

4.3.2 Production of diapause larvae

Diapause larvae are reared individually in polythene test tubes by the same method as is described for *Epiphyas postvittana* larvae with the following change:

- Use a photoperiod of 12 hours to induce diapause

General

Use a very tight cotton wool plug to prevent escape of mature larvae. Most larvae cocoon about 30 days after being inoculated onto the diet. Survival to cocooned fifth instar is about 98%.

4.4 Sex determination

- Adult females have a large ovipository pore and are generally larger than males
- Female pupae have 4 ventral abdominal segments. Males have 5.

4.5 Rearing schedule

a. Days 1-7
   - Check developing eggs for contamination
   - Surface-sterilize eggs on days 4-6 if necessary

b. Days 8-10
   - Check diet condition
   - Cut eggsheets into strips and place on diet
   - Check progress of diet colonisation by first instars

c. Day 11
   - Remove sheets from diet and discard

d. Days 8-37
   - Periodically check diet condition and insect development. Development can be gauged indirectly by checking size of accumulating frass mounds.

e. Days 38-44
   - Set up mating and oviposition containers
   - Collect adults either individually or phototaxically into mating and ovipostion containers
   - Collect and incubate eggs

4.6 Insect quality

Insect quality is checked periodically by measuring adult fecundity, egg fertility, larval development, pupal weights and susceptability of wild and laboratory reared larvae to fumigants.

a. Adult fecundity

Fresh females (30 pairs/container) lay a mean of 97 eggs each. A minimum of about 60 eggs/female is acceptable. Eggs are easier to count toward the end of their development.

b. Egg fertility

Egg fertility should lie between 95 and 100%.

c. Larval development

The duration (days) of each larval stadium and head capsule measurements (mm) should lie within the following ranges:

	Larval development (days)			Head capsule measurements (mm)		
Instar	Minimum	Maximum	Mean	Minimum	Maximum	Mean
1	4	5	4	0.295	0.375	0.330
2	2	3	2	0.425	0.575	0.495
3	3	4	3	0.715	0.850	0.800
4	3	6	4	1.020	1.300	1.145
5	10	13	11	1.420	1.630	1.521

d. Pupal weights

Pupae with an average weight of 36.5 ± 3.8 mg for females and 29.9 ± 3.0 mg for males are considered of high quality with respect to most user-group research objectives in our laboratory.

e. Larval susceptibility to fumigants

During post-harvest disinfestation programmes, LD50 values after methyl bromide fumigation are compared between laboratory-reared and field-collected larvae. Larvae are considered of high quality when there are no significant differences in LD 50s.

4.7 Special problems

a. Dietary deficiency

The diet rarely produces any deformed insects. It is good practice, however, to date all dietary ingredients when they arrive in the laboratory. Prepared, unused vitamin mixture kept at 2-4°C should be discarded after one year. Ascorbic and linoleic acids must be kept in dark bottles.

b. Contamination of diet

Bacterial and fungal infection of the diet can occur if high hygiene standards are not maintained. Freshly-prepared diet should not be left exposed to the air for prolonged periods.

c. Human health

The inhaling of moth scales may produce an allergic or other unpleasant respiratory reaction. The use of a protective particle mask is advised, particularly at egg collection.

Formalin is carcinogenic and must always be used under a fume hood.

5. HOLDING INSECTS AT LOWER TEMPERATURES

5.1 Adults

Moths may be stored at 15°C in polystyrene tubes for up to 10 days. However, fecundity will begin to drop if moths are held for more than 3-4 days.

5.2 Eggs

Eggs may be held at 10-15°C provided a humidity of 60-75% is maintained. Where control rooms or incubators have no humidity control, a piece of lightly moistened filter paper can be placed under cellulose wadding in the bottom of each egg storage container to maintain higher humidity.

PRECAUTIONS

(i) Containers should be checked daily for moisture build-up. Any visible moisture on the insides of containers should be wiped away to discourage growth of contaminants.
(ii) Do not hold 1-2 day old eggs at temperatures lower than 10°C since some desiccation may occur.
(iii) Sterilize eggs if mould develops.

5.3 Larvae

Neonate larvae may be held for up to 36 h in egg storage containers at 15°C. Although larvae can be held on diet at 15°C for their entire larval period, this is not advised since development rates will vary considerably and some may go into diapause. A minimum temperature of 20°C is suggested. Diapausing fifth instar larvae may be held for 6-18 months at 15°C under conditions of no light and ambient humidity.

PRECAUTIONS

(i) Larval development is more satisfactory if first instars are allowed to establish at 25°C.
(ii) Maintain a photoperiod of 18 h or longer at lower temperatures to avoid diapause.

5.4 Pupae

Pupae may be held in rearing containers at 15°C for 10-14 days.

PRECAUTION

Pupae held at temperatures lower than 15°C may desiccate if they were reared at 20-25°C as larvae.

6. LIFE CYCLE DATA

Developmental data for individually-reared insects are given in the chapter on insect rearing management (Vol. I)

Life cycle and survival data are given below for optimum rearing conditions of 25 ± 1°C, 60 ± 5% RH and an 18 h photoperiod.

6.1 Developmental data

Stadium	Number of days		
	Min	Mean	Max
Eggs	6.0	6.5	7.0
Larvae	22.0	24.0	31.0
Pupae	7.0	8.0	9.0
Adult emergence (from inoculation)	29.0	32.0	40.0
*Total development	37.0	40.5	49.0

* Egg to egg including pre-mating and pre-oviposition periods.

6.2 Survival data

Egg hatchability	95%
First instar survival	60-70%
Established neonate larva to pupa	>95%
Established larva to adult	>90%

6.3 Other data

Pre-mating period: 1 day
Pre-oviposition period: 1 day
Peak egg-laying period: 2-3 days after mating
Mean fecundity: 97 eggs/female
Larval development: See 4.6 c.
Quality of pupae: See 4.6 d.

7. INSECT YIELD

The production of eggs and adults by this method results in a high insect yield for little investment in time and labour resources. Although the mortality of first instar larvae is high (30-40%), these larvae are an easily obtainable resource.

Each rearing container can produce 550-1000 moths with a mean of about 770 moths/container. Thirteen kilograms of diet (sufficient for 3 rearing containers) is the standard amount used for each colony per generation in our laboratory. Two colonies (each of the same generation) are kept with the beginning of adult emergence of one colony occurring two weeks after the end of emergence of the other. Thus about 4600 moths are produced for each generation. These moths are capable of producing more than 220,000 eggs if required. The time required to collect these moths individually (although it is not usually necessary to collect all of them) is about 24 h for one person with a further 3.5-4 h needed to place them in mating and oviposition boxes. Phototaxic collection requires only the systematic changing of mating and oviposition boxes and using this system, one person can maintain both colonies in 5-7 days/generation.

8. PROCEDURES FOR SUPPLYING INSECTS

8.1 General prerequisites

- Establish a healthy and disease-free colony
- Adopt a standard rearing method and routine
- Maintain a laboratory maintenance colony of at least 2000 adults per generation
- Maintain more than 1 colony if necessary to facilitate user-group requests
- Ensure that the duration of all life cycle stages under various conditions of temperature, humidity and photoperiod is known
- Make a habit of informing user groups about colony status and ensure that requests for insects are placed well in advance
- Organise adequate material and personnel resources
- Maintain a back-up colony away from the main laboratory colony to avoid total loss in case of equipment failure

8.2 Shipping

C. pomonella may be shipped as eggs, larvae or pupae within the country of origin, but are most safely shipped as larvae for exportation overseas. The regulations regarding importation should be investigated if shipping is international. All containers should be well marked with their contents. Weekend arrivals should be avoided if possible.

a. Eggs

Eggs should be collected and cut out as large batches to save space, and packed into polystyrene Petri dishes lined with cotton wool. The dishes should be taped together, surrounded with packing material, and placed in a suitable sized box. Time between dispatch and arrival should not exceed 7-8 days.

b. Larvae

Larvae should be sent individually packed in polystyrene tubes with an adequate supply of freshly-dispensed diet. Tubes should be tightly bundled together and packed between cushions of packing material such as cotton wool, cellulose wadding or styrofoam beads. Larvae should be dispatched as first or second instars and the mode of shipping should be expeditious to prevent diapause.

c. Pupae

Pupae may be packed between cotton wool cushions in Petri dishes or rolled between 12-layer absorbent paper (Baumhover *et al.* 1977). They should be sent as soon as they have hardened. It is advisable to increase humidity by placing a small piece of lightly moistened Whatman No. 1 filter paper in the bottom of each Petri dish. Time between dispatch and arrival should not exceed 8 days.

REFERENCES

Ashby, M.D.; Hancock, D.F.; Singh, P. 1982. A container for rearing codling moth. N.Z. J. Zool. 9: 515-518.

Baumhover, A.H.; Cantelo, W.W.; Hobgod, J.M.; Knott, C.M.; Lam, J.J. Jr. 1977. An improved method for rearing the tobacco hornworm. USDA Bull., ARS-S-167, 13 pp.

Brinton, F.E.; Proverbs, M.D.; Carty, B.E. 1969. Artificial diet for mass production of the codling moth, *Carpocapsa pomonella* (Lepidoptera: Olethreutidae). Can. Entomol. 101: 577-584.

Butt, B.A. 1975. Bibliography of the codling moth. ARS W-31, USDA, 221 pp.

Singh, P. 1977. Artificial diets for insects, mites and spiders. Plenum Press, New York, 594 pp.

Singh, P. 1983. A general purpose dietary mixture for rearing insects. Insect Sci. Appl. 4: 357-362.

ACKNOWLEDGEMENT

The authors gratefully acknowledge Melody Tapene who typed the manuscript.

CYDIA POMONELLA

D. K. REED and N. J. TROMLEY

USDA-ARS , Fruit and Vegetable Insect Research Laboratory, P. O. Box 944, Vincennes, IN 47591

THE INSECT

Scientific Name: Cydia (Laspeyresia) pomonella (L.)
Common Name: Codling moth (CM)
Order: Lepidoptera
Family: Olethreutidae

1. INTRODUCTION

The codling moth is a key pest affecting pome fruit production in all major growing regions of the U.S. and the world, with the exception of some areas in Asia. Without adequate control measures, this insect can destroy 100% of the apple crop and even with current control, up to 10% of the fruit is lost annually (Schwartz and Klassen, 1981). The codling moth not only attacks apples and pears but is also a serious pest of apricots, English walnuts, quince, crab and many wild Rosaceous plants.

The codling moth overwinters as a fully mature larva within a cocoon under loose bark on tree trunks or under other shelters within the orchard environment. Packing sheds in the area are also sometimes used as overwintering sites. These larvae are resistant to temperatures as low as -35°C. The pupal stage in the field lasts from 2-6 weeks. Adult females may deposit 100 or more eggs under optimum field conditions, normally on the upper side of leaves or on twigs and fruit spurs. Oviposition normally takes place after trees have bloomed, so that larvae hatching 1-3 wk later may find young apples to enter, either through the calyx cup at the blossom end or through the surface. Such larvae usually tunnel to the core and may feed on developing seeds. After 3-5 wk, full grown larvae exit the apples and either drop to the ground or crawl down the trunk to pupate. There are 2 or more generations in the U.S. with 1 in more northern areas.

Until 1930, codling moth research conducted only during the summer, when adequate supplies of feral insects were available. Farrar and Flint (1930) changed this when they devised a system where relatively large numbers of insects could be reared on apples in the laboratory. It is interesting that they used wax paper for oviposition, which is still utilized today. Dickson (1949) determined the role of photoperiod in governing diapause, making continuous rearing possible. Using apple thinnings, Dickson et al. 1952, produced 60 insects per tray of 130 apples.

The first work on artificial diets for codling moth was done by Theron (1947) who reared larvae on a corn meal medium, but he was unable to overcome the diapause problem and yields were low. Redfern (1960) obtained satisfactory numbers of insects with a semisynthetic diet which was based on apple seeds, dried apple and English walnut meats. Rock (1967) successfully used a casein diet. By 1975 a number of diets had been developed, most of which were reported by Singh (1977). Butt (1975) classified most of the diets into 1 of 2 types, the sawdust type (Brinton et al. 1969) or wheat germ type (Howell, 1970, 1971). The bean diet of Shorey and Hales (1965) and more recently the corn meal diet of Guennelon et al. (1980) has also given good results. Each diet has it own merits and short-comings. The sawdust diet has fewer problems with fungi, but virus has been a problem. The larvae pupate in the sawdust, bean, and corn meal diets because of dryness, but with wheat germ diets the larvae leave the diet to pupate elsewhere, as in corregated strips provided. Many of the rearing methods used in our laboratory were reported by Hamilton and Hathaway (1966). A laboratory colony can be established from field collected specimens or by obtaining insects from an established colony.

2. FACILITIES AND EQUIPMENT REQUIRED

2.1 Facilities

Larvae and pupae are reared in chambers or controlled environment rooms maintained at 27.8-28.8 ° C, 65-70% RH and an LD of 16.5:7.5 or a continuous light photoperiod furnished by fluorescent lighting. Oviposition is carried out within an incubator maintained at 27.8 ° C. Lighting intensity is not critical but should be subdued, with a range of 3-5 fc adequate.

2.2 Equipment and materials required for insect handling

2.2.1 Adult oviposition

- Cardboard cylinder (salt box) 7.6 cm dia-15 cm length with caps at both ends. (Fig. 1)
- 25x52 mm, 4 dr (15.6 g) glass vials
- Absorbent cotton wool
- Wax paper
- Scissors
- Dowel rods or pearwood 1-2 cm dia x 15 cm
- Cheesecloth
- Rubberband

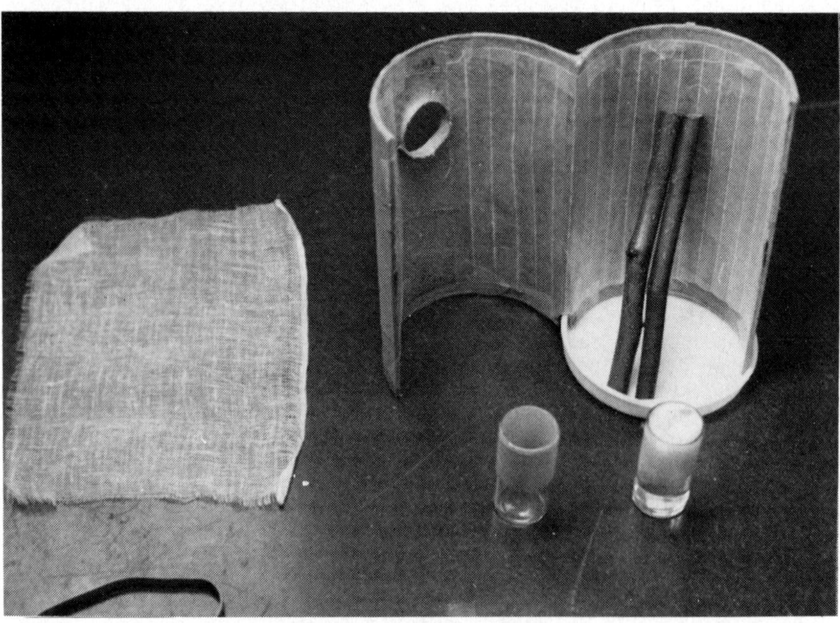

Fig. 1. Oviposition chamber for codling moth.

2.2.2 Egg hatch and larval inoculation of diet
- 25x52 mm, 4 dr (15.6 g) glass vials
- Desk lamp or other light source
- Fine camel hair paint brushes (size 000)

2.2.3 Larval rearing containers
- 1 oz (29.5 ml) plastic cups
- Plastic coated paper lids for cups
- Dinner trays 52x39 cm

2.2.4 Adult collection
- Screen cage
- Fluorescent lamp
- Collecting device made from a portable vacuum cleaner, No. 5 size rubber stopper and 6 mm ID copper tubing (15 cm and 35 cm) (Fig. 2)

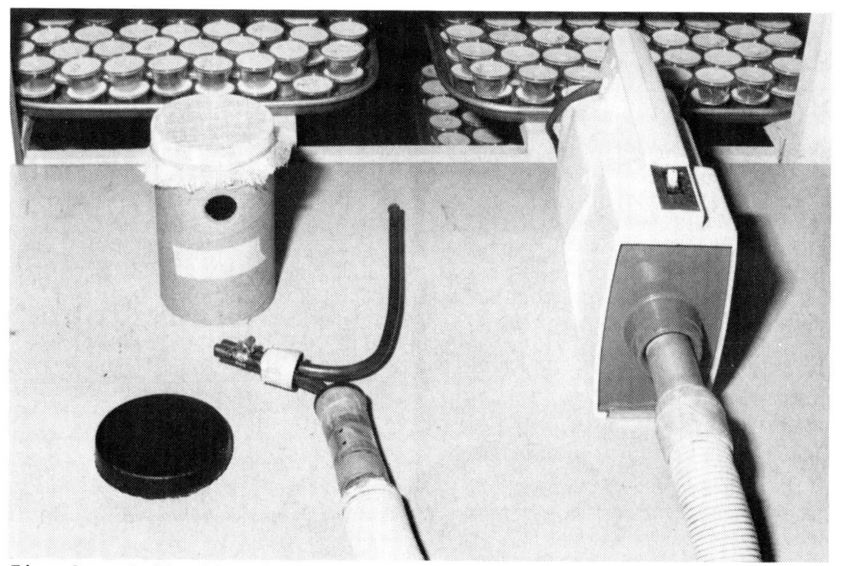

Fig. 2. Collection device for codling moth adults.

2.3 Diet preparation equipment
- Stainless steel basins 1.5-3. liter (1 per batch)
- Colander
- Blender
- Top loading balance (0.1-1000 g)
- 10 ml graduated cylinder
- Hot plate
- Stirrer
- 1000 ml beaker
- 4 oz (113 ml) plastic container
- Plastic catsup dispensers
- Large rubber spatula
- Small rubber spatula
- Thermometer (0-100 C)
- Asbestos gloves
- Fan
- Terricloth towels

3. ARTIFICIAL DIET

3.1 Composition

Ingredient	Amount for 700 cups	% of total wt.
Dry pinto beans	457.09 g	23.6
Dried brewers yeast	64.09	3.5
L. ascorbic acid	6.49	0.4
Methyl-p-hydroxybenzoate	4.09	0.2
Sorbic acid	2.09	0.1
Formalin (37%)	4.0 ml	0.2
Agar	25.69 g	1.4
Distilled water	1280.0 ml	70.6
Total Diet	1813.0 ml	100.0

3.2 Names and sources of ingredients

Brewers yeast, sorbic acid, agar-granulated, (ICN Nutritional Biochemicals, Inc. 26201 Miles Rd., Cleveland, OH 44128); L ascorbic acid, formalin solutions, (Fisher Scientific Co., P. O. Box 12405, St. Louis, MO 63132); methyl-p-hydroxybenzoate (Tenneco Chemicals, Inc. Organics and Oplymers Div. Turner Place, Box 365, Pictaway, NJ 08854); Pinto beans (Bemhen and Co. c/o Southwestern Sales Assoc., P. O. Box 2948, Jacksonville, FL 32203). Other sources may be obtained from the "Frass" Newsletter 7(1) 1981.

3.3 Names and sources of non-dietary items

Plastic cups (Fill-Rite Corp. 49-55 Liberty St., Newark, NJ 07102, Premium Plastics, 465 W. Cermak Rd., Chicago, IL 60616, Bio-Serv. Inc., P. O. Box BS, Frenchtown, NJ 08825), Salt cans which are kraft paper cans w/paper ends (Star Packaging, Inc., 1028 N. Illinois St., Indianapolis, IN 46204); Vials-Kimble 60965-L (Fisher Scientific Co., P. O. Box 12405, St. Louis, MO 63132)

3.4 Diet preparation procedure

- Soak beans overnight in 2X volume of water
- Drain
- Add 1/2 the total required volume of distilled water (640 ml)
- Blend at high speed
- Add other ingredients with exception of agar when pasty consistency is reached.
- Dissolve agar in 1/2 the remaining distilled water at 88.8°C with stirring. (This is accomplished while other materials are being blended).
- Add agar to rest of blended materials and blend until smooth

3.5 Diet dispensing procedure

- Pour hot diet into plastic squeeze bottles
- Dispense 3-4 g into each plastic jelly cup. Less than 3 g will not provide sufficient food for maturation and will dry out. More than 4 g increases moisture levels beyond optimum and can increase larval life span
- Dry the diet for 2 h using a fan to circulate air above the cups. Diet should attain a light crust.
- If diet is not to be immediately infested, it can be held for 4-5 days at room temperature if the trays containing filled diet cups (uncapped) are covered tightly with towels to prevent desiccation.

PRECAUTIONS

Diet preparation room should be scrupulously clean. Counters should be disinfected with Roccal or other anti-microbial agent. Equipment and glassware should be washed with hot soapy water, rinsed with distilled water and covered until used.

4. HOLDING INSECTS AT LOWER TEMPERATURES

4.1 Adults

Moths can be stored individually within the diet cups for 2-3 days within the rearing environment until mating but fecundity is best if used within first 1-2 days after eclosion.

4.2 Eggs

The wax paper containing eggs can be stored for up to 2 wk at 11 ° C within the cardboard oviposition chambers which are capped with cardboard lids.

4.3 Larvae

Larvae are not ordinarily stored, but are utilized soon after hatching.

5. REARING AND COLONY MAINTENANCE

The colony was established from field collected insects and reared continuously for over 60 generations. Ideally, the colony should be renewed after 20-30 generations on artificial diet. Care should be taken not to introduce pathogens such as a granulosis virus into the colony when starting it up.

5.1 Egg collection

Eggs are not ordinarily handled in the rearing procedure but may be utilized by cutting out the required number from the wax paper substrate with scissors.

5.2 Larval collection and diet inoculation

- Remove oviposition chambers from incubator after 5 days when many eggs have reached the "black-head" stage.
- Remove water vial and replace it with an empty vial.
- Orient vials toward a bright light source such as a desk lamp at room condition (27-28°C, 60-55% RH).
- Positively phototrophic neonate larvae will move into the lighted vials and may be transferred to diet cups with a brush.

5.3 Adults

The sex ratio is approximately 1:1

5.3.1 Emergence period ranges from 23-28 days from egg with peak emergence on the 25th day. Some are extended beyond 28 days but for practical purposes these may be discarded.

5.3.2 Collection

- Check cups for adults 1-2 days before time for eclosion and then on 1-2 day basis.
- Open cups containing adults and place them in screen cage near bright light source.
- Collect directly into oviposition chambers using the collection device. A random selection of moths (100/carton) is made. Vacuum should be reduced as low as possible to avoid damage to adults.
- For small colonies, collections may be made with shell vials.

5.3.3 Oviposition

This is done within a cabinet with controlled conditions or may be done in a room if lighting and environmental conditions are satisfactory. Females will survive for 6-7 days if provided with water and will oviposit ca 90-100 eggs each.

- Line salt cans with wax paper 25x13 cm, which is tightly pleated to provide roughened surfaces for oviposition.
- Place a vial of water with cotton stopper into hole in can.
- Place 2-3 dowel rods or pear wood sections in can for resting sites.
- Cover top of can with cheesecloth secured with a rubberband.

5.4 Sex determination

- Female adults tend to be larger than males
- Female pupae have 3 dark segimental bands across caudal margins on ventral surface while males have 4.
- Testes of males are visible through 7-8th dorsal segments of mature larvae.

- Moths immobolized by cold can be examined to observe the tubular structure of the female genitilia.

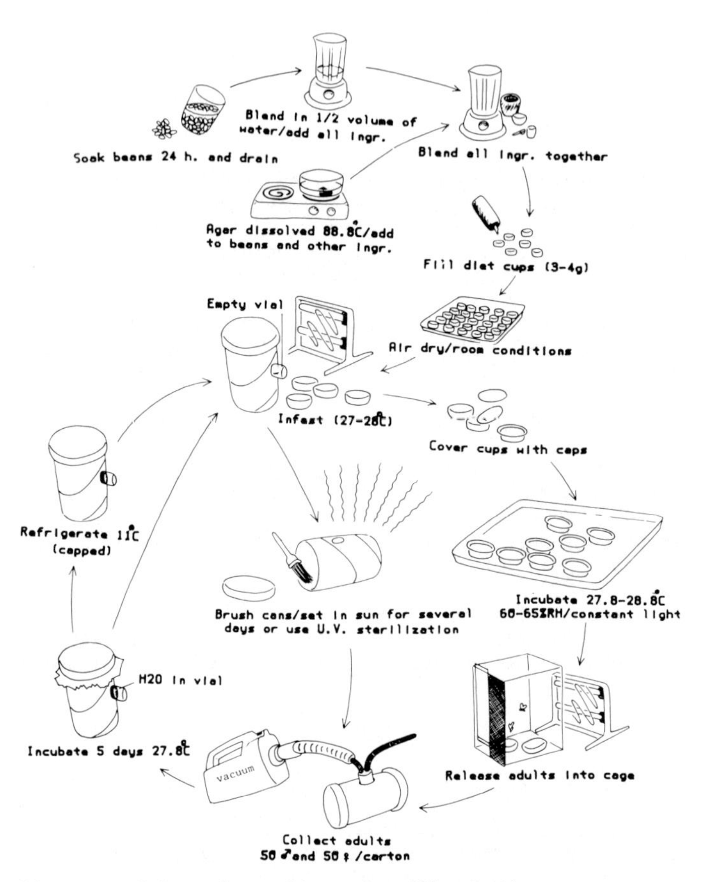

Fig. 3. Schematic outline of codling moth

5.5 Rearing schedule

- Monday	- Remove oviposition containers from incubator and refrigerate. (Discard adults) - Open diet cups and release adults in cage. - Mix diet for the week - Collect moths into oviposition containers and place in incubator.
- Tuesday	- Put oviposition containers under lights with vials in holes for egg hatch. - Infest diet with 2 neonatal larvae each.
- Wednesday	- Same as Monday, no diet mixing - Infest diet
- Thursday	- Same as Tuesday.
- Friday	- Same as Wednesday. - Soak beans (in refrigerator to prevent mold).

This schedule may be arranged differently if desired. Infestation may be done any day that neonate larvae are available (Fig. 5.1)

5.6 Insect quality

Insect quality is monitored continuously by keeping records of mating percentage, average oviposition, percent hatch, pupal weights etc. Routine observations of larval development and mortality and adult behavior should also be made.

5.7 Special problems

5.7.1 Diseases

- Granulosis virus infection of larvae can be a problem, particularly in newly established colonies. This disease may be recognized by observing individual larvae periodically. Infected insects show an opaque ventral integument with a milky appearance. Bodies tend to become sac-like and soft, then turn progressively darker until at death they appear entirely black.

  This may be prevented by autoclaving diet containers and materials, surface sterilization of eggs with formaldehyde and the dipping of larval transfer brushes into alcohol between transfers. Isolation of larvae within diet cups make it possible to disinfect diseased larvae by autoclaving, without having to discard the whole collection. Care should be taken when examining cups so that those containing diseased larvae do not contaminate the others.
- Bacterial and fungal diseases may occur if strict sanitation within the diet preparation room and larval holding room is not maintained.

5.7.2 Human health

- Formalin should be used with care when mixing it in the diet and if used as a disinfectant, work should be done in a hood.
- Exposure to moth scales can be hazardous. Use a suitable mask if working with adults in confined quarters. If an allergic reaction is suspected, consult a physician immediately.

6. LIFE CYCLE DATA

Life cycle and developmental data are given for normal rearing environment

6.1 Developmental data

Stage		Number of Days Min	Mean	Max
Egg		4	5	7
Larva		14	15	18
Pupa	♂		9	
	♀		7	
Adult emergence	♂	21	24	31
	♀	22	22	33
Total development from egg to egg			30	

6.2 Survival data (%)

- Egg viability 85% (70-94)
- Neonate larvae to adult 97%

6.3 Other data

- Preoviposition period 2 days
- Mean fecundity 107 eggs/♀ with pear wood
  99 eggs/♀ with dowel rod
  82 eggs/♀ with no support
- Mean percent mating ♀ - 98
- Mean pupal wt - ♂ 23.3 mg, ♀ 35.8 mg
- Mean larval length last instar 2.0 cm (1.9-2.1 cm)

7. PROCEDURES FOR SUPPLYING INSECTS

No special procedures have been developed to supply insects since the colony is essentially for research. Small numbers of insects may be supplied upon request with adequate lead time.

8. REFERENCES

Brinton, F. E., Proverbs, M. D., and Carty, B. E., 1969. Artificial diets for mass production of the codling moth, Carpocapsa pomonella (L.) (Lepidoptera:Olethreutidae). Can. Entomol. 101:577-584.

Butt, B., 1975. Survey of synthetic diets for codling moths. In "Sterility Principle for Insect Control 1974." International Atomic Energy Agency, Vienna. IAEA-SM-186/14. p. 565-78.

Dickson, R. C., 1949. Factors governing the induction of diapause in the oriental fruit moth. Ann. Entomol. Soc. Amer. 42:511-537.

Dickson, R. C., Barnes, M. M., and Turzan, C. L., 1952. Continuous rearing of the codling moth. J. Econ. Entomol. 45:66-68.

Farrar, B. D. and Flint, W. P., 1930. Rearing codling moth larvae throughout the year (Carpocapsa pomonella). J. Econ. Entomol. 23:41-44.

Guennelon, G., Audemard, H., Fremond, J., et. Ammari, E. I., 1980. Progres realises dans l'elevage permanent de Carpocapse (Laspeyresia pomonella L.) sur milieu artificiel.

Hamilton, D. W., and Hathoway, D. O., 1966. Codling moths. p. 339-354. In C. N. Smith (ed.) Insect Colonization and Mass Production. Academic Press, NY.

Howell, J. F., 1970. Rearing the codling moth on an artificial diet. J. Econ. Entomol. 63:1148-1150.

Howell, J. F., 1971. Problems involved in rearing the codling moth on diet in trays. J. Econ. Entomol. 64:631-636.

Redfern, R. E., 1960. Concentrate medium for rearing the codling moth. J. Econ. Entomol. 57:607-608.

Rock, G. C., 1967. Aseptic rearing of the codling moth on synthetic diets: Ascorbic acid and fatty acid requirements. J. Econ. Entomol. 60:1002-1005.

Singh, P., 1977. Artificial diets for insects, mites and spiders. IFI/Plenum Publ. Co., New York, 594 pp.

Schwartz, P. H. and Klassen, W., 1981. Estimates of losses caused by insects and mites in agricultural crops. P. 15-77. In Pimental, D. (ed.) CRC Handbook of Pest Management in Agriculture. CRC Press, Boca Raton, FL 597 p.

Shorey, H. H. and Hales, R. L., 1965. Mass-rearing of the larvae of nine noctuid species on a simple artificial medium. J. Econ. Entomol. 58:522-524.

Theron, P. P. A., 1947. Studies on the provision of hosts for the mass-rearing of codling moth parasites. Union S. Afr. Dep. Agr. Sci. Bull. 262:1-39.

Mention of a commercial product in this paper does not constitute an endorsement of this product by the USDA.

DIATRAEA GRANDIOSELLA

G. MICHAEL CHIPPENDALE and KATHERINE CASSATT

Department of Entomology, University of Missouri, Columbia, MO 65211, USA

THE INSECT

Scientific Name: *Diatraea grandiosella* Dyar
Common Name: Southwestern corn borer
Order: Lepidoptera
Family: Pyralidae

1. INTRODUCTION

The southwestern corn borer is present in Mexico and the United States between about 18° and 38°N latitude. The larvae feed on various wild and cultivated grasses. Corn (*Zea mays*) is the primary host plant and sorghum (*Sorghum bicolor*) and sugar cane (*Saccharum officinarum*) are secondary hosts (Davis *et al.*, 1933; Rolston, 1955).

Since *D. grandiosella* first migrated into the United States early in this century (Chippendale and Reddy, 1974), it has been responsible for losses of corn and sorghum in 14 states of the southern corn belt, ranging from Arizona to Alabama. Early instar larvae feed externally on the host plant, whereas late instar larvae tunnel into the stalk. In the fall, fully grown larvae may girdle the plant above ground level before entering their overwintering cells in the base of the stalk (Davis *et al.*, 1933).

The insect exhibits a facultative mature larval diapause induced by exposure of the sensitive larval instars to short days and low temperature. Throughout its range in the United States the southwestern corn borer appears to be bivoltine or trivoltine (Takeda and Chippendale, 1982a).

Keaster and Harrendorf (1965) first described an artificial diet and procedures to maintain a laboratory colony. Subsequently, several nutritional requirements of the larvae were determined and continuous mass rearing was achieved (Chippendale, 1972, 1979; Davis, 1976). Procedures for preparing artificial media and for handling the life stages have been described (Jacob and Chippendale, 1971; Davis, 1976; Chippendale, 1975, 1979; Yin and Peng, 1981). The development of procedures for rearing the southwestern corn borer on an artificial diet has facilitated the use of the insect in laboratory and field research (e.g., Yin and Chippendale, 1976; Takeda and Chippendale, 1982b; Barry and Darrah, 1978; Whitworth and Poston, 1979).

2. FACILITIES AND EQUIPMENT EMPLOYED

2.1 Facilities

The colony of southwestern corn borers is reared in an insect physiology laboratory equipped to investigate the developmental physiology of plant-feeding insects. Moderate numbers of nondiapausing and diapausing southwestern corn borers are maintained for our research program. Larvae are held in incubators equipped with daylight fluorescent tubes controlled with 24-h timers. Larvae receive at least 240 microwatts cm^{-2} of light energy, and the temperature is regulated to within $\pm$ 1°C. Nondiapausing larvae are maintained at 30°C and LD 16:8. Pupae are transferred to adult oviposition cages maintained at 25°C and LD 10:14.

Fig. 1. Equipment for handling the life stages of a research colony of southwestern corn borers. Left, covered stainless steel pan (5 x 24 x 34 cm) to receive hot medium; wooden frame with a 2.5-cm grid (5 x 26 x 35 cm) to prepare cubes from slab of medium; 32-ml clear plastic cups (1 oz) with laminated cardboard lids to rear larvae; 950-ml jar (1 quart Mason®) to hold egg-laden waxed paper. Right, cylindrical oviposition cage (19 cm diam x 30 cm high) in aluminum pan (5 x 22 x 22 cm). In use, this cage is wrapped with waxed paper.

2.2 Equipment and material required to handle the insects

2.2.1 Oviposition cage (Fig. 1)

- Wire mesh cage of no. 4 hardware cloth (19 x 30 cm)
- Aluminum pan (5 x 22 x 22 cm) lined with paper towels
- Waxed paper, masking tape, elastic band
- Petri dish (1.7 x 9 cm) lined with filter paper

2.2.2 Egg collection (Fig. 1)

- Jar, 950 ml, to hold egg-laden waxed paper
- Moistened filter paper (9 cm diam)

2.2.3 Larval rearing containers (Fig. 1)

- Disposable clear plastic cups, 32 ml (1 oz) (van Brode 410, Fill-Rite Inc., 49-55 Liberty St., Newark, NJ 07102, USA)
- Disposable plastic-coated cardboard caps (Standard Cap and Seal Inc., P. O. Box 80336, 3695 Longview Dr., Chamblee, GA 30341, USA)
- Clear plastic show boxes (9 x 17 x 31 cm) (Max Klein Co., 715 Lynn Ave., Baraboo, WI 53913, USA)

2.2.4 Pupal collection

- Pointed forceps (18 cm)
- Disposable polyethylene gloves

2.3 Equipment for diet preparation

- Autoclave to generate, 1.2 kg/cm^2 (17 psi), 121°C
- Refrigerator, freezer

- Bacteriological glove box, fiberglass (Labconco Corp., 9911 Prospect, Kansas City, MO 64132, USA)
- Fume hood
- Top loading balances (0.01 to 99 g, and 0.1 to 1200 g)
- Analytical balance (0.001 to 99 g)
- Commercial Waring® blender (5+ liter capacity)
- Hot plate (30 x 30 cm)
- Double boiler (5 liter capacity)
- Motorized stirrer
- Jar mill (1800 ml capacity)
- Stainless steel pans with lids (5 x 24 x 35 cm) (Fig. 1)
- Desiccator
- Erlenmeyer flasks, polypropylene (2 liter capacity)
- Graduated cylinders (10, 25, and 2000 ml)
- Wire mesh grid on wooden frame (5 x 26 x 35 cm) (Fig. 1), paper towels
- Insulated gloves and safety goggles

3. ARTIFICIAL DIET

3.1 Composition

Ingredients to prepare 5.3 kg diet	Amount	%
Agar solution		
Agar, fine ground (g)	108	2.04
Distilled water (ml)	3,948	74.53
Dry components (g)		
Casein, 30-40 mesh	262	4.95
Sucrose	172	3.25
Wheat germ, whole	148	2.79
Salt mixture [a]	50	0.94
L-ascorbic acid	27	0.51
Cellulose powder	24	0.45
Potassium sorbate	10.7	0.20
Beta-sitosterol, practical	10	0.19
Methyl paraben	2.4	0.05
Vitamin mixture [b]	3	0.06
Liquid components (ml)		
Distilled water	497	9.38
Wheat germ oil	25	0.47
Choline chloride (50%)	10	0.19
Total	5,297.1	100.00

[a] Salt mixture: Tumble mix in jar mill for 4 h 1140.1 g of the following salts (g/100 g); K_2HPO_4, 52.36; KH_2PO_4, 18.95; $MgSO_4$, 16.84; $CaHPO_4$, 5.26; NaCl 4.21; $FeSO_4$, 1.16; $MnSO_4.H_2O$, 0.53; Zn $(C_2H_3O_2)_2.2H_2O$, 0.53; and $CuSO_4$, 0.17.

[b] Vitamin mixture: Tumble mix in jar mill for 4 h 607.4 g of the following vitamins (g/100 g): inositol, 75.73; DL Ca pantothenate, 14.0; niacin, 5.43; p-amino-benzoic acid, 2.64; riboflavin HCl, 0.79; pyrodoxine HCl, 0.61; thiamine HCl, 0.46; folic acid, 0.18; biotin, 0.13; B_{12}, 0.03.

3.2 Sources of ingredients

Agar (Moorehead & Co., Inc., 14801 Oxnard St., Van Nuys, CA 91401, USA); casein (Milk Specialties Co., P. O. Box 278, Dundee, IL 60118, USA); sucrose (local retail); wheat germ, cellulose powder, methyl paraben, vitamins, except B_{12}, wheat germ oil, choline chloride (ICN Nutritional Biochemicals, 26201 Miles Rd., Cleveland, OH 44128, USA); salts (Fisher Chemical Co., 124 Ambassador Blvd., St. Louis, MO 63132, USA); L-ascorbic acid (Roche Chemical Div., Hoffmann La Roche, Inc., Nutley, NJ 07110, USA); potassium sorbate, beta sitosterol, B_{12} (Sigma Chemical Co., P. O. Box 14508, St. Louis, MO 63178, USA)

3.3 Procedure to prepare diet

a. Mixing ingredients

- Weigh out dry ingredients, excluding the agar, into plastic bags and store sealed at -20°C, or weigh out ingredients directly into blender
- Weigh out agar and transfer sterilized distilled water into double boiler
- Stir with mechanical stirrer over boiling water for 1.25 h or until the solution is translucent
- Add wheat germ oil, choline chloride solution, and sterilized distilled water (497 ml) to the dry ingredients in the blender at least 20 min before the hot solution of agar is added (this procedure permits the water to soak into the dry ingredients)
- Pour the hot agar solution into the blender and mix at high speed for 1 min

b. Dispensing and storing the diet

- Pour the hot diet immediately to a depth of 2 to 2.5 cm into 5 stainless steel pans located in glove box
- Allow 30 min for the diet to gel and cool, cover the pans, and store at +5°C
- Use the diet between 1 and 7 days of its preparation

PRECAUTIONS

(i) Store stocks of wheat germ, wheat germ oil, and casein at -20°C, working solutions of choline chloride and wheat germ oil at +5°C, and salt and vitamin mixtures in a desiccator.
(ii) Sterilize distilled water in the 2-liter flasks.
(iii) Minimize exposure to powdered ingredients by dispensing in fume hood.

4. MAINTAINING A COLONY

It is convenient to start a laboratory colony in the fall by collecting about 200 diapausing larvae from their overwintering cells. Larvae should be placed individually into vials containing moist paper strips. Diapause can be terminated relatively quickly by holding larvae at 30°C under constant illumination. Under these conditions 50% pupation is reached in about 40 days (Chippendale, 1979). Our present colony originated thus, and has been reared continuously on artificial diet for > 30 generations. For our research program we maintain nondiapausing and diapausing generations by rearing the larvae under different photoperiods and temperatures as shown in Table 1.

TABLE 1. Regimen used to maintain nondiapausing and diapausing generations of D. grandiosella in culture.

State	Newly hatched larvae	Diapause initiated (days, posthatch)	50% Pupation (days, posthatch)
Nondiapause	30°C LD 16:8	--	17
Diapause	23°C LD 12:12	40	190

4.1 Egg collection

- Transfer oviposition cage to fume hood to remove egg-laden waxed paper and clean (this transfer minimizes exposure to air-borne moth scales)
- Fold egg-laden waxed paper lengthwise, roll, and place with its folded edge down in jar containing moistened filter paper (when incubation is carried out at 25°C and LD 10:14 eggs hatch in about 4 days and larvae survive up to 2 days in the absence of food)

4.2 Transferring larvae onto the diet

- Transfer newly hatched larvae from jar to cubes of artificial diet in clear plastic cups in glove box as follows:

Remove slab of diet from each pan using disposable gloves and press through a 2.5-cm wire grid back into the pan to produce about 112 cubes (This procedure results in each cube having some rough surfaces to provide larval feeding sites)
- Dispense diet cubes required onto sterilized paper towels (store remain ing cubes in pan at +5°C for use within 5 days)
- Air dry cubes for about 10 min before placing them in clear plastic cups followed by 2-3 larvae/cup using a small soft-bristled brush (large groups of larvae cannot be reared in a single container because late instar larvae are cannibalistic)
- Seal each cup with 2 laminated cardboard lids and place 56 cups in shoe box in incubator.
- Minimize microbial contamination by sterilizing the stainless steel pans and wire grid (Fig. 1), and by using a glove box to pour and cut the diet and to set up newly hatched larvae on the diet (if these precautions are followed and the diet contains antimicrobial agents, additional sterile techniques should not be required)
- Transfer prediapausing larvae onto fresh medium at about 18 days of age (This procedure insures that larvae have sufficient food to complete their growth before diapause begins at about 40 days of age and available moisture for the time spent in diapause. A single larva should be placed into each cup to prevent cannibalism in late instars.)

4.3 Pupae and adults
- Remove pupae (> 24 h post-ecdysis) from cups using forceps and disposable gloves (after removing pupae return cups which still contain larvae to culture)
- Expected yield is 1 pupa/cup (our records show that 2470 pupae were obtained from 2481 cups over a 3-month period)
- Add up to 100 pupae to a filter-paper-lined petri dish held in the oviposition cage (adults do not require food)
- Remove dead adults and empty pupal cases from the oviposition cage twice a week

5. REARING SCHEDULE FOR NONDIAPAUSING AND DIAPAUSING GENERATIONS

Table 2 presents a weekly timetable of the operations necessary to maintain a colony of the southwestern corn borer for our studies. This schedule has proved to be convenient for our purposes, but it could be modified easily for other uses of the insect. The procedures described here produce about 200 diapausing larvae and 500 nondiapausing larvae per week. This level of production requires about 3 person-hours per day of a 5-day week.

TABLE 2. Weekly schedule used to maintain nondiapausing and diapausing generations of *D. grandiosella* in culture.

Day	Procedure
1	(a) Harvest pupae and exchange waxed paper on oviposition cage (b) Set up larvae (50-60 cups, 2-3 larvae/cup) and incubate at 23°C, LD 12:12 to induce diapause
2 or 3	(a) Set up larvae (160-170 cups, 2-3 larvae/cup) and incubate at 30°C LD 16:8 to maintain colony
4	(a) Harvest pupae and exchange waxed paper on oviposition cage (b) Prepare 5 pans of medium
5	(a) Exchange diet of prediapausing larvae (1 larva/cup) which had been set up 18 days earlier

6. LIFE CYCLE DATA

6.1 Larvae

Growth and development characteristics have been determined for a colony of D. grandiosella originally obtained from southeast Missouri in 1969. Larvae were found to pass through 5 or 6 instars (Yin and Chippendale, 1976). A sex dimorphism was observed beginning in the 4th instar of nondiapausing larvae. When larvae were reared at 30°C and LD 12:12, both males and females ecdysed into the 4th instar at 7 and 8 days (posthatching), whereas males ecdysed into the 5th instar at 10 days and females ecdysed into the 5th instar at 10 and 11 days. Most of the females then entered a 6th instar and thus pupated later than did males. Fifty percent pupation was reached by 16 days for males, compared with 18 days for females.

Prediapausing larvae reared at 23°C and LD 12:12 had a markedly lower rate of growth than did nondiapausing larvae reared at 30°C (Chippendale and Yin 1976) Six larval instars were observed in both sexes. Larvae ecdysed into the 2nd, 3rd, 4th, 5th, and 6th instars at 5, 10, 14, 18 to 19, and 24 to 28 days, respectively. A sex dimorphism was detected beginning on the 24th day. The head capsules of 6th instar females were significantly larger than were those of 6th instar males. After reaching maturity around 40 days, each diapausing larva undergoes a stationary ecdysis into a pigment-free "immaculate" morph.

6.2 Pupae and adults

The pupal-pharate adult period lasts about 7 days at 30°C. The following mean weights of pupae and adults have been obtained: 138-mg female, 107-mg male pupae and 36-mg female, 29-mg male adults. Although the nonfeeding adults may live up to 10 days, life spans of 4 days for males and 5 days for females are more common (Davis et al., 1933; Rolston, 1955; Gifford et al., 1961). Mating usually takes place within 24 h of emergence and each female lays from 100 to 400 eggs in overlapping masses.

7. CONCLUSION

Since 1969 we have maintained colonies of southwestern corn borers for physiological research using these methods. It is important to periodically check findings obtained from a laboratory colony in insects collected from a wild population. For example, we have observed that photoperiodic responses of larvae diminished after they had been continuously colonized under a standardized photoperiod and temperature (Takeda and Chippendale, 1982b). It appears necessary, therefore, to periodically reestablish a laboratory colony to maintain photoperiodic responses similar to those found in the natural population. For this reason our present laboratory colony was founded from diapausing larvae collected in September 1980 from Scott County, southeast Missouri (36.9°N latitude).

These rearing procedures for the southwestern corn borer rely upon antimicrobial agents incorporated into the diet and sanitation to control microbial contamination. If those procedures are employed, or appropriately modified for other laboratory environments, a successful rearing program should be assured.

8. REFERENCES

Barry, D. and Darrah, L. L., 1978. Identification of corn germ plasm resistant to the first generation of the southwestern corn borer. J. Econ. Ent. 71: 877-879.

Chippendale, G. M., 1972. Composition of meridic diets for rearing plant-feeding lepidopterous larvae. Proc. North Centr. Branch Ent. Soc. Am. 27: 114-121.

Chippendale, G. M., 1975. Ascorbic acid an essential nutrient for a plant-feeding insect, Diatraea grandiosella. J. Nutrition 105: 499-507.

Chippendale, G. M., 1979. The southwestern corn borer, Diatraea grandiosella: case history of an invading insect. Missouri Agric. Exp. Sta. Res. Bull. 1031, 52 p.
Chippendale, G. M. and Reddy, A. S., 1974. Diapause of the southwestern corn borer, Diatraea grandiosella: low temperature mortality and geographical distribution. Environ. Ent. 3: 233-238.
Chippendale, G. M. and Yin, C.-M., 1976. Endocrine interactions controlling the larval diapause of the southwestern corn borer, Diatraea grandiosella. J. Insect Physiol. 22: 989-995.
Davis, E. G., Horton, J. R., Gable, C. H., Walter, E. V., Blanchard, R. A., and Heinrich, D., 1933. The southwestern corn borer. USDA Tech. Bull. 388, 61 p.
Davis, F. M., 1976. Production and handling of eggs of southwestern corn borer, Diatraea grandiosella, for host-plant resistance studies. Agr. For. Exp. Sta., Mississippi State Univ. Tech. Bull. 74, 11 p.
Gifford, J. R., Walton, R. R., and Arbuthnot, K. D., 1961. Sorghum as a host of the southwestern corn borer. J. Econ. Ent. 54: 16-21.
Jacob, D. and Chippendale, G. M., 1971. Growth and development of the southwestern corn borer, Diatraea grandiosella, on a meridic diet. Ann. Ent. Soc. Am. 64: 485-488.
Keaster, A. J. and Harrendorf, K., 1965. Laboratory rearing of the southwestern corn borer, Zeadiatraea grandiosella, on a wheat germ medium. J. Econ. Ent. 58: 923-924.
Rolston, L. H., 1955. The southwestern corn borer in Arkansas. Arkansas Agr. Exp. Sta. Bull. 533, 40 p.
Takeda, M. and Chippendale, G. M., 1982a. Phenological adaptations of a colonizing insect: the southwestern corn borer, Diatraea grandiosella. Oecologia 53: 386-393.
Takeda, M. and Chippendale, G. M., 1982b. Environmental and genetic control of the larval diapause of the southwestern corn borer, Diatraea grandiosella. Physiol. Ent. 7: 99-110.
Whitworth, R. J. and Poston, F. L., 1979. A thermal-unit accumulation system for the southwestern corn borer. Ann. Ent. Soc. Am. 72: 253-255.
Yin, C.-M. and Chippendale, G. M., 1976. Hormonal control of larval diapause and metamorphosis of the southwestern corn borer, Diatraea grandiosella. J. Exp. Biol. 64: 303-310.
Yin, C.-M. and Peng, W.-K., 1981. A simplified soybean and wheat germ diet for rearing the southwestern corn borer, Diatraea grandiosella Dyar (Lepidoptera:Pyralidae). Ann. Ent. Soc. Am. 74: 425-427.

ACKNOWLEDGMENT

We thank S. Y. Young of the University of Arkansas for his helpful comments about the manuscript. This article is a contribution from the Missouri Agricultural Experiment Station, journal series no. 9516.

DIATRAEA SACCHARALIS

E. G. KING and G. G. HARTLEY

Agricultural Research Service, USDA, P. O. Box 225,
Stoneville, Mississippi 38776

THE INSECT

Scientific Name: *Diatraea saccharalis* (Fabricius)
Common Name: Sugarcane borer
Order: Lepidoptera
Family: Pyralidae

1. INTRODUCTION

The sugarcane borer (SCB), *Diatraea saccharalis* (F.), is the most important stalk borer of sugarcane in the Western Hemisphere (Long 1969). It is the only species of moth borer attacking sugarcane in Louisiana and is also a major pest in Cuba, Peru, Puerto Rico, Jamaica, Trinidad, Mexico, Florida, and Texas. Damage is characterized by stand reduction, tunneling within stalks with consequent reduction in (stalk) weight and juice quality. The cane crop is susceptible to red rot caused by *Physalospora tucamanensis*.

SCB has been successfully reared in the laboratory for parasite production, insecticide testing and development of varieties resistant to damage by SCB. Singh (1977) provides a general review of diets and techniques. More successful diets have been modifications of the casein-wheat germ diet reported by Adkisson et al. (1960), Hensley and Hammond (1968).

We maintained a SCB colony at this laboratory for 10 years. During this period several million insects were produced. Many aspects of this rearing program are described in King et al. (1979).

2. FACILITIES AND EQUIPMENT REQUIRED

2.1 Facilities

Rearing of larvae and pupae is conducted in controlled environment rooms maintained at 26° ± 1°C and 60-70% RH and a LD 14:10 photoperiod. Adults are maintained for oviposition in controlled environment rooms held at 24°C ± 1°C and a LD 14:10. Cool, white fluorescent lighting of medium intensity is recommended.

2.2 Equipment and materials required for insect handling

2.2.1 Adult oviposition

- 4 liter cardboard buckets
- White polyester-cotton cloth
- Wax paper
- Vermiculite
- Absorbent cotton
- 50 ml glass or plastic vial
- 1.25 cm masking tape
- Scissors

2.2.2 Egg sterilization

- Plastic pan (30 x 40 cm)

- Fumehood
- Rubber gloves

2.2.3 Egg hatch and larval inoculation of diet
- 2 liter Erlenmeyer flasks
- Absorbent cotton
- Aluminum foil
- 1.25 cm masking tape
- Fine camel hair brushes
- Absorbent paper towels

2.2.4 Larval rearing containers
- 22.5 ml clear plastic cups
- Wax coated paper cup lids
- Cell pak containers or other suitable containers for holding cups

2.2.5 Pupal collection and holding
- Forceps
- 4 liter cardboard buckets
- Vermiculite
- Polyester-cotton cloth

2.2.6 Insect holding
- Environmental room maintained at 24°C ± 1°C and 70-80% RH

2.3 Diet preparation equipment
- Top loading balance - 2000 g capacity
- Portable concrete mixer (optional)
- Aluminum scoop
- 3.8 liter blender
- 7.6 liter stainless steel garden sprayer - modified to dispense artificial diet
- Electric or gas hot plate
- 1 liter cardboard buckets
- 3 liter plastic measuring pitcher
- 6 liter aluminum or stainless steel boiler
- Kitchen pot-holder gloves
- Freezer
- Laminar flow table (Agnew & Higgins Inc., Garden Grove, CA)
- Large stainless steel spatula

3. ARTIFICIAL DIET

3.1 Composition

This diet is also suitable for rearing *Heliothis virescens* (Fabricius), *Heliothis zea* (Boddie), *Anticarsia gemmatalis* (Hubner), and *Pseudoplusia includens* (Walker) (with addition of 25 ml of raw linseed oil).

a. Formula for 1 liter of prepared diet

- Soybean flour (Nutri-Soy Flour #40)	41 g/liter
- Wheat germ	35 g/liter
- Wesson salt	10 g/liter
- Sugar	41 g/liter
- Vitamin mix	9.5 g/liter
- Agar	22.4 g/liter
- Methyl paraben	1 g/liter
- Aureomycin	1 g/liter
- Sorbic acid	1 g/liter
- Water	930 ml

3.2 Sources of ingredients

Nutri-Soy Flour #40, (Flavorite Lab Inc., Memphis, TN); vitamin mix, (Roche Chemical Division, Nutley, NJ - Specify mix used at Stoneville, MS by USDA); agar, (Perny Inc., Ridgewood, NJ); aureomycin, (American Cyanamid Co., Wayne, NJ); sorbic acid, wheat germ, wesson salt, methyl paraben, (Nutritional Biochemicals Corp., Cleveland, Ohio); and sugar (local grocery).

3.3 Diet preparation procedure

- Weigh and place all dry ingredients required for up to 100 liters of finished diet into a portable concrete mixer. Ingredients can be weighed in quantities to prepare 3.8 liters of diet and placed directly into blender.
- Mix in concrete mixer for two hours.
- Weigh out 616 g of dry mix and add to 3535 ml of boiling water.
- Blend dry mix in boiling water for 4 minutes.
- Pour blended diet into diet dispenser and dispense hot diet into 22.5 ml cups (10 to 15 ml per cup).
- Cover cups of diet with wax paper and allow to cool under a laminar flow table.
- Dry mix has a shelf life of approximately six months when stored at 0°C.
- Keeping prepared diet in storage more than two weeks is not recommended.

PRECAUTION

To avoid a dusty work area cover the top of the concrete mixer with a sheet of plastic.

4. DEVELOPMENT OF INSECT STAGES AT OPTIMAL TEMPERATURES

4.1 Eggs

Temperature	Incubation time (days)	% hatch
26°C	5.7	98.6

4.2 Larvae

Temperature	Larval period (days)		% larval mortality
28°C	18.8	20.7	3.3

4.3 Pupae

Temperature	Pupal period (days)		Pupal weight (mg)		% pupal mortality
26°C	8.2	8.3	96.5	163.5	9.5

4.4 Adults

Temperature	Adult longevity (days)
24°C	7.0

5. REARING AND COLONY MAINTENANCE

The sugarcane borer founder colony was established from larvae collected in sugarcane fields in Louisiana during 1972-1974. This colony was maintained continuously on artifical diet through 1981.

5.1 Egg collection

- Remove wax paper lining containing eggs from the 4 liter buckets starting the 2nd day after moth emergence.
- Replace wax paper lining in the 4 liter buckets.

5.2 Egg sterilization

- Wash the wax paper lining with eggs in formaldehyde (3%) active ingredient for ten minutes.
- Rinse the wax paper lining with eggs in distilled water for 15 minutes and place under a laminar-flow hood until almost dry.
- Place the egg covered wax paper linings into sterile 2 liter Erlenmeyer flasks containing moist cotton.
- Plug flasks with sterile cotton, wrap in aluminum foil, and hold until larvae hatch.

5.3 Diet inoculation with neonate larvae

- Remove the positively phototactic larvae from the top of the aluminum foil wrapped flasks with a fine camel hair brush and transfer them to cups of diet.

- Rear sugarcane borer larvae in 22.5 ml plastic cups containing about 10 ml of the soybean flour-wheat germ diet. Use wax impregnated lids to cover the cups as larvae frequently chew through plastic lids.
- Rear larvae at the rate of 3 per cup.

5.4 Pupal collection

- Remove pupae from the cups and disinfect in formaldehyde (4%) active ingredient for 2 mintues.
- Rinse pupae thoroughly in distilled water.
- Separate pupae by sex.
- Place 200-300 pupae in each 4 liter cardboard container containing moist vermiculite.
- Replace the top of the 4 liter cardboard container with nylon organdy or white polyester/cotton cloth.

5.5 Adults

- Transfer moths (40 pair per container) as they emerge to 4 liter cardboard containers lined with wax paper, which serves as an oviposition surface.
- Place a water-soaked piece of cotton on the cloth top of the 4 liter container. This serves as a water source.
- Discard moths after 4 nights as egg production slows and moth mortality is high.

5.6 Sex determination

- For rearing purposes pupae were sexed by size (females larger than males). This proved 90% accurate.

5.7 Rearing schedule

- For rearing purposes, eggs and larvae are held at 28°C and 80% relative humdity in complete darkness. Harvested pupae and moths are held at 24°-26°C and 80% RH on LD 14:10 photoperiod. Under these conditions, the eggs require 5-6 days, the larvae 19-21 days, and the pupae ca. 8 days to complete development. The eggs are stored at 15.6°C to prevent hatching over weekends.

5.8 Insect quality

- Production is monitored on a daily basis. Checks include egg production, egg hatch, larval survival and emergence of adults from the pupae and adult survival. Thus, quality control was appropriate for parasite production. However, when insects are used for other purposes such as host plant resistance or chemical control research, other criteria should also be used. The average quality of each check is as follows: egg production per female, 729.8; egg hatch, 98.6%; larval survival, 96.7%; emergence of adults from pupae, 95%; and adult survival, 7 days.

5.9 Special problems

a. Diseases

A bacterium, *Serratia marsescens*, occasionally caused mortality among nonparasitized larvae. However, this was most prevalent among larvae that had been parasitized by maggots of the tachinid, *Lixophaga diatraeae* (Townsend). Mortality was prevented by sterilizing the maggots with formalin and maintaining only one host larvae per cup.

b. Microbial contamination of diet

The artificial diet served as a good medium for bacteria, yeast, and fungi growth. This growth was prevented through the use of antimicrobial agents in the diet and exposure of the diet only to clean air hoods. All equipment and work areas were thoroughly disinfected with sodium hypochlorite or ammonia type compounds. If possible, equipment used in rearing was autoclaved.

c. Human health

(i) All work with formalin is conducted under a fume hood.

(ii) This insect is not highly active, so under small-scale rearing conditions, moth scales are not a major problem. However, a protective mask is worn while removing moths from emergence containers and pairing them in oviposition containers. This operation could also be done under a fume hood for greater ventilation.

6. LIFE CYCLE DATA

6.1 Developmental data

- Duration of the egg, larval, pupal, and adult stages generally decrease with each increment of temperature increase within limits of 15.6 to 30-32°C. However, at 33-34°C, duration of the egg, larval, and pupal stages are lengthened, and at 34°C no developmental time has been determined for larvae due to high mortality. Egg hatch and pupal survival are also severely reduced at temperatures of 34 and 36°C, respectively. Within parameters of 22 to 31°C, optimal egg production is obtained at 24°C. Production decreases about 2.3-fold at 22°C and about 1.7, 3.8, 5.2, and 26.5-fold, respectively, at 26, 28, 30, and 31°C. Our data indicate that for purposes of rearing the sugarcane borer on artificial diet in the laboratory at constant temperatures 26, 28, and 26°C are optimal for eggs, larvae, and pupae, respectively. For best egg production, moths should be held at 24°C.

6.2 Other data

Developmental stage	Temperature C	
Adult	24	Survival Time = 7 days No. eggs/moth = 729.8
Egg	26	Incubation time = 136.1 hours % Hatch = 98.6
Larval	28	Developmental time = = 18.8 days; = 20.7 days % survival = 96.7
Pupal	22-26	Developmental time = 8.24 - 13.4 days

7. PROCEDURES FOR SUPPLYING INSECTS

Pupae can be shipped (up to 200) in 0.47 liter cylindrical cardboard containers filled with moist vermiculite. The top of each container is perforated for ventilation and is taped on with 2.5 cm masking tape to prevent contents from spilling. The container is then placed in corrugated cardboard boxes (15 x 10 x 10 cm) equipped with four 2.5 cm diameter ventilation holes.

Eggs can be shipped while still attached to the wax paper oviposition sheets using the same procedure as described for shipping pupae; although a moist piece of cotton is substituted for the vermiculite in the 0.47 liter container.

8. REFERENCES

Adkisson, P. L.; Vanderzant, E. S.; Bull, D. L.; and Allison, W. E. 1960. A wheat germ medium for rearing the pink bollworm. J. Econ. Entomol. 53: 759-762.

Hensley, S. D.; and Hammond, A. M. 1968. Laboratory techniques for rearing the sugarcane borer on artificial diet. J. Econ. Entomol. 61: 1742-43.

King, E. G.; Hartley, G. G.; Martin, D. F.; Smith, J. W.; Summers, T. E.; and Jackson, R. D. 1979. Production of the tachinid *Lixophaga diatraeae* on its natural host, the sugarcane borer, and on an unnatural host, the greater wax moth. U.S. Dept. of Agric., SEA, AAT-S-3/April 1979.

Long, W. H. 1969. Insecticidal control of moth borers of sugarcane. pp. 149-61. *In* J. R. Williams, J. R. Metcalfe, R. W. Montgomery, and R. Mathes (*Eds*.), Pests of Sugarcane. Amsterdam: Elsevier. 568 pp.

Singh, P. 1977. Artificial diets for insects, mites, and spiders. 594 pp. Plenum Press, New York.

EPIPHYAS POSTVITTANA

PRITAM SINGH, GRAEME K. CLARE and MICHAEL D. ASHBY

Entomology Division, Department of Scientific and Industrial Research,
Private Bag, Auckland, New Zealand.

THE INSECT

Scientific Name : *Epiphyas postvittana* (Walker)
Common Name : Lightbrown apple moth (LBAM)
Order : Lepidoptera
Family : Tortricidae

1. INTRODUCTION

The lightbrown apple moth is a leafroller species that is indigenous to Australia and has been introduced into New Zealand, Hawaii, New Caledonia and England. It is a major orchard pest in Australia and New Zealand. Within New Zealand it occurs throughout the cultivated lowland areas, and has a large range of recorded hosts. The common hosts include apples, pears, black and red currants, citrus, red and white clover, hops, lucerne, lupin, tree lupin, other weed species and flowering plants, various shrubs, and young conifers (Thomas 1974).

The larvae cause fruit and foliage damage characteristic of leafrollers. Early instar larvae feed on tissue beneath the upper epidermis of leaves. Larger larvae migrate from these positions to construct feeding niches between adjacent leaves and between leaves and fruit causing the typical leaf roll to develop. The late stage larvae feed on all leaf tissue except main veins.

There are three generations annually in Australia and in the central New Zealand region with no winter resting stage, and in some years there is a partial fourth generation. Adults produced by the overwintering larval generation emerge during October and November. These give rise to the first summer generation, in which final instar larvae mature between January and mid-February. Second generation larvae reach maturity during March and April and these adults provide third generation eggs. Eggs are laid in clusters of 3-150 on leaves or fruit (Thomas, 1974; Danthanarayana, 1975).

E. postvittana has been reared on several artificial diets by Dunwoody and Hooper (1967), Thomas (1968), Bartell and Shorey (1969), King (1972) and Singh (1974 a,b). In our laboratory it has been reared on artificial diet for more than 100 generations over a 10 year period. The insects are used for virus and parasite production, and post-harvest disinfestation research. A laboratory colony can be established either from field collected specimens or by obtaining insects from an already established laboratory colony.

2. FACILITIES AND EQUIPMENT REQUIRED

2.1 Facilities

Rearing of larvae and pupae is carried out in controlled environment rooms maintained at 25 ± 1°C, 50-60% RH and an 18 h photoperiod. Oviposition occurs in the laboratory at 18-20°C under natural light conditions.

2.2 Equipment and materials required for insect handling

2.2.1 Adult oviposition

Two methods are used

a. Single pair mating (see Fig. 1)
- Open-ended perspex or glass tubing, 120-150 mm long and 30 mm internal diameter
- 0.005-grade 150 x 110 mm transparent polythene sheets
- Scissors
- Absorbent cotton wool
- 75 x 12 mm polystyrene test tubes

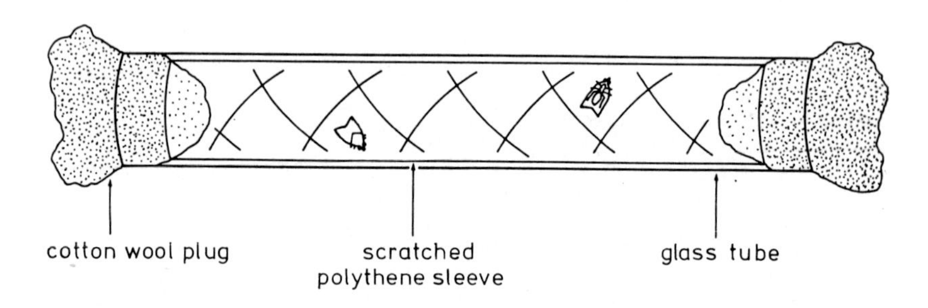

Fig. 1 Single pair oviposition container

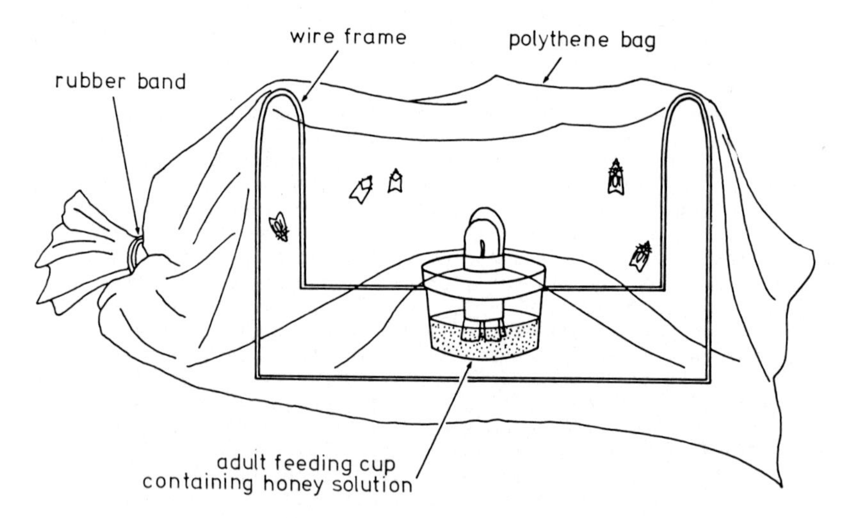

Fig. 2 Group oviposition container

b. Group mating (see Fig. 2)
- Rubber band
- Transparent polythene bag (250 x 375 mm, 35 gauge)
- Galvanised wire frame
- Feeding cup (see Fig. 3) consisting of:
  - Absorbent cotton wool dental rolls (75 mm)
  - Small clear plastic vial (60 mm diameter x 50 mm high)
  - Plastic tubing, 25 mm outer diameter

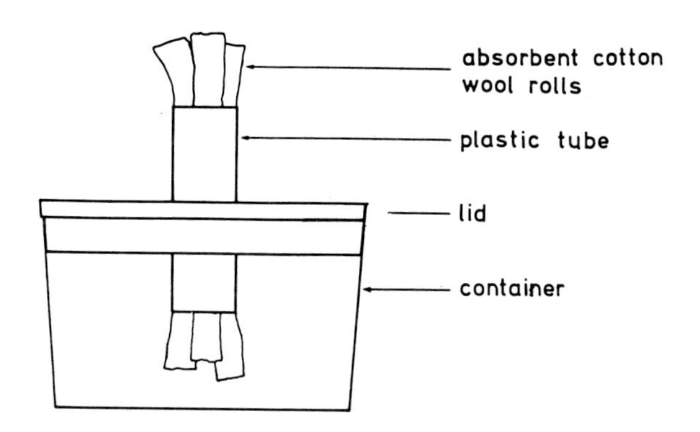

Fig. 3 Adult feeding cup

2.2.2 Egg collection and storage (see Fig. 4)

- Round, transparent, plastic container (Clearseal®) with airtight lid (112 mm diameter x 63 mm high) (Adelphi Plastics Ltd, N.Z.). [Also used as pupal storage container]
- Whatman No.1 filter paper 42.5 mm diameter
- Absorbent, bleached cellulose wadding
- Scissors

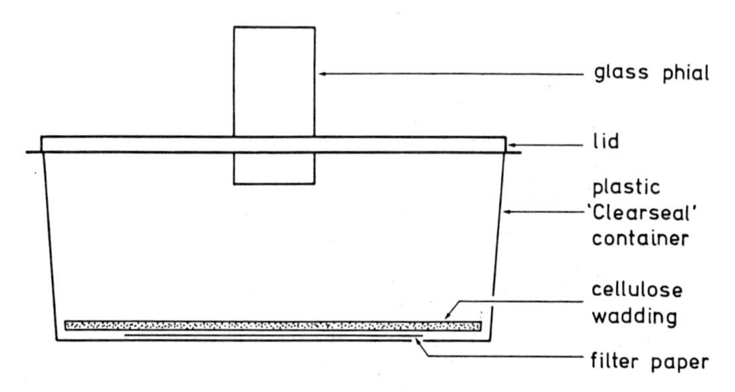

Fig. 4 Egg and pupal storage container

2.2.3 Egg sterilization

- Container as used for egg storage except lid with 25 mm diameter hole
- Plastic containers 200 mm diameter x 200 mm high with lids
- 1 liter container of 5% formalin (2% formaldehyde)
- 2 liter container of sterile water
- Soft tissue paper (sterile)
- Glass vial (25 x 75 mm)
- Fine nylon gauze bag
- Soft-nosed forceps
- Long metal forceps
- Fume cupboard or fume hood

2.2.4 Inoculation of diet with larvae

- Fine camel hair paint brush (size 000)
- 75% ethanol
- Sterile tissue paper

2.2.5 Larval rearing containers
a. Individual rearing
- Polystyrene test tubes (75 x 12 mm)
- Cotton wool
- Container or test tube rack for polystyrene tubes
b. Group rearing
- Wax coated No. 75, 250 ml Lily® cups with lids (Frank M. Winstone Ltd, Auckland, N.Z.)
- Paraffin wax
- 500 ml glass beaker for melting wax
- Bunsen burner, tripod and gauze
- Tongs for holding beaker of wax

2.2.6 Pupal collection and storage
- Soft-nosed forceps
- Wire hook 115 mm long, 1 mm diameter
- Egg storage container (see 2.2.2 and Fig. 4)

2.2.7 Insect holding
- Incubator set at 15°C

2.3 Diet preparation equipment

2.3.1 Dry mix
- Stainless steel basin (500 x 300 x 150 mm)
- Top loading balance (0.1-1000 g)
- Waring Blendor
- 1 liter glass beaker
- 50 ml measuring cylinder
- Plastic scoop
- Large flat stainless steel spatula (250 x 40 mm)
- Fume hood
- Non-toxic particle mask
- Self-sealing polythene bags (450 x 600 mm)

2.3.2 Finished diet
- Round stainless steel mixing bowl (230 mm diameter x 230 mm high)
- 1 liter measuring cylinder
- 50 ml measuring cylinder
- Aluminium foil
- Autoclave or pressure cooker
- Large flat stainless steel spatula
- Small stainless steel spatula for weighing
- Small rubber spatula
- 0.0001-10 g and 0.1-1000 g balances
- 1 liter glass beaker
- Thermometer (0-100°C)
- Magnetic stirrer with magnetic stirring bar
- Heat-resistant gloves
- Automatic diet dispensing machine (Singh *et al.*, 1983) or hand operated diet plugger (Ashby and Singh, 1983)
- Gladwrap® cellophane food wrapper (Union Carbide N.Z. Ltd)
- Refrigerator at 5°C
- Self-sealing polythene bags (450 x 600 mm)

3. ARTIFICIAL DIET

3.1 Composition

a. Dry mix to prepare 6.5 kg finished diet

Ingredients	Amount	Approx. %
Agar	150 g	12.32
Casein	210 g	17.24
Cellulose powder	600 g	49.25
Wesson's salt mix	60 g	4.93
Wheatgerm	180 g	14.78

Cholesterol	3 g	0.25
Linoleic acid	15 g	1.23
Dichloromethane*(evaporates)	50 ml	
Total	1218 g	100.00%

b. Finished diet

Dry mix	1218 g	18.56
Distilled water	4290 ml	65.33
4N KOH	30 ml	0.46
Vanderzant vitamin mixture	120 g	1.83
Sucrose	180 g	2.74
Glucose	30 g	0.46
Streptomycin sulphate BP*	900 mg	0.01
Penicillin*	900 mg	0.01
Distilled water	600 ml	9.15
Mould inhibitor	90 ml	1.37
Total diet approx.	6558 g	100.00%

c. Preparation of stock mould inhibitor solution

Nipagin (methyl *p*-hydroxybenzoate)	37.5 g
Sorbic acid	50.0 g
95% Ethyl alcohol	425 ml

d. Preparation of stock 4N KOH solution
56 g KOH added carefully to 250 ml distilled water

*Not included in total weight

3.2 Brand names and sources of ingredients

Agar (Davis Gelatine N.Z. Ltd, Christchurch); casein (N.Z. Co-op Dairy Co., Casein Milling Station, Penrose, N.Z.); cellulose powder, Wesson's salt mix, Vanderzant vitamin mixture (Bio-Serv Inc., P.O. Box B.S., Frenchtown, New Jersey 08885, U.S.A.); wheatgerm (natural raw), glucose powder, sucrose (Sanitarium Health Food Co., Auckland, N.Z.); dichloromethane BP (Scientific Supplies Ltd, P.O. Box 14-454, Auckland); streptomycin sulphate BP, benzylpenicillin (Sodium BP) (Glaxo Laboratories Ltd, Greenford, England); linoleic acid (L.R.), potassium hydroxide, sorbic acid (B.D.H. Chemicals Ltd, Poole, England); cholesterol (Sigma Chemicals Co., P.O. Box 14508, St Louis, M.O. 63178, U.S.A.); methyl *p*-hydroxybenzoate (Koch-Light Labs Ltd, Bucks, England).

3.3 Diet preparation procedure

a. Dry mix
- Place agar, casein, cellulose powder and Wesson's salt mixture into stainless steel basin
- Grind wheatgerm into fine powder using blender
- Mix all of the above together thoroughly with plastic scoop
- Combine cholesterol and linoleic acid and dissolve in dichloromethane in 1 liter glass beaker
- Add cholesterol-linoleic acid solution to dry ingredients and mix thoroughly with large metal spatula
- Leave in fume cupboard overnight to allow solvent to evaporate
- Weigh and dispense dry mixture into self-sealing polythene bags and store at 5°C until required

b. Finished diet
- Place dry mix into stainless steel bowl
- Add distilled water and KOH
- Mix well using large spatula
- Cover top of bowl tightly with aluminium foil

- Place in autoclave for 20 minutes at 115 kPa pressure (small pressure cooker can be used for 1-2 kg of diet)
- Place Vanderzant vitamin mixture, sucrose, glucose, streptomycin and penicillin into 1 liter glass beaker
- Add distilled water
- Mix solution thoroughly using magnetic stirrer
- Cool autoclaved diet to 60°C
- Add vitamin solution and mould inhibitor, mix thoroughly
- Pour and scrape into diet dispenser hopper using rubber spatula
- Use dispensing machine to dispense 2.5 g hot diet into test tubes or 80 g into waxed cups, or manually dispense into waxed cups or trays
- Condition diet overnight to allow excess moisture to evaporate
- Cut diet from trays into blocks and wrap in cellophane. Seal waxed cups and test-tubes of diet in polythene bags. Store diet at 5°C until required
- Prepared diet has a shelf life of approximately 1 month at 5°C

PRECAUTIONS

(i) A non-toxic particle mask should be worn when mixing dry ingredients.
(ii) Dichloromethane should be used in fume cupboard.
(iii) All equipment should be sterilized.

4. REARING AND COLONY MAINTENANCE

4.1 Adult oviposition

This is done under natural light conditions at room temperature (mean 18-20°C) for best results.

The pre-oviposition period is about 2 days at room temperature. Peak egg laying occurs 3-4 days after pairing. Discard adults after 8 days.

a. Single pair mating in perspex tubes (see Fig. 1)
- Cut 150 x 110 mm polythene sheet to fit around the outside of the perspex tubing
- Score several grooves in the polythene by using the pointed ends of a pair of scissors
- Insert the sheet into the tube to form an internal sleeve
- Seal one end of the tube with a tight cotton wool plug moistened in 10% honey solution
- Place one pair of adults into the tube

PRECAUTIONS

(i) Honey-moistened cotton wool should be changed daily as mould may develop.
(ii) Avoid direct sunlight.

b. Group mating in polythene bags (see Fig. 2)
- Place wire frame inside bag
- Assemble feeding cup containing honey solution, plastic tubing and dental roll wick as shown in Fig. 4
- Cut a hole just large enough for the plastic tube with wick to fit through in one side of the polythene bag
- Seal the open end of the bag with a rubber band
- Place 10 pairs of moths into the bag through the hole in the side
- Turn bag over and place plastic hose and wick of feeding cup through hole in bag

PRECAUTION

Same as for a.

4.2 Egg collection

a. From perspex tubing
- Prepare new oviposition container (See 4.1 (a))
- Transfer moths from tube containing eggs to new container
- Remove sleeve containing egg masses from tube
- Carefully cut around individual egg masses using a pair of scissors
- Set up an egg storage container (See Fig. 4)
- Lightly moisten filter paper with sterile water
- Place egg batches face up into container using soft-nosed forceps
- Incubate at 25°C
- Collect eggs daily

b. From polythene bags
- Prepare new bag (see 4.1 (b))
- Remove rubber band from bag containing eggs and carefully transfer moths into new bag and set up for oviposition. (Moths may be placed at 5°C for 10-15 min to slow them for easy handling)
- Cut open bag containing eggs
- Cut around individual egg batches carefully using scissors
- Place in egg storage container
- Incubate at 25°C

PRECAUTION

Eggs should be collected daily as adults often lay a new batch on top of an older one, thus killing the bottom layer.

4.3 Egg surface sterilization

- Use fume hood
- Place eggs (4-5 days old) into fine nylon gauze bag
- Place bag into 1 liter container of 5% formalin solution using forceps for 20 min with periodic agitation
- Remove bag and dip several times in 2 liter container of sterile water
- Place in sterile water for 20 minutes
- Remove bag and place eggs between two layers of sterile soft tissue paper for 30 minutes
- Air-dry at room temperature
- Return eggs to storage container
- Change the lid to one with a 25 mm hole in it and place glass vial in top
- Darken container by covering sides with black paper

PRECAUTIONS

(i) Eggs less than 4 days old should not be sterilized.
(ii) Equipment and tissue paper should be sterilized.
(iii) Do not use formalin without a fume hood.

4.4 Diet inoculation with neonate larvae

a. In test tubes
- Condition diet by allowing to stand at rearing temperature overnight
- Diet can be placed into bottom of test tubes by dispenser or by using hand operated diet plugger or cork borer to cut 25 x 8 mm plugs from blocks of diet
- Remove vial containing larvae from top of egg holding container and replace with another vial for larval collection (neonate larvae migrate towards the light into the vial)
- Transfer 1 larva per tube very carefully onto the diet using a paint brush
- Plug test tube tightly with cotton wool
- Place test tube, cotton wool end upright, in container or test tube rack

b. In waxed cups
- Condition dietary cups by allowing to stand at rearing temperature overnight

- Remove vial containing larvae from top of egg holding container and replace with another vial for larval collection
- Scratch the surface of the diet several times with sterilized forceps
- Transfer 30 larvae into each cup
- Place lid on and seal with hot liquid paraffin wax

PRECAUTIONS

(i) Paint brush must be sterilized by dipping in 75% ethanol. Rinse in distilled water and then dry with sterile tissue paper.

(ii) Extreme care must be taken when inoculating diet with larvae to reduce mortality due to handling.

(iii) Cotton wool plug placed in test tube must be a close fit to prevent desiccation and larval escape.

4.5 Pupal collection

a. From test tubes

- Remove any diet or frass necessary to uncover pupa with wire hook
- Carefully dislodge pupa from any webbing or diet. Pull out using wire hook, separate sexes and transfer pupae to storage container using soft-nosed forceps

b. From waxed cups

- Carefully uncover pupae from diet or webbing using soft-nosed forceps and transfer to storage container

PRECAUTIONS

(i) Pupae are green and soft when newly formed and the cuticle must be allowed to harden for at least 1 day before collecting.

(ii) Check moisture content in pupal containers to prevent desiccation and mould.

4.6 Adults

4.6.1 Emergence

a. In test tubes

The first adult male will usually emerge 28 days after inoculation with the first female a day or two later. Adult emergence is complete in 41 days with a peak occurring on day 31. Adult survival from neonate larvae is 90%.

b. In Waxed cups

The first adult male emerges about 29 days after inoculation with the first female a day or two later. Emergence is complete in 44 days with several peaks occurring. Adult survival from neonate larvae is 61%.

4.6.2 Collection and maintenance

Collection is done individually on day of emergence using polystyrene test tubes. Adults are fed 10% honey solution. The sex ratio is usually 1:1.

4.6.3 Sex determination

- Female pupae and adults are usually larger than males
- The female pupa has four ventral abdominal segments, the male has five
- Female moths are distinguished by a large ovipository pore.

4.7 Rearing schedule

a. Daily

- Collect newly emerged moths
- Set up new mating containers
- Feed moths
- Collect eggs
- Check moisture content in egg and pupal containers
- Sterilize eggs when 4-5 days old if necessary
- Discard old moths
- Check previously sterilized eggs for larval hatching and inoculate if necessary

b. Alternate days

- Collect pupae

c. Friday afternoon

- Place eggs and pupae 5 days and older at 15°C for weekend storage (if required)

4.8 Insect quality

Insect quality is checked periodically by keeping a record of pupal weight, adult fecundity and egg viability (see section 6.2 and 6.3). Insects from each generation are also checked visually for growth and size.

4.9 Special problems

a. Diseases

Nuclear polyhedrosis virus infection of the larval stage is prevented by egg sterilization.

b. Microbial control

Microbial infection of the diet can occur if sterile conditions are not maintained. Diet should not be exposed to air for long periods. If infection occurs very strict hygiene control must be enforced. All equipment and benches must be disinfected with a concentrated solution of Chlorodux® or other disinfecting agent. All infected material should be destroyed by autoclaving.

c. Human health

(i) All work with formalin and dichloromethane should be carried out in a fume cupboard as they are carcinogenic.

(ii) Exposure to scales of moths can be hazardous. A suitable protective mask should be worn when scales are evident in large amounts, or when working in confined spaces.

(iii) If an allergic reaction is suspected, consult a doctor immediately.

5. HOLDING INSECTS AT LOWER TEMPERATURES

5.1 Adults

Moths can be stored individually at 15°C in 75 x 12 mm test tubes plugged with honey-moistened cotton wool up to 2 or 3 days until mating.

PRECAUTION

Do not hold moths more than 2-3 days as fecundity may drop.

5.2 Eggs

These can be also held at 15°C if necessary. For best results store only 2-3 day old eggs or older. Eggs are placed in storage containers containing lightly moistened filter paper and cellulose wadding (Fig. 4). The lid is airtight to prevent the eggs from drying out. Containers must be checked daily for build-up of excess moisture, and wiped out if necessary.

PRECAUTIONS

(i) Keep humidity at 55-70% using moist filter paper if necessary to prevent egg desiccation.

(ii) Too much moisture will encourage mould development.

(iii) Eggs must be sterilized if mould develops.

5.3 Larvae

Neonate larvae can be held at 15°C safely for one day in the egg storage container. Larvae that have been inoculated on diet in test tubes can be held at 15°C for their entire larval period.

PRECAUTION

Avoid excess moisture in containers to prevent larval drowning and mould growth.

5.4 Pupae

These can be held for 1-2 weeks at 15°C in pupal storage containers.

PRECAUTIONS

As for eggs (see 5.2).

6. LIFE CYCLE DATA

Life cycle and survival data during development of various stages under optimum rearing conditions of 25°C, 50-60% RH and an 18 hour photoperiod are given.

6.1 Developmental data (individual rearing)

Stage		Number of days Min	Mean	Max
Eggs		6.0	6.5	7.0
Larvae	♂	19.0	22.0	30.0
	♀	21.0	25.0	30.0
Pupae	♂	7.0	9.0	11.0
	♀	7.0	8.0	11.0
Adult emergence	♂	26.0	31.0	41.0
(from inoculation)	♀	28.0	33.0	41.0
Total development (egg to egg,	♂	34.0	39.5	50.0
inclusive pre-oviposition)	♀	36.0	41.5	50.0

6.2 Survival data (%)

a. Individual rearing
- Egg viability 60-70
- Neonate larva to pupa 97
- Neonate larva to adult 90

b. Waxed cups
- Neonate larva to adult 61

6.3 Other data

Preoviposition period, 2 days
Peak egg laying period, 3-4 days after emergence
Mean fecundity, 305 eggs/♀ in polythene bags; single pair oviposition yields higher
Mean adult life span, 12 days
No. of larval instars, 5♂, 6♀
Mean pupal weights (mg), ♂34 (min 26, max 40), ♀56 (min 48, max 68)

7. PROCEDURES FOR SUPPLYING INSECTS

7.1 General prerequisites

- Research and development of rearing procedures must be completed
- The founder colony should be robust and disease free
- The duration of all life cycle stages at various temperatures, humidities and light must be known
- The best diet or food should be used
- Standard rearing method(s) should be adopted
- Adequate equipment and labour should be organised
- The rearing facility should have several environmental control units

7.2 Insect yield

The investment in labour for the production of insects by group rearing is considerably less than that required for individual rearing. Although insects are usually produced for colony maintenance by group rearing, they develop less uniformly than those that are reared individually and adult yields are considerably lower (61% compared to 90%).

The laboratory colony should be maintained at at least 300 pairs/generation. Under group rearing this requires about 20 hours/week (30 Lily cups @ 20 moths yield/cup). Allowing 100 usable eggs/female as an absolute minimum, 30 group mating and oviposition bags with 10 pairs in each will produce 30000 eggs. At a viability rate of about 65%, 19500 first instar larvae should be available for inoculation. About 1000 (5.1%) of these larvae should be used for colony maintenance. The remainder may be used to produce new colonies for research projects, or may be used for supply to user groups.

7.3 Shipping

Most countries have strict regulations regarding the importation of live insects and these should be investigated if shipping is international. *E. postvittana* is sent within New Zealand and sometimes to Australia as eggs or pupae, but large numbers of larvae are also sent in individual containers for post-harvest fumigation research. The time required to ship the insects from their point of origin to their destination may dictate the stage at which they should be sent. It is also advisable to ship orders early in the week to avoid weekend delays at the other end.

a. Eggs

Eggs are collected and transferred to polystyrene Petri dishes lined with lightly moistened filter paper to prevent desiccation. If time between despatch and arrival permits, they may be checked for fertility and surface-sterilized (see 4.3) prior to packing. A light layer of cotton wool is placed over the eggs and covered with a Petri dish lid. The dishes should be taped together and then packed into a suitably sized box lined with cellulose wadding or cotton wool.

b. Larvae

Larvae may be sent at any stage of development. They should be individually placed in 12 x 75 mm polystyrene test tubes with an adequate supply of diet for the journey. Diet that can move about in the tubes may cause injury and for this reason it should be freshly dispensed (and allowed to set in the tubes) immediately prior to shipping. The tubes should be tightly bundled together with tape in groups of about a dozen and then packed between cushions of cotton wool or multi-layered cellulose wadding in a suitably sized container.

c. Pupae

Pupae should be shipped as soon as they have hardened. They may be packed in close proximity between cotton wool cushions inside Petri dishes as for eggs, or rolled between 12-layer absorbent paper cushion as in Baumhover *et al.* (1977). The dishes or rolls should then be placed in a suitable container to prevent their movement.

REFERENCES

Ashby, M.D.; Singh, P. 1983. A simple dispenser for agar-based insect diets. N.Z. Entomol. 7(4): 469-470.

Bartell, R.J.; Shorey, H.H. 1969. A quantitative bioassay for the sex pheromone of *Epiphyas postvittana* (Lepidoptera) and factors limiting male responsiveness. J. Insect Physiol. 15: 33-40.

Baumhover, A.H.; Cantelo, W.W.; Hobgod, J.M.; Knott, C.M.; Lam, J.J. Jr. 1977. An improved method for mass rearing the tobacco hornworm. USDA Bulletin, ARS-S-167, 13 pp.

Danthanarayana, W. 1975. The bionomics, distribution and host range of the light brown apple moth, *Epiphyas postvittana* (Walk.) (Tortricidae). Aust. J. Zool. 23: 419-437.

Dunwoody, J.E.; Hooper, G.H.S. 1967. An artificial medium for rearing *Epiphyas postvittana*. J. Econ. Entomol. 60: 1753-1754.

King, P.D. 1972. Studies on the toxicity of insecticides to the light brown apple moth *Epiphyas postvittana* (Walker) (Lepidoptera: Tortricidae). Unpublished Masters Thesis, Lincoln College, N.Z. 178 pp.

Singh, P. 1974a. Aseptic rearing of *Epiphyas postvittana* (Lepidoptera: Tortricidae) on a meridic diet. N.Z. J. Zool. 1: 111-117.

Singh, P. 1974b. A chemically defined medium for rearing *Epiphyas postvittana* (Lepidoptera: Tortricidae). N.Z. J. Zool. 1: 241-243.

Singh, P.; Ashby, M.D.; Hunter, J.R. 1983. A semi-automatic volumetric filling machine for dispensing insect diets. N.Z. J. Zool. 10: 413-418.

Thomas, W.P. 1968. An artificial diet for rearing the light brown apple moth *Epiphyas postvittana* (Walk.) (Lepidoptera: Tortricidae). N.Z. Entomol. 4: 31-32.

Thomas, W.P. 1974. Light brown apple moth, *Epiphyas postvittana* (Walk.), life cycle. D.S.I.R. Information Series No. 105/3 N.Z.

ACKNOWLEDGEMENTS

We thank Mrs Judith Matheson for useful discussions in the preparation of this manuscript, and Mr Bruce Philip for drawing the diagrams. We are also very grateful to Melody Tapene for typing and processing the manuscript.

ESTIGMENE ACREA

P. V. VAIL and D. K. COWAN

USDA Agricultural Research Service , Horticultural Crops Research Laboratory, 5578 Air Terminal Drive, Fresno, CA 93727 USA

THE INSECT

Scientific Name: *Estigmene acrea* (Drury)
Common Name: Saltmarsh caterpillar
Order: Lepidoptera
Family: Arctiidae

1. INTRODUCTION

The Saltmarsh caterpillar (SMC) is a common pest throughout the United States and parts of Europe. The larvae damage beans, cotton and sugar beets and are often found on other vegetables and forages, flowers, grasses and weeds (Essig, 1958). Heavy infestations may defoliate entire plantings. The white or yellowish colored eggs are deposited in masses on the undersides of leaves, turning grey-black 18 - 24 hours prior to hatching. During development larvae may molt from 6 to 9 times, and attain a length of 3.5 to 4.5 cm. Often called "woolyworms", the caterpillars normally pupate by spinning their cocoons on or in the soil at the base of the host plant. Adults emerge as black-speckled, white moths with a wing span of from 4 to 6 cm for the females and somewhat less for the males. The hind wings of the females are white and the males' hind wings are orange.

The Saltmarsh caterpillar is relatively easy to rear in the laboratory. Colonies have been used for bioassay of chemical and microbial agents, physiology and parasitoid studies, and large-scale production of insect pathogens and parasites (Young and Sifuentes, 1959). All stages are relatively large and easy to handle, and the larvae can be reared together. Larvae do not burrow into the rearing media in preparation for pupation. The pupal and adult stages are easy to sex. The moths do not require food to reproduce.

The SMC has been reared on various host plants (Dunn et al. 1964) and semisynthetic diets. However, only rearing of the insects on semisynthetic diets will be discussed in this paper.

2. FACILITIES AND EQUIPMENT NEEDED

2.1 Facilities

Rearing of larvae is carried out in incubators or environmental chambers maintained at 27°C, ambient humidity and a LD of 16:8 photoperiod. Pupae and ovipositing moths are maintained in the laboratory at ca. 23°C under prevailing light conditions.

2.2 Equipment and materials required for insect handling.

2.2.1 Adult oviposition

- 3.8 liter cardboard carton or glass jar cages
- Cotton muslin for lid
- Rubber bands to hold muslin to lip of carton or jar
- Paper towelling oviposition substrate
- Shredded paper towel to absorb excrement

2.2.2 Egg collection and storage

- Additional paper towelling
- 2 liter glass or cardboard containers
- Forceps

- Incubators (10°C)
- Petri dishes or 8 oz. (236 ml) or larger paper cups

2.2.3 Egg sterilization
- 1 liter beaker
- Fume hood
- Forceps
- Formalin solution (4% formaldehyde AI)
- Running tap water
- Sterile cloth towels

2.2.4 Egg hatch and larvae infestation
- Neonate larvae
- Prepared diet in selected containers or greenhouse flats
- Fine camel's-hair paint brush (size 1)
- Sterile water for dipping paint brush
- Lids for prepared diet cartons

2.2.5 Larval rearing containers

One or more of the following containers may be used:
a. 30 ml plastic cups with lids
b. 236 and 472 ml paraffin coated Lily cups with lids
c. Mason glass jars with filter paper replacing lid insert may be used
d. Two plastic storage box containers (15.7 x 26.7 cm) with 12 x 20 cm opening in lid, and bottom replaced with hardware cloth plus two, 12.5 cm x 12.5 cm fiber trays (wood fiber, styrofoam, etc.) for diet and sterile paper towelling.
e. Porcelain coated steel trays approximately 50 x 33 x 6.3 cm deep, chickenwire, tempered glass cover and felt or plastic foam gasket material; sterile paper towelling

2.2.6 Pupal collection and storage
- Large blunt forceps
- 1 liter glass container
- 0.03% AI sodium hypochlorite
- Paper towels
- 3.8 liter glass or cardboard containers
- Fume hood
- Running tap water

2.2.7 Insect holding
- Incubator set at 27°C

<u>2.3 Diet equipment and preparation</u>

2.3.1 Dry mix for artificial diets
- Balance (capable of weighing from 0.1 to 120 g or more)
- Analytical balance for weighing vitamins
- Spatulas
- Weighing paper
- 4 liter mixing bowl
- Electric food or commercial blender of appropriate size for number of batches of diet to be made
- 1 and 10 ml pipettes
- Graduated cylinders of appropriate size for amount of diet to be prepared
- Plastic bags
- Fume hood
- Six-500 ml screw cap Erlenmeyer flasks

2.3.2 Finished diet
- Bunsen burner, hot plate or autoclave
- 4 liter stainless steel pot
- Spatulas
- Asbestos gloves

- 50 ml and 1 liter graduated cylinders
- 5 and 10 ml pipettes
- Electric food or commercial blender of appropriate size
- Large spoons
- Refrigerator (to hold dry and liquid ingredients)
- Thermometer (0-100°C)
- Plastic squeeze bottles for dispensing diet into 30 ml cups
- 1,000 ml beakers for dispensing diet into large cups
- Sterile cotton sheets (1 x 2 m)
- Laboratory coat or smock

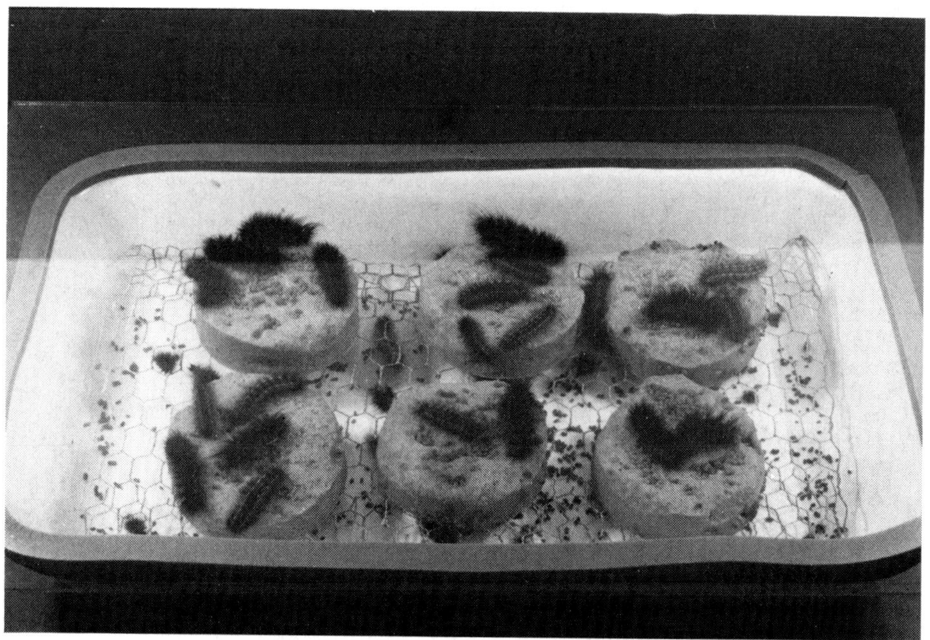

Fig 1. Saltmarsh caterpillar rearing tray showing chicken wire support for diet and placement of styrofoam insulation material for sealing.

3. ARTIFICIAL DIET

Artificial diets provide an easy, although not necessarily economical, means of efficiently rearing the SMC in relatively large numbers. However, the ingredients may not be available at some locations and care must be exercised as the diets provide an excellent substrate for the growth of microbial contaminants even though anti-microbial agents are included in the diets. Three semisynthetic diets have been commonly used to rear the SMC. One batch of diet will feed 10-25 larvae for 1-2 weeks.

3.1 Composition

a. Ingredients

	Diet		
	Wheat Germ [a/]	Lima Bean [b/]	Cabbage Looper [c/]
	Grams/batch of diet		
Wheat germ, raw	120.0	-	28.5
Ground baby lima beans	-	166.8	-
Alfalfa leaf meal	-	-	14.3
Brewer's yeast	-	31.8	-
Agar	15.0	-	23.7
Gelcarin [d/]	-	10.0	-
Casein	25.0	-	33.2
Sucrose	-	-	33.2
Wesson salt mix	8.0	-	9.5
Ascorbic acid	-	3.8	3.9
Alphacel	-	-	4.3
Vitamin mixture	10.0 [d/]	-	3.3 ml
Choline chloride (10% AI, aqueous)	-	-	9.0 [e/]
Potassium hydroxide	-	-	4.5 ml [e/]
Formaldehyde	-	1.5	0.4 ml
Aureomycin	0.5	0.6	0.13
Methyl-p-hydroxybenzoate	1.0	3.3	2.01 ml [e/]
Sorbic acid	2.0	1.15	2.1 ml [e/]
Deionized water (preferably sterile)	800.0	850.0	810.0

a/ Bell et al., 1981.
b/ Patana 1969 and 1977.
c/ From the diet of Henneberry and Kishaba, 1966; Vail et al., 1967.
d/ Mix #28262 also available from: Roche Chemical Co., 340 Kingsland ST. Nutley, N.J. 07110
e/ Added as stock solutions/suspensions; other diets added dry

b. Preparation of stock vitamin solution for cabbage looper diet:

Stock = 600 mg each of niacin and calcium pantothenate; 300 mg riboflavin; 150 mg each of thiamine (HCl), pyroxidine and folic acid; 12 mg biotin; and 1.2 mg vitamin B_{12} in 100 ml sterile deionized water.

c. Preparation of stock mold inhibitor solutions for cabbage looper diet: 23.9 g methyl-p-hydroxybenzoate + 38 ml 95% ethanol; same formula for sorbic acid.

3.2 Brand names and sources of ingredients.

Wheat germ (raw), Brewer's yeast, agar, casein (Industrial grade, 80% protein, high nitrogen), Wesson salt mix, ascorbic acid, alphacel, Vanderzant Modification Vitamin Mixture for Insects, choline chloride, methyl-p-hydroxybenzoate (p-hydroxybenzoic acid), and sorbic acid can be ordered through ICN Nutritional Biochemicals, 26201 Miles Road, Cleveland, Ohio 44128; Aureomycin and alfalfa meal can be ordered through Bio Serv, Inc., P.O. Box B5, Frenchtown, N.J. 08825; formaldehyde, reagant grade, Van Waters and Rogers Scientific Inc., P.O. Box 3200, San Francisco, CA 94119, or the nearest VWR outlet; Gelcarin, Marine Colloids Division, FMC Corporation, 360 Civic Drive, Suite G, Pleasant Hill, CA 94523.

3.3 Diet preparation procedure.

Sterilize all utensils and materials prior to use. Also, surface sterilize countertops, etc. with 0.5% AI sodium hypochlorite solution prior to mixing dry ingredients, preparing or pouring diet. Always arrange preferred diet containers on bench top prior to preparation of diet and cover with sterile cotton cloth until ready to pour diet.

a. Wheat germ diet.

- Have wheat germ, casein, Wesson salt mix, vitamin mix, aureomycin, methyl-*p*-hydroxybenzoate and sorbic acid preweighed or measured and ready to add to hot agar.
- Mix 15 g agar slowly into 800 ml water and bring to boil. Check that agar gels properly.
- Transfer hot agar solution to motorized blender. Let cool to 70°C.
- Add all dry ingredients and other components to agar and blend for 2 min or until thoroughly mixed.
- Dispense into prepared containers approximately half full while still hot using plastic squeeze bottle, 1 liter beaker, or automatic dispensing device.
- Cover diet while cooling with sterile cotton sheet to prevent airborne contamination.
- Allow covered diet to sit out overnight at room temperature to allow excess moisture to evaporate.
- Cover with appropriate lids the following day and infest within 3 days.

b. Lima bean diet.

- Weigh out and combine dry ingredients.
- In 50% (425 ml) of total water, blend ground baby lima beans, brewer's yeast, ascorbic acid, formaldehyde, dry aureomycin, methyl-*p*-hydroxybenzoate and sorbic acid and blend for about 5 min or until temperature is about 44°C. The higher temperature is necessary to keep the diet liquid so that when the hot Gelcarin is added, it will mix better and dispense readily.
- Add Gelcarin to the other 425 ml water and bring to a boil.
- Add to premixed dry ingredients in blender and blend for 1 min
- Dispense while hot into preferred containers approximately half full using plastic squeeze bottle, 1 liter beaker, or automatic dispensing device.
- Cover diet containers while cooling with sterile cotton sheet to prevent airborne contamination.
- Allow diet to sit overnight at room temperature to allow excess moisture to evaporate.
- Cover with appropriate lids and infest within 3 days.

c. Cabbage looper diet.

- Weigh out dry ingredients and prepare and measure out wet ingredients.
- Blend dry ingredients in 30% (243 ml) of total water.
- Add liquid ingredients; blend again. Add KOH solution last and blend again.
- Slowly add agar to 70% (567 ml) of water and bring to boil.
- Allow hot agar solution to cool to 70°C, add to previously mixed ingredients and blend for 1 min
- Dispense into preferred containers using plastic squeeze bottle, 1 liter beaker, or automatic dispensing device.
- Cover with sterile cotton sheet while cooling to prevent airborne contamination.

- Allow diet to sit overnight at room temperature in order to allow excess moisture to evaporate.
- Cover with appropriate lid and infest within 3 days.

PRECAUTIONS

(i) Preferably use sterile distilled or deionized water.

(ii) Some diet ingredients may be incompatible in certain uses, e.g., formaldehyde in the diet reduces the activity of nuclear polyhedrosis viruses (Vail, et al., 1968).

(iii) Check for drying of diet. Diet will need to be replaced after 1-2 weeks (depending on relative humidity) due to loss of moisture and/or larval feeding.

4. HOLDING INSECTS AT LOWER TEMPERATURES

Only the egg and pupal stages have been routinely stored at lower temperatures in order to either synchronize the stages for various experiments or adjust the developmental rates to coincide with projected needs for the colony.

4.1 Eggs

- After sterilization, eggs can be held for up to a week at 15°C without any apparent effect on viability.

4.2 Pupae

After removal from their cocoons and surface sterilization, heavily melanized and sclerotized pupae have been held at 15°C for 1 week.

5. REARING AND COLONY MAINTENANCE

5.1 Egg collection

SMC moths are very quiescent during the day and there is no need for CO_2 or cold immobilization.

- Discard first eggs laid as the hatch will be low.
- Remove paper towelling every 1-2 days after oviposition commences.
- Replace with fresh paper towelling.
- Cut out individual egg masses and discard excess paper towelling.

5.2 Egg sterilization

- Use fume hood
- Place eggs in 1 liter beaker, cover with formalin solution, and agitate occasionally for 20 min
- Decant off formalin solution.
- Place beaker containing eggs under running tap water for 20 min
- Remove eggs and place on one side of sterile towel; fold over other side to protect eggs from airborne contaminants.
- When dry, remove with forceps and place in large petri dish or Lily cup.
  Eggs can also be placed in a Lily cup containing diet so that time of infesting does not become critical.
- Place in incubator at 27°C until hatch.

5.3 Diet infestation

In all cases use sterile utensils and surface sterilize the work area using a 0.5% sodium hypochlorite solution. If the diet has been refrigerated, allow sufficient time for the diet to come to ambient temperature. If infested immediately, cold diet will sweat when placed in an incubator and drown the newly infested larvae. All containers should be incubated at 27°C.

a. 30 ml plastic cups

- Infest each plastic cup with one neonate larvae.

- Place lid on cup and incubate. Note: 30 ml cups will not support larvae through complete development because of both space and diet limitations.

b. 236 and 472 ml paraffin coated paper cups.
- Infest small cups with 5 larvae; large with 10 larvae.
- Place lids on container.
- Incubate at 27°C.
- Check consumption and condition of diet every week.
- If diet is consumed, transfer larvae to new diet containers until pupation occurs.
- Because of possible cannibalism, pupae should be collected daily once pupation begins.

c. Box type rearing container
- Prepare box(es). The box type container (15.7 x 26.7 x 36.8 cm) utilizes 2, 12.5 cm x 12.5 cm fiber trays inside, arranged in vertical tiers. Each tray contains a 5 x 5 cm block of artificial diet (Patana, 1977). The changeable fiber trays enable easy removal of frass, thus decreasing the incidence of mold and facilitating the replacement of diet. A 12 x 20 cm opening is cut in the center of the box lid and covered with window screen to promote air circulation. The bottom of the plastic box is replaced with 6.35 mm mesh hardware cloth. This first box is set inside another plastic box of the same size providing a 5 cm deep pupation space between their bottoms filled with shredded, sterilized paper. This method also decreases cannibalism.
- Infest with 50 neonate larvae.

d. Tray type rearing.
- Add shredded paper towels to bottom of porcelain tray.
- Insert chicken wire diet support in bottom of tray so that surface is 1.3 cm above bottom.
- Autoclave completed tray.
- After sterilization, insure that foam rubber or felt gasket is intact.
- Place 6-8 discs of diet from 236 ml Lily cups on top of chicken wire.
- Infest trays with 125 neonate larvae.
- Replace spent diet as required or about once a week until development is completed.

5.4 Pupae collection
- Carefully remove pupae in cocoons from rearing container(s) with blunt forceps.
- When melanized and sclerotized (approx. 3 days post pupation) carefully remove pupae from cocoons. Note: if pupae are surface sterilized too late in development, emergence may be reduced.
- Under fume hood, place pupae in 1 liter glass container with 0.03% sodium hypochlorite and agitate for 5-10 min to remove foreign material.
- Rinse several times with tap water.
- Blot pupae dry on paper towelling.
- Place pupae in 3.8 liter glass or cardboard emergence cartons with paper towelling in the bottom to absorb excrement, etc. or previously prepared oviposition cage (20 pupae).

5.5 Adults

The sex ratio is approximately 1:1.

5.5.1 Adult emergence

Adult emergence will commence approximately 10 days after pupation at 27°C.

5.5.2 Adults can be collected daily.

5.5.3 Oviposition

Oviposition cages can be held under natural light conditions and at room temperature (Approx 23°C). Adults commence mating 2-3 days after eclosion, therefore eggs laid the first two days after emergence should be discarded.
Adults do not require a sugar source for oviposition (Dunn et al., 1964) although dilute honey (Patana, 1977) has been used.

- Wrap paper towelling oviposition substrate around inside surfaces of 3.8 liter cardboard carton or glass jar.
- Place shredded paper towelling in bottom of container to absorb excrement.
- Place 10 pair of moths or 20 pupae in container.
- Remove and discard oviposition substrate 2 days after first adult emergence.
- Thereafter, collect eggs for 3-4 two day intervals and replace oviposition substrate

5.6 Sex determination

- SMC can be easily sexed in the adult stage.
- Rear wings of the female are white.
- Rear wings of the male are orange.

6. LIFE CYCLE

Developmental time and fecundity data of SMC(from Fye and Surber, 1971 and Fye and McAda, 1972) reared at different temperatures are provided.

6.1 Development and fecundity data for SMC.

	Developmental time[a]							
		Larval-pupal (days)		Adult stage (days)		Fecundity[c]	Days after emerge when 100%	
Temp. °C	Eggs (hr)					(# eggs/ female)	females ovipositing	eggs laid
20	183.2	48.4	48.2	11.2	9.5	1,577.2±500.4	5	14
25	109.2	31.4	32.0	8.2	8.1	862.2±363.7	6	12
30	85.2	29.0	28.8	14.9	10.2	636.6±236.6	7	7
33	90.4[b]	25.5	26.0	4.6	5.7	80.0± 99.1	6	8

a/ Larvae fed on diet of Patana (1969), maintained at 50% RH; the data presented in this table will vary with the rearing methods used and the genetic makeup of each *E. acrea* culture.
b/ Egg survival very poor at 33°C.
c/ ± standard deviation

6.2 Other data

- Percent emergence is generally over 90% for both sexes.
- Preoviposition period, 2-3 days at room temperature (23°C)
- The number of eggs laid/female can vary from 400 to over a 1,000 (avg about 500)
- The number of larval instars may vary from 6 to 9 instars, 6-7 being normal
- Adult longevity is about 10 days at normal room temperature (23°C)
- Eggs are very sensitive to high temperature
- The majority of eggs are laid by the 8th day after emergence at temperatures of 20-25°C

- Percent egg hatch usually exceeds 80%

7. PROCEDURES FOR SUPPLYING INSECTS

7.1 Routine colony maintenance and schedules

The number of adults required to maintain a colony depends on the number of insects utilized for purposes other than colony maintenance. If, for example, 200 larvae are needed for bioassays each generation, and they must be from eggs collected on the same day, then a total of 10 pairs of healthy adults should be more than adequate. If one female lays 400 eggs over 4 collection periods (100 eggs/period), then a total of 1,000 eggs would be collected from 10 females. Egg production of the first collection period will be low, because of the variation in developmental times among adults, but the second through the fourth should yield an excess for other uses as well as continuing the colony.

To yield 10 pairs of moths, 40 neonate larvae should be infested on fresh plants or artificial substrates. This insures recovering at least 30 healthy pupae which will yield at least 25-30 adults, approximately one-half of which should be females. Dunn et al. (1964) recovered 10,000 to 12,000 eggs weekly from the moths of 150 to 200 caterpillars reared on red kidney bean plants (*Phaseolus vulgaris* var. *rubra*).

8. REFERENCES

Bell, R. A.; Owens, C. D.; Shapiro, M.; Tardif, J. R. 1981. Development of mass rearing technology. In C. C. Doane (ed.): The gypsy moth; research towards integrated pest management. U. S. D. A. Tech. Bull. 1584.

Dunn, P. H., Hall, I. M.; Snideman, M. L. 1964. Bioassay of *Bacillus thuringiensis*-based microbial insecticides III. Continuous propagation of the saltmarsh caterpillar, *Estigmene acrea*. J. Econ. Entomol. 57: 374-377.

Essig, E. O. 1958. Insects and mites from western North America. The Macmillan Co., New York. 1050 pp.

Fye, R. E.; Surber, D. E. 1971. Effects of several temperature and humidity regimens on eggs of six species of lepidopterous pests of cotton in Arizona. J. Econ. Entomol. 64: 1138-1142.

Fye, R. E.; McAda, W. C. 1972. Laboratory studies on the development, longevity and fecundity of six lepidopterous pests of cotton in Arizona. U. S. D. A. Tech. Bull. 1454: 48-57.

Henneberry, T. J., Kishaba, A. N. 1966. Cabbage loopers. 461-478 In C. N. Smith (ed.), Insect colonization and mass production. Academic Press, Inc., New York. 618 pp.

Patana, R. 1969. Rearing cotton insects in the laboratory. U. S. D. A. Prod. Res. Rpt. 108, 6 pp.

________. 1977. Rearing selected western cotton insects in the laboratory. U. S. D. A., ARS W-51, 8 pp.

Vail, P. V.; Henneberry, T. J.; Pengalden, R. 1967. Artificial diets for rearing the saltmarsh caterpillar, *Estigmene acrea* (Lepidoptera:Arctiidae), with notes on the biology of the species. Ann. Entomol. Soc. Am. 60: 134-138.

Vail, P. V.; Henneberry, T. J.; Kishaba, A. N.; Arakawa, K. Y. 1968. Sodium hypochlorite and formalin as antiviral agents against nuclear polyhedrosis virus in larvae of the cabbage looper. J. Invert.

Pathol. 10: 84-93.
Young, W. R.; Sifuentes, J. A. 1959. Biological and control studies on *Estigmene acrea* (Drury), a pest of corn in the Yaqui Valley, Sonora, Mexico. J. Econ. Entomol. 52: 1109-1111.

9. Footnote

Mention of a commercial (or proprietary) product in this paper does not constitute an endorsement of this product by the USDA.

EUXOA SCANDENS and EUXOA MESSORIA

S. BELLONCIK, C. LAVALLÉE, and I. QUEVILLON

Virology Res. Center, Institut Armand-Frappier, Laval, Québec, Canada H7N 4Z3

THE INSECTS

Scientific Name: *Euxoa scandens* (Riley)
Common Name: White cutworm
Order: Lepidoptera
Family: Noctuidae

Scientific Name: *Euxoa messoria* (Harris)
Common Name: Darksided cutworm
Order: Lepidoptera
Family: Noctuidae

1. INTRODUCTION

Euxoa scandens found in sandy soils was described as causing damage to tobacco plants in Québec (Meloche *et al.*, 1980), vegetable crops in Ontario and Manitoba (Beirne, 1971) and asparagus in Michigan (Lampert *et al.*, 1982). In southern Québec, *E. scandens* lays its eggs during July. The larvae are present in the field from mid July to early June of the following year. They pass through seven instars and overwinter at the fifth or sixth one. The pupae are found in June and the adults emerge in July.

Euxoa messoria is a pest common in the United States and Canada. This insect was found in California, Oregon, Washington, and British Columbia on wild grasses and on other crops, as well as on young orchard trees in Michigan, Indiana, and Illinois (Chapman and Lienk, 1981). In eastern Canada, it is a major pest of tobacco and other cultivated crops (Cheng, 1971). In Québec, the flight period is from the beginning of August to the first week of September. The eggs overwinter and seven larval stages occur from mid May to mid July. The pupae are found during July and the first half of August.

The best way to start a new generation of: 1) *E. scandens* is to collect adults from light traps or to collect advanced instar larvae from an infested area. Two males and two females or 100 larvae are adequate to start the culture. 2) *E. messoria* is to collect mature larvae (approx. 200) from an infested region or to ask for eggs (approx. 1000) from specialized laboratories.

In our laboratory the insects are reared for the production of viruses and cell culture initiations.

2. FACILITIES AND EQUIPMENT REQUIRED

2.1 Facilities

The rearing of the insects is done in a controlled environment room at 25°C, 40% RH, and an LD 16:8 photoperiod (cool white neon tube, Sylvania F20T12-CW).

2.2 Equipment and materials required for insect handling

2.2.1 Adult oviposition

a. single-pair mating

- Polystyrene disposable soup bowls (100 mm diam, 60 mm high)
- Plastic tubing (75 mm long x 12 mm diam)

- Plastic petri dish (100 x 15 mm)
- Wire-netting (opening 6.5 mm)
- Fine sand (less than 355 micrometers)
- Cheese cotton
- Dental cotton
- Scissors
- Elastic bands
- Soft-nosed forceps

b. Group mating
- Wood cages (380 x 240 x 250 cm) made with wood, wire screen and plastic; each cage can accomodate 10 to 15 pairs
- Metal plate (395 x 270 mm)
- Plastic dish for shade (200 mm diam)
- Feeding cup
- Dental cotton
- Fine sand (less than 355 micrometers)
- Soft-nosed forceps
- Plastic tubing (75 mm long x 12 mm diam)

2.2.2 Egg collection and storage
- Small plastic container (35 mm diam, 60 mm high)
- Metallic sieve (42 mesh)
- Petri dishes (49 x 9 mm)
- Fine, squirrel-hair paintbrush (no. 2)
- Whatman no. 1 and no. 2 filter papers (90 mm and 125 mm diam, respectively)
- Scissors
- Distilled water squirt bottle
- Cold room or refrigerator (4°C)

2.2.3 Egg sterilization
- Vacuum pump
- Buchner funnel
- Erlenmeyer flask (500 ml)
- Rubber bung (7-1/2)
- Filter paper (55 mm diam)
- Squirt bottle
- Sodium hypochlorite (0.2% AI)
- Sharp-pointed forceps
- Fine hair brush (no. 2)
- Small plastic container (60 x 30 mm diam)
- Plastic dishes (49 x 9 mm)

2.2.4 Egg hatch, larval inoculation of diet and rearing
- Flat wood stick (115 x 5 mm)
- Soft-nosed forceps
- Scissors
- Fine hair brush (no. 2)
- Petri dishes (49 mm diam)
- Filter paper (55 mm) divided in four
- Plastic trays for rearing of larvae, 50 cavities (29 x 92 x 10 mm each cavity), purchased from Nu Trend Container Division, USA, Corrigan Inc., P. O. Box 2833, Jacksonville, FL 32203, USA; these trays are used for the entire larval period
- Heat sensitive Mylar® sheet (Champion Package, 15 Spinning Wheel Road, Hensdale, IL 60521, USA) used to cover the trays
- Disposable hypodermic needles (20 G x 1)
- Moderate heating iron

2.2.5 Pupal collection and storage
- Soft-nosed forceps
- Cylindrical, plastic covered container (60 x 30 mm diam) with 6 ml of 2% agar
- Aluminum paper placed on the agar surface
- Paper strips (100 x 20 mm)

Fig. 1. Some important material and equipment used for rearing larvae: (a) 50 cavity tray divided in half; (a*) E. scandens larvae in cavities; (b,c) 96 and 24 cavity trays used, respectively, for virus assays; (d) Mylar plastic sheet, self adhesive after iron (e) heating; (f) pupal container; (g) group-mating moth cage, the bottom contains sand for deposition of eggs by the adults; (h) single-pair moth cage; (i) sieve for egg straining; (j) vacuum installation used for egg drying after disinfection; (k) stereomicroscope for sex differentiation of pupae.

- Stereomicroscope with 7 to 40X magnification

2.2.6 Diet preparation equipment
- Waring® blender (3.78 liter)
- Hot plate
- 0.1 to 1000 g balance
- 10-ml pipettes
- Pyrex® flask (4 liter)
- 2-liter plastic containers
- 1-liter measuring cylinder
- Magnetic stirrer with magnetic bar
- Asbestos gloves
- Rubber spatula
- Cellophane sheet
- Cold room (4°C) or refrigerator

3. ARTIFICIAL DIET

3.1 Composition

The quantities given are enough for complete development of approx. 250 larvae.

Ingredient	Amount	% Estimated
a. Pinto beans (soaked for 12 h and drained)	854 g	30.2
b. Wheat germ	200 g	7.1
c. Brewer's yeast	128 g	4.5
d. Ascorbic acid	13 g	0.5
e. Sorbic acid	4 g	0.1
f. Nipagine (methyl-p-hydroxybenzoate)	8 g	0.3
g. Inositol	2.82 g	0.1
h. Formalin* = 40 % solution of formaldehyde	8 ml	0.3
i. Benomyl**	750 mg	0.03
j. Corn oil	10.8 ml	0.4
k. Agar	40 g	1.4
l. Distilled water	1560 ml	55.1

* Replaced by 160 mg of tetracycline if the larval use is intended for virus production.

** Used if microsporidial infection is suspected.

3.2 Brand names and source of ingredients

(a) through (c) and (j): purchased from local grocery stores. (d) through (h) and (k): purchased from Fisher Scientific Co. (8505) Devonshire Road, Montreal, Quebéc, Canada, H4P 2L4).

3.3 Diet preparation procedure

Grind (a) in the blender with 300 ml of distilled water and add together items (b) through (j), and 460 ml of cold water and mix again (Mixture A).

Dissolve the agar in 800 ml of boiling distilled water in Pyrex flask (Mixture B). Mix well with the magnetic stirring bar.

Add Mixture B to Mixture A, while B is still hot, and blend.

Distribute the warm liquid diet in rearing trays or in the plastic containers. Wait until it sets and store at 4°C; the containers are covered with cellophane sheet.

4. HOLDING INSECTS AT LOWER TEMPERATURES

- Eggs of <u>E. messoria</u> can be kept at least 1 year at 4°C
- Rearing of <u>E. scandens</u> and <u>E. messoria</u> can be slowed down by keeping 4th- and 5th-instar larvae at 10°C for 3 weeks

5. REARING AND COLONY MAINTENANCE

5.1 Egg collection and sterilization

The sand inside the adult mating cage is strained, every 2 to 3 days or weekly to collect the eggs.

The eggs are disinfected in the small plastic container with 0.2% sodium hypochlorite for 10 min, and then rinsed 3 times in distilled water. The eggs are placed on a filter paper in a Buchner funnel that is fixed on an Erlenmeyer flask with a rubber bung. The Erlenmeyer is connected to a pump and the eggs are dried by vacuum. The eggs are then removed from the filter paper by holding the paper with the sharp-pointed forceps and collecting the eggs with the fine brush.

- *E. scandens* eggs are placed on a filter paper in a petri dish and immediately incubated at 25°C, for larvae rearing (section 5.2)
- *E. messoria* eggs are stored in a plastic container at 25°C for 3 to 4 weeks to permit the embryonic development with a moist filter paper that does not touch the eggs to prevent desiccation. During this time the eggs must be checked 4 times a week for moisture and mould control. Thereafter, the eggs are mixed with moist sand and stored at 4°C for at least 1 month in a sterile petri dish containing a filter paper to absorb the excess humidity. This process is required for the eggs to diapause. After the incubation period the moist sand containing the eggs is rubbed against dry sand to free the eggs from the sand that sticks on it. The sand is then sieved to recover the clean eggs.

5.2 Larval rearing

The eggs of *E. scandens* or those of *E. messoria* after diapause are placed in a petri dish containing filter paper and incubated at 25°C. Two or 3 days later, the neonate larvae in groups of 10 are transferred on a filter paper in a petri dish using a fine brush. A cube (1 cm^3) of diet is added. Approximately 2 weeks later, the larvae are individually transferred to the cavities of the rearing trays. The trays are then covered by a Mylar sheet and sealed with a warm iron. Holes are made in the sheet with needles. Thereafter diet is added every 3 days.

5.3 Pupae and moth handling

Twenty-four hours after pupation, the pupae are sexed under a stereomicroscope (Fig. 2) and placed in the individual plastic containers and returned to an incubator for emergence.

The emerging adults are transferred every 2 days into adult mating cages. They are fed a 10% honey solution absorbed in dental cotton. After a period of 4 days, the eggs are laid in the sand on the bottom of the cages.

5.4 Rearing schedule

The frequency of operations indicated depends upon the quantity of insects.

Monday, Wednesday, and Friday
- Collect and disinfect eggs
- Transfer neonate larvae to rearing trays
- Transfer adults from pupae containers to cages

Monday
- Determine sex of pupae

Tuesday
- Feed larvae

Thursday
- Prepare diet and do the miscellaneous work

5.5 Insect quality

Insect quality can be checked by evaluation of pupal weight and length, adult size, fecundity, and egg viability.

5.6 Special problems

- *E. messoria* adults are very difficult to catch by light traps
- After 4 laboratory generations both species must be regenerated by the addition of field-collected insects; otherwise a loss of vigour and fertility occurs and adult malformations appear

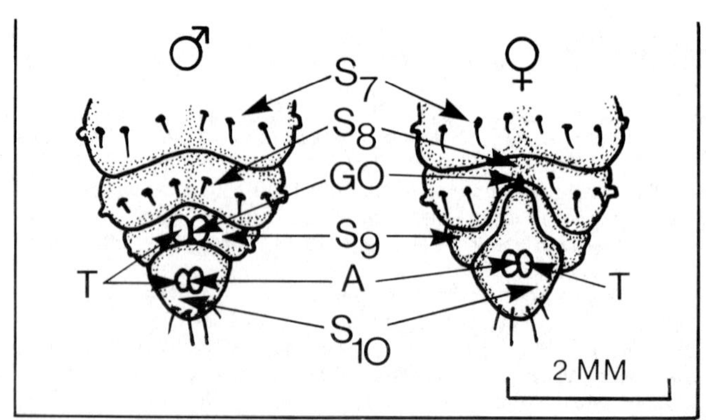

Fig. 2. Sex differentiation of pupae of E. scandens and E. messoria: (A) anal opening; (GO) genital opening; (S) abdominal segment; (T) tubercle

6. LIFE CYCLE DATA

	E. scandens	E. messoria
Egg incubation	4 to 7 days	1 to 3 days
Number of larval instars	7	7
Length of larval period	1-1/2 months	2 months
Average weight of pupae	230 mg (male) 242 mg (female)	294 mg (male) 272 mg (female)
Pupal period	2 weeks	2 weeks
Preoviposition period	3 to 4 days	3 to 4 days
Oviposition period	10 to 13 days	10 to 13 days
Fecundity (no. eggs/female)	approx. 1000	approx. 1000
Adult longevity	2 to 2-1/2 weeks	2 to 2-1/2 weeks
Diapause	no	eggs (4°C) 7th stage
Larval cannibalism	yes	yes

7. INSECT YIELD

The insect rearing by this method is efficient and provides a high yield of insects with a low percent of mortality.

Each rearing container of 50 cavities can produce 40 to 50 moths with a mean of about 46 moths per container. Each recipe of diet is the standard amount used for 250 larvae per generation in our laboratory. Thus, about 230 moths are produced for each generation. These moths are capable of producing more than 50,000 eggs, if required. The time required to place the moths in mating cages is about 10 h for 1 person.

8. PROCEDURES FOR SUPPLYING INSECTS

8.1 Shipping

E. scandens and E. messoria may be shipped as eggs, larvae or pupae within the country of origin but are most safely shipped as eggs or pupae for exportation overseas. The regulation regarding importation should be investigated if shipping is international. All containers should be well marked with their contents. Weekend arrivals should be avoided if possible.

a. Eggs

- For E. scandens, the eggs may be shipped immediately after oviposition in a container with pieces of lettuce or rye

- For *E. messoria*, the eggs may be shipped before or during the 1st diapause (at 25°C), or at the end of the 2nd diapause (at 4°C) in a container with an icepack and some pieces of lettuce or rye

b. Larvae

- Larvae should be sent individually packed in polystyrene tubes with an adequate supply of freshly dispensed diet. Tubes should be tightly bundled together and packed between cushions of packing material such as cotton wool, cellulose wadding or styrofoam beads. Time between dispatch and arrival should not exceed 7 to 8 days.

c. Pupae

- Pupae should be shipped in hermetic containers with peat moss; time between dispatch and arrival should not exceed 10 to 14 days

9. REFERENCES

Beirne, B. P., 1971. Pests insects on annual crop plants in Canada. Part I. Lepidoptera; II. Diptera; III. Coleoptera. Ent. Soc. Canada, Ottawa. 78: 123 p.

Chapman, P. J. and Lienk, S. E., 1981. Flight periods of adults of cutworms, armyworms, loopers and others (family Noctuidae) injurious to vegetable and field crops. Search Agriculture. 14: 39.

Cheng, H. H., 1971. Assessment of tobacco losses caused by the dark-sided cutworm *Euxoa messoria* (Lepidoptera: Noctuidae), Delhi, Ontario. Can. Ent. 103: 534-541.

Lampert, E. P., Haynes, D. L. and Cress, D. C., 1982. White cutworm: bionomics and evaluation of larval sampling schemes in asparagus. Environ. Entomol. 11: 21-28.

Meloche, F., Pilon, J. G., Mailloux, G. and Thierry, V., 1980. Inventaire des problèmes entomologiques et nématologiques dans les plantations de tabac jaune au Québec. Ann. Soc. Ent. Québec. 25: 81-89.

ACKNOWLEDGEMENT

The authors express their appreciation to D. Rouleau for her excellent technical assistance and to Drs. G. Lussier (Institut Armand-Frappier) and G. Mailloux (Agriculture Québec) for reviewing the manuscript and H. H. Cheng for helpful advice. The work is supported by the Department of Education of the Province of Québec (FCAC EQ-60), Agriculture Canada and Québec and by the Natural Sciences and Engineering Research Council Canada.

GALLERIA MELLONELLA

E. G. KING and G. G. HARTLEY

Agricultural Research Service, USDA , P. O. Box 225,
Stoneville, Mississippi 38776

THE INSECT

Scientific Name: *Galleria mellonella* (Linnaeus)
Common Name: Greater wax moth
Order: Lepidoptera
Family: Pyralidae

1. INTRODUCTION

The greater wax moth, *Galleria mellonella* (L.), is a major pest of the honey bee, *Apis mellifera* L. It feeds on wax and pollen stored in combs of active honey bee colonies (Milam 1970). The species is used extensively for studies in insect pathology and physiology; is mass reared for fish bait; and is a host for parasites of several lepidopterous pests. We developed a rearing program for the greater wax moth as a host to rear the tachinid *Lixophaga diatraeae* (Townsend). Procedures developed are suitable for smaller scale rearing as reported in this paper.

Rearing procedures were modified and expanded from those reported by Nielsen and Lambremont (1976). They obtained their colony and based their cereal diet and rearing procedures on those reported by Dutky et al. (1962) which Petersen (1953) had modified from Haydak (1936). Diets were reviewed for rearing the greater wax moth (Marston and Campbell 1973), and Marston et al. (1975) reported on an inexpensive diet containing a whey-grown yeast product (Wheast). However, Wheast is no longer available and a comparable product is too expensive. Specifically we summarize here a greater wax moth production system we reported in 1979 (King et al.). Hartley et al. (1977) reported on oviposition chambers, a moth scale collector, and larval harvesting while King et al. (1979) reported a highly nutritious diet for rearing larvae for parasite production.

2. FACILITIES AND EQUIPMENT REQUIRED

2.1 Facilities

Rearing of larvae and pupae is done in controlled environment rooms maintained at 30°C and 50% RH in complete darkness. Adults oviposit at 26°C and 70 to 80% RH with a photoperiod LD 14:10. Cool white fluorescent lighting of medium intensity is recommended.

2.2 Equipment and materials required for insect handling

2.2.1 Adult oviposition

- Aluminum screen wire
- 1 liter glass jar with mason type lid
- Oviposition chamber (constructed of 15 cm diameter PVC plastic pipe and 6.4 mm thick Plexiglas - Hartley et al. 1977).
- Scale collector (constructed from a blower - Grainger #3M254, 25 x 25 x 2.5 cm filter, 5.0 cm diameter plastic pipe, 12.6 mm hose insert fittings and 12.7 mm flexible hose - Hartley et al. 1977).

- Wax paper
- Paper stapler and staples

2.2.2 Egg collection and storage
- Staple remover
- Small plastic pan

2.2.3 Egg sterilization
- Small plastic pan or beaker
- Fine mesh sieve
- Absorbent paper towels
- Fume hood
- Formaldehyde

2.2.4 Egg hatch and larval inoculation of diet
- 5 ml vials
- Laboratory balance

2.2.5 Larval rearing containers
- Aluminum screen wire
- 4 liter plastic or glass jars
- Lids for 4 liter jars with center cut out for ventilation
- Filter paper

2.2.6 Pupal collection and storage
- 1 liter glass jars with mason type lid
- Large pan

2.3 Diet preparation equipment
- Top loading balance (2000 g capacity)
- Heating plate with magnetic stirrer
- 4 liter flask
- Aluminum scoop
- Portable concrete mixer

3. ARTIFICIAL DIET

3.1 Composition and mixing procedure

Ingredient	Amount (g/kg)	Source
Gerber mixed cereal	187 g	Gerber Products Co., Manufacturers Fremont, Mich.
Gerber high-protein cereal	187 g	do.
Wheat germ	120 g	Nutritional Biochemicals Corp., Cleveland, Ohio
Sucrose	148 g	
Glycerol	219 g	Dow Chemical Co., Midland, Mich.
Vitamin mix (Poly-Vi-Sol)	1.2 ml	Mead Johnson & Co., Evansville, Ind.
Water	138 g	

Mixing instructions for 1 kg of diet: dissolve sucrose in heated water; add glycerol; cool mixture and add vitamin mix; add liquid to dry ingredients and mix about 1 minute. For larger batches of diet a portable concrete mixer is used to mix diet.

4. INSECT HOLDING AT LOWER TEMPERATURES

4.1 Eggs
- Eggs can be stored at 16°C for up to 5 days with little or no reduction in percent hatch. Eggs are placed in containers which allow air exchange, such as a cardboard ice cream cup with a perforated top, and provided with moisture by placing a moist paper towel inside the container.

4.2 Larvae

- Larvae can also be held at 16°C for up to 10 days with no apparent damage. Larvae are usually stored in rearing containers filled with artificial diet.

4.3 Pupae

- Pupae can be stored at 16°C for up to 5 days. Late stage larvae are placed in 1-liter glass jars (up to 200 per jar) covered with aluminum window screen. After larvae pupate, the jar is transferred to 16°C storage area.

4.4 Adults

- Storage of moths is not recommended.

5. REARING AND COLONY MAINTENANCE

5.1 Egg collection

- Female moths deposit their eggs in accordion-folded strips of wax paper inserted into the oviposition chamber through slots cut in the chamber top
- Collect eggs by removing the folded strips of wax paper, removing the staples from each end of the wax paper strips, and gently scraping the eggs from the sheets of wax paper.
- Collect eggs three times each week.
- Insert new sheets of folded wax paper into the oviposition chambers each time eggs are collected.

5.2 Egg sterilization

- Disinfect eggs in formalin (3.3%) active ingredient for 10 minutes followed by a 5-minute rinse in distilled water
- Place eggs on paper towels to air dry.
- Weigh dry eggs (48 mg = 1500 eggs) and placed in 5-ml vials until they are transferred to jars containing diet.

5.3 Diet inoculation with eggs

- Place eggs directly into 4-liter jars containing 750 grams of artificial diet (eggs are placed directly on diet).
- The mouths of the jars used for larval rearing are covered with aluminum window screen and filter paper and are held in place by the jar lid with the center removed. The filter paper prevents the escape of the newly hatched larvae.
- Transfer the jars containing diet and eggs to 30°C and 50% RH. After about 37 days larvae move to the top of the jar to spin cocoons (this period includes six days for the eggs to hatch).

5.4 Pupal collection

- Remove larvae as they move to the top of the 4-liter rearing jars, and place (200 per jar) in 1-liter glass jars.
- Replace 1-liter glass jar lids with aluminum window screen to prevent larval escape and provide ventilation. (Good ventilation is required to allow evaporation of moisture accumulating from the emptying of larval gut tracts before pupation.
- Remove screen jar lid after larvae pupate and place in a chamber which serves for adult emergence, mating, and oviposition (Hartley et al. 1977).

5.5 Adults

- The sex ratio is about 1:1
- Adult survival from eggs is about 65%
- First adults begin to emerge about 47-50 days after egg is oviposited

5.5.1 Oviposition

- Moths oviposit best in controlled environment rooms maintained at 24°C and 70-80% RH. A LD 14:10 photoperiod is maintained. Fluorescent cool white lighting of medium intensity is recommended.

- Large amounts of loose scales occur during oviposition; because of this a special oviposition chamber and scale collector system is used (Hartley et al. 1977).
- Moths oviposit the second night after mating.
- Discard moths seven days after mating.
- Each female moth deposits about 700 eggs over a seven-day period.

5.6 Sex determination

- For rearing purposes, Galleria mellonella was not sexed. Approximately a 1:1 sex ratio was achieved as larvae were collected and placed in 1 liter glass jars for the oviposition cycle.

5.7 Insect quality

- Insect quality is checked by recording number of eggs per female, % egg hatch, number of larvae harvested per larval rearing container, larval size, and adult longevity. Average quality should be as follows: percent egg hatch, 90%; eggs per female, 700; number larvae per container, 1000; larval size, 250 mg; and adult longevity, 12 days (adults discarded after 7 days).

5.8 Special problems

a. Human health

In rearing Galleria mellonella serious respiratory problems can occur among workers if precautions are not taken to collect the large number of wing scales present during the oviposition cycle. Workers should always wear protective masks and some type of scale collector should be installed where moths are held.

b. Microbial contamination of diet

Diet should be mixed and infested under sanitary conditions. All larval rearing containers and work instruments should be sanitized with a suitable disinfectant.

6. LIFE CYCLE

6.1 Optimal rearing conditions

- Holding adults for mating and oviposition - 26°C; 70-80% RH
- Holding larvae for development on artificial diet - 30°C; 50% RH
- Holding pupae for adult emergence - 30°C; 50% RH

6.2 Longevity of various stages at optimal rearing conditions

- Adults - moths are held for 7 days; each female deposits about 700 eggs over this period. Moths will survive longer but most egg production is completed after 7 days
- Larvae - from egg hatch until larvae begin to spin cocoons just prior to pupation averaged 32-34 days
- Pupae - Beginning when larvae start to spin cocoons until moth emergence averages 9-12 days

7. PROCEDURES FOR SHIPPING INSECTS

Persons interested in establishing a colony of Galleria mellonella are advised to contact either the Insect Pathology Laboratory of the Plant Protection Institute, Beltsville Agricultural Research Center, ARS, USDA, Beltsville, MD or the USDA, ARS, Bee Breeding and Stock Center Laboratory at Baton Rouge, LA.

8. REFERENCES

Dutky, S. R.; Thompson, J. V.; and Cantwell, G. E. 1962. A technique for mass rearing the greater wax moth. Proc. Entomol. Soc. Wash. 64: 56-58.

Hartley, G. G.; Gantt, C. W.; King, E. G.; and Martin, D. F. 1977. Equipment for mass rearing of the greater wax moth and the parasite, Lixophaga diatraeae. U.S. Agric. Res. Serv. ARS-S-164, 4 pp.

Haydak, M. H. 1936. A food for rearing laboratory insects. J. Econ. Entomol. 29: 1026.
King, E. G.; Hartley, G. G.; Martin, D. F.; Smith, J. W.; Summers, T. E.; and Jackson, R. D. 1979. Production of the tachinid Lixophaga diatraeae on its natural host, the sugarcane borer, and on an unnatural host, the greater wax moth. U.S. Dept. of Agric., SEA, AAT-S-3, 16 pp.
Marston, N.; and Campbell, B. 1973. Comparison of nine diets for rearing Galleria mellonella. Ann. Entomol. Soc. Am. 66: 132-136.
Marston, N.; Campbell, B.; and Boldt, P. E. 1975. Mass-producing eggs of the greater wax moth, Galleria mellonella (L.). U.S. Dep. Agric. Tech.
Milam, V. G. 1970. Moth pests of honey bee combs. Glean. Bee Culture 68: 424-428.
Nielsen, R. A.; and Lambremont, E. N. 1976. Radiation biology of the greater wax moth: inherited sterility and potential for pest control. U.S. Dept. Agric. Tech. Bull. 1539, 31 pp.
Peterson, A. 1953. A manual of entomological techniques. Edwards Bros., Ann Arbor, Ich., p 47.

9. FOOTNOTE

Mention of a trademark, proprietary product or vendor does not constitute a guarantee or warranty of the product by the U.S. Department of Agriculture and does not imply its approval to the exclusion of other products or vendors that may also be suitable.

GRAPHOLITHA MOLESTA

D. J. PREE

Research Station, Agriculture Canada, Vineland Station, Ontario LOR 2EO

THE INSECT

Scientific Name: Grapholitha molesta (Busck)
Common Name: Oriental fruit moth
Order: Lepidoptera
Family: Olethreutidae

1. INTRODUCTION

The Oriental fruit moth (OFM) was introduced into the U.S., probably from Japan, near Washington D.C. about 1913 (Chaudhry 1956). By 1925, infestations were reported from over most of eastern North America (Dustan 1960) and by 1943 it had reached California (Smith and Summers 1948). In the northeastern U.S. and eastern Canada, the OFM is the most important pest of peach, but quince, apricot, pear, and to a much lesser degree apple, cherry and plum are also attacked (Dustan 1960). There are 3-4 generations/year. The larvae damage peach in two ways: by boring into terminal and lateral shoots for several inches causing death of the growing tip; and by entering and feeding inside the fruit. Larvae prefer twigs as long as they are succulent and a single larva may attack 3 - 4 twigs before it matures. As twigs harden, larvae may leave and enter fruit. Damaged twigs may become infected with Leucostoma canker and this problem may be serious in more northerly growing areas.

The Oriental fruit moth has long been cultured on green thinning apples (Peterson 1930, Garman 1933, Daniel et al 1933). Artificial diets for Oriental fruit moth have bee developed by Tzanakakis and Phillips (1969), Laing and Hagen (1970) and more recently by Szocs and Toth (1982), but culturing on green thinning apples remains the most practical and efficient method to this date. Current uses of the culture range from studies of comparative toxicity of pesticides (Pree 1979) to use as a food source for predaceous insects.

2. FACILITIES AND EQUIPMENT REQUIRED

2.1 Facilities

Rearing of all stages is carried out in a controlled environment room maintained at 20°C ± 1° C, 65 ± 5:RH and a 16:8 photoperiod.

Light is from F40 cool white incandescent tubes, and ranges from 20-30 lux over egg laying and larval rearing containers to 1200 lux directly below the lights.

2.2 Equipment and materials required for insect handling

2.2.1 Oviposition cages

- Screw top glass jars 18 x 8 cm inside diameter
- Cotton wool dental rolls 75 mm
- 8 x 2 cm shell vials
- Metal jar lids, partially cut away, screened with 10 mesh/cm - saran screen (Barrday Div. of Wheelabrator Corp. of Canada, Galt, Ontario)
- Glue; 3M Scotch Grip® plastic adhesive
- Spatula
- Waxed paper Rap Rite® Brand, Perkins Papers Ltd., Laval, Quebec

- Refrigerator at 2-4° C
- corks 75 mm

2.2.2 Preparation of ovipositon cages
- With metal sheers cut away ca 1/3 of surface of metal lid
- Glue saran screen in place with plastic adhesive
- Cut additional 75 mm hole in metal top to fit cork
- Cut hole in cork to fit dental roll

2.2.3 Egg collection and storage
- Absorbent cotton
- 4 liter plastic container with snap lids (Plastipak Containers, Toronto, Ontario, Canada)
- Water squirt bottle

2.2.4 Egg sterilization
- 4 liter plastic containers
- Household bleach (Javex®, 6% sodium hypochlorite)
- Metal forceps
- Household fans 15 cm blades

2.2.5 Egg hatching and larval rearing containers
- 4 liter plastic containers or (preferrable) 24 liter plastic pails with lids
- lids 20 cm diam circle cut out, screened with saran

2.2.6 Host
- Thinned apples ca 3.5-5 cm diam. preferably cultivars Wealthy or Sandow Spy

2.2.7 Preparation of fruit
- Apples are harvested ca June 25-July 4 (in Ontario) from orchards sprayed only with fungicides
- Apples are held in cold storage for up to one year
- After 6 months, apples are sorted monthly, and rotted fruit discarded
- Sprayed fruit may be used if treated for 5 minutes with 2% Na_3PO_4, followed by a 5 minute water wash

2.2.8 Pupal collection and storage
- Corrugated carboard strips, B flute, 1.5 cm x 1 m masking tape
- 4 liter plastic containers with snap lids

3. REARING AND COLONY MAINTENANCE

The colony was established from field collected adults and has been maintained for ca 25 years. In the past 5-6 years, limited additions of newly collected OFM from the field have been added.

3.1 Egg production

3.1.1 Preparation of oviposition containers
- Cut strips of waxed paper to line inside of glass oviposition cage, overlapping paper ca 2-3 cm
- Place in jar
- Seal waxed paper tube by scratching with edge of spatula

3.1.2 Transfer of moths
- Moths in emergence or oviposition containers are held 1/2 h or longer in refrigerator to slow activity
- Moths are tipped from previous containers to oviposition jar inside waxed paper tube
- Replace metal lid, fit damp dental wick into hole in lid, placing proximal end inside jar ca 3 cm
- Place distal (outside) end of dental wick in water filled 8 cm shell vial
- Change oviposition cage daily
- Wash wicks daily to remove moth scales and debris, discard weekly
- Oviposition cages and other glassware are washed in a dishwasher and

dry sterilized at 200°C between uses

3.1.3 Comments
- Egg production begins after 3-4 days of adult emergence
- Egg production peaks after 3-4 days of egg laying
- Adults should be discarded after 6-7 days
- Field collected moths lay fewer eggs than those conditioned to the laboratory
- Choice of waxed paper affects numbers of eggs laid, Rap Rite® waxed paper works best
- To avoid excess water build up inside egg laying container, dental roll should not contact waxed paper

3.1.4 Handling and storage of eggs
- Rinse waxed paper egg sheets with eggs under the tap in lukewarm water for ca 30 seconds to remove loose scales
- Hang "egg sheets" to dry in rearing room
- Eggs may be stored in 2 liter covered plastic containers containing a wad of damp cotton in a refrigerator for 7-10 days
- At 22° C eggs reach 'blackhead' stage after 4 days
- Hatch occurs 24 h later

3.1.5 Comments
- After 7-10 days storage in refrigerator eggs do not develop
- Blackhead stage eggs hatch ca 12 h after removal from cold storage
- If 1st stage larvae are required, sheets containing blackhead eggs are held in sealed plastic bags overnight in the rearing room

3.1.6 Larval rearing
- In 24 liter plastic pails, place layer of thinned apples (to cover bottom)
- Place 2 egg sheets (1000-1500 eggs) on surface of apples
- Place 1 sheet facing up, other opposite add rest of 4 liter container of apples gently to cover egg sheets (to about 1/3 full)
- Cover pail with screened lid
- Place in rearing room at 23° C
- Air movement in rearing room is essential-use 1-2 small fans to circulate air

3.1.7 Comments
- Apples tend to deteriorate in storage. In April, May and June fruit may break down during larval life. During this period it is necessary to add (after ca 7-10 days) additional layers of apples to rearing pails as the fruit breaks down
- Larvae migrate to fresh apples as older food sources deteriorate
- Rearing efficiency which is 50-60% from July-January deteriorates to ca 25% as fruit begins to rot

3.1.8 Pupal collection
- After 7 days in rearing room strips of cardboard 1.5 cm x 100 cm are taped horizontally with narrow strips of masking tape to side of larval rearing pails
- Usually 3-4 strips/pail are necessary
- After 10-12 days larvae leave fruit and crawl up sides of container, pupating inside cardboard strips
- Remove strips every 2 days until larvae cease to appear (usually 2-3 x)
- Place cardboard strips containing spun-up larvae in 2 liter plastic containers with screened lids (4-6 strips/container)

3.1.9 Adults
- Adults emerge from pupae in bands inside 2 liter holding containers

- after ca 7 days
- To transfer adults to egg laying containers place emergence containers in 2°C refrigerator for ca 1/2 h
- Sex ratio is ca 1:1
- Emergence occurs over ca 5 days
- Discard cardboard strips
- Wash containers in hot water, dry before reuse

4. REARING SCHEDULE

4.1 Daily (except Sunday)

- Prepare fresh wax paper lined egg laying containers
- Collect newly emerged moths
- Change egg laying containers
- Rinse "egg sheets", hang to dry
- Place "egg sheets" from previous day in 2-liter plastic tubes with moist cotton

4.2 Alternate days

- Collect pupae in cardboard bands
- Replace bands

4.3 Twice weekly

- Set up fresh pails of larvae

5. INSECT QUALITY

Quality is checked 2-3x per year by assessment of toxicity of 2-3 insecticides (representing organophosphorous, carbamate and pyrethroid insecticides). Mortality data are compared with previously established dose-response curves (Pree 1979).

5.1 Special problems

5.1.1 Diseases

Nuclear polyhedrosis virus infections are prevented by sterilization of eggs. This can be accomplished by dipping waxed paper sheets with eggs attached into dilute, 0.5% sodium hypochlorite solution for 30 sec, then rinsing with lukewarm tap water.

5.2 Human health hazards

Scales of moths may be hazardous or at least an irritant. A suitable mask should be worn when working with moths.

Scales in the air can be removed by circulating air through a filteration system.

6. LIFE CYCLE DATA

6.1 Development data (at 20°C; mean days duration)

- Eggs	4
- Larvae	12
- Pupae: Male	7
Female	8
- Preoviposition period	2 days
- Duration of egg production	7 days
- Peak egg production	3-4th day
- Larval instars	4
- Complete life cycle (egg-egg)	25-26 days

6.2 % Survival

- Egg viability	95+
- Larvae + pupae	30-60
- Pupae + adult	90

7. REFERENCES

Chaudhry, G.U. 1956. The development and fecundity of the Oriental fruit moth, Grapholitha molesta (Busck) under controlled temperatures and humidities. Bull. Ent. Res. 46:869-898.

Daniel, D.M., Cox, J., and Crawford, A. 1933. Biological control of the Oriental fruit moth. New York State Agr. Exp. Sta. Bull. 635. 27p.

Dustan, G.G. 1960. The Oriental fruit moth Grapholitha molesta (Busck) in Ontario. Proc. Entomol. Soc. Ont. 91:215:227.

Garman, P. 1933. Notes on breeding Macrocentrus ancylivorus from reared hosts. J. Econ. Entomol. 26:330-334.

Laing, D.R. and Hagen, K.S. 1970. A xenic, partially synthetic diet for the Oriental fruit moth, Grapholitha molesta. Can. Ent. 102:250-252.

Peterson, A. 1930. A biological study of Trichogramma minutum Riley as an egg parasite of the Oriental fruit moth. USDA Tech. Bull. 215. 21 p.

Pree, D.J. 1979. Toxicity of phosmet, azinphosmethyl and permethrin to the Oriental fruit moth and its parasite Macrocentrus ancylivorus. Environ. Entomol. 8:969-972.

Smith, L.M. and Summers, F.M. 1948. Propagation of the Oriental fruit moth under central California conditons. Hilgardia 18:369-387.

Szocs, G. and Toth, M. 1982. Rearing of the Oriental fruit moth Grapholitha molesta Busck. on simple semisynthetic diets. Acta Phytopathol. Acad. Sci. Hung. 17:295-299.

Tzanakakis, M.E. and Phillips, J.H.H. 1969. Artificial diets for larvae of the Oriental fruit moth. J. Econ. Entomol. 62:879-882.

Acknowledgements

I thank Mrs. D. Archibald and D. R. Marshall for helpful discussions and reviews of the manuscript.

HELIOTHIS PUNCTIGER

R. E. TEAKLE and J. M. JENSEN

Entomology Branch, Department of Primary Industries, Indooroopilly, Queensland, Australia, 4068.

THE INSECT

Scientific Name: *Heliothis punctiger* Wallengren
syn. *Heliothis punctigera*, *Helicoverpa punctigera*
Common Name: Native budworm
Order: Lepidoptera
Family: Noctuidae

1. INTRODUCTION

Heliothis punctiger is an endemic Australian species and occurs throughout this continent on a wide range of host plants. In southern Australia, it is the predominant or only *Heliothis* species, but in northern Australia it often occurs in association with *Heliothis armiger*, and may give way to *H. armiger* as the season progresses.

H. punctiger is a destructive pest of a wide range of commercial crops. Field crops, such as cotton and tobacco, and oil seeds, such as linseed, sunflower, safflower and sesame; and legumes such as lucerne (= alfalfa), lupins, peas and peanuts are attacked. Additionally, horticultural crops such as tomatoes, strawberries, apples, watermelons, gladioli and carnations may be damaged. While reproductive parts and growing points are preferentially attacked, extensive damage to foliage may also occur.

The species generally overwinters as diapausing pupae. There are up to 7 generations between August and May. Eggs are laid singly on leaves, growing tips or flowers in the field, and on rough, vertical surfaces in the laboratory. Eggs hatch after about 80 hours at 25°C. Field-collected larvae usually undergo 6 instars although 7 instars may be recorded (Hardwick 1965). Larvae in our laboratory culture normally complete development in 5 instars, probably owing to continuous selection for rapid larval development (see Section 4.4). This takes about 17 days at 25°C. The pupal period is normally spent in the soil within a small, silk-lined chamber. The duration at 25°C is about 16 days, and is slightly shorter for females than males, which tend to emerge later. Adults mate and commence oviposition usually within 3 to 5 days of emergence.

Pupal diapause is likely to occur at short, reducing day-lengths (e.g. LD 12:12), particularly if coupled with low temperatures (e.g. below 20°C).

The preferred stage from the field for starting the colony is gravid females, which can be collected at a light trap. These offer a better chance of avoiding disease and parasites than, for example, eggs or larvae.

We have continuously reared *H. punctiger* in the laboratory, without additions to the colony, for 10 years, mainly for studies of pathogens. One technician part-time is required for colony maintenance based on the production of 2 batches of 60 moths per week. This involves larval diet preparation, harvesting and surface-sterilizating of eggs and pupae, transfer of batches of newly-hatched larvae to diet and later to individual rearing containers, and transfer of moths to mating cages and then to oviposition cages.

2. FACILITIES AND EQUIPMENT REQUIRED

2.1 Facilities

All stages are maintained in a constant temperature room at 25 ± 1°C and ambient (40-70%) RH with an LD 16:8 photoperiod. Natural lighting is obtained via a clear glass window. The rearing room contains a bench measuring 1070 x 580 mm and 380 x 300 mm sink with a goose-neck cold water tap.

2.2 Equipment and materials required for insect handling

2.2.1 Adult mating and oviposition

- Mating cage of at least 0.05 m^3, composed of non-corrodible metal with mesh sides or clear perspex with cloth mesh ends
- Vacuum cleaner
- Aspirator (Fig. 1)
- Oviposition cage (after P.H. Twine*, personal communication) constructed of aluminium, in which paper towelling constitutes the vertical walls (Figs. 2a and 2b). The floor is made of expanded aluminium mesh (BD Expamet No. 401A, Comalco Aluminium Supply, Acacia Ridge, Queensland, Australia) to allow the moth scales to pass through and collect in a dish of water. Feeding cups containing sucrose solution (see Section 3.5) are set into the floor. The paper is drawn through a 3 mm gap, which retains the moths in the cage.
- 1 ounce (28 ml) clear portion cups for sucrose solution. (Premier Distributors Pty. Ltd., Bulimba, Queensland, Australia). The cups have plastic lids in which 8 mm holes are punched and plugged with dental roll wicks.

2.2.2 Egg collection and storage

- Plastic bags (400 x 255 mm)
- Rubber bands
- Waterproof marking pen

2.2.3 Egg sterilization and hatch

- Photographic developing tray (440 x 430 mm)
- 10% commercial sodium hypochlorite (Nightingale Chemicals, Eagle Farm, Queensland, Australia)

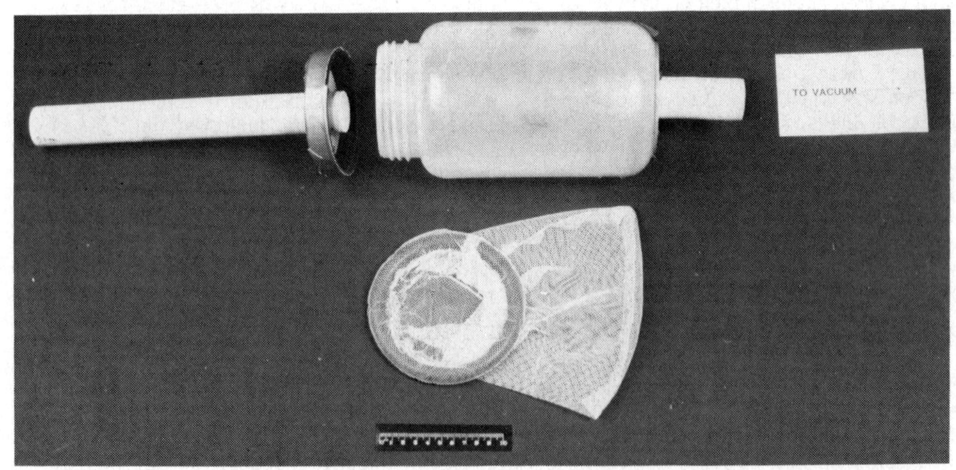

Fig. 1. Aspirator with mesh bag insert for transferring moths. (Scale 10 cm).

* Department of Primary Industries, Toowoomba, Queensland, Australia

Fig. 2a. Oviposition cage for *H. punctiger*. Paper towelling from the roll at the left forms the walls of a cylindrical cage at the right. The paper protrudes through a 3 mm slit, and is withdrawn to harvest the eggs. (Scale 10 cm).

Fig. 2b. Cage exposed to show circular paper guides and expanded mesh floor with sugar solution dispensers set into holes. The clear perspex lid is suspended from an aluminium strip.

- Buchner funnel and flask, and vacuum source
- Absorbent tissues(for household use)
- Plastic bags (400 x 255 mm)
- Interval timer graduated in minutes
- Paper towelling
- Rubber bands
- 50 ml measuring cylinder (polythene)
- 2 liter measuring cylinder (polythene)

2.2.4 Larval inoculation of diet

- Brushes (Roymac 3550 Golden Sable imitation Size 0 (fine))
- 1.0% sodium hypochlorite for sterilizing brushes
- Metal trays (about 520 mm x 360 mm)

2.2.5 Larval rearing containers

a. Group rearing

- Plastic cups Lily P125-77 (70 mm diameter x 50 mm high), (Lily Plastics, Sydney, N.S.W., Australia)

b. Individual rearing

- Plastic portion control cups (Lily P100) with waxed cardboard lids
- Forceps, 15 cm curved or straight, blunt
- Bunsen burner

2.2.6 Pupal collection and storage

- Blunt forceps
- 1600 ml of 0.25% sodium hypochlorite
- 2 liter polythene measuring cylinder
- 50 ml polythene measuring cylinder
- Plastic strainer (170 mm diameter, with handle) (Fig. 3)
- Plastic container, 220 mm diameter x 90 mm high (Fig. 3)
- Wide (90 mm)-neck glass jar with bakelite lid with 70 mm diameter hole cut out
- Paper towelling

2.3 Artificial diet preparation

2.3.1 Equipment

- 2 Basins (enamel or stainless steel, 220 mm top diameter, 125 mm high) with lids
- 2 Basins (enamel or stainless steel, 205 mm top diameter, 110 mm high)
- Top loading balance (0.1 - 3000 g)
- Blender (e.g. "Waring" blendor, 1 gallon) (3.78 liter)
- 1 liter measuring cylinder, polythene
- Stainless steel spatula, 25 mm wide x 155 mm long plus a wooden handle
- Stainless steel spatula, 35 mm wide x 200 mm long plus a wooden handle
- Trays (230 x 230 x 45 mm) for diet storage
- Plastic bags (500 x 300 mm)
- Thermometer (0-100°C)
- Waterbath (70°)
- Refrigerator (5°C)

3. ARTIFICIAL DIET

The following diet was adopted as suitable for both *H. punctiger* and *H. armiger*. Formerly the diet of Shorey and Hale (1965), supplemented with wheat germ at the rate of 10% of the weight of navy bean, was used successfully for *H. punctiger*. Cooper (1979) used a modified Sender's diet (Griffith and Smith, 1977).

3.1 Composition (per 1000 ml diet)

		Amount	%
a.	Navy bean (or equivalent)	86 g	8.6
	Tap water	470 ml	47.0
b.	Agar	12.5 g	1.25
	Tap water	303 ml	30.3

Fig. 3. Pupae in plastic sieve for surface-sterilizing in sodium hypochlorite solution.

	Amount	%
c. Dry mix		
Wheat Germ (natural, raw)	60 g	6.0
Yeast (Torula or Brewers)	50 g	5.0
ℓ Ascorbic acid	3 g	0.3
Nipagin M (methyl p-hydroxy benzoate)	3 g	0.3
Sorbic acid	1 g	0.1
Pollen (ex honey bee)	10 g	1.0
d. Formaldehyde (40%)	1 ml	0.1

3.2 Brand names and sources of ingredients

- Agar, (Davis Gelatin (Australia) Co., Acacia Ridge, Q. Australia)
- Yeast, (Sanitarium Health Food Co., Brisbane, Q. Australia)
- ℓ Ascorbic acid, (Sigma Chemical Co., P.O. Box 14508, St. Louis, Mo. 63178, U.S.A.)
- Nipagin M, (Nipa Laboratories, South Wales, U.K.)
- Sorbic acid, (Sigma Chemical Co., P.O. Box 14508, St. Louis, Mo. 63178, U.S.A.)
- Pollen (Pollen Processers, P.O. Box 35, Willetton, Western Australia)

3.3 Diet preparation procedure

The navy beans are weighed into a large bowl and washed with several changes of tap water. The volume of water added finally is determined by weighing (to the required total weight of water plus beans) to allow for water absorbed by the beans during washing.

- Steam beans a and agar b separately in covered basins for 1.5 hours.
- Cool beans and agar to about 70°C.
- Mix the beans and dry ingredients and 1 ml formaldehyde in the blender.
- Add the agar and remix.
- Dispense the diet, either directly into individual containers using a plastic sauce dispenser or into a tray, to set at room temperature.
- Put tray of diet in a plastic bag after the diet has set.
- Refrigerate diet at 5°C, at which temperature it will keep for at least 10 days.
- Cut the diet in the tray with a knife and remove as required.

Fig. 4. Larvae hatching from eggs sandwiched between paper.

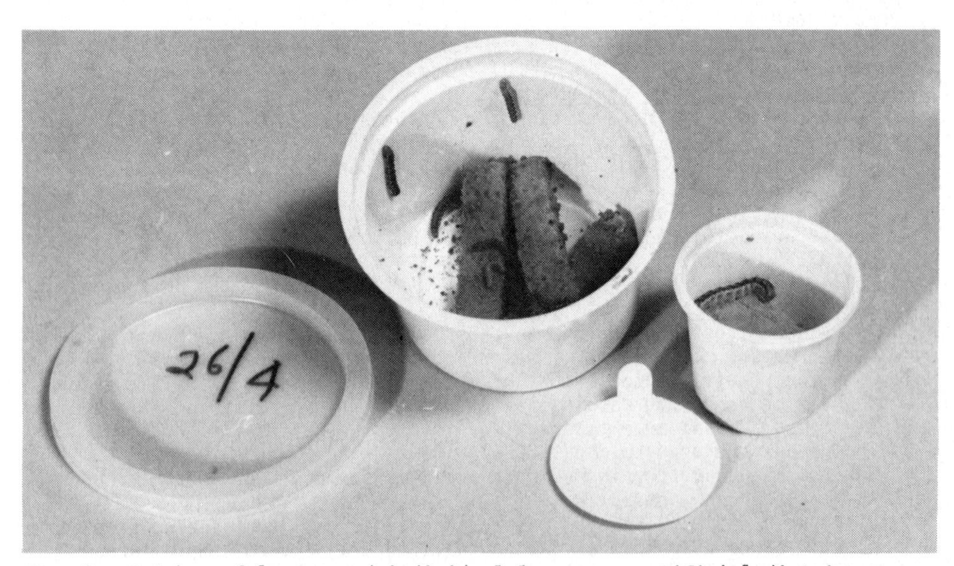

Fig. 5. Batches of larvae and individual larvae on artificial diet in disposable plastic containers.

3.4 Natural food

French bean pods (*Phaseolus vulgaris*) constitute the most convenient natural food (Kirkpatrick, 1962; Cullen, 1969).

- Wash the pods in 70% ethanol and dry.
- Dispense 3-cm lengths into autoclaved glass jars about 45 mm diameter x 60 mm high.
- Rear larvae individually in the jars.
- Provide late-instar larvae with 1 cm layer of insecticide-free sawdust to absorb excess moisture and provide a substrate in which to pupate.

3.5 Sugar solution for moths

- Sugar 100 g
- Deionised water 1000 ml
- Autoclave at 120°C for 15 min and cool to room temperature. Add ascorbic acid at 6 g per liter. Add 1 ml per liter of a stock solution of streptomycin sulphate* of 1 g in 25 ml water, which is stored frozen. This provides a final concentration of 40 mg/liter. *(Glaxo Australia Pty. Ltd., Boronia, Victoria, Australia)

4. REARING AND COLONY MAINTENANCE

Our laboratory culture was established from moths collected at light in December, 1972, and has been maintained continuously without further introductions from the field.

4.1 Egg collection

- Draw the paper towelling with attached eggs through the moth-excluding slit in the oviposition cage and tear off past the last eggs seen.
- Wash off the moth scales collected in the dish below the cage and refill with water.

4.2 Egg surface-sterilization (after Ignoffo, 1965)

This is preferably done on the day before the larvae are due to hatch.

- Tear the paper towelling into 10-cm lengths.
- Place eggs downward in a developing tray and add 2 liters of 0.2% sodium hypochlorite.
- Agitate the tray for 8 min, during which the eggs become detached and settle to the bottom, then remove the paper towelling.
- Pour off most of the liquid.
- Pour the remaining liquid containing the eggs into a Buchner funnel lined with 2 absorbent tissue papers.
- Wash the eggs several times with tap water.
- Remove the tissue papers with eggs and wrap in paper towelling.
- Store moist in a sealed plastic bag until the eggs hatch (80 hours at 25°C) (Fig. 4).

PRECAUTION

Eggs desiccate if not kept moist after sodium hypochlorite treatment.

4.3 Diet inoculation

- Dispense 3 pieces of diet each about 50 mm x 10 mm x 5 mm per cup (Lily P125-77). At least 12 cups are set up on a metal tray.
- Using brushes previously sterilised for 10 min in 1% sodium hypochlorite, transfer newly hatched larvae to diet, 10 larvae per cup. Apply lids.
- Label with date and stack trays for incubation (Fig. 5).

PRECAUTIONS

(i) Preferably handle larvae by suspending threads to avoid damage to the insects.

(ii) Larvae may attack and cannibalise other larvae. The large diet surface minimises contact between larvae.

(iii) Avoid contamination of diet with non-sterilised instruments or hands.

4.3.1 Individual rearing of larvae (after 10 days at 25°C)
- Dispense 6 cm of diet into Lily P100 cups.
- Transfer the 5 most advanced larvae per Lily P125-77 cup to individual Lily P100 cups using flamed forceps (Fig. 5). Discard the remaining larvae.
- Seal cups with waxed cardboard lids. These allow some ventilation and drying of the diet before the larvae pupate in the frass and remaining diet at the bottom of the container.

PRECAUTIONS
(i) Final-instar larvae may die of asphyxiation in non-ventilated containers.
(ii) Some field-collected larvae may chew through cardboard lids. Those showing this tendency should be transferred to containers with plastic lids plugged with a dental roll wick for ventilation.

4.4 Pupal collection (less than 1 month after larval hatch)
- Carefully extract pupae from frass and diet in the cups using blunt forceps.
- Transfer to a plastic sieve (Fig. 3).
- Immerse sieve and pupae in 1600 ml or more of 0.25% sodium hypochlorite for 10 min with occasional agitation.
- Wash pupae under running tap water, drain.
- Transfer to a wide-neck jar lined with paper towelling and seal, using a lid with the centre removed and lined with paper towelling or 9 cm diameter filter paper.

PRECAUTIONS
(i) Avoid stabbing pupae with forceps.
(ii) Discard any deformed pupae.
(iii) Paper should extend well up the sides of the jar to provide a support for the emerging moths.

4.5 Adults
- Set up a mating cage (on 4 supports, each standing in paraffin oil, if ants are a problem) and line with absorbent paper on the bottom to absorb meconium.
- At the first sign of adult emergence, add the jar of pupae with the lid removed and at least 4 sugar solution dispensers, and maintain the moths in the cages for approximately 3 days at 25°C.

PRECAUTIONS
(i) Avoid any direct contact of the cage with the bench to exclude ants.
(ii) Female adults emerge before males. Females have brown or buff-coloured forewings, whereas males have lighter-coloured, grey forewings.
(iii) Sodium hypochlorite treatment of pupae appears to make adult emergence more synchronous.
(iv) Cage must be either steamed or soaked in 1% sodium hypochlorite before reuse.

4.5.1 Oviposition
- Set up the oviposition cage with feeding cups with 10% sucrose. (Use paraffin oil ant excluders as for the mating cage if ants are a problem).
- Remove clumps of moth scales from the mating cage using a vacuum cleaner.
- Transfer moths to the oviposition cage using the aspirator.

PRECAUTIONS
(i) Make moth transfers in the early morning when they are least active.
(ii) Avoid inhalation of moth scales, which can be allergenic. Wear a dust mask filter.

4.6 Sex determination

- Pupae. This is based on the position of the gonopore and the shape and size of the terminal segment (see Kirkpatrick, 1961).
- Female - The gonopore occurs midway between the second and third last segments (segments 8 and 9). The terminal segment is larger than in the male and tends to protrude into the second-last segment.
- Male - The gonopore is positioned in the second-last segment and edged by lateral swellings.
- Adults. These can be distinguished by the colour of the forewings (Kirkpatrick, 1961).
- Female - Forewings reddish brown to light brown, darker than males.
- Male - Forewings greyish, lighter than females.

4.7 Rearing schedule

Two weekly collections of eggs are sufficient to maintain the culture. Prepare 2000 ml of diet weekly and sugar solution as required.

- Clear eggs on Monday and Thursday and discard if not otherwise required.
- Collect eggs on Tuesday and Friday for the culture.
- Treat eggs with sodium hypochlorite on Thursday (Tuesday's collection) and Friday (same day collection) and these will hatch on Friday and Monday, respectively, on the third day after collection.
- Transfer 12 batches of 10 neonates to diet on both Monday and Friday.
- After 10 days, transfer the 60 most advanced larvae to individual containers, and discard the remainder.
- In less than 1 month after larval hatch, harvest pupae and surface-sterilize in sodium hypochlorite.
- At approximately 4.5 weeks, when adults start to emerge, transfer moths and the rest of the pupae to the mating cage.
- 3 days after adult emergence, transfer the moths to an oviposition cage.
- After a week or less, destroy the moths by autoclaving the cage (minus perspex lid and paper towel roll).

4.8 Special problems

a. Diseases.

Infection by a nuclear polyhedrosis virus (Teakle, 1973) can be very destructive and must be excluded by sodium hypochlorite-treatment (sterlization) of eggs and pupae and sterilization of working surfaces and instruments used for handling the insects.

A bacterial disease of the moths has been apparently controlled by the addition of streptomycin sulphate to the sugar solution.

b. Microbial contamination of diet.

Appropriate hygiene precautions to control disease also help to exclude contaminants from the diet. Development of contaminants can also be retarded by maintaining a low RH (less than 60%) to discourage fungal growth and by allowing the diet to dry out in the individual larval rearing cups.

c. Development of allergy to moth scales.

This has occurred with heavy exposure to the scales. Scale inhalation can be minimised by the wearing of a protective face mask when handling the moths, and the removal of scales by vacuum or trapping in water.

5. LIFE CYCLE (25°C, LD 16:8 photoperiod)

Preoviposition period	3 to 5 days	Eggs to pupae	20 to 21 days
Eggs hatch in	3.3 days	Eggs to adults	34 to 35 days

6. PROCEDURES FOR SUPPLYING INSECTS

Eggs constitute the most readily available and convenient stage for dispatch. These can normally be supplied from established cultures, or field-collected moths, and reared to the required stages by the recipient.

Within Australia the insects are usually forwarded in insulated boxes. Delivery via Priority Paid Post is usually within 1 or 2 days.

7. OTHER SPECIES

This rearing method is also used for *Heliothis armiger* (Hübner) (syn. *Heliothis armigera*, *Helicoverpa armigera*).

REFERENCES

Cooper, D.J., 1979. The pathogens of *Heliothis punctigera* Wallengren. PhD thesis, University of Adelaide, South Australia.

Cullen, J.M., 1969. The reproduction and survival of *Heliothis punctigera* Wallengren in South Australia. PhD thesis, University of Adelaide, South Australia.

Griffith, I.P. and Smith, A.M., 1977. A convenient method for rearing lepidopteran larvae in isolation. J. Aust. ent. Soc. 16 : 366.

Hardwick, D.F., 1965. The corn earworm complex. Memoirs of the Entomological Society of Canada, No. 40.

Ignoffo, C.M., 1965. The nuclear-polyhedrosis virus of *Heliothis zea* (Boddie) and *Heliothis virescens* (Fabricius) II. Biology and propagation of diet-reared *Heliothis*. J. Invertebr. Pathol. 7 : 217-226.

Kirkpatrick, T.H., 1961. Comparative morphological studies of *Heliothis* species (Lepidoptera : Noctuidae) in Queensland. Queensl. J. Agric. Sci. 18 : 179-194.

Kirkpatrick, T.H., 1962. Methods of rearing *Heliothis* species, and attempted cross-breeding. Queensl. J. Agric. Anim. Sci. 19 : 565-566.

Shorey, H.H. and Hale, R.L., 1965. Mass-rearing of the larvae of nine noctuid species on a simple artificial medium. J. Econ. Entomol. 58 : 522-524.

Teakle, R.E., 1973. A nuclear-polyhedrosis virus from *Heliothis punctigera* Wallengren (Lepidoptera : Noctuidae) Queensl. J. Agric. Anim. Sci. 30 : 161-177.

HELIOTHIS VIRESCENS

E. G. KING and G. G. HARTLEY

Agricultural Research Service, USDA, P. O. Box 225,
Stoneville, Mississippi 38776

THE INSECT

SCIENTIFIC NAME: *Heliothis virescens* (Fabricius)
COMMON NAME: Tobacco budworm
ORDER: Lepidoptera
FAMILY: Noctuidae

1. INTRODUCTION

The tobacco budworm overwinters as a mahogany colored pupa in the top 5 to 10 cm of soil in or around cotton fields and other areas where host plants occur. The moths emerge in the spring and deposit their eggs on plants such as clover, vetch, geranium, evening primrose and alfalfa. The second generations infest cotton in mid-June, after earlier hosts become less numerous. There, the moths feed on plant exudates and females oviposit at night. Terminal buds are preferred sites for oviposition, but lateral buds, expanded leaves, bracts of squares and bolls are often used. A female moth is capable of depositing up to 1000 eggs during her life span when reared in the laboratory. (Anonymous 1978).

Several rearing methods have been used to successfully rear the tobacco budworm. The method described here is one of the simpler ones, and can be adapted by most any laboratory desiring to rear 1000 to 2000 insects per week. For laboratories requiring more insects, it is suggested that they investigate the use of a form-fill-seal machine (Harrell et al. 1974) or the use of multicellular rearing units (Raulston and Lingren 1972) and (Hartley et al. 1982).

2. FACILITIES AND EQUIPMENT REQUIRED

2.1 Facilities

Tobacco budworm are reared in controlled environment rooms. For oviposition purposes, adult rearing requires 25° C ± 1°, 70-80% RH, with a LD 14:10 photoperiod. Larval rearing requires 29.5° C ± 1°, 60-70 RH, with a LD 14:10 photoperiod.

2.2 Equipment and materials required for insect handling

2.2.1 Adult oviposition

- 3.8 liter cardboard buckets
- White polyester-cotton cloth
- 40 ml vial
- Nonsterile cotton
- Rubber gloves
- 5% sucrose solution

- Scale collector (constructed from a blower (Grainger #3m254) 25x25x2.5 cm Filter, 5.0 cm diameter plastic pipe and 12.6 mm hose insert fittings (Hartley et al. 1977)
- Face masks 3-M No. 8500 non-toxic particle mask (Laminaire Corp., Rahway, NJ)
- Hair covers, utility cap 512 (Laminaire Corp., Rahway, NJ)

2.2.2 Egg collection and sterilization
- Plastic pan (30x35x8.5 cm)
- 40% formaldehyde solution
- 2000 ml flask
- Nonsterile cotton
- Aluminum foil
- White polyester-cotton cloth

2.2.3 Larval rearing
- Disposable 22.5 ml plastic cups with waxed paper lids (Comet Products, Inc., Chelmsford, MA)
- Camel hair brush (size 6/128)
- Containers for holding cups

2.2.4 Pupal collection, sexing, and holding
- Forceps
- Stereoscope (American Optical Model 569)
- White polyester cotton cloth
- 3.8 liter cardboard buckets

2.3 Diet preparation equipment
- Top loading balance - 2000 gram capacity (Mettler P-2000)
- Portable concrete mixer (optional) (Sears Roebuck and Co.)
- Scoop
- 3.8 liter blender (Waring Model 91-215)
- 7.6 liter stainless steel garden sprayer - modified to dispense artificial diet (Stadelbacher & Brewer 1974)
- Electric or gas hot plate
- Plastic bags (used to store pre-mixed diet)
- 3 liter measuring pitcher
- 5 liter boiler
- Insulated gloves (for holding hot items)
- Freezer
- Laminar flow table (Agnew & Higgins Inc., Garden Grove, CA 92640)
- Spatula

3. LARVAL DIET

3.1 Composition (formula for 1 liter of diet)

Ingredients	g/liter
Soybean flour (Nutri-Soy® Flour # 40)	41
Wheat germ	35
Wesson salt	10
Sugar	41
Vitamin mix	9.5
Agar	22.4
Methyl paraben	1
Aureomycin®	1
Sorbic acid	1
Water	930 ml

3.2 Sources of ingredients

Nutri-soy Flour® #40, (Flavorite Lab Inc., Memphis, TN); vitamin mix, (Roche Chemical Division, Nutley, NJ - Specify mix used at Stoneville, MS by USDA); agar (CM-100), (Perny Inc., Ridgewood, NJ);

Aureomycin® (concentrate soluble powder), (American Cyanamid Co., Wayne, NJ); sorbic acid, wheat germ, Wesson salt, methyl paraben, (Nutritional Biochemicals Corp., Cleveland, Ohio); and sucrose (purchased from local markets).

3.3 Diet preparation procedure

- Weigh and place all dry ingredients required for as many as 100 liters of finished diet into a portable concrete mixer. Ingredients can be weighed in quantities to prepare 3.8 liters or less of diet and placed directly into blender.
- Mix in concrete mixer for 2 hours
- Weigh 616 g of dry mix for each 3.8 liters of diet; add to 3535 ml of boiling water
- Blend dry mix in boiling water for 4 minutes
- Pour blended diet into diet dispenser and dispense hot diet into 22.5 ml cups (10 to 15 ml per cup)
- Cover cups of diet with wax paper and allow to cool under a laminar flow table
- Dry mix has a shelf life of approximately 6 months when stored at 0° C
- Storing prepared diet more than 2 weeks is not recommended

PRECAUTION To avoid a dusty work area, cover the top of the concrete mixer with a sheet of plastic.

4. HOLDING INSECTS AT LOWER TEMPERATURES

Eggs, larvae, and pupae can be stored at 16°C for 7 days with no noticeable detrimental effects. Stored larvae should be provided with food. Storage of adults is not recommended.

5. REARING AND COLONY MAINTENANCE

5.1 Egg collection

- Remove polyester-cotton cloth, containing eggs, from the 3.8 liter buckets starting the 2nd day after mating.
- Place fresh polyester-cotton cloth in the 3.8 liter buckets

5.2 Egg sterilization

- Wash the polyester-cotton cloths with eggs in formaldehyde (3% active ingredient) for 10 minutes.
- Rinse the polyester-cotton cloths with eggs in distilled water for 15 minutes (at least five changes of water) and place under a laminar-flow hood until almost dry.
- Place the egg covered polyester-cotton cloths into sterile 2 liter Erlenmeyer flasks containing moist cotton
- Plug flasks with sterile cotton wrapped in polyester-cotton cloth, wrap bottom of flask with aluminum foil, and hold until larvae emerge.

5.3 Diet inoculation with neonate larvae

- Remove the positively phototropic larvae from the top of the aluminum foil wrapped flasks with a fine camel hair brush and transfer them to cups of diet.
- Rear tobacco budworm larvae in 22.5 ml plastic cups containing about 10 ml of the soybean flour-wheat germ diet. Use wax impregnated lids to cover the cups, since larvae frequently chew through plastic lids.
- Rear tobacco budworm larvae at the rate of 1 per cup, since they are cannibalistic.
- Move cups with larvae to the 29.5° C - 60-70% RH room and hold for 14 days.

5.4 Pupal collection

- Remove pupae from the cups; disinfect them in sodium hypochlorite solution (0.5% active ingredient) for 2 minutes.
- Rinse pupae thoroughly in distilled water.
- Separate pupae by sex.
- Place 200 pupae in each 3.8 liter cardboard container.
- Replace the top of the 3.8 liter cardboard container with white polyester cotton cloth.

5.5 Adults

- As they emerge, transfer moths (40 pair per container) to 3.8 liter cardboard containers which have had the top replaced with white polyester-cotton cloth and a 7.5 cm wide polyester-cotton cloth strip suspended from the container top edge to the bottom.
- Place a 7.5 cm square piece of cotton, saturated with 5% sucrose solution, on the cloth top of the 3.8 liter cardboard container. This serves as a food source.
- Discard moths after 9 nights; if they are kept longer, egg production decreases rapidly and moth mortality is high.

5.6. Sex determination

- The pupae can be sexed with the aid of a stereoscope. The eighth segment of the female pupa is divided by the ninth segment on the ventral median, and the opening of the bursa copulatrix is visible in this segment; the eighth segment of the male pupae is not divided, and the ninth segment has two rounded pads, one on each side of the middle ventral line.

5.7 Rearing schedule

a. Daily tasks
- Prepare artificial diet
- Mate newly emerged adults - (7 days per week)
- Feed adults - (7 days per week)
- Remove and disinfect egg cloths
- Infest cups of diet with first instar larvae
- Harvest pupae from cups of diet
- Separate pupae by sex
- Sanitize rearing areas

b. Schedule

This schedule can be adapted to a Monday-Wednesday-Friday routine although adults should be mated and given sucrose solution daily including Saturday and Sunday.

5.8 Insect quality

- Quality is monitored daily. Checks include egg production, egg hatch, larval survival, emergence of adults from the pupae, and adult survival. These checks have proven adequate to allow for production of over five million TBW pupae during the past 5 years.

5.9 Special problems

(a) Microbial contamination of diet

The soybean flour-wheat germ diet is an excellent medium for the growth of bacteria, yeast, and fungi. This growth is prevented by using antimicrobial agents in the diet and exposing it to the air only under clean air hoods. All work areas and equipment are thoroughly disinfected with sodium hypochlorite solution or ammonia type compounds.

If possible, equipment used in rearing is autoclaved.

(b) Human health

(i) All work with formalin is conducted under a fume hood

(ii) Moth scales are a continuous problem when rearing the tobacco budworm; workers quickly develop skin and respiratory allergies when they are exposed to wing scales. Containers holding adult moths are hooked to scale collectors at all times; workers are required to wear hair covers, face masks, and rubber gloves when handling adults

6. LIFE CYCLE

6.1 Optimum rearing conditions

Eggs

Temperature	Incubation period (days)	% hatch
29.5°C	3	96

Larvae

Temperature	Larval period (days)	% larval mortality
29.5°C	13	3

Pupae

Temperature	Pupal period (days)	Pupal wt. (mg)	% mortality
25°C	6-8	300	3

Adults

Temperature	Adult longevity (days)
25°C	12

6.2 Longevity of various stages at optimum rearing conditions

- Adults - moths are held for 9 nights; each female deposits between 500-1000 eggs during this period. Adults will survive longer but most egg production is completed after 9 nights.
- Larvae - Beginning when larvae hatch from eggs and lasting until larvae begin pupation average 12 to 14 days.
- Pupae - Beginning when larvae pupate, lasting until adult moths emerge; averages 6 to 8 days.

7. PROCEDURES FOR OBTAINING INSECTS

The Southern Field Crop Insect Management Laboratory within USDA-ARS at Stoneville, MS, can supply researchers with tobacco budworm pupae and/or eggs. This is possible through a cooperative program supported by The Cotton Foundation and USDA, designed to accelerate development of control measures for the tobacco budworm. There is a small charge by the Cotton Foundation for all pupae and eggs to help defray expenses of the program.

8. OTHER INSECTS REARED WITH THIS PROCEDURE

Heliothis zea (Boddie)

9. REFERENCES

Anonymous. 1978. Tobacco budworm distribution program for R & D. The Cotton Foundation, Memphis, TN. 10 pp.

Harrell, E. A., Sparks, A. N., Perkins, W. D., and Hare, W. W. 1974. Equipment to place insect eggs in cells on a form-fill-seal machine. U.S. Dep. Agric., Agric. Res. Serv. -S-42. 4 pp.

Hartley, G. G., King, E. G., Brewer, F. D., and Gantt, C. W. 1982. Rearing of the *Heliothis* sterile hybrid with a multicellular larval rearing container and pupal harvesting. J. Econ. Entomol. 75: 7-10.

Hartley, G. G., Gantt, C. W., King, E. G., Martin, D. F. 1977. Equipment for mass rearing of the greater wax moth and the parasite, *Lixophaga diatraeae*. U.S. Dep. Agric., Agric. Res. Serv. ARS-3-164, 4 pp. *

*(Not cited in text but recommended as a reference)

Raulston, J. R., and Lingren, P. D. 1972. Methods for large-scale rearing of the tobacco budworm. U.S. Dep. Agric. Prod. Res. Rep. 10 pp.

Stadelbacher, E. A., and Brewer, F. D. 1974. Dispensers of insect diets made from a stainless steel sprayer. J. Econ. Entomol. 67: 312-313.

Mention of a trademark proprietary product or vendor does not constitute a guarantee or warranty of the product by the U.S. Department of Agriculture and does not imply its approval to the exclusion of other products or vendors that may also be suitable.

HELIOTHIS ZEA / HELIOTHIS VIRESCENS

RAYMOND PATANA

USDA-ARS , Biological Control of Insects Laboratory, 2000 East Allen Road, Tucson, AZ 85719 USA

THE INSECTS

Scientific Name: *Heliothis zea* (Boddie)
Common Name: Bollworm/corn earworm/tomato fruitworm

Scientific Name: *Heliothis virescens* (Fabricius)
Common Name: Tobacco budworm
Order: Lepidoptera
Family: Noctuidae

1. INTRODUCTION

Colonies of the bollworm (*Heliothis zea*) and the tobacco budworm (*Heliothis virescens*) have been continuously maintained at the Tucson laboratory since 1963. The cultures have been maintained on the lima bean diet developed by Shorey (1963) and later modified by Patana (1969). For several years the larvae of both species were reared individually in plastic cups. A new method of rearing tobacco budworm using dry diet flakes was reported by Patana and McAda (1973). The same method was also adopted for rearing the bollworm.

The *Heliothis* cultures at the Tucson laboratory are "selected laboratory strains", and have been maintained for over 19 years without outside introduction. This factor has proven valuable when cultures have been started at other laboratories throughout the country with Tucson stock. Colonies can be started at a new location without the processes of "screening out" of non-adaptable individuals required when a new culture is started from wild stock. The methods described here can be used in maintaining cultures in large or small operations. This system can be adapted to fit particular needs and does not require sophisticated or expensive equipment.

2. FACILITIES AND EQUIPMENT REQUIRED

2.1 Facilities

Rearing of larvae and pupae is done in controlled temperature rooms maintained at 29°C in a 14:10 photoperiod. Lighting for the 4.6 x 6.1 m room is supplied by 6 fluorescent fixtures (four 1.2 m, 40-watt lamps per fixture). No attempt is made at controlling humidity in the rearing rooms. It is felt that humidity within the various containers used in rearing processes serves a more useful purpose.

2.2 Equipment and materials required for insect handling

2.2.1 Adult oviposition

- 3.79 liter glass jars (1 gal), wide mouth
- Lids for jars (with center cut out to form a jar ring)
- Single ply paper toweling (household kitchen towel)
- Saran screen or nylon screen
- 500 ml plastic squeeze bottle
- Glass screw cap vials 2.5 ml
- Plastic bags (8.9 x 12.7 x 33 cm)

2.2.2 Egg collections and sterilization
- Storage container for bleach solution (3 to 18.9 liter)
- Buchner funnel
- Vacuum pump for drawing off bleach and rinse solutions
- Hardware cloth trays, 31.8 x 40.6 x 5.1 cm (0.64 mesh)

2.2.3 Egg hatch and larval inoculation of diet
- Plastic tubs for egg holding, 3.97 or 1.89 liter (1 gal or 1/2 gal, Lily® 10MP or 5MP) (Lily, Division of Owens-Illinois, Toledo, OH 43666)
- Fine camel hair brushes, 1.27 cm (1/2 in)

2.2.4 Larval rearing
- Plastic utility boxes (UB-200), 8.9 x 25.4 x 34.3 cm (Sterling Products Co. Inc., St. Paul, MN 55118)
- Tissue, triple thickness wipes, Kaydry® (Kimberly-Clark Corporation, Roswell, GA 30076)

2.2.5 Pupal collection
- Sink spray hose
- Collection funnel for surface sterilization and draining
- Forceps
- Tea strainer

2.3 Diet preparation equipment

2.3.1 Pre-mix, mixing, and dispensing

a. Small batch
- Top loading balance (1-4000g)
- 3.79 liter (1 gal) blender capacity
- 2000 ml measuring cylinder
- 5.05 liter (5 qt) cooking pot
- Wooden handled rubber spatula
- Diet dispenser (Patana, 1967)
- Refrigerator(s) for holding diet
- Dial thermometer (30°-100°C)

b. Large-scale production
- Vertical cutter mixer, 90.9 liter (20-1/2 gal capacity), (VCM-80E, Hobart Mfg. Co., Troy, OH 45374)
- Dial thermometer (30°-100°C)
- Top loading balance (1-10000g)
- Steam kettle, 75.7 liter (20 gal) (Mod. RR-K-193B Groen Mfg., Elk Grove Village, IL)
- Hammermill (Model No. 10, C. S. Bell Co., Hillsboro, OH 45133)
- Cooking pot, 18.9 liter (5 gal)
- Measuring cylinder, 250 ml
- Measuring container, 8000 ml
- Air powered stirrer (Model 4 AM air motor, Gast Mfg. Co., Benton Harbor, MI 49022)
- Walk-in refrigerator (for diet storage)
- Diet pump (Patana, 1977)

c. Diet flakes
- Drying chamber (Patana and McAda, 1973)
- Drying frames, 57.8 x 182.9 x 5.1 cm (wood with hardware fabric cover (0.64 cm mesh)
- Wet waxed paper, 45.7 cm wide
- Smooth paint roller, 22.9 cm wide
- Scissors

3. ARTIFICIAL DIET

3.1 Composition

a. Partial mix for small batch	Amount	% estimated
- Beans, dry ground (baby lima)	630 g	15.53
- Brewer's yeast	120 g	2.96
- Ascorbic acid	15 g	0.37
- Methyl-p-hydroxy benzoate	12 g	0.30

- Scorbic acid	3 g	0.07
- Aureomycin	3 g	0.07
- Formalin (37% USP)	4 ml	0.09
- Water (tap) with above ingredients	1630 ml	40.17
- Mix separately		
- Gelcarin®	40	0.98
- Water (hot tap) with Gelcarin	1600	39.44
Total	4057 g	99.98%
b. Partial mix for large mixer		
- Beans, dry ground (baby lima)	10080	15.46
- Brewer's yeast	1920	2.94
- Ascorbic acid	250	0.38
- Methyl-p-hydroxy benzoate	200	0.31
- Sorbic acid	70	0.11
- Aureomycin	35	0.05
- Formaldehyde	64	0.09
- Water (tap) with above ingredients	28 liters	42.93
- Mix separately		
- Gelcarin	600 g	0.92
- Water	24 liters	36.80
Total	65219	99.99%

3.2 Brand names and sources of ingredients

Ascorbic acid (Merck & Co. Inc., Merck Chemical Division, Rahway, NJ 07065); Aureomycin (American Cyanamid Company, Agricultural Division, Princeton, NJ 08540); baby lima beans, No. 1 grade (The Trinidad Bean & Elevator Co., 215 Market Street, San Francisco, CA 94126); Brewer's yeast (U.S. Biochemical Corp., P.O. Box 22400, Cleveland, OH 44122); Formaldehyde, USP 37%; Gelcarin (Marine Colloids Inc., 2 Edison Place, Springfield, NJ); Methyparosept (Tenneco Chemicals Co., Intermediates Division, Turner Place, P.O. Box 2, Pescatawny, NJ 08854); Sorbic acid (Pfizer Inc., Chemicals Division, New York, NY 10017).

3.3 Diet preparation procedure

3.3.1 Dry mix (small and large scale)

- Pre-weigh Gelcarin and keep separately.
- Other ingredients can be pre-weighed and combined.
- Set up diet containers

a. Mixing 3.79 liter (blender capacity) batch

- Mix Gelcarin and measured hot tap water for 1 to 2 min in blender.
- After mixing, pour into cooking pot and bring to a boil on hot plate.
- In blender container add remaining dry ingredients, formaldehyde and water, mix about 6 min or until mixture reaches 48°C.
- After Gelcarin comes to a boil, add to other ingredients, mix until combined in a smooth mixture - about 1 minute.
- Diet can be dispensed into desired container while hot (40°-60°C).
- Pour thin layer around sides of plastic utility boxes to a depth of about 0.6 cm on the bottom, yield ca 9 boxes.
- Refrigerate to store until needed.

b. Mixing 90.9 liter (mixer capacity) batch

- Mix Gelcarin and 24 liter hot water in steam kettle, stir with air-powered stirrer, bring to low boil.
- Mix dry ingredients and formaldehyde in vertical cutter mixer with 28 liter water.
- Adjust water temperature so that dry ingredients will be 48°C after 5 min mixing at high speed.
- Add boiling Gelcarin to ingredient mixture, mix 1 min at low speed.
- Remove from mixer with diet pump and dispense into desired containers (Patana, 1977).
- Yield one batch, ca 130 plastic utility boxes.
- Cool to room temperature, then store in walk-in refrigerator until needed, up to 3 weeks.

c. Diet flakes

- Hot diet is spread to a depth of about 0.3 cm on a sheet of waxed paper 45.7 x 182.9 cm, placed on a drying frame 57.8 x 182.9 x 5.1 cm, with a smooth paint roller in successive layers.
- The frames with the diet are placed in the drying chamber and allowed to dry overnight.
- The dried flakes are then collected and stored in the refrigerator until needed.

4. REARING AND CULTURE MAINTENANCE

The original laboratory culture was established on diet in 1965 and has been reared continuously since. The most important factor in maintaining these cultures over prolonged periods of time at this laboratory can be attributed to the practice of gathering eggs and implanting new hatched larvae daily, 7 days a week. This provides a continuous series of individuals which serve as a back-up in the event of loss at any stage during the life cycle. The daily implanting of larvae also possibly contributes to the genetic diversity of the culture. The newly hatched larvae used daily are picked randomly from a group of eggs gathered on a particular day, but which come from different aged moths. This allows eventual cross mating between progeny from several different day's moths. This mixing process was not intentionally built into the system but since the method has endured, this is a possible reason for its success.

4.1 Larval stage

Preferably ca 24 h prior to use, add a layer of dried flakes (1.5 to 2.0 cm deep) to boxes with poured diet. Placing the dry diet flakes on the diet surface absorbs any excess surface moisture present. Additionally, the diet flakes later serve to isolate larvae and to serve as additional food as they become moistened.

- The diet boxes with flakes are implanted with ca 75 newly hatched larvae which are sprinkled over the surface using a small brush.
- The box is then covered with a three-ply wiping tissue which is then covered with the lid.
- The boxes are then held at 29°C.

4.2 Pupal stage

- Harvest pupae after 14 to 16 days.
- Wash diet boxes to float off pupae into a screen sieve.
- Surface sterilize pupae with a 0.03 % (AI) sodium hypochlorite solution, then rinse with water.
- Place pupae into 3.79 liter glass jars, lined with plastic bags (8.9 x 12.7 x 33 cm) blown up inside of the jars to fit the extremities of the jars. Date bag before placing it into the jar.
- Numbers used per jar is dependent on need or size of the culture, 20 to 30 are generally used.
- Hang a paper toweling strip 5 cm wide x 28 cm long down inside the jar and cover the top of the jar with a 12.7 cm square of paper toweling. Use a cutout jar lid to hold it in place.
- Hold the jars at 26°C for emergence.
- Sex ratio is 1:1 so sexing is not required.
- Use of different colors of paper toweling to quickly distinguish between species.

4.3 Moth emergence

- Moth emergence begins in about 2-4 days.
- When first moth emerges, a feeding vial (2.5 ml screw top) with the neck is inserted through a hole in a 3 cm^2 piece of plastic screen is placed over a hole in the toweling top of the jar. The vial is filled with 10 percent honey solution.
- Females emerge usually one day before males and oviposition begins 2 to 3 days after emergence.

4.4 Oviposition

a. Egg collection
- Oviposition jars are checked daily after emergence for eggs.
- When eggs are noted, toweling strips and top are changed and watering vials are filled daily.
- Moths are kept for 4-6 days after oviposition begins, the insect bag is then removed, tied and discarded.

b. Egg sterilization
- Egg-bearing toweling strip and tops are placed in Buchner funnel.
- The toweling is covered with a 0.03 percent sodium hypochlorite solution for 5 min.
- The bleach solution is drawn off with a vacuum pump.
- The eggs are rinsed with water and this is then drawn off.
- The toweling strips are then spread on a wire tray, covered and allowed to dry overnight in a ventilated cabinet.
- Eggs from previous day are placed into 3.79 liter plastic containers (Lily 10MP) held at 26°C and allowed to hatch.

4.5 Rearing schedule

a. Daily
- Implant newly hatched larvae in diet boxes.
- Discard excess larvae.
- Feed moths.
- Collect and sterilize eggs.
- Discard old moths.

b. Monday, Wednesday, Friday
- Wash pupae (14, 15, 16 day old).
- Surface sterilize pupae.
- Put up pupae into jars for emergence.

4.6 Special problems

a. Mold contamination

No attempt is made to control humidity. Rearing containers that are used tend to lose moisture as larvae develop. This "drying out" tends to produce a less favorable environment for mold development.

b. Larval overcrowding

If excessive numbers of larvae are implanted, they tend to consume all of the diet flakes, thus eliminating the isolation barrier and increasing cannabalism.

5. LIFE CYCLE DATA

Life cycle data and survival data under normal rearing conditions, photoperiod LD 15:9. For additional development and oviposition data at different temperatures, see Fye and McAda (1972).

5.1 Developmental data (± SD)

Stage (Temp. 25°C)	Number of days: *H. zea*	Number of days: *H. virescens*
Egg (hatch)	3 days	3 days
Larvae to pupation	17.3 ± 1.3 days	15 days
Pupal stage	14.4 days	13 days
Number egg/♀	173.1 ± 42.6	1247 ± 536
Pupal wt ♂ mg	426.7 ± 26.3	283 ± 35
Pupal wt ♀ mg	431.1 ± 24.6	284 ± 38

5.2 Survival data (%)	*H. zea*	*H. virescens*
- Egg hatch	85-95	85-95
- Pupal yield	65	85
- Moth emergence	95-98	95-98

5.3 Other data - 25°C both species
- Preoviposition period - 3-4 days
- Peak oviposition period - 4-7 days after emergence

6. REFERENCES

Bryan, D. E.; Jackson, C. G.; Patana, R.; Neemann, E. G. 1971. Field cage and laboratory studies with Bracon kirkpatricki, a parasite of the pink bollworm. J. Econ. Entomol. 64: 1236-1241.*

Bryan, D. E.; Fye, R. E.; Jackson, C. G.; Patana, R. 1973. Releases of parasites for suppression of pink bollworm in Arizona. USDA, ARS, W-7. 8 p.*

Bryan, D. E.; Fye, R. E.; Jackson, C. G.; Patana, R. 1976. Non-chemical control of pink bollworms. USDA, ARS, W-39. 26 p. *

Fye, R. E.; McAda, W. C. 1972. Laboratory studies on the development, longevity, and fecundity of six lepidopterous pests of cotton in Arizona. USDA, Tech. Bull. 1454. 73 p.

Patana, R. 1967. A pressure paint tank modified for use as a dispenser of insect diets. J. Econ. Entomol. 60: 1755-1756.

Patana, R. 1969. Rearing cotton insects in the laboratory. USDA, ARS, Prod. Res. Rep. 108. 6 p.

Patana, R. 1977. Rearing selected western cotton insects in the laboratory. USDA, ARS, W-51. 8 p.

Patana, R.; McAda, W. C. 1973. Tobacco budworms: use of dry diet flakes in rearing. J. Econ. Entomol. 66: 817-818

Shorey, H. H. 1963. A simple artificial rearing medium for the cabbage looper. J. Econ. Entomol. 56: 536-537.

Mention of a commercial (or proprietary) product in this paper does not constitute an endorsement of this product by the USDA.

*(Not cited in text but recommended as references)

HYALOPHORA CECROPIA

LYNN M. RIDDIFORD

Department of Zoology, University of Washington, Seattle, WA 98195, USA

THE INSECT

Scientific Name: *Hyalophora cecropia*
Common Name: Cecropia moth
Order: Lepidoptera
Family: Saturniidae

1. Introduction

The wild silkmoth *Hyalophora cecropia* is found throughout the United States east of the Rocky Mountains and is highly prized by collectors. The larvae are polyphagous and, depending on the area, are found on a wide variety of trees (for example, wild cherry, willow, birch, maple, and dogwood) (Scarbrough *et al*. 1974). These nonfeeding moths are nocturnal with male flight periods just after lights-off and near dawn, the latter of which coincides with the time of sex pheromone release in the female. Virgin females are relatively inactive although they show some flight activity during the first hour after lights-off; this activity is greatly enhanced and prolonged after mating as the female oviposits.

The larvae feed and pass through 5 larval instars, growing to a final weight of about 25-30 grams before spinning a cocoon. The resultant pupae undergo an obligatory diapause which can be terminated by chilling for 3 or more months (Williams 1956). Adult development usually ensues when the pupae are returned to 25°C. Under field conditions in Illinois, about 8% of the adult population eclose in late May whereas the remainder eclose about a month later. This polymorphism has a genetic basis and is due to a difference in the sensitivity of chilled pupae to initiate adult development in response to a temperature rise (Willis *et al*. 1974).

The importance of *H. cecropia* to the scientific world lies in its seminal role in the development of insect endocrinology in the pioneering experiments of C. M. Williams. For many years the larvae for experimental purposes were reared outdoors on leaves, usually that of wild cherry (*Prunus virginiana* or *P. serotina*), apple (*Malus pumila*), or willow (*Salix babylonica*), under nylon-mesh nets (Telfer 1967). This paper describes 2 laboratory rearing methods.

2. FACILITIES AND EQUIPMENT REQUIRED

2.1 Facilities

Larvae can be reared in normal laboratory rooms but we have had much better success rearing them in a controlled photoperiod (either LD 12:12 or 16:8) at 26-27°C and 60% RH. An even higher survival rate is obtained when larvae are reared at 32°C, LD 16:8, provided that the relative humidity does not exceed 75%.

Either natural lighting conditions at room temperature or a LD 16:8 photoperiod (fluorescent light) are sufficient for mating and oviposition. The eggs oviposited in paper bags should be kept in the rearing room at 25-27°C and 60-70% RH.

2.2 Equipment and materials required for insect handling

2.2.1 Adult emergence and mating
- 30 x 30 x 30-cm mesh cage or any screened container
- Alternatively, large (#46; 30.5 x 43 x 17.5 cm) paper bags

2.2.2 Oviposition
- Large (#46) or small (15 x 28 x 9.5 cm) paper bags

2.2.3 Egg hatch and placement on medium

a. For leaf rearing
- Scissors
- Branches of wild cherry (*P. virginiana*); see Scarbrough *et al.* (1974) for other common food plants

b. For diet rearing
- Scissors
- 100 x 20-mm plastic petri dishes (Falcon®, sterile)
- Wooden applicator sticks (sterilized by autoclaving)

2.2.4 Larval rearing containers

a. For leaf rearing
- Narrow-mouth bottles (either polyethylene or old soft drink bottles)
- Paper towels
- Small polyethylene sprayer that provides a fine spray

b. For diet rearing
- Wooden applicator sticks (sterilized by autoclaving)
- Soft-nosed forceps
- 100 x 20-mm plastic petri dishes (Falcon, sterile) for 1st- and 2nd-instar larvae
- 237-ml (8 oz) plastic dessert cups with lids for 3rd- and 4th-instar larvae
- 473-ml (16 oz) paper cups with lids or polyethylene sheeting (such as Baggie®) held with a rubber band for 5th-instar larvae
- Tray for holding dishes and cups
- Alternatively, plastic boxes (29 x 18 x 13 cm) fitted with hardware cloth (coarse mesh) so pellets drop through

2.3 Diet preparation equipment

2.3.1 Preparation of ingredients and mixing
- Top loading balance (0.1 to 1000 g)
- Plastic weighing cup
- Metal scopula or spoon for weighing materials
- Hot plate
- Waring® blender
- One 1- or 2-liter Pyrex® glass beaker
- Two 1-liter and three 25-ml graduated cylinders
- One 1-ml and one 5-ml graduated pipet plus pipettor
- Thermometer (0 to 100°C)
- Rubber spatula

2.3.2 Finished diet
- 22-cm (9 in.) square cake pan or foil lined box of similar dimensions
- Aluminium foil
- Polyethylene bags, 4 liter size (1 gal) and twist tie
- Refrigeration at 2 to 5°C

2.3.3 Antibiotic preparation for leaves
- Erlenmeyer flask (1 liter)
- Analytical balance (0.001 to 1 g)
- Weighing paper

3. ARTIFICIAL DIET (modified from Riddiford 1968)

3.1 COMPOSITION

a. Dry ingredients (per 850 ml distilled water) to prepare about 1 kg

Ingredients	Amount	% Final diet
- Fine agar	25.0 g	2.4
- Wesson's salt mixture	10.0 g	1.0
- Casein, high nitrogen	35.0 g	3.4
- D-sucrose (table sugar, fine)	35.0 g	3.4
- Kretschmer wheat germ (toasted, vacuum packed)	30.0 g	2.9
- Cholesterol, U.S.P.	1.0 g	0.1
- Alphacel (powdered cellulose)	5.0 g	0.5
- Kanamycin sulfate	0.14 g	0.01
- Aureomycin (chlorotetracycline HCl)	1.0 g	0.1
- Sorbic acid	2.2 g	0.2
- Methylparaben, U.S.P.	1.6 g	0.16
Total	145.94 g	

b. Finished diet

- Dry mix	145.94 g	
- Distilled water	850.0 ml	82.9
- Linseed oil (raw)	2.0 ml	0.19
- Linolenic acid (55%)*	2.0 ml	0.19
- 10% potassium hydroxide	10.0 ml	1.0
- Formaldehyde (standard 37% solution)	1.0 ml	0.01
- Vitamin mixture	15.0 ml	1.5

* Use if adults do not spread wings properly on eclosion (see Discussion)

c. Preparation of 10% potassium hydroxide
- 10 g KOH made to 100 ml distilled water

d. Preparation of vitamin mixture (may also use purchased Vanderzant vitamin mixture)

Ingredients	Amount
- Alpha-tocopherol	8 g
- Ascorbic acid	270 g
- Biotin	20 mg
- Calcium pantothenate	1 g
- Choline chloride	50 g
- Folic acid (crystalline)	250 mg
- Inositol	20 g
- Niacinamide	1 g
- Pyridoxine HCl	250 mg
- Riboflavin	500 mg
- Thiamine-HCl	250 mg
- Vitamin B-12 trituration in mannitol	2 g
- Distilled water	1 liter

3.2 Brand names and sources of ingredients

All the ingredients except kanamycin sulfate, potassium hydroxide, linseed oil and formaldehyde (BioServ Inc., P. O. Box 100, Frenchtown, NJ 08825, USA); toasted wheat germ (BioServ or Kretschmer brand toasted, vacuum-packed wheat germ obtained at local grocery store); kanamycin sulfate, nonsterile form [Bristol Laboratories, Syracuse, N. Y. (potency approx. 785 micrograms/g) obtained upon request for experimental use only]; potassium hydroxide and formaldehyde (Mallinckrodt); linseed oil (Hains, cold-pressed; Hains Pure Food Co., Los Angeles, CA 90061, USA).

3.3 Diet preparation procedure

a. Dry mix
- Weigh out agar
- Weigh and mix remainder of dry ingredients in a plastic bag and store at 50°C until needed

b. Finished diet
- Bring 850 ml distilled water to a boil, pour into Waring blender

- Add agar and immediately blend until agar is dissolved
- Add bag of remaining dry ingredients and blend
- Let cool to 60°C
- While it is cooling, measure out liquid ingredients; shake vitamin mixture well to resuspend
- Add liquid ingredients when temperature is 60°C, then reblend
- Pour into pans or foil-lined containers; refrigerate overnight
- Store in closed polyethylene food bags in refrigerator for no longer than 2 weeks

4. PREPARATION OF ANTIBIOTIC MIXTURE FOR LEAVES (Riddiford 1967)

Ingredients	Amount
- Aureomycin (chlorotetracycline HCl)	2.8 g
- Kanamycin sulfate	0.16 g
- Distilled water	1 liter

- Mix and store in refrigerator; sources of these 2 antibiotics are listed in Section 3.2

5. REARING AND COLONY MAINTENANCE

5.1 Egg collection

- Transfer mated females daily (in morning) to fresh paper bag
- Incubate eggs at 25-27°C and 60-70% RH

5.2 Placement on food

- On day 8 of incubation, cut out groups of eggs and place as follows

5.2.1 Leaf rearing

- Place freshly cut branches immediately into water in a narrow-mouth jar, binding the branches together with a paper towel to plug the remainder of the jar opening. Small branches with few leaves are sufficient for 1st- and 2nd-instar larvae which are gregarious and often feed in groups on a leaf. Larger branches are necessary for the later stages which feed under solitary conditions and are stressed by crowding.
- Spray the leaves with the antibiotic mixture until the surface is wet; let dry before placing eggs or larvae on leaves
- Fix paper containing eggs or old leaves on which larvae are feeding onto the branches bearing fresh leaves; to avoid stress do not handle the larvae themselves if at all possible
- Change branches at least every 3 days or sooner if wilted; during the final (5th) larval instar, fresh leaves may have to be added twice a day as the larva is a voracious feeder
- Maintain leaves in upright position so the larvae can move freely; as long as a plentiful food supply is present, they will remain in the branches
- Set the jars on metal or plastic trays which should be cleaned daily of frass and larvae which have fallen off the branches (these larvae are usually those that later become sick and should be discarded without handling)
- Place large 5th-instar larvae, which have evacuated their gut and are beginning to wander either individually into small paper bags that are then clipped shut or in groups in nylon mosquito-netting bags that are then securely tied. Evacuation of the gut is signaled by the cessation of hard fecal pellets and the appearance of a reddish liquid dripping from the anus accompanied by a characteristic posture in which the larva is immobile and anal prolegs and sometimes more anterior abdominal prolegs are no longer clasping the branch. This behavior continues for a few hours followed by an intense period of wandering behavior during which the larva usually crawls off the branches.

5.2.2 Diet rearing

a. 1st- and 2nd-instar larvae

- Place 10 eggs (still attached to paper) in 100 x 20-mm plastic petri

dish
- On the morning that the larvae hatch, place 2 small chunks of diet on either end of 3 broken wooden applicator sticks to make dumbbells; larvae tend to cling to sticks while feeding and molting
- Remove the paper and any unhatched eggs after 1 day
- Add 2 or 3 free-rolling applicator sticks on which the larvae can spin their molting pads
- Change diet and sticks every 3 days as described below

b. 3rd- and 4th-instar larvae
- Place five 3rd-instar larvae in 237-ml cups on upright dumbbells of diet (pieces at top and bottom of applicator sticks), 1 per caterpillar plus 2 extra per cup at this stage
- Cover with top that has 1 or 2 small holes to prevent moisture accumulation and condensation
- Change diet at least every 3 days or sooner, if dried; remove feces at this time

c. 5th-instar larvae
- Place individually in tall 473-ml paper cups and cover with lid or polyethylene sheet with a few small holes to allow moisture to escape; fasten with rubber band
- Put diet dumbbells upright; use several sticks so they may be changed often without disturbing the larvae
- Check daily and add diet as needed
- Remove feces every other day
- Alternatively, if disease problems have not occurred (see Section 6), rear these larvae in groups of 5 to 10 in plastic boxes with diet on hardware cloth (to allow the frass to fall through); change diet and clean cage frequently and separate if too crowded

d. Spinning larvae
- Check 5th-instar larvae daily for signs of gut evacuation (liquid feces); at this time they will also appear somewhat shiny
- Place larvae into small (15 x 28 x 9.5 cm) paper bag
- Close bag with clip
- Discard cup and contents immediately as these feces rapidly grow mold
- After 10 days the cocoons may be removed and spread out in trays

e. Changing of diet
- Do every 3 days or sooner if eaten or dried out
- Remove all old diet, except for that part on which larvae are feeding
- Discard all applicator sticks, except those on which larvae have spun their molting pad
- Move feeding larvae to new sticks by allowing them to walk onto the sticks
- Try not to disturb larvae during this change
- Add new chunks of diet; place old pieces containing larvae next to them

PRECAUTIONS

(i) Autoclave, then thoroughly dry applicator sticks before use.

(ii) Use new sticks whenever the diet is changed to prevent mold growth.

(iii) Do not touch larvae with either hands or forceps, as they are very subject to stress. Therefore, to remove old sticks, let animals move to new stick of their own accord.

(iv) Do not disturb molting larvae as they depend on the silk pad to hold the old skin when they ecdyse.

(v) Discard all dead, dying, and potentially sick larvae (those that are not feeding or are not growing) into animal waste outside the rearing room. Do not open container if it has several dead or dying animals to remove others; discard all.

(vi) Be scrupulously clean, cleaning surfaces and utensils with 70% ethanol before and after diet changing.

5.3 Pupal collection

- Remove cocoons from bags 11 days after spinning at which time the pupa will be formed
- Spread out in boxes at 25°C, LD 12:12 or 16:8, for 30 days to allow them to enter diapause
- Placing at 5°C for a minimum of 3 months before returning to 25°C will break diapause (Williams 1956)
- Maximal storage time at 5°C is ca 10 months to 1 year

5.4 Adults

5.4.1 Emergence

- Adults eclose about 3 to 5 weeks after removal from 5°C, if they have been chilled 3 or more months; the longer they are chilled, the sooner they initiate development upon return to 25°C and LD 16:8 photoperiod
- Removal from the cocoon by a careful slitting with scissors allows one to assess the later stages of adult development
- If desired, assess progress of adult development by observing the appearance of various morphological characteristics of the adult underneath the pupal cuticle after swabbing 70% alcohol on the cuticle and viewing the insect with incident light with a polarizing filter [see Schneiderman and Williams (1953) for a timetable]. By so doing, one can then synchronize eclosion of males and females by placing faster developing pupae at 18-20°C for a time until the other ones catch up.
- Males usually eclose about 1-2 days earlier than females if removed from the cold at the same time
- Place in screened cages or large (#46) paper bags as soon as you can readily see pigment on pharate adults
- Keep males and females separate, 3 or 4 per large bag (they emerge during the morning or early afternoon and may be used for mating that evening)

5.4.2 Mating

- Place 2 females and 3 males into 30.5-cm^3 cage just before lights-off (they should mate around lights-on, depending on the photoperiod)
- After a pair has mated, remove extra unmated animals (do not disturb mating pair)
- Just before lights-off the following night, take mated pair and put in paper bag for oviposition (if male falls off, remove him)
- Both sexes mate most readily the first 3 nights after emergence (the males can be used for a second mating, if necessary)

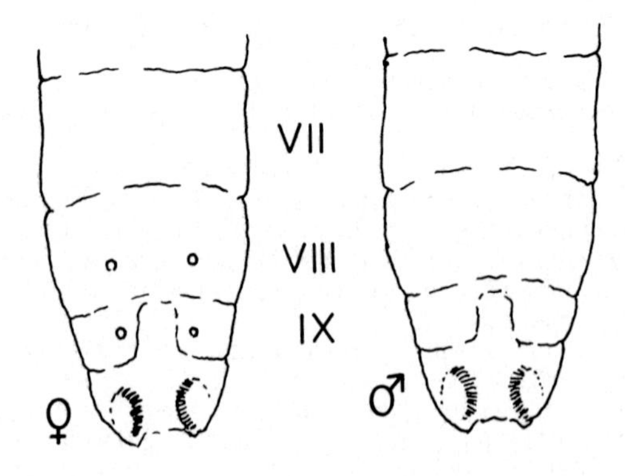

Fig. 1. Drawing of ventral side of 8th and 9th abdominal segments showing location of translucent spots indicating Ishiwata's glands in a female Cecropia larva.

5.5 Sex determination

5.5.1 Larvae

- Ishiwata's glands appear as 2 pairs of translucent spots on the ventral side of the 8th and 9th abdominal segments on the female (Fig. 1)
- No spots are seen in the male (Fig. 1)

5.5.2 Pupae

- Male pupae have 2 small raised protuberances on the 9th abdominal segment in the vental midline
- Female pupae have a line on the 8th abdominal segment in the ventral midline

5.5.3 Adults

- Male has large feather-like antennae
- Female has barbed, yet narrow antennae

5.6 Special problems - diseases

Daily care is necessary for rearing Cecropia larvae since diseases often occur. Most of the diseases seem to be related to stress of handling, overcrowding, and starvation due to inadequate or old food (either leaves or diet). Since most are very contagious and can rapidly spread through the entire stock, we rear the larvae on diet individually in the final instar. Dead, dying, and potentially sick larvae should not be handled; their containers should be discarded outside of the rearing room.

The prevalence of some disease among larvae is maternally related. While routine surface sterilization of the eggs with 0.05% sodium hypochlorite (1% Clorox ®) helped decrease mortality, it was not as effective as the use of breeding stock collected in the wild. From these latter adults, usually over 80% of the 1st-instar larvae which fed could be reared to the pupal stage when the diet was frequently changed and the larvae were not handled.

Also, the rearing room and utensils and space used for handling diet and changing the larvae should be kept scrupulously clean. A germicidal detergent such as I Stroke Ves-phene® (Vestal Laboratories) is excellent as a general cleaning agent for the room. Utensils are washed with dish detergent unless an outbreak occurs; then sterilization is necessary.

Disease problems are minimized by rearing at 32°C. Common problems encountered in H. cecropia rearing on either leaves or diet are:

- Failure to eat in the 1st instar (larvae usually die in ca 2-3 days); affliction is not transmissible
- A shrivelling affliction which occurs normally in the 2nd or 3rd instar (this happens very quickly but does not seem to be highly contagious)
- Extremely slow growth (those that lag behind their siblings for 2 or 3 days in any of the early larval instars should be discarded because they will never reach maturity and will eventually show the intestinal or viral diseases described below)
- A disease of the gut which is characterized by cessation of feeding and a discharge from the gut. This normally occurs in the 4th or 5th instar but may occur even as early as the 2nd instar. When this disease is discovered, the larva should be immediately discarded outside the rearing room, and the remaining larvae transferred to a clean container as it is highly contagious. This disease is not as prevalent in diet-reared as in leaf-reared animals and seems to be kept under control by the antibiotic mixture.
- A viral disease which usually strikes suddenly in the 2nd or 3rd instar and causes cessation of feeding, then shrinkage and death in a pool of liquid within 2 days. It is highly contagious. If 1 larva in a petri dish has it, the unopened dish (including the remaining larvae) should be discarded immediately outside the rearing room.
- The other main source of death seems to be due to a failure to molt properly. Usually the larva has been knocked off its molting pad or has spun this pad on a wet stick or on the diet. Therefore, it cannot pull itself out of its old cuticle. This problem is most often found in conditions of overcrowding and of high humidity. Therefore, adhering

strictly to the above numbers, keeping the amount of diet small but adequate, especially in the first few instars, and maintaining a good supply of clean, dry sticks should practically alleviate this problem.

PRECAUTIONS

(i) Moth scales are allergenic. Therefore, wear a protective mask when handling the adults for mating and when cutting eggs out of paper bags.

(ii) Some people also show an allergic reaction to the larvae, especially to their setae and prolegs. If you find hives or any itchiness after handling them, use gloves.

6. LIFE CYCLE DATA

6.1 Life cycle at 26 ± 1°C, 60-70% RH, and either LD 12:12 or 16:8 photoperiod.

Stage	Number of days
Egg	9-10
Larvae	
1st-4th	5 per stage
5th	10-15
Spinning	2
Spinning-pupa	10
Pupa	
25°C (to allow entrance in diapause)	30
5°C	3-9 months
After return to 25°C to adult emergence	21-34
Adult	10-14

6.2 Survival data (%)

a. Egg viability (90)

b. Leaf rearing

- Feeding larva to pupa (80)
- Pupa to adult (100)

c. Diet rearing (optimal)

- Feeding larva to pupa (60)
- Pupa to adult (90)

6.3 Summary of characteristics of each larval instar [see Riddiford (1970)]

- 1st: black
- 2nd: yellow with black head, tubercles, and spots
- 3rd: green with black spots; blue prothoracic, lateral abdominal and posterior abdominal tubercles; red or orange meso- and metathoracic tubercles, and yellow dorsal abdominal tubercles
- 4th: like 3rds but lacking the black spots and modified setae
- 5th: green with modified tubercles and setae
- Those reared on the artificial diet are pale blue with pale red and yellow tubercles, apparently due to the lack of yellow plant-derived pigments, especially xanthophyll in the diet (Clark 1971; personal observation)

7. DISCUSSION

The methods outlined above have been used for successful rearing of *H. cecropia* larvae in the laboratory. The adults obtained from diet-reared larvae eclose and mate normally and produce fertile eggs. We have not reared successive generations on the diet although we have taken the second generation successfully through to adult. Waldbauer and Sternburg (1979) have found a severe depression of egg hatchability from the 1st brother-sister mating of moths from leaf-reared larvae. Therefore, mass rearing successive generations in the laboratory without careful attention to outbreeding may not be successful.

Besides the disease problems often encountered in rearing H. cecropia larvae in the laboratory, the long life cycle enforced by the obligatory pupal diapause makes breeding successive generations on diet to select a strain better adapted to laboratory conditions a formidable task. Although Mansingh and Smallman (1966) reported that rearing these larvae under long day (17 h light: 7 h dark) in the final 2 larval instars prevented the pupal diapause, I have not been able to repeat their findings with larvae derived from H. cecropia pupae collected in the wild in either Nebraska or North Dakota. Nor was I able to find any set of photoperiod conditions for maintenance of eggs and larvae to prevent the pupal diapause.

The same diet and rearing methods described here can be successfully used to rear Philosamia cynthia. The addition of 2 ml 55% linolenic acid to prevent wing deformities and 10 microliters trans-2-hexenal as a phagostimulant with the liquid ingredients given in Table 1 makes this diet also satisfactory for raising Antheraea polyphemus, A. mylitta, and Actias luna larvae. These and other modifications including the addition of ground oak leaves, however, were unsuccessful in stimulating the feeding and subsequent development of Antheraea pernyi larvae.

8. REFERENCES

Clark, R. M., 1971. Pigmentation of Hyalophora cecropia larvae fed artificial diets containing carotenoid additives. J. Insect Physiol. 17: 1593-1598.

Mansingh, A. and Smallman, B. N., 1966. Photoperiod control of an 'obligatory' pupal diapause. Can. Entomol. 98: 613-616.

Riddiford, L. M., 1967. Antibiotics in the laboratory rearing of cecropia silkworms. Science 157: 1451-1452.

Riddiford, L. M., 1968. Artificial diet for cecropia and other saturniid silkworms. Science 160: 1461-1462.

Riddiford, L. M., 1970. Effects of juvenile hormone on the programming of postembryonic development in eggs of the silkworm, Hyalophora cecropia. Devel. Biol. 22: 249-263.

Scarbrough, A. G., Waldbauer, G. P. and Sternburg, J. G., 1974. Feeding and survival of cecropia (Saturniidae) larvae on various plant species. J. Lepid. Soc. 28: 212-219.

Schneiderman, H. A. and Williams, C. M., 1953. Physiology of insect diapause. VII. Respiratory metabolism of the Cecropia silkworm during diapause and development. Biol. Bull. 105: 320-334.

Telfer, W. H., 1967. Cecropia. In: Methods in Developmental Biology (Eds., F. H. Wilt and N. K. Wessells), pp. 173-182. Crowell, New York.

Waldbauer, G. P. and Sternburg, J. G., 1979. Inbreeding depression and a behavioral mechanism for its avoidance in Hyalophora cecropia. Amer. Mid. Nat. 102: 204-208.

Williams, C. M., 1956. Physiology of insect diapause. X. An endocrine mechanism for the influence of temperature on the diapausing pupa of the cecropia silkworm. Biol. Bull. 110: 201-218.

Willis, J. H., Waldbauer, G. P. and Sternburg, J. G., 1974. The initiation of development by the early and late emerging morphs of Hyalophora cecropia. Ent. Exp. Appl. 17: 219-222.

ACKNOWLEDGEMENTS

I thank Prof. Carroll M. Williams for introducing me to H. cecropia; the undergraduate student helpers, graduate students, and technicians that I have had over the years who have helped raise these animals and aided in the evolution of these rearing methods; Dr. James Truman and Dr. Masami Sasaki for reading the manuscript; and NSF for grant support over the years which enabled us to continue rearing these larvae when needed.

LAMBDINA FISCELLARIA

DAIL GRISDALE

Department of the Environment, Canadian Forestry Service, Forest Pest Management Institute, Sault Ste. Marie, Ontario, Canada, P6A 5M7.

THE INSECT

Scientific Name: *Lambdina fiscellaria fiscellaria* (Guenée)
Common Name: Eastern hemlock looper
Order: Lepidoptera
Family: Geometridae

1. INTRODUCTION

The eastern hemlock looper is distributed in Canada from Newfoundland to Saskatchewan and occurs in the United States from Maine to Georgia and westward to Minnesota. The insect is an important pest in mature forest stands where balsam fir and eastern hemlock predominate. Outbreaks causing extensive tree mortality have been recorded in Newfoundland, Nova Scotia, Quebec and Ontario in Canada, while in the United States infestations have been reported in Ohio, Wisconsin, Michigan, New York, Maine, Connecticut, Massachusetts, New Hampshire and Vermont.

The life cycle has been described by Carroll and Waters (1967). Moths appear from mid-August to early October. Eggs are usually laid singly but occasionally in groups of two or three in a variety of locations including moss on the forest floor, under lichens and bark scales on the tree trunk and on twigs and branches. Eggs overwinter and hatch about the first week of June. Young larvae feed on current year's foliage and older larvae on older needles. They feed wastefully and needles are seldom entirely consumed. Larvae feed for about 50 days, then pupate in late July or early August, emerging about 20 days later. Carroll (1956) has reported four larval instars in Newfoundland while de Gryse and Schedl (1934) observed five larval instars in Ontario populations.

Until the mid-1960s, looper larvae were reared on their natural hosts during the summer months. Carroll (1956) reported that first stage larvae could not survive when restricted to a diet of old needles of balsam fir although larvae that had been fed new foliage of balsam fir, white spruce or the leaves of white birch, red maple and mountain maple throughout the larval period were reared successfully to pupation. Danard (1968) reared western hemlock looper *Lambdina fiscellaria lugubrosa* (Hulst) during winter months on forced American larch foliage with 94% surviving the first instar. Eastern hemlock foliage used as a control showed only 25% survival past the first instar. Grisdale (1975) described methods for rearing the looper that had been used at the Forest Pest Management Institute since 1968. During these early rearings at the Institute, larvae were fed an artificial diet (McMorran 1965) until they reached the second instar, when they would accept dormant balsam fir foliage. Larvae could also be reared through to pupation on the diet in 22 ml plastic cups but because of a high incidence of cannibalism only one larvae could be reared in each cup.

In 1972, rearing tests were carried out using the low cost CSM diet (Burton 1970). This diet gave good results but was too viscous to facilitate automatic dispensing and dried out too quickly. After modification (Grisdale

1975) it proved excellent for rearing hemlock looper. Larvae complete development in approximately 50 days and individual rearing to avoid cannibalism was unnecessary. A high percentage of first instar larvae reached the pupal stage resulting in well formed adults producing the normal number of viable eggs.

2. FACILITIES AND EQUIPMENT NEEDED

2.1 Facilities

Egg incubation after cold storage and rearing of larvae and pupae are carried out in environmental control cabinets (130 x 69 x 190 cm) maintained at 22 ± 1°C, 50-60% RH and a LD 18:6 photoperiod. Lighting is provided by three 30 watt daylight florescent tubes, 117.5 cm in length. Mating and oviposition are at 20°C, 80-90% RH and LD 16:8 photoperiod. A cooler environment of 18°C is used to reduce overnight larval activity at time of egg hatch, for holding male pupae and for holding eggs before cold storage. Eggs are stored in a walk-in cold room at 2°C.

2.2 Equipment and materials required for insect handling

2.2.1 Adult oviposition

- One gallon (4.5 litre) cylindrical cardboard container (Metripac Nested Food Containers, Purity Packaging, Montreal, Quebec, Canada)
- Cheesecloth
- Paper towelling
- Plastic or glass vial (22 x 70 mm)
- Plug of absorbent cotton
- Feather weight forceps
- Scissors
- Distilled water squirt bottle

2.2.2 Egg collection and storage

- HEPA filtered exhaust hood used when large numbers of mating cages are handled (Canadian Cabinets Co. Ltd., P.O. Box 11336 Station H, Nepean, Ontario K2H 7V1, Canada)
- Particle face mask for handling small numbers
- Clear plastic bag (13 x 22 cm)
- Twister

2.2.3 Egg sterilizaton

- 300 ml glass jar fitted with a screened lid
- Rubber gloves
- Fume hood with outside venting
- Distilled water
- Formalin (37% Formaldehyde)
- Paper towelling
- Clear plastic round box (15 diam. x 6.3 cm) with friction fit lid (Althor Products, Div. American Hinge Corp., 496 Danbury Road, Wilton Con., U.S.A. 06897)

2.2.4 Egg hatch and larval innoculation of diet

- Camel hair paint brush (No. 1)
- Distilled water spray bottle

2.2.5 Larval rearing container

a. Diet rearing

- Translucent, ribbed cups (22 ml) sold as coffee creamers by Portion Packaging Ltd., 26 Tidemore Ave., Rexdale, Ontario, M9W 5H4
- Disposable pressed board cafeteria trays (46 x 36 cm) to hold cups (Keyes Fibre, Waterville, Maine USA 04901)
- Spray bottle of distilled water

b. Foliage rearings

- Clear plastic box (28 x 38 x 15 cm) with a vented lid (Althor Products)
- Paper towelling
- Paint brush (No. 1)
- Feather weight forceps
- Pruners
- Spray bottle of distilled water

2.2.6 Pupal collection
- Feather weight forceps
- Magnifier illuminator work station, Model 3775 (Cole-Palmer, Chicago, Ill., 60626, USA)
- Clear plastic box (28 x 38 x 15 cm) with lid and ends well ventilated (Althor Products)
- Paper towelling

2.2.7 Insect holding
- Incubator set at 18° C

2.3 Diet preparation equipment

2.3.1 Finished diet
- 4000 ml flask
- 50 ml beaker
- 1000 ml graduated cylinder
- 10 ml pipet
- Autoclave, small pressure cooker or hot plate
- Waring blendor, 1 gal (4.5 liter)
- Balances; 0.0001 to 10 g and 0.1 to 3000 g
- Absestos gloves
- 1000 ml stainless steel beaker with pouring spout and handle
- Spoon
- Metal scoop
- Rubber spatula
- Plastic bag (66 x 91 cm)
- Marker pen and masking tape

3. ARTIFICIAL DIET

3.1 Composition

Ingredients to prepare approximately 4 litres of finished diet.

	Ingredients	Amount	% (estimated)
a.	Water	625 ml	15.69
	Wheat germ	50 g	1.25
	Torula yeast	36 g	.90
	Ascorbic cid	14.4 g	.36
	Sorbic acid	2.7 g	.07
	Methyl paraben	5.4 g	.14
	Formalin	1.8 ml	.05
b.	Water	2752	69.07
	Agar	75 g	1.88
c.	CSM	422 g	10.59
	Total	3984.3 g	100.00%

3.2 Brand names and sources of ingredients

Agar, sorbic acid, methyl paraben, torula yeast (Bio-Serv. Inc., P.O. Box R.S. Frenchtown, New Jersey, 08884, USA), ascorbic acid, (Hoffman-LaRoche Inc., 1000 Roche Blvd., Vaudreuil, Quebec J7Z 6B3, Canada); formalin (Canlab, 80 Jutland Road, Toronto, Ontario, M8Z 2H4, Canada); wheat germ (fresh stock from local health food store). CSM was developed as a complete diet for human consumption in protein deficient countries. Its ingredients are: 63.8% yellow corn meal gelatized; 24.2% soy flour, toasted; 5.0% non-fat dried milk; 5.0% soy oil, refried; 1.9% mineral premix; and 0.1% vitamin premix. Companies manufacturing CSM include: Krause Milling Co., 4222 W. Burnham St., Milwaukee, Wis. 53246; Lauhoff Grain Co., Danville, Ill. 61832; Archer Daniels Midland Co., 4666 Faries Parkway, Decatur, Ill. 62521.

3.3 Diet preparation procedure

a. Finished diet
- Weigh or measure diet ingredients
- Mix agar and water (Group B) in a large flask and place in an autoclave or heat in a double boiler until agar is liquified. (94-95° C)

- The ingredients from group A are added in order listed to a running blender.
- Remove liquified agar from the autoclave using asbestos gloves
- Pour agar mixture slowly into running blendor, increase blendor speed and mix for an additional 2 or 3 minutes.
- Add CSM (Group C) to running blender and blend for 2 or more minutes to ensure thorough mixing of all ingredients.
- Pour hot diet into stainless steel pouring beaker.
- Manually dispense about 15 ml of diet into each plastic cup (cups are positioned on cafeteria trays before diet is prepared. Each tray will hold about 90 cups).
- Place trays of diet into plastic bags, record diet specifications on bag and refrigerate at 2° C.

PRECAUTIONS

(i) Occasionally, perhaps because of a lower mixing temperature or variation in agar quality, the diet may appear too thick to dispense smoothly into the cups. When this occurs a small amount of extra hot tap water may be added and blended into the mixture.

(ii) Cups of diet should not be surface sprayed with anti-fungal solutions as heavy mortality of newly hatched larvae may occur.

4. HOLDING INSECTS AT LOWER TEMPERATURE

4.1 Eggs

Once eggs have been removed from cold storage diapause is broken. Eggs received or shipped to initiate a rearing culture may be again placed in the refrigerator and a portion removed weekly to implement rearings. Such eggs are not in a true diapause and after a month or so the length of time to hatch will increase and hatching rate will sharply decrease.

4.2 Larvae

Mid-to-late larvae reared on balsam fir foliage in plastic boxes have been held for one week at a temperature as low as 6° C.

4.3. Pupae

Pupae may be stored in the refrigerator (10° C) for 2 or 3 days. To synchronize emergence male pupae are initially held at a temperature 4° C lower than female pupae.

5. REARING AND COLONY MAINTENANCE

A laboratory colony is usually started from field collected larvae. To avoid excessive parasitism it is best to collect early instar larvae and to rear them through to the pupal stage on host foliage.

5.1 Eggs

- Eggs may be removed from cold storage as early as 3 months and as late as 9 months.
- Sterilize eggs, rinse and dry.
- Line the bottom of a round plastic box (15 diam. x 6.3 cm) with paper towelling.
- Place eggs in box and cover with a tight fitting lid.
- Incubate eggs at 22 + 1° C and relative humidity of 70%.
- Eclosion occurs in approximately 11 days and most of the eggs that have been in storage 4-7 months will hatch over a 3 day period.

PRECAUTIONS

(i) In low humidity environments eggs should be slightly moistened three or four times before hatch.

(ii) During periods when hatching larvae cannot be transferred to food (after work hours) hold the dish in a darkened area at 18° C or lower and lightly spray the sides of the dish with water. These conditions will reduce larval activity and cannibalism.

5.2 Egg sterilization

- Use fume hood.
- Place cheesecloth containing eggs into a jar with a screened lid.
- Pour 3% formalin solution into jar to cover cloth, replace lid.
- After 15 minutes pour out formalin solution and rinse eggs with tap water at room temperature for about 10 minutes.
- Drain off tap water and rinse thoroughly with distilled water.
- Air dry at room temperature.
- Place eggs in round plastic box for hatch.

5.3 Diet innoculation with neonate larvae

a. Diet reared

- Transfer 10-20 larvae to diet-filled cups with a paint brush.
- Cap cups and place in an upright position on cardboard trays.

Fig. 1 Plastic box with vented lid used for rearing loopers on foliage.
Fig. 2 Well vented plastic box used for adult eclosion.
Fig. 3 Mating and oviposition cage.

- Larvae will reach the third instar in about 2 weeks and are transferred to fresh diet at four per cup. The number of larvae per cup must be thinned at this time as excessive crowding will result in cannibalism.
- Some larvae will pupate without additional food but usually another change is required when food in cups becomes too dry or the quantity insufficient.

PRECAUTIONS

First instar loopers are strongly photopositive and active. When several hundred are present in a hatching dish they will quickly crawl away from the dish and work area when the lid is removed. All stages of looper larvae drink readily and this wandering can be controlled by spraying the dish and immediate area with a fine spray of distilled water.

b. Foliage reared

- To obtain insects physiologically more compatible with an outdoor environment larvae required for field testing are reared on balsam fir foliage. Larvae required for the maintenance of stock are also reared on foliage.
- Select looper infested cups about 10 days old (most should be in the second instar).
- Place pieces of balsam fir in a 27 x 20 x 10 cm plastic box (Fig. 1)
- Transfer about 200 larvae from cups to the plastic box using either featherweight forceps or a paint brush.
- Tape the vented lid to the plastic box to ensure an escape-proof fit.
- Spray foliage lightly with water every 3 or 4 days, move often in a low humidity environment.
- Clean boxes and transfer larvae to fresh foliage after 2 weeks.
- Water daily, add fresh foliage as required and clean boxes weekly until pupation.

PRECAUTIONS

Periodically an intestinal bacterial problem may result in larval mortality when rearing on foliage. When this occurs an antibiotic solution comprised of 2.8 g aureomycin (chlortetracycline hydrochloride) and 0.155 g kanamycin sulphate dissolved in a litre of distilled water is used to spray the foliage instead of distilled water (Riddiford 1967). The solution has proven to be an excellent control for bacterial infection and can be used routinely as a preventative measure.

5.4 Pupal collection

- Remove pupae from cups or plastic boxes with featherweight forceps.
- Separate sexes and place in well ventilated 27 x 20 x 10 cm plastic boxes. Line the sides of the box with paper towelling to ensure that at the time of eclosion adults will have a suitable area for wing drying and development (Fig. 2)
- Synchronize adult emergence by initially holding male pupae at a temperature about 4° C lower than for females.

5.5 Adults

- The sex ratio is approximtely 3♂:2 ♀.

5.5.1 Emergence period

Foliage reared adult males require an average of 59 days to develop from neonate larvae. Females require an average of 68 days.

Diet reared adult males require an average of 64 days and females require an average of 68 days to develop from neonate larvae.

5.5.2 Collection

Collect adults daily by clasping wings with featherweight forceps. Place plastic boxes with newly eclosed adults in the refrigerator to facilitate handling of moths.

5.5.3 Oviposition

The pre-oviposition period is about 3 days. Females lay about 100 eggs which are usually deposited between the layers of cheesecloth and few if any are found elsewhere in the chamber.

When first laid, eggs are green but if fertile change to a brownish colour in a few days. Remove the cheesecloth in two weeks or more when most of the adults are dead. Hold the eggs for about one month after laying at 18 ± 1° C, moisten slightly, place in a plastic bag and move to cold storage to satisfy diapause requirements.

a. Mating and oviposition cage (Fig. 3)

- Remove and discard central portion of the lid from a 4.5 liter carton.
- Line bottom of carton with paper towelling.
- Cut a hole 25 cm in diameter midway up the side of the carton.
- Place a piece of cheesecloth (at least 6 layers thick) on top of the carton.
- Force lid rim over the top of the cheesecloth to hold in place. Tape if necessary.
- Transfer up to 40 pairs of adults to the chamber through the hole in the side of the carton.
- Plug hole by using the bottom of an appropriately sized vial.
- Give adults a liberal supply of water daily by spraying the cheesecloth.
- Larger (20 liter fibre drums) or smaller (400 ml paper cartons) containers may be used to construct mating chambers and will give excellent results.

PRECAUTIONS

(i) It has been observed that decrease in fecundity will occur if water is not available. During periods when daily watering may not be possible, the vial used to plug the hole in the side of the mating chamber may be filled with water and fitted with a cotton wick. Position of the vial is reversed so that enclosed adults can imbibe water from the wick.

(ii) A light-dark period is a specific mating requirement and continuous light will result in a very high percentage of infertile eggs.

5.6 Sex determination

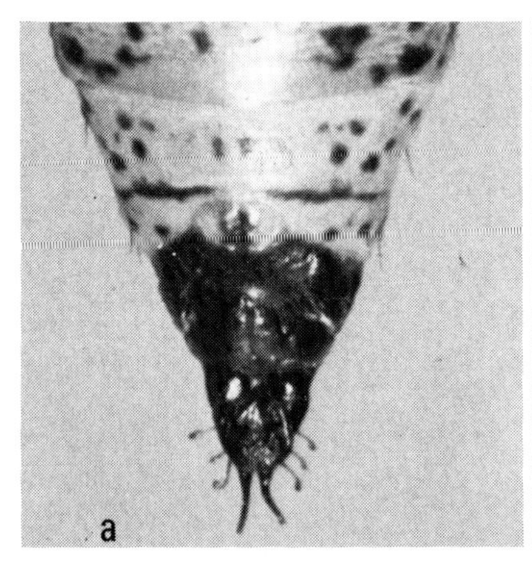

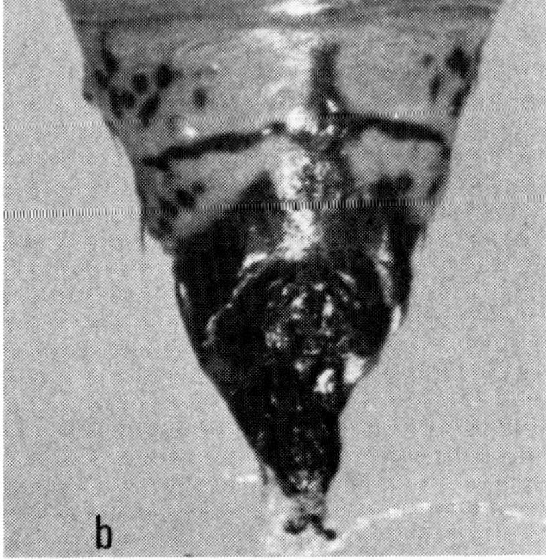

Fig. 4a Male pupa terminal abdominal segment.
Fig. 4b Female pupa terminal abdominal segment.

5.7 Rearing schedule

a. Daily

- Collect newly eclosed moths and place in mating chamber.
- Spray mating chamber, incubating eggs, and foliage in larval rearing boxes with water.
- Infest diet cups with neonate larvae.

b. Weekly (Establish a routine by assigning a specific day for each task).

- Remove eggs from cold storage and sterilize.
- Move eggs ready for storage into cold room.
- Remove eggs from mating chambers.
- Remove second instar larvae from diet cups and place on foliage.
- Thin diet reared third instar larvae to four per cup and place on fresh diet.
- Clean plastic boxes, harvest pupae and give larvae fresh foliage.
- Sex pupae and place in plastic boxes for adult eclosion.

5.8 Special problems

a. Diseases

The larval bacterial infection and its control using antibiotics has been discussed in 5.3 b.

b. Human Health

Work with formalin should be carried out in a fume hood.

Exposure to moth scales can be hazardous. Using a particle mask or biological containment hood is recommended.

6. LIFE CYCLE DATA

Life cycle and survival data during development of various stage under rearing conditions of 22 ± 1° C, 50-60 RH and LD 18:6 photoperiod are given. Comparative data for loopers reared on balsam fir foliage and on artificial diet is also given.

6.1 Developmental data

Stage	Average number of days Spent in stage	
	Diet Reared	Foliage Reared
1st instar	8	8
2nd instar	6	4
3rd instar	7	6
4th instar	7	5
5th instar	8	7
6th instar	13	15
Pupa	15	15
Adult	*ca* 14	*ca* 14

6.2 Survival data (%)

Egg viability 75

Neonate larvae to pupae, 88 diet - 84 foliage

Neonate larvae to adults, 82 diet - 80 foliage

6.3 Other data

Mean pupal weight for diet reared (μg) ♂ 84 (min 78, max 93) ♀ 103 (min 93, max 114)

Mean pupal weight for balsam fir reared (mg) of ♂ 98 (min 89, max 110) ♀ 133 mg (min 112, max 145)

7. PROCEDURES FOR SUPPLYING INSECTS

7.1 Pre-requisites

- Rearing colony should be of a known quality.
- The duration of developmental stages at various temperatures, humidity and light must be known.
- Standard laboratory insect rearing equipment should be available and daily and weekly work routines organized.

7.2 Placing an insect order

About two months notice is the maximum lead time required to supply adult moths. Neonate larvae could be supplied as early as 11 or 12 days.

7.3 Planning the order

An order for 2000 new fifth instar larvae weekly over several months would be handled in the following way.

- Each Thursday remove about 3000 diapause satisfied eggs from cold storage. The extra 1000 eggs is to compensate for overwintering egg mortality and for those insects that don't develop according to schedule. Eggs are removed on Thursday to ensure hatching does not occur on a weekend.
- Ideally new fifths should be made available for supply early in the week. Adjust temperature in the rearing cabinets to increase or decrease develop ment so that insects will be available on a specific day.
- About one month before the order is to terminate, do not remove eggs from storage.

7.4 Conclusion

A central rearing unit is in operation here at the Institute to serve the needs of approximately 40 research scientists and support staff. As many as 20 insect species are reared routinely in numbers that may exceed 5,000,000 annually. Because of the volume of work, cultures are usually initiated weekly and experimental insects for research projects are also supplied weekly. If specific stages are required more often, we manipulate temperature to increase or delay development.

8. REFERENCES

Burton, R.L. 1970. A low-cost artificial diet for the corn earworm. J. Econ. Entomol. 63: 1969-1970.

Carroll, W.J. 1956. History of the hemlock looper, *Lambdina fiscellaria fiscellaria* (Guen.) (Lepidoptera:Geometridae) in Newfoundland, and notes o its biology. Can. Ent. 88: 587-599.

Carroll, W.J., Waters, W.E. 1967. In A.G. Davidson and R.M. Prentice (eds). Important forest insects and diseases of mutual concern to Canada the United States and Mexico. Queen's Printers, Ottawa, pp. 121-122.

Danard, A.S. 1968. Newly hatched western hemlock looper larvae reared on forced foliage of American larch. Can. Forestry. Serv. Bi-Monthly Res. Notes 24: 7.

de Gryse, J.J., Schedl, K. 1934. An account of the eastern hemlock looper on hemlock with notes on allied species. Sci. Agr. 14(10): 523-539.

Grisdale, D.G. 1975. Simplified rearing of eastern hemlock looper. Can. Forestry Serv. Bi-Monthly Res. Notes 31: 19-20.

McMorran, A. 1965. A synthetic diet for spruce budworm *Choristoneura fumiferana* (Clem.), (Lepidoptera-Tortricidae). Can. Ent. 97: 58-62.

Riddiford, L.M. 1967. Antibiotics in the laboratory-rearing of cecropia silkworm. Science 157: 1451-1452.

LYMANTRIA DISPAR

THOMAS M. ODELL, CAROL A. BUTT, and ARTHUR W. BRIDGEFORTH

USDA Forest Service , Northeastern Forest Experiment Station, Center for Biological Control of Northeastern Forest Insects & Diseases, 51 Mill Pond Rd., Hamden, Connecticut 06514 USA

THE INSECT

Scientific Name: *Lymantria dispar* L.
Common Name: Gypsy moth
Order: Lepidoptera
Family: Lymantriidae

1. INTRODUCTION

The gypsy moth is a forest pest found in the United States, Europe, along the northern coast of Africa, across the Eurasian continent, and throughout Japan. Within the United States it is a major threat to commercial forests, parks, natural areas, and residential communities. The number of host plants exceeds 300 species of trees and shrubs, with species of oak ranked among the most favored foods. Conifers, when mixed with favored host species, are fed on by large larvae and in outbreaks may be completely stripped. Some less favored species, such as red maple, are fed on when larval population densities are high and favored foliage is scarce. Under such conditions larvae will consume nearly any foliage available, with only a few species such as tulip poplar and ash spared.

Larvae are phototropic and thus tend to initiate feeding in the upper crown of trees. Feeding by young larvae gives leaves a frayed, shot-hole appearance. Older larvae do the most damage; 90% of the defoliation occurs during the last 2 larval instars. In outbreaks whole forests can be denuded to look like winter. Although most trees will refoliate, the depletion of nutritional reserves makes the tree susceptible to attack by other insects and disease.

The gypsy moth is univoltine; eggs are laid in light brown setae-covered masses; embryonation is completed within approximately 6 weeks, after which time the apparently fully developed larva begins its diapause period. During diapause eggs require chilling (-10°C to 8°C) and over 50% RH for greatest survival. Egg hatch generally coincides with budbreak of oaks, particularly those in the black and red oak groups. First stage larvae appear to be particularly discriminating in selection of host plants, establishing well on the young, tender leaves of oak, gray birch, and apple as well as other hardwood species. They are not physically able to establish on conifers, but large larvae readily feed on hemlock, pine and spruce. Males and females normally have five and six instars, respectively, which range in size from approximately 3.5 mm in the first stage to approximately 50 mm in the sixth stage. Female pupae are characteristically larger than male pupae. The pupal stage lasts about 2 weeks. The adult male emerges first, followed several days later by the adult female; males and females live only 3-5 days. The female, which does not fly in the United States but does in Asia and some parts of Europe, attracts the male with a sex lure. Mating usually occurs on the day of eclosion, and shortly thereafter eggs are deposited. Each egg mass contains 200-800 eggs.

The univoltine life cycle, the large larval size, and the relatively long developmental period have been the major obstacles in the evolution of a biologically sound and economical rearing methodology. Rollinson (1964) began the first attempt to develop a uniform diet by modifying a wheat germ diet originally developed for spruce budworm (McMorran and Grisdale, 1963). This ultimately led to the development of a diet and rearing technique (ODell and Rollinson, 1966) which, with slight modification, is still being used for small scale laboratory rearing. Other diet formulations for rearing gypsy moth were subsequently developed by Leonard and Doane (1966), Magnoler (1970), Vasiljevic and Injac (1971), and Ridet (1972). Bell et al. (1981) developed a high wheat germ diet which is now used for mass production as well as smaller scale rearing for research.

Several strains have been established in the laboratory, the oldest was established in 1967 at the USDA Animal and Plant Health Inspection Service's (APHIS) Otis Gypsy Moth Methods Development Laboratory, Otis Air Force Base, Massachusetts, and has been reared continuously for 24 generations. Presently the strain is available at all times of the year, and is used for mass production of virus, parasites, and sterile males. In our laboratory this strain, and others, are used for research involving plant-pest interaction, development of microbial controls, parasite-host relationships, and quality control for colonized insects.

2. FACILITIES AND EQUIPMENT REQUIRED

2.1 Facilities

Rearing of larvae and pupae is carried out in controlled environmental rooms maintained at 25±1° C, 50-60% RH, and a LD 16:8 photoperiod. Air movement within the chamber circulates at 2 to 3 cubic feet per min. Lighting is provided by high output type fluorescent lamps with minimum intensity of 70 foot candles, 102 cm above the floor. Oviposition and embryonation prior to egg diapause are carried out at the same environmental parameters.

2.2 Equipment and materials required for insect handling

2.2.1 Adult oviposition

Two methods are used:

a. Individual mating
- 236.5 ml unwaxed cup with plastic cover
- Labels
- Forceps
- Disposable gloves
- Hepa exhaust hood with inward flow of air <u>away</u> from the operator

b. Group mating (2-15 pairs)
- 0.95 liter round, rolled-rim cardboard container
- Delicatessen paper liner
- Labels
- Scissors
- Hepa exhaust hood with an inward flow of air <u>away</u> from the operator
- Disposable gloves
- Forceps

2.2.2 Egg collection and storage

- Cold chamber (refrigerator or equivalent) maintained at 8-9° C, 80-90% RH, and an LD 16:8 photoperiod. Two chambers located in different places avoids loss of colony if one chamber fails.
- Scissors (only for individual method)
- 0.95 liter round, rolled-rim cardboard container
- Central vacuum system
- Hepa exhaust hood with an inward flow of air <u>away</u> from the operator
- Labels

- Disposable gloves

2.2.3 Egg sterilization
- 2-liter autoclavable plastic pitcher
- 3.7% formaldehyde solution *Precaution (fumes)
- Fume hood
- Sterile forceps
- Screening (nylon or plastic, 20 squares/2.54 cm)
- Stapler
- Scissors
- Disposable gloves
- Hardware cloth platform (0.63-cm gauge)
- Hardware cloth basket (0.63-cm gauge) with removable lid (7.62 cm high x 7.62 cm diam)
- Disinfectant: Zephiran chloride aqueous solution 0.13%. AI = benzalkonium chloride.
- Hepa exhaust hood with an inward flow of air <u>away</u> from the operator

2.2.4 Egg hatch
- Plastic box with lid; 304 x 152 x 95 mm
- Galvanized wire screening, 0.63-cm gauge (292 x 146 x 19 mm)
- 60 ml water
- 15 x 100 mm plastic petri dishes
- Marking crayon
- Environmental conditions (25°C, 100% RH)

2.2.5 Larval set-up on diet
- Fine camel hair paint brushes (size 3)
- Hood
- Zephiran chloride (0.13% solution)
- Beaker (100 ml, for Zephiran)
- Paper toweling
- Spatula
- Alcohol lamp
- Forceps
- Small beaker of 95% alcohol with <u>lid</u> *Precaution (alcohol fires)

2.2.6 Larval rearing containers

a. Rearing for bioassay
- First and second instars use 15 x 100 mm plastic petri dishes
- Third instars use 177.4 ml fluted plastic cups with lids (Sweetheart Plastics, Willmington, MA 01887)
- Fourth, fifth and sixth instars use uncoated 471.2 ml cardboard cups with plastic lids, or 177.4ml plastic fluted cups with cardboard lids
- Trays

b. Maintaining colony (stock replacement)
- 177.4 ml Sweetheart plastic fluted cups with uncoated cardboard lids
- Trays

2.2.7 Pupal collection and holding for eclosion, mating and oviposition
- Forceps
- Disposable plastic gloves
- 0.95 liter round, rolled-rim cardboard containers
- Labels
- Delicatessen paper
- Hepa exhaust with an inward flow of air <u>away</u> from the operator
- Scissors

<u>2.3 Diet preparation equipment</u>

2.3.1 Equipment for weighing dry ingredients (same for all diets)
- Plastic bags
- Twist ties
- Non-toxic particle mask
- Freezer
- Labels
- Metal scoop
- Top-loading balance (0.1-1200 g)

2.3.2 Equipment for making finished diet

a. 8-12 liters of diet
- 2000 ml autoclavable measuring cylinder
- Large rubber spatula
- Direct steam table mixer, 2-20 liter capacity (Groen Division, Dover Corp., Elk Grove, Illinois)
- Disposable plastic gloves
- Aluminum foil
- Autoclave
- Autoclavable pitchers
- Asbestos gloves
- Thermometer
- Zephiran chloride (0.13% concentration) (disinfectant)
- Containers: 29.6 ml plastic cups (Fabri-Kal Corp., Hazelton, PA); 177.4 ml fluted cups and lids (Sweetheart Plastics, Inc., Wilmington, MA); 15 x 100 mm plastic petri dishes (Falcon Labware Div., Becton-Dickinson, Oxnard, CA)

b. 2-6 liters of diet
- Same as for 8-12 liter batch except for substitution of a 5 liter blender with stainless steel bowl and stainless steel overflow pan (229 x 330 x 51 mm) (Hobart Corp., Troy, OH) for steam kettle

c. 1 liter or less
- Same as for 8-12 liter batch except substitute a small (Osterizer) blender for the steam kettle and reduce size of spatula
- Stainless steel bowl and stainless steel overflow pan used in 2-6 liter batches

3. ARTIFICIAL DIET

3.1 Composition

3.1.1 Bell diet

a. Dry ingredients to prepare 10 liters of finished High Wheat Germ Diet

Ingredients	Amount (g)	%
Wheat germ	1200	77.42
Casein	250	16.13
Salt mix	80	5.16
Sorbic acid	20	1.29
Total	1550	100.00

b. Finished diet

Dry mix	1550 g	14.339
Tap water	9000 ml	83.256
Methyl paraben	10 g	0.093
Vitamin premix	100 g	0.925
Agar (fine mix, 80-100 mesh)	150 g	1.388
Total	10,810 g	100.001

3.1.2 Prepackaged, commercially available diet (available from BioServ, Inc., Frenchtown, NJ)

Composition for 1, 2, 8, and 10 liter batches of BioServ diet (see BioServ catalogue for ingredients)

	Amount			
Composition	1 liter	2 liters	8 liters	10 liters
Water	850.00 ml	1700.00 ml	7600.00 ml	9500.00 ml
Agar	18.75 g	37.5 g	150.0 g	187.5 g
Dry mix	130.25	260.5	1042.0	1302.5
Fats and sugars	20.00	40.0	160.0	200.0
Vitamins	1.29	2.57	10.28	12.85

3.2 Brand names and sources of ingredients

3.2.1 Bell diet

Wheat germ (Mennel Milling Co., Fostoria, OH); Casein (New Zealand Milk Products, Petaluma, CA); Salt mix, Wesson's #902851 (ICN Nutritional, Cleveland, OH); Sorbic acid (ICN Nutritional, Cleveland, OH); Vitamin Premix #26862 (Hoffman-LaRoche, Inc., Salisbury, MD); Agar (Moorehead & Co., Van Nuys, CA); Methylparaben (Kalama Chemical, Inc., Garfield, NJ)

3.2.2 Prepackaged, commercially available diet

a. Gypsy moth rearing media: Bio-Mix #722-A

b. Field collected gypsy moth diet: Bio-Mix #909-A

3.3 Diet preparation procedure

3.3.1 Bell diet

a. 8-12 liter batch

- Flip switch on steam generator
- Wipe down work tables and area with Zephiran
- Remove preweighed ingredients from freezer for acclimation to room temperature
- Mix all dry ingredients together (except vitamins) in plastic weighing bags; set aside
- Put agar in steam kettle. Add distilled water
- Set out required diet containers
- Turn steam kettle valve on with beater speed set at #5
- When agar starts to boil, slowly add all ingredients except vitamins; let boil for 12 sec, then turn steam valve off
- Beat for 8 min at speed #7
- Continue to beat until well-mixed
- Dispense into containers

b. 1 liter or less

- Wipe down work tables and area with Zephiran
- Remove preweighed ingredients from freezer for acclimation to room temperature
- For 1 liter batch, place 15 g agar in mixing bowl
- Add 800 ml cool water to agar, cover bowl with foil, set bowl into overflow pan and autoclave
- While agar mixture is in autoclave, set out required diet containers
- Remove agar from autoclave and return to the mixing bowl any water that may have collected in the overflow pan
- Add other ingredients
- Put mixture in Osterizer
- Blend until well mixed
- Dispense into containers

3.3.2 Commercial diet (BioServ, Inc.)

a. 8-12 liter batch

- Flip switch on steam generator
- Wipe work tables and area with Zephiran
- Remove preweighed BioServ ingredients from freezer for acclimation to room temperature
- Put agar in steam kettle, add cool distilled water
- Set up required containers (as per order)
- Turn steam kettle valve on with beater speed set at #5
- When agar starts to boil, slowly add all ingredients except vitamins; let boil for 12 sec, then turn steam valve off
- Beat for 8 min at speed #7
- Add vitamins
- Continue to beat until well mixed
- Dispense into containers

b. 1 liter or less

- Wipe down as above
- Remove preweighed ingredients from freezer for acclimation to room temperature
- Place agar (18.75 g) in mixing bowl
- Add 850 ml cool distilled water, cover with foil, set bowl into overflow pan and autoclave with other utensils
- While agar mixture is in autoclave set up required diet containers
- Remove agar from autoclave and return to the mixing bowl any water that has collected in the overflow pan
- Mix in other ingredients
- Put mixture in Osterizer and blend well
- Dispense into containers

3.3.3 Ideas and recommendations

- Note scale-up in mixing container size when increasing diet quantity from 1 liter (section 2.3.2).
- Diet mix, excluding vitamins, should be brought to boil (blanched) for 12-15 sec to eliminate enzymatic processes. This will stabilize diet and reduce darkening.
- The amount of water added is not necessarily derived directly from ingredient scale-up. The size of container opening will influence amount of water lost through evaporation which must be considered in deriving final moisture content of diet.
- Presoaking agar in a portion of the water often eliminates lumping and makes blending easier.

4. REARING AND COLONY MAINTENANCE

4.1 Egg collection and storage

From cardboard mating containers

- Clean work surface of hood with Zephiran and turn on just prior to use
- When eggs are 4-6 weeks old transfer containers to hood, remove oviposition sleeve, cast pupal skins, moths, and vacuum excess wing scales off of oviposition sleeves
- Count and record number of egg masses per infest date
- All oviposition sleeves from the same infest date are placed together (maximum of 6) in a 0.95 liter rolled rim container and stored at 8-9° C or 4-5°C and a minimum 50% RH for the 180-day diapause period
- After completion of diapause period, egg masses are cut from the oviposition sleeve as needed

PRECAUTIONS

(i) Viability of eggs will be extended by storing at lower temperatures; minimum 2-3°C, storage up to 320 days

(ii) Use new, sterile containers to store eggs in chill to prevent mold

4.2 Egg sterilization and incubation

- Use fume hood or appropriate OSHA-approved mask
- Place enclosed eggs into a 2 liter container with 1 liter of formaldehyde solution. Egg masses, on paper or in screen packets, are placed in a hardware cloth basket and submerged in the solution for 60 min
- Remove egg masses and submerge in running tap water for 60 min
- Using sterile forceps, egg masses are removed from water, placed on a hardware cloth platform (previously rinsed in Zephiran chloride) and dried in a hood for a minmum of 60 min
- Egg masses are then placed individually in 15 x 100 mm plastic petri dishes
- Dishes are put in a 304 x 152 x 95 mm incubation box containing approximately 60 ml of tap water in the bottom, and incubated at 25° C and a LD 16:8 period. Dishes containing eggs are kept above the water

using empty petri dishes or some other platform

PRECAUTIONS

(i) Eggs in chill for more than 230 days may be killed by sterilization procedure

(ii) Incubation time shortens as eggs age, and is mediated by diapause and incubation temperature. Thus, eggs held for 180 days at 2-3°C or 8-9°C take 5 days, and 2-3 days to hatch, respectively. Eggs held for only 150 days at the same temperatures may take 8 and 4-6 days to hatch.

(iii) Eggs can be placed directly on diet, after removal from egg mass and disinfection. Hatch time is variable as is development time, but in mass rearing this technique is time and cost efficient.

4.3 Diet inoculation with larvae

PRECAUTIONS

Aging neonates

(i) Eggs which have been held at 8-9°C for 180 days begin to hatch within 24 h of initiation of incubation at 25°C

(ii) The majority hatch between 24 h and 96 h

(iii) Neonates can be left in eclosion dish for 48-72 h without apparent loss of viability

(iv) Neonates held without diet for 24-48 h often establish on diet sooner than newly hatched larvae

(v) All inoculation should be done under the hood

(vi) For colony, use neonates from one egg mass for each 5 cups inoculated for a good genetic mix. One tray will contain 30 cups with neonates from 6 different egg masses

a. In 177.4 ml cups - colony
- Air cool freshly poured diet (80 ml/cup) in clean area for 2 h (or until room temperature) prior to covering
- Transfer 8 neonates to each cup using a soft camel hair paint brush which has been disinfected in Zephiran chloride
- Cover cup with cardboard lid and mark with inoculation date, strain, order number, and other appropriate identifying information
- Place cups in rearing chamber on trays
- Do not place cups directly on top of each other; maintain full "breathing" surface of cover

b. In 177.4 ml cups - orders for 3rd, 4th, 5th and 6th instars
- Air cool freshly poured diet in clean area for 2 h prior to covering. Use the following diet and insect number per cup for instar 3: 20 ml diet, 50 neonates; instar 4: 40 ml diet, 15 neonates; instar 5: 60 ml diet, 12 neonates; instar 6: 80 ml diet, 12 neonates
- Approximate time to 60% attainment for 3rd instars is 7 days; 4ths, 11-12; 5ths, 14-17; 6ths, 19-21
- Larvae are removed from the cups with sterile forceps and placed in requested container for delivery
- If age requirement for larvae is critical, an earlier instar is delivered

c. In petri dishes
- Air cool freshly poured diet (15 x 100 mm petri dish) in clean area for 20-30 min (high wheat germ or BioServ diet) before covering
- Diet is cut into cubes, approximately 0.5 cm^3
- Three 0.5 cm^3 pieces of diet are evenly placed on bottom of 15 x 100 mm plastic petri dish
- Transfer 50 neonates to dish containing diet (using disinfected brush) invert dish and place in appropriate environmental chamber
- After each molt, larvae are transferred to new sterile dishes using a small camel hair brush for instar I, a small spatula for instars II and III, and forceps for instars IV, V, and VI. Combinations of larvae and

found optimal for petri dishes were: instar II, 20-25 larvae, 3 diet cubes approximately 1 cm^3 each; instar III, 10 larvae, same diet; instar IV, 5 larvae, same diet; instar V, 2-3 larvae, same diet; instar VI, 1 larva, 1 diet cube approximately 1 cm^3

- Approximate time to 60% attainment for 2nd instars is 4 days. The approximate time to 60% attainment for 3rd through 6th instars is the same as in item #2 of 4.3b.

d. In 471.2 ml unwaxed cup (Dixie/Marathon Products, Greenwich, CT)

- Diet is poured into 29.6 ml plastic cups, allowed to cool for 20-30 min and stored in plastic bag at 3-4°C
- Larvae are transferred to 471.2 ml cups from petri dishes at instar III
- To rear to pupation, 20 instar III larvae are transferred to cups containing 2, 29.6 ml cups of diet (3/4 full)
- Diet is replaced as needed, but at least every 5 days
- Use clear lids for 471.2 ml container (Sweetheart Plastics, Inc., Willmington, MA).

PRECAUTIONS

(i) Inspection of containers for development should be minimal to avoid contamination

(ii) Clear tops will allow inspection without disturbing microclimate

(iii) Label containers to be used for checks so others will not be opened by mistake

4.4 Pupal collection

a. Laboratory strain, F_{24}, from 177.4 ml rearing cups, petri dishes, or 471.2 ml Dixie cups

- Transfer rearing containers to hood approximately 34 days after inoculation
- Carefully remove pupae from container by hand (use disposable, sterile gloves), and transfer to mating containers lined with delicatessen paper, 10 males and 10 females per container

b. Newly colonized strains (P_1, F_1, F_2)

- Developmental variability may require daily monitoring to determine pupation date. At 25°C and 50% RH, 80% pupation occurs at approximately 36 days from inoculation

PRECAUTIONS

(i) Pupae are green and soft when newly formed and are easily punctured. These should be harvested 4-8 h later

(ii) Do not hold pupae in air-tight cups or in densities which require pupae to be stacked in more than 2 layers

(iii) For standard strain mate males from one tray with females from another tray (different egg mass source) to prevent inbreeding

(iv) Schedule is relative to this particular strain and generation. This may change from strain to strain and with generation.

4.5 Adults

4.5.1 Eclosion period

a. Colony-F_{24}

Colony development is relatively uniform so males and females emerge in approximately 10 days after harvest, with all eclosion occurring within 4 days. Adult survival from neonate larvae is approximately 80%

b. Newly colonized strains-P_1, F_1, F_2

- Eclosion is variable with male eclosion beginning 2-3 days earlier than females
- If pupae are separated by sex, adults can be removed and paired as they eclose

4.5.2 Collection

- Collection and pairing of adults should be done under hood

- Females, which do not fly, are removed by grasping the forewing near the thorax and lifting gently
- Males can be similarly transferred to containers with females. It is usually advantageous to cool them for 5 min at 5°C to avoid escape and/or injury
- Colony adults emerge in the 0.95 liter cardboard containers, mate, and oviposit. Moths are not removed until eggs are harvested

4.5.3 Oviposition

Females and males will mate within 3 h after eclosion. Oviposition is initiated soon after completion of mating. Females mate only once.

a. In 0.95 liter mating containers
   - Cut delicatessen paper to line the mating container
   - Place 10 pupae of each sex, of similar age, into container
   - Mark container with strain, generation, chill temperature, and dates of inoculation, pupal harvest and chill initiation
   - Males and females will emerge within 12 days and mate. Females which have crawled on to the paper liner initiate oviposition soon after mating and complete the process within 72 h (at 25°C). Eggs are then held for completion of embryonation (25°C, 30 days) and diapause (8-9°C, 180 days).

b. In 471.2 ml Dixie cups
   - Cut paper towel to fit around inside of cup and attach to lip of cup in several places with tape
   - 5 adults of each sex can be accommodated easily in this mating arena

PRECAUTIONS

(i) A male can mate successfully with at least 5 females so they can be placed, when necessary, with more than one female, or can be transferred between cups containing females

(ii) All work involving adults and egg masses should be conducted under a hood with negative air flow; i.e. conducts insect dust away from worker

4.6 Sex determination

- Female pupae and adults are usually larger than the males
- Pupae antennal features: female antennal region is flat, male antennal region is convex
- Female moths are white with black markings, males are brown with black markings
- A urogenital slit on the last abdominal segment of the female is the most reliable morphological feature for sex determination

4.7 Rearing schedule

a. Colony-F_{24}--Eggs have been stockpiled for continuous rearing if required

b. Scheduling for 600 egg masses per month

Monday - eggs sterilized and incubated
(10-20 egg masses randomly selected)
Thursday - neonates transferred (720)
Tuesday - egg sheets collected
Wednesday - pupae collected and placed in mating container

4.8 Insect quality

4.8.1 Production-colony

Growth rate: Determined by percent pupation 34 days after inoculation. Standard = 80% pupation for colony. A lower percent could be caused by a change in temperature and/or humidity, a difference in diet quality, sublethal microbial infection, inadvertent genetic selection, or a combination of these. Each would be investigated.

Pupal weight: On day of pupal harvest, the pupae from each of 30 randomly selected cups are weighed, by cup, and the number of males and females per cup is determined. Pupal weight, by cup, is then compared statistically with a previously determined standard which sets quality limits based on sex ratio per cup and total weight per cup. If more than 5% of the pupal weights/cup fall below the confidence limit procedures are initiated to determine cause.

Egg viability: Each month 10 egg masses, 183 days in chill, are set out to hatch. First instars are counted by egg mass and recorded. The remainder of the egg mass is cleaned; embryonated and nonembryonated eggs are separated, counted, and recorded. Fecundity, percent embryonated and percent hatch are determined. These are compared with a standard.

4.8.2 Product performance

a. Adult males, used in the sterile male technique (SMT), are tested for response to pheromone in a flight tunnel.

- Periodicity of flight is determined using an actograph.
- A standard is developed by comparing the performance of wild and SMT insects in field plots and laboratory bioassays (flight tunnel and actograph). The standard reflects periodicity of peak flight and varies with geographic location (photoperiod, temperature cycle, and RH).

b. Microbial bioassay of *Bacillus thuringiensis* (Bt) and nucleopolyhedrosis virus (NPV). A standard LD_{50}, with confidence limits is developed using control data from previous bioassays. Control data from subsequent bioassays are then compared statistically against this standard. Larval weight is considered in the analysis.

c. Rearing and bioassay of entomophagous insects.
Continuous records are maintained for parasite size, longevity and fecundity. Factors associated with host rearing which can effect significant changes in these quality control characteristics include host diet, host density, microbial infection of host, and change in environmental parameters.

4.9 Special problems

a. Diseases:

Nucleopolyhedrosis virus infection of the larval and pupal stages is greatly reduced by egg sterilization. Contents of containers with probable infections should be destroyed. Sterile techniques should be used in all areas where rearing procedures are taking place.

Contamination by other pathogenic organisms can be contained by strict adherence to proper use of laboratory coats, gloves, and other laboratory apparel.

Maintaining a clean work environment by daily vacuuming (use in-house vacuum system), washing floors and misting with disinfectant, and routine disinfection of work surface, will greatly reduce potential for colony contamination.

b. Microbial control of diet:

Bacterial and fungal infection of diet can occur if sterile conditions are not maintained. Diet should not be exposed to air for more than 2 h. Samples from each day's diet production should be incubated to determine incidence of occurrence of bacteria and fungus.

Dietary ingredients should be tested for microbial contamination when it is received, and then stored in a freezer to prevent contamination and reduce potential for degradation.

c. Human health:

All work with formalin should be carried out under a fume hood as it is a carcinogen. Larval setae, moth scales and egg mass hairs can cause serious allergenic reactions. All work should be carried out under hoods which carry material away from worker, and protective masks, lab coats, and gloves should be worn.

d. Further discussion of quality control in rearing gypsy moth can be found in ODell et al., 1984.

5. LIFE CYCLE DATA

Life cycle and survival data during development of various stages under optimum rearing conditions of 25°C, 50-60% RH, and a LD 16:8 photoperiod are given.

5.1 Developmental data (when reared in 177.4 ml fluted cups with no transfer)

Developmental period	Number of days
Egg embryonation	28 ± 7
In chill (diapause)	
Minimum	120
Optimum	180
Maximum	220
Egg hatch	
120 days at 8-9°C	10
180 days at 8-9°C	3
220 days at 8-9°C	2
Larval (neonate to pupa)	
Male	28 ± 3
Female	32 ± 3
Pupal	
Male	12 ± 2
Female	12 ± 2
Adult eclosion (neonate to adult)	
Male	40 ± 5
Female	44 ± 5
Total development (egg to egg)	51

Note: - When reared in petri dishes or 471.2 ml cups, total development takes 5-6 days longer.
- Rearing schedule may differ from strain to strain and with generation.

5.2 Survival data (%)

a. 177.4 ml cups, no transfer
- Egg viability, $\overline{X}$=85, minimum=73, maximum=95
- Neonate larva to pupa, $\overline{X}$=85, minimum=76, maximum=96
- Pupal mortality = less than 3

b. When rearing in petri dishes and/or 471.2 ml cups, survival is approximately the same as in 177.4 ml cups

5.3 Other data and information

- Preoviposition period, 4 h
- Egg laying period, 4-36 h after eclosion
- Once unfertilized females begin to lay eggs, they will not mate
- Females will scatter eggs if not mated, or disturbed too much
- Mean fecundity: mean=900, minimum=675, maximum=995
- Adult longevity: males=3-5 days, females=2-3 days
- Number of larval instars: normally 5 for males, 6 for females (see below)
- Mean pupal weights (grams): males=0.70, females=2.14
- Sex ratio: All hatch - 1:1. Note: sex ratio can be distorted by shortening chill period (120 days); sex ratio of first 25% hatch will be biased toward males; after 180 days of chill first 25% of hatch will be biased toward females

- Through inadvertent selection, 97% of the females of the F_{24} laboratory strain now have only 5 instars. Selection probably occurred by continual harvesting of pupae at earliest date possible.

6. PROCEDURES FOR SUPPLYING INSECTS

6.1 Prerequisites

- Research and development of rearing procedures must be completed.
- The founder colony should be robust and disease free.
- The duration of developmental stages at a particular temperature, humidity, and light must be known.
- The best diet should be determined through research of the gypsy moth's nutritional needs, making certain several generations are observed for long term effects of malnutrition.
- Standard rearing methods should be adopted including sterile technique and sanitation control.
- Adequate equipment and labor should be organized, with special attention to protective clothing and safety procedures.
- The rearing facility should have several environmental control units.
- Colony should be maintained to produce weekly a batch of eggs that has been chilled 180 days at 8-9°C.

6.2 Placing an insect order

- Insect orders are filled out in duplicate for each separate order. (One copy is returned with the order and one filed in the rearing facility.)
- Order cards are filled out in advance depending on the insect stage desired.

Eggs	1 day
Instar I	1 week
Instar II	2 weeks
Instar III	2-1/2 weeks
Instar IV	3 weeks
Instar V	4 weeks
Instar VI	5 weeks
Pupae	6 weeks

- Insect Order Form

INSECT ORDER

NAME: Dr. A. B. Smith DATE: June 6, 1983
SPECIES: *Lymantria dispar* L. STRAIN: New Jersey
STAGE: 2nd instar
QUANTITY: 4000
DIETARY REQUIREMENTS: Bell Diet
SPECIAL HANDLING; Deliver insects 100/cup (471.2 ml)
PICK-UP DATE: June 20, 1983

Order filled by: AWL
On: June 20, 1983
Log #: 171 (Julian date)
Comments:

6.3 Planning the order

a. When an order is received, all necessary steps in filling the order are recorded on a master schedule calendar, starting with the pick-up date and working backwards. Example: Dr. Smith would like 4000 instar II gypsy moths on June 20, 1983. The chronological schedule would be as follows: June 9 - Order the amount of diet needed to fill the order (this gives the diet technician one week's notice). In this example, order 12 petri dishes of

Bell diet (yield: 30 cubes of diet per dish). June 13 - Remove 12 egg masses from 8-9°C (180-day chill), sterilize and incubate; an underestimation of 500 neonates from each egg mass allows for sufficient neonates to fill the order even if egg hatch is at a minimum. June 16 - Each of 120 petri dishes are supplied with 3 diet cubes and 50 neonates (total of 6,000 neonates transferred to allow for a 66% rate of molt to instar II in 4 days). June 20 - Remove instar II insects at a rate of 100 per 471.2 ml cup and package the order in a plastic bag with a copy of the order prominently displayed for easy identification (the bottom right hand corner of the order card contains information which allows the order to be traced in the event Dr. Smith has questions concerning the order).

b. Calculation of insect numbers

- Section 4.3 and Table 5.1 can be used as timing guides for optimum numbers of each instar and stage.
- The 66% attainment figure is variable and experience is the best tool for estimating numbers.
- Planning schedules are recorded on the back of each order card that is filed in the rearing facility for easy reference in adjusting successive orders.

6.4 Conclusion

This method of planning the order necessitates the weekly availability of eggs that have been chilled for 180 days. In a fairly consistent ordering situation, the colony may be maintained to produce a constant number of egg masses each week. Approximately 8.5 month's notice must be given for extraordinarily large orders so that the size and production of the colony can be increased accordingly.

REFERENCES

Bell, R.A.; Owen, C.; Shapiro, M.; Tardif, J.R. 1981. Development of mass-rearing technology. pp. 599-655 In C.C. Doane and M.L. McManus (eds.). The gypsy moth: Research toward integrated pest management. U.S. Dep. Agric. Tech. Bull. 1584.

Leonard, D.E.; Doane, C.C. 1966. An artificial diet for the gypsy moth, Porthetria dispar. Ann. Entomol. Soc. Am. 59: 462-464.

Magnoler, A. 1970. A wheat germ medium for rearing the gypsy moth, Lymantria dispar L. Entomophaga 15: 401-406.

McMorran, A.R.; Grisdale, D.G. 1963. Rearing insects on artificial diet. (Mimeo). 24 April 1963. Insect Pathol. Res. Inst., Sault Ste. Marie, Ontario.

ODell, T.M.; Rollinson, W.D. 1966. A technique for rearing the gypsy moth, Porthetria dispar (L.), on an artificial diet. J. Econ. Ent. 59: 741-742.

ODell, T.M.; Bell, R.A.; Mastro, V.C.; Tanner, J.A.; Kennedy, L.F. 1984. Production of gypsy moth for research and biological control. In E.G. King and N. C. Leppla (eds.). Advances and Challenges in Insect Rearing. U.S. Dep. Agric. Tech. Bull. 306 + xvi pages, 80 illus. (In press).

Ridet, J. 1972. Etude des conditions optimales d'elevage et d'alimentation de Lymantria dispar L. Ann. Soc. Entomol. Fr. (N.S.). 8:653-668.

Vasiljevic, L.J.; Injac, M. 1971. Artificial diet for the gypsy moth (Lymantria dispar L.). Zast. Bilja. 22: 389-396.

MALACOSOMA DISSTRIA

DAIL GRISDALE

Department of the Environment, Canadian Forestry Service, Forest Pest Management Institute, Sault Ste. Marie, Ontario, Canada, P6A 5M7

THE INSECT

Scientific Name: *Malacosoma disstria* Hübner
Common Name: Forest tent caterpillar
Order: Lepidoptera
Family: Lasiocampidae

I. INTRODUCTION

The forest tent caterpillar appears at irregular intervals throughout the forested area of Canada from the Maritimes to British Columbia and north to the limits of its principal host tree. In the United States this insect is distributed across the continent and as far south as Louisiana. Trembling aspen is the principal host species, but larvae also feed on sugar maple, red oak, basswood, willow and alder and in the southern states on bottomland gums.

Hosts, descriptions of damage and life history were described by Sippell and Evan (1967). The moths appear in late June or early July, and eggs are laid in bands that encircle small twigs on the host trees. Each egg band usually contains about 150 eggs. Fully-formed larvae develop within the eggs in about three weeks and the insect overwinters in this form. Eclosion usually coincides with the swelling and breaking of buds in early spring. With favourable weather and development of foliage the larvae feed voraciously, and mature by mid-June. Cocoons are spun among the leaves of trees, shrubs or other vegetation, and the moths emerge about 8 to 12 days later.

Unlike other species of tent caterpillar, the larvae do not form a tent but feed openly, often in large groups or clusters. Although tree mortality is rare, heavy defoliation causes appreciable losses in radial growth of host trees. Because of the nuisance created by migrating larvae, the insect can have a serious impact on the camping and tourist industry.

Rearing the forest tent caterpillar on artificial food was reported by Grisdale (1963) and a rearing method was described by Addy (1969). Grisdale (1976) described a rearing method that had been in use at the Forest Pest Management Institute since 1966. Our laboratory colony is largely maintained by field collected eggs. For the past 20 years infestations have occurred in various areas of Ontario and each fall host trees are felled and egg bands removed from the mid and upper crowns. Because of the incidence of nuclear polyhedral virus in field populations, eggs for larval rearings must be sterilized.

2. FACILITIES AND EQUIPMENT NEEDED

2.1 Facilities

Incubation of eggs after cold storage and rearing of larvae and pupae are carried out in an environmental control cabinet maintained at 24 ± 1°C, 50-60% RH and an LD 18:6 photoperiod. Lighting is provided by three 30 watt daylight fluorescent tubes 117.5 cm in length. Oviposition is carried out in a controlled environment room at a temperature of 20°C, 80-90% RH and an LD 16:8 photoperiod.

2.2 Equipment and materials required for insect handling

2.2.1 Adult oviposition

Two methods are used

a. Individual mating

- 1 pint (0.57 litre) cylindrical cardboard container (Metripac Nested Food Containers, Purity Packaging, Montreal, Quebec, Canada)
- Petri dish covers (100 x 15 mm)
- Short section of an aspen branch tip (other hosts or non-host suitable)
- Pruners

b. Group mating

- Screened cage (35 x 35 x 25 cm) fitted with a sliding glass door
- Paper towelling
- Pruners
- Aspen branches

2.2.2 Egg collection and storage

- Pruners
- Cardboard mailing tube
- HEPA filtered exhaust hood (Canadian Cabinets Co. Ltd., P.O. Box 11336, Station H, 25F Northside Road, Nepean, Ontario K2H 7V1, Canada) when large numbers of cages are handled
- Particle mask when small numbers are handled

2.2.3 Egg sterilization

- 300 ml glass jar fitted with a screened lid
- Household bleach with 6% available sodium hypochlorite
- Scissors
- Paper towelling
- Petri dish (100 x 20 mm)
- Rubber gloves

2.2.4 Egg hatch and larval innoculation of diet

- Soft nosed forceps
- Surgical scissors

2.2.5 Larval rearing containers

a. General laboratory rearing

- Translucent, ribbed cups (22 ml) sold as creamers by Portion Packaging Ltd., 26 Tidemore Ave., Rexdale, Ontario, Canada M9W 5H4
- Unwaxed paper lids for 22 ml cup (also sold by Portion Packaging)
- Disposable pressed board cafeteria trays (46 x 36 cm) to hold cups (Keyes Fibre, Waterville Maine, USA 04901)

b. Group larval rearings

- Materials as in 2.2.5. (a)
- Plastic box (28 x 38 x 15 cm) with lid and ends ventilated with plastic screening (Althor Products, Div. American Hinge Corp., 496 Danbury Road, Wilton, Connecticut, USA 06897)
- Paper towelling
- Aluminum foil
- Soft nosed forceps

2.2.6 Pupal collection (Chemical separation of pupae from cocoons)

- Fume hood or particle mask
- 2 plastic dishpans (35 x 30 x 14 cm)
- 3.5 litre stainless steel beaker
- Cylindrical wire basket (20 x 15 cm) made from galvanized hardware cloth
- Cheesecloth
- Plastic twister
- Household bleach (6% available sodium hypochlorite)

2.2.7 Insect holding

- Incubator set at 18°C
- Refrigerator set at 10°C
- Walk-in cold room set at 2°C

2.3 Diet equipment and preparation

2.3.1 Finished diet

- 4000 ml flask

- 50 ml beaker
- 1000 ml graduated cylinder
- 250 ml graduated cylinder
- 10 ml pipet
- Autoclave, small pressure cooker or hotplate
- Waring blendor
- Fume hood
- 0.0001-10 g to 3000 g balances
- Asbestos gloves
- 1000 ml stainless steel beaker with pouring spout and handle
- Large plastic bags (green garbage)
- Plastic spray bottle
- Metal scoop
- Spoon
- Rubber spatula

3. ARTIFICIAL DIET

3.1 Composition

Ingredients to prepare approximately 3.6 kg finished diet

a. Ingredients

Ingredients	Amount	% (estimated)
Granulated agar	60 g	1.68
Distilled water to blendor	792 ml	22.10
Distilled water to dissolve agar	2201 ml	61.41
Casein, vitamin free	126 g	3.52
4 N KOH	18 ml	.50
Alphacel	18 g	.50
Salt mixture W	36 g	1.00
Sucrose	126 g	3.52
Wheat germ (toasted)	108 g	3.02
Choline chloride	3.6 g	.10
Vitamin solution	36 g	1.00
Ascorbic acid	14.4 g	.40
Formalin (37% formaldehyde)	1.8 g	.05
Methyl paraben	5.4 g	.15
Aureomycin powder (5.5% available chlortetracycline hydrochloride)	20 g	.56
Raw linseed oil	18 ml	.50
Total	3584	100.00%

b. Preparation of fungus spray solution

Scorbic acid	1.5 g
Methyl paraben	0.6 g
95% ethyl alcohol	100 ml

c. Preparation of 4 N KOH solution
Potassium hydroxide pellets 224 g in 1 litre distilled water (always add pellets to water)

d. Preparation of B-vitamin solution
In 100 ml distilled water add 100 mg niacin, 100 mg calcium pantothenate, 500 mg riboflavin, 25 mg thiamine hydrochloride, 25 mg pyroxdine hydrochloride, 25 mg folic acid, 2.0 mg biotin, and 0.2 mg vitamin in B-12

3.2 Names and sources of ingredients

Agar, alphacel, choline chloride, methyl paraben, sorbic acid, (Bio-Serv Inc., P.O. Box B.S., Frenchtown, New Jersey, USA, 08884); casein, (National Casein (Canada) Ltd., 450-6 Tapscott Rd., Scarborough, Ontario, Canada, M1B 1Y4); Wesson salt mixture, B-vitamins, (ICN Pharmaceuticals 26201 Miles Road, Cleveland, Ohio, 44128, U.S.A.); ascorbic acid, (Hoffman-La Roche Inc., 1000 Roche Blvd., Vaudreuil, Quebec, J7Z 6B3, Canada); aureomycin product no. 8742-65, (Cyanamid of Canada Ltd., Animal Products Division, 2031 Kennedy

Road, Scarborough, Ontario M1P 2N4, Canada); wheat germ, (fresh stock from local health food store); white sugar, raw linseed oil, (local stores); potassium hydroxide pellets, formalin, (Canlab, 80 Jutland Road, Toronto, Ontario, Canada M8Z 2H4).

3.3 Diet preparation procedure

a. Finished diet

- Pour 2201 ml cold water into 4000 ml flask and add the agar, mix well.
- Place agar mixture in autoclave and steam to liquify agar.
- To the blendor add 792 ml water and the KOH solution.
- Measure other ingredients, add to blendor in the order given in 3.1, blend for 20 or 30 seconds after each addition.
- Remove liquified agar (93-94°C) from autoclave using asbestos gloves. pour agar mixture slowly into running blendor, increase blendor speed and mix for an additional 2 or 3 minutes.
- Pour hot diet into stainless steel pouring beaker.
- Manually dispense about 10 ml diet into each plastic cup. Cups are positioned on cafeteria trays before diet is prepared. Each tray will hold about 90 cups.
- Pour diet into shallow metal or plastic trays (Pieces of this diet will be used to feed late larvae in plastic boxes).
- Move trays of cups to fume hood and spray with the anti fungal solution. Alcohol evaporates rapidly leaving a thin film of anti-fungal agents on all exposed surfaces of the cups.
- Place trays of diet into plastic bags and refrigerate at 2°C.

PRECAUTIONS

(i) If several batches of diet are made daily a particle mask should be used to weigh and mix dry ingredients.

(ii) A fume hood must be used when spraying diet cups with the fungal solution.

4. HOLDING INSECTS AT LOWER TEMPERATURES

4.1 Adults

Moths can be stored in capped vials in the refrigerator (10°C) for 2 or 3 days.

PRECAUTIONS

Best mating success will occur if female moths are mated as soon after eclosion as possible.

4.2 Eggs

When eggs have been removed from storage and a delay of hatch becomes necessary, eggs may be held in the refrigerator (10°C) for a week or so before being returned to normal conditions. Hatching can be delayed one or two days by holding at 18°C.

4.3 Larvae

Pupae may be held in the refrigerator (10°C) for 2 or 3 days. Holding at 18°C will delay eclosion for 2 or 3 days.

5. REARING AND COLONY MAINTENANCE

Rearing is started with eggs collected from host trees in the fall of the year. Every effort should be made to collect from areas of new infestation where disease and parasitism are at a low level. Do not store egg bands when they are wet as fungus may develop on them during storage. Field collected eggs are used to start rearing during the months of January to July. After this period, eggs are over a year old and hatch is unreliable. If year-round rearings are required eggs must be produced in the laboratory to supply rearing material during the months of August to December, when field collected eggs are unsuitable. These eggs should be produced during the months of March to May.

5.1 Eggs

- Remove eggs from cold storage (Fig. 1). Fall collected eggs taken out in January (3-4 months) will hatch in 5 or 6 days. Optimum time for storage is 6-9 months; eggs so treated hatch in 2 or 3 days. Eggs may be stored as long as 12 months but as many as 40% of the bands will not hatch . For those bands that do hatch incubation time is increased to 6 or more days, the hatch rate is poor and eclosion occurs over several days.

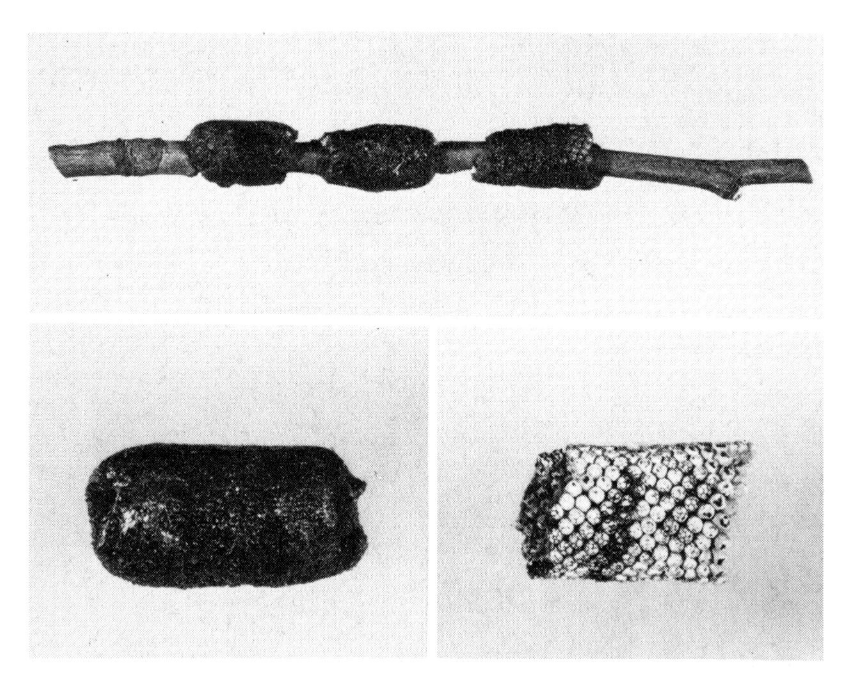

Fig. 1 Field collected egg band on an aspen twig.
Fig. 2 Egg band removed from twig.
Fig. 3 Egg band after sodium hypochlorite treatment.

5.2 Egg sterilization

- Remove egg ban intact from twig (Fig. 2). This is not always possible; if not, cut close to band
- Pour about 100 ml household bleach into a jar fitted with a screened lid.
- Place egg bands in jar, replace lid and agitate to ensure bands are fully immersed for 1½ minutes or until the spumilin covering has been removed (Fig. 3)
- Rinse for 3 or 4 minutes under the cold water tap to flush away the bleach.
- Place eggs on paper towelling in a covered petri dish for incubation and eclosion.
- Cut egg bands into two or more pieces if smaller numbers per cup of diet are required.

5.3 Diet innoculation with neonate larvae

Two rearing methods are used

a. General laboratory rearing (for colony maintenance and to supply early instar larvae for experimentation)

- Place a hatching egg band or one near hatch into a cup of diet.
- Cap with a paper lid.
- Invert cup so that the egg band falls onto the lid and away from the surface of the diet. Newly eclosed larvae leave the egg band, climb the wall of the cup and readily establish on the artificial food.
- After about 12 days remove third instar larvae from cups.
- Transfer about 10 larvae to each cup of fresh diet and again place in an inverted position. Cups need not be checked again until larvae reach the late 5th or early 6th stage.
- Transfer last instar larvae to new food at the rate of one per cup.
- Place cups in an upright position (larvae usually spin their cocoon on the underside of the lid).

b. Group larval rearing (late larvae used for pathogen production or for experimentation requiring larvae of a specific size)

- Follow first five steps as in 5.3a.
- Line bottom of a large plastic box with paper towelling.
- Place pieces of diet on small piece of aluminium foil on the bottom of the box.
- Transfer about 300 late larvae from the cups to the plastic box (Fig. 4)
- Feed fresh diet daily. Do not give more food than can be consumed in 24 hours.

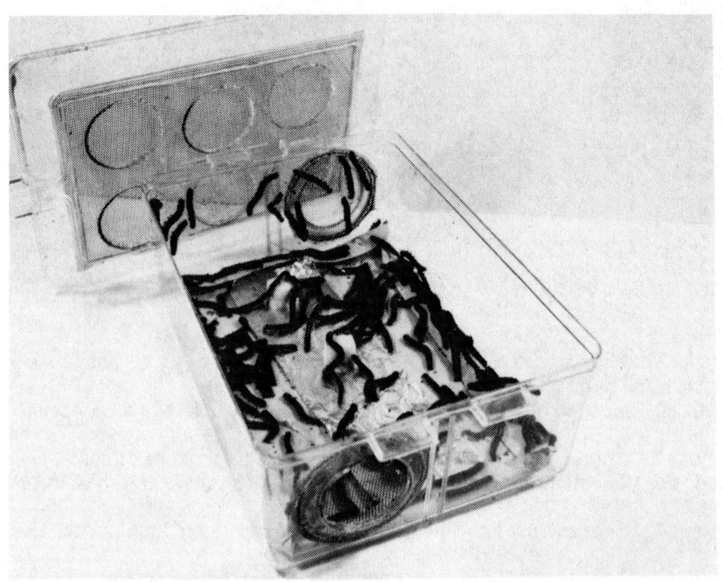

Fig. 4 Vented plastic box for rearing late instar larvae.

PRECAUTIONS

(i) During the first three instars larvae appear to thrive and develop better under crowded conditions; however, fungus may develop if crowding is excessive and if high humidity cannot be controlled.

(ii) The paper lid on the rearing cup allows for some transfer of air and moisture from the cup. If inverted cups are placed on a smooth surface this transfer of gases is eliminated and excessive moisture and mould will develop in the cup.

(iii) Plastic boxes must be very well vented on the sides and lid to prevent moisture buildup and fungus growth on the bottom. If frass and uneaten diet haven't dried after 24 hours, the boxes should be vented more or the environmental humidity lowered.

5.4 Pupal collection

- Remove lids with attached cocoons from cups.
- Remove pupae from cocoons manually if only a small number of pupae present.
- Separate cocoons from lids using sodium hypochlorite when large numbers of cocoons (300+) are present.
- Place cocoons on a piece of cheesecloth, draw ends together and tie to form a loose bag.
- Place an equal amount of household bleach and cool tap water in a 3.5 liter beaker.
- Immerse the cheesecloth bag in the solution for 30 seconds or until cocoon silk is dissolved and free pupae are observed in the bag.
- Place immediately into dish pan under running tap water and rinse thoroughly.
- Transfer the bags of pupae to another dish pan half filled with water and open the bags.
- Collect and discard cheesecloth bags and lids leaving only pupae, cast skins and excrement.
- Place cylindrical wire basket in a dish pan with running tap water
- Pour contents of the dish pan (pupae etc.) into wire basket. Gently shake contents under the tap water to remove as much debris as possible. Manually remove debris floating at the surface of the basket.
- Place pupae on paper towelling to dry.
- Separate pupae by sex (See Section 5.6).
- Transfer dry pupae to well-vented plastic boxes with floor and sides lined with towelling to facilitate adult eclosion.

PRECAUTIONS

(i) Several workers have become allergic to the yellow powder found in the inner layer of the cocoon and if pupae are removed manually a hood or particle mask should be used.

(ii) Use rubber gloves when handling sodium hypochlorite.

(iii) The sodium hypochlorite solution looses its strength after treating 2 or 3 thousand cocoons.

5.5 Adults

The sex ratio is approximately 1:1.

5.5.1 Emergence period

Adult eclosion occurs 12-14 days after pupation or about 45 days from egg hatch.

5.5.2 Collection

Adult moths, particularly males, are active flyers and boxes of newly eclosed adults should be placed in the refrigerator for a few minutes to facilitate handling. Moths are collected individually using a glass vial for transfer to mating and oviposition cages.

5.5.3 Oviposition

This is done in a controlled environmental room as described in 2.1. Females will mate soon after eclosion, often before the wings are dry. Eggs are usually laid within one day of mating. Males will mate more than once.

a. Cardboard containers (Fig. 5)

- Place a short section of a branch tip in a pint container. Wedge the twig off the floor to allow female moths room to encircle the twig when depositing eggs.
- Place 2 pairs of adults in the mating chamber and cover with a petri dish.

b. Screened cage (Fig. 5)

- Line cage floor with paper towelling.
- Place portion of branches with leaves removed and with numerous lateral twigs and shoots on the bottom of the cage.
- Position branches off the floor in a manner that will present as many oviposition sites as possible.
- Transfer up to 100 pairs of adults to the cage or place a similar number of sexed pupae on the floor of the cage to await eclosion.

Fig. 5 Screened cage for mating large numbers of adults and cylindrical cardboard container for mating one or two pairs of adults.

5.6 Sex determination

- Pupae may be sexed as described by Muggle (1974).
- Female pupae and adults are usually larger than males.
- Male moths have a slender body, the antennae are plumose and they are more active than female moths.

5.7 Rearing schedule

a. Daily

- Collect newly emerged moths.
- Set up new mating containers.
- Feed large larvae in plastic boxes.

- Clean plastic boxes if necessary.
- Select larvae of appropriate age and size and supply for experimentation.

b. Weekly (establish a routine by assigning a specific day for each task)
- Remove eggs from cold storage and sterilize.
- Remove eggs from mating cages.
- Store laboratory produced eggs.
- Place hatching egg bands in rearing cups.
- Remove third instar larvae from crowded cups, reduce numbers and place in new cups of food.
- Transfer last instar larvae from cups to plastic boxes.
- Harvest cocoons.
- Separate pupae from cocoons using bleach.

5.8 Insect quality

We have not placed emphasis on production in-line testing for insect quality except for regular sampling for the presence of the protozoan parasite *Nosema disstriae* (Thom.). Areas where field eggs are collected are carefully chosen.

5.9 Special problems

a. Diseases

Nuclear polyhedrosis virus is very common in wild forest tent caterpillar populations and is transferred from contaminated eggs to laboratory rearings. All eggs must be sterilized.

Nosema disstriae (Thom.) causes high mortality in larval rearings and also reduces the quality of surviving insects. To avoid this problem we collect a small egg sample from proposed egg collecting areas and diagnose eggs for the presence or absence of microsporidian spores. First instar larvae are removed from eggs, crushed, and examined microscopically. Eggs are not collected from areas where the parasite is prevalent.

b. Fungal control of diet

Fungal development in synthetic diet can create very serious problems unless careful sanitation practices, rearing techniques and recommended environmental rearing conditions are closely followed. Periodically, in spite of all precautions, fungus infection of diet in cups will occur. When this happens discard the cup without removing the lid. If insects are particularly valuable then cups must be opened in the biological containment hood.

c. Human health

(i) Exposure to wing scales can be hazardous. Some workers have also shown an allergic reaction to the yellow powder found in the inner layer of the cocoon. A protective mask should be worn when working with small numbers of insects. Large numbers of adults or cocoons should be processed in a biological containment hood.

(ii) If an allergic reaction is suspected, consult a doctor immediately.

6. LIFE CYCLE

Life cycle, survival data, and development of various stages under rearing conditions of 24 ± 1° C, 50-60% and an LD 18:6 photoperiod are given.

6.1 Development data

Stage	Number of Days	
	To reach stage	Spent in stage
1st instar	-	3
2nd instar	4	5
3rd instar	9	5
4th instar	14	5
5th instar	19	6
6th instar	25	10
pupal	35	10

6.2 Survival data (%)

Egg viability 82
Neonate larva to pupa - 67
(most mortality occurs during the third and early fourth instar)

6.3 Other data

An average of 168 eggs per band and an average hatch of 82% was recorded at 18 locations in Ontario and Manitoba in 1952 and 1953 (Blais et al. 1955).
Female moth requires about 3 hours to lay egg mass
No. of larval instars, 6
Mean pupal weight (mg) ♂ 286 ♀ 490

7. PROCEDURE FOR SUPPLYING INSECTS

7.1 Pre-requisite

- Rearing colony should be of a known quality.
- The duration of developmental stages at specific temperatures must be known.
- The best food should be used and a standard rearing equipment should be available and daily and weekly routines organized.

7.2 Placing an insect order

About two months notice is the maximum time required to supply any developmental stages of the forest tent caterpillar once egg diapause requirements have been satisfied.

7.3 Planning the order(s)

a. Present orders demand a supply of 400 second instar larvae every second week for protozoology studies, 500 third instar and 500 sixth instar weekly for mycology studies and 300 - 500 pupae weekly for sex pheromone studies. Larval requests are for the months of February to June and pupae are required throughout the year.

b. Calculation of insect numbers
To meet these demands, eggs are removed from cold storage each Monday so that all stages are available weekly about two months after the first eggs are incubated. For planning purposes the number of eggs per band are underestimated at 100. Though there is a maximum demand for 1900 insects weekly we initiate weekly rearings with 30 egg bands. Excess egg bands are a cushion against unexpected rearing losses. Extra insects can be discarded when about the fourth larval instar and before food and handling demands increase.

7.4 Conclusion

A central rearing unit is in operation here at the Institute to serve the needs of approximately 40 research scientists and support staff. As many as 20 insect species are reared routinely in numbers that may exceed 5 million annually. Because of the volume of work, cultures are usually initiated weekly and experimental insects for research projects are also supplied weekly. If specific stages are required more often, we manipulate temperature to increase or delay development.

8. REFERENCES

Addy, N.D. 1969. Rearing the forest tent caterpillar on an artificial diet. J. Econ. Ent. 62: 270-271.

Blais, J.R., Prentice, R.M., Sippell W.L., Wallace D.R. 1955. Effects of weather on the forest tent caterpillar *Malacosama disstria* Hon., in central Canada in the spring of 1943. Can. Ent. 87: 1-8.

Grisdale, D.G. 1963. Rearing insects on artificial diet. Interim Research Report of the Insect Pathology Research Institute, Sault Ste. Marie, Ontario, Canada.

Grisdale, D.G. 1976. Laboratory methods for rearing the forest tent caterpillar. Can. Forestry Serv. Bi-monthly Res. Notes. 32: 1.

Muggle, J.M. 1974. Sex identification of *Malacosoma disstria* pupae (Lepidoptera: Lasiocampidae) Ann. Entomol. Soc. Am. 67: 521-522.

Sippell, W.L., Evan, H.E. 1967. Important forest insects and diseases of mutual concern to canada, the United States and Mexico. pp. 127-129 Queen's printers, Ottawa, Canada.

MAMESTRA BRASSICAE

BRIAN O. C. GARDINER

ARC Unit of Insect Neurophysiology and Pharmacology, Department of Zoology, University of Cambridge, Downing Street, Cambridge, CB2 3EJ, England

THE INSECT

Scientific Name: *Mamestra brassicae* Linnaeus
Common Name: Cabbage moth
Order: Lepidoptera
Family: Noctuidae

1. INTRODUCTION

The cabbage moth (*Mamestra brassicae* L.) is a medium-sized noctuid moth of widespread Palearctic distribution. It has frequently been reported in the literature as a minor pest, particularly of cabbage crops. The larvae feed on many plants, but have a marked preference for *Cruciferae* (Stokoe, 1948). There are normally 1-3 broods in a year. Diapause in an established culture is facultative and light controlled. It is therefore possible to rear it continuously for about 6 generations per year.

Several workers have experienced difficulty in the past in preventing diapause (see Dusaussoy 1964 for summary). This should be borne in mind when starting with a fresh wild-collected stock. Both the quality of the larval food and the strain have been implicated and it seems likely that there is a genetic factor involved for univoltism, particularly with northern strains.

M. brassicae is less prone to disease than are armyworms. It is large enough to be useful as an experimental insect, but is not so large as to require an extensive amount of space and expensive food. It may be maintained either on one of its natural foods, a *Brassica* cultivar, or an artificial diet.

The use of *M. brassicae* for the study of insecticides was first investigated by Way *et al.* (1951). They used *Brassica* cultivars as the food and had trouble breaking the diapause. At 24°C they found the life cycle, egg to adult, to be 53 days. This work was followed in Japan by a number of authors who were principally interested in the diapause requirements. Otuka and Santa (1955) showed that all nondiapause pupae were produced under rearing conditions of 20°C and 18 h light per day for the larvae. Also in Japan, the larvae, reared under these conditions, were extensively used for insecticide testing by Ishikura and Ozaki (1958) who showed that a density of 10 larvae per petri dish produced heavier larvae than higher densities or more elevated temperatures. In France, Dusaussoy (1964) used similar techniques as Ishikura and Ozaki who also maintained a colony for testing against insecticides and pathogens. Finally, David and Gardiner (1966) were successful in breeding and maintaining a colony for several years on artificial diet and it is basically this technique, with later modifications, that is described.

These instructions allow the maintenance of a culture of *M. brassicae* capable of producing a weekly surplus of 15,000-20,000 eggs.

2. FACILITIES AND EQUIPMENT REQUIRED

2.1 Facilities

- A constant temperature room maintained at 20°C and 60% RH, LD 18:16 (tungsten or fluorescent) for the larvae; the adults may either be kept in another room, or in a subdivision (a large cupboard arrangement for

instance) of the same constant temperature room. They must be given 12-h light using a 60-100W bulb at a distance of 3 to 6 ft and a 12-h night using a similarly distanced 7.5W bulb.
- Usual facilities of a laboratory (small glassware items, balances, workspace, washing-up facilities etc.) are needed

2.2 Equipment
- Diet preparation, rearing, cleaning and storage should be carried out in separate rooms
- Six adult cages are needed, each of 1 cubic ft capacity (30.5 cm^3); a wooden frame covered with strong synthetic fabric is ideal
- The floor should be of formica and one side should consist of a removable perspex (leucite) sheet to enable the inside of the cage to be readily cleaned
- The perspex sheet should be fitted with a conventional sleeve through which adult moths can be easily introduced and removed.
- Plastic vials (60 and 150 ml)
- Petri dishes 90 mm diam
- Plastic boxes 280 x 150 x 85 mm (with a 30-mm fabric-covered ventilation hole in the lid)
- Plastic containers (75 mm diam)
- Sterile and insecticide free sawdust
- Saucepan (1-10 litre capacity depending on size of culture;
- Kenwood® (1 litre), Waring® (5 litre), or Hobart® (20 litre) blender, depending on size of culture
- Whatman® No. 3 chromatography paper (570 x 470 mm)
- Red or blue cotton wool
- Sucrose (as a 10% solution)
- Refrigerator
- Deepfreeze set at -25°C

3. DIET AND ITS PREPARATION

3.1 Ingredients
- Diet (1.1 litre)

Group	Ingredients	Weight/volume	%
A	Agar, powdered	27.0 g	2.4
	Water	600 ml	53.0
B	Casein, light white	37.8 g	3.3
	Wheat germ (Bemax®)	32.4 g	2.9
	Wessons® salts	10.8 g	1.0
	Sucrose (commercial sugar)	37.8 g	3.3
	Cellulose (Whatman CF 11)	5.4 g	0.5
	Dried cabbage powder	16.4 g	1.5
	Water	330 ml	29.0
C	Choline chloride (10% in water)	10.8 ml	1.0
	Methyl paraben (15% in 95% ethyl alcohol)	10.8 ml	1.0
	Linseed oil	3.0 ml	0.26
	Formalin (10% solution)	4.5 ml	0.4
	Vitamin mix (see below)	1.8 ml	0.16
D	Ascorbic acid	4.5 g	0.4

- Vitamin supplement is as follows:

Nicotinic acid	5.0 g
Calcium pantothenate	5.0 g
Riboflavine	2.5 g
Aneurine HCl	1.25 g
Pyridoxine HCl	1.25 g
Folic acid	1.25 g
D-biotin	0.1 g
Cyanocobalamine	0.01 g

Weigh 2 grams of the above vitamins and put into distilled water (50 ml) and ethyl alcohol 98% (50 ml). Shake well before use since some vitamins do not dissolve.

3.2 Sources of supply

All the ingredients required for the diet may be purchased from chemical drug suppliers, British Drug Houses Ltd. in England, ICN Nutritional Biochemicals in the United States, or from the various suppliers listed in Frass newsletter vol. 6, no.1.

The dried cabbage leaf powder is prepared by taking the outer green leaves of various Brassica cultivars, preferably in late summer or in early autumn, drying them to crispness in an oven at 105°C and then reducing them to a fine powder in a conventional powder mill. This powder has been stored 14 years without deterioration.

The aureomycin used is a veterinary grade powder [25 g aureomycin per 454 g (lb)] distributed by the Cyanamid Company. The actual concentration is 120 mg pure aureomycin per 1 litre diet.

3.3 Preparation

- Prepare fresh diet each week; weigh 10 to 12 aliquots of the diet ingredients at one time and store in vials ready for future use
- To save time make up occasional bulk supplies to be kept at -25°C
- Combine all solid ingredients (group B) after weighing; sufficient liquid ingredients should be prepared to make a similar number of batches of diet and stored at 4°C
- To prepare a batch of diet the agar is added to cold water in a saucepan; bring to a boil and simmer for 10 to 15 min; remove from the heat and cool to 70°C and stir occasionally
- Blend the solid ingredients into their aliquot of water
- Add liquid ingredients (group C) one by one
- Add the 70°C agar
- Add the ascorbic acid and aureomycin
- Dispense into vials to a depth of 20-30 mm

For quantities of 2 litres or less, pour the diet out of the blender back into the agar saucepan which is standing in the outer saucepan (at about 70°C). The adept can pour straight from this, stirring from time to time to keep the mixture from premature surface cooling. For larger quantities, a 20-litre capacity, hot water jacketed kettle that can be maintained at 50°C is required. As each vial is filled it should be tipped to an angle of 45°C and completely rotated. A quantity of diet should also be dispensed into the 90-mm petri dishes. Vials to be stored in the deep freeze should first be allowed to cool and set. Small vials and dishes can be put into plastic bags in convenient numbers but large vials are best individually capped before storage.

PRECAUTION

The equipment used and spillages should be cleaned at once since dried diet could lead to infection from bacterial or fungal spores.

3.4 Storage

- The diet has a storage life of at least 2 years at -25°C
- Storage life is only 2-3 weeks at 4°C

4. ADULTS AND OVIPOSITION

Weekly

- Place 100 pupae of approximately the same age in a plastic box, the sides of which are lined with Whatman No. 3 chromatography paper
- Place box in 1 of the adult cages
- Place a petri dish containing red or blue cotton wool soaked in 10% sucrose solution into cage
- Discard remaining moths or pupae in the adult cage after 7 days; wash cage in hot water containing any proprietary brand soap powder together with a disinfectant and a little washing soda

Daily
- Inspect and, if necessary, replenish sugar solution
- Remove mating pairs to individual plastic vials (150 ml), lined with paper
- Remove dead adults

5. MAINTENANCE OF LARVAE

Daily
- Use fresh diet or, if frozen, remove diet from freezer 24 h earlier

Weekly
- Put 30-40 newly hatched larvae into each of 20 small vials containing diet
- Cap with Whatman filter paper (or finely woven cloth) and lay on their sides
- Transfer 15 to 20 3rd instar to larger vials
- Discard (or use for supply) surplus larvae
- Place one of the diet filled petri dishes on top of plastic boxes half filled with moist sawdust
- When full grown, the larvae start to burrow into the diet in the vials; transfer 75-100 larvae from the vials to the boxes for pupation
- Discard any laggard larvae
- Harvest pupae after one week by tipping out onto large sheet of paper; lay the pupae on top of layer of sawdust in another box lined with the Whatman no. 3 paper
- Wash and clean boxes

6. LOGISTICS

It is estimated that 10 to 12 h per week should ensure the maintenance of a _Mamestra_ culture which can produce a weekly surplus of 15,000-20,000 eggs. At the time of writing, 1982, the cost of the ingredients to make up a litre of diet is $3.20 based on moderate quantities bought from British Drug Houses Ltd. To maintain the culture as described requires 1.5 litres of diet per week. At current English prices the cost of the vials is $500 per year.

7. LIFE CYCLE DATA

7.1 _Developmental_
- At 20°C the mean time, in days, of the various stages is as follows
  - (a) Egg (10)
  - (b) Larval instars (1st, 5; 2nd, 4; 3rd, 4; 4th, 4; 5th, 11-13)
  - (c) Pupa (25-30)
  - (d) Adult (7)

7.2 _Survival_
- Egg viability, 90-100%
- Larval viability, 80-90%
- Pupal viability, 90%
- Adult longevity, 5-10 days

7.3 _Other_
- Preoviposition period, 0-3 days
- Oviposition after mating, 1 day
- Maximum oviposition after 3-4 days
- Fecundity, 400-900 eggs
- Pupal weight, 350-420 mg

8. SUPPLYING

- The culture produces double the number of insects required for maintenance and a very large surplus of eggs; therefore, requests for small numbers can be met the same or following day

- Orders for large numbers are met by setting up more larvae as they hatch from the eggs and from the data given in 7.1.
- Diapause stages can be reared under a light regimen of LD 9:15

9. REFERENCES

David, W.A.L. and Gardiner, B.O.C., 1966. Rearing Mamestra brassicae (L.) on semisynthetic diets. Bull. ent. Res., 57: 137-142.

Dusaussoy, G., 1964. L'elevage permanent de Mamestra (Barathra) brassicae L. comme insecte test de preparations pathogenes ou insecticides. Ann. E'piph., 15: 171-192.

Ishikura, H. and Ozaki, K., 1958. Continuous mass rearing of the cabbage armyworm Barathra brassicae Linnaeus and the studies on the resistance of the reared armyworm to certain insecticides. Bull. Nat. Inst. Agric. Sci., Tokyo (C), 10: 1-42.

Otuka, M. and Santa, H., 1955. Studies on the diapause in the cabbage armyworm, Barathra brassicae L. III. The effect of the rhythm of light and darkness on the induction of diapause. Bull. Nat. Inst. Agric. Sci., Tokyo (C), 5: 49-56.

Stokoe, W. J., 1948. The caterpillars of the British moths. Vol. 2. F. Warne and Co., London. 408 p.

Way, M. J., Smith, P. M. and Hopkins, B., 1951. The selection and rearing of leafeating insects for use as test subjects in the study of insecticides. Bull. ent. Res., 42: 331-354.

MANDUCA SEXTA

A. H. BAUMHOVER

USDA-ARS , Tobacco Research Lab., Route 2, Box 16 G, Oxford, NC 27565

THE INSECT

Scientific Name: *Manduca sexta* (L.)
Common Name: Tobacco hornworm
Order: Lepidoptera
Family: Sphingidae

1. INTRODUCTION

The tobacco hornworm (TBH) is indigenous to North and South America and various Caribbean Islands. It is a major pest of tobacco and occasionally tomatoes. Wild hosts such as jimson weed and horse nettle are restricted to the plant family *Solanaceae* (Madden and Chamberlin, 1945).

Five to ten 5th instar larvae can consume the leafy areas of a mature tobacco plant in 3 to 4 days leaving only the stalk and leaf ribs. Until the 1950's the TBH was considered the "king" of tobacco insect pests. In recent years, however, the tobacco budworm, *Heliothis virescens* (Fabricius) and the green peach aphid, *Myzus persicae* (Sulzer) frequently have been the major insect pests.

In the mid-Atlantic states there are 3 generations per year. Larvae developing after mid-August enter diapause as pupae and comprise the first brood the following year and about 1/3 of the second brood. In tropical climates reproduction may be continuous (up to 9 generations per year), except for aestivating populations during periods of drought. In Florida emergence of the moths begins about May 1st while in the more northern states it is delayed several weeks. Eggs are laid singly, usually on the underside of the leaf within an inch of the leaf's edge. Most eggs are deposited on the upper third of the plant. Six or more eggs per plant would cause an outbreak unless predators are active (unpublished reports, USDA Tobacco Research Laboratory, Oxford, NC).

Initially the TBH was reared on field grown tobacco (Hoffman and Lawson, 1964; Allen and Kinard, 1969). However, this limited production to the tobacco growing season and presented problems with disease, predators and parasites. Yamamoto (1968, 1969) was the first to rear the TBH on an artificial diet en masse, while Bell and Joachim (1976) reared the TBH individually on a similar diet. Yamamoto's diet was used initially but has been modified to improve quality and reduce cost (Baumhover et al, 1977). Our colony has been reared for 17 years or approximately 170 generations. Up to 30,000 TBH have been reared weekly for sterile male releases (Snow et al, 1976). More recent efforts have been devoted to egg production (50 million annually) for rearing the spined stilt bug (*Jalysus wickhami* (Van Duzee) = *spinosus*), an efficient predator for budworm and hornworm eggs (Lam et al, in press). We also supply TBH for sex attractant studies and various other physiological studies in other laboratories. A laboratory colony can be established readily from our culture, by trapping adults in blacklight traps or by collecting larvae from tobacco fields. The latter technique may be hazardous, particularly in late season, since 90% or more may be parasitized by *Apanteles congregatus*, a Braconid wasp. It is also available commercially from Carolina Biological Supply, 3700 York Road, Burlington, NC 27215.

2. FACILITIES AND EQUIPMENT NEEDED

2.1 Facilities

Starting containers are held at 28 ± 1°C and 50-60% RH. Lighting is continuous and must be uniform to prevent the newly hatched larvae from orienting to the brightest light source and becoming entangled with each other, preventing migration to the diet at the top of the container. Third instar larvae are transferred to large containers held in racks under conditions similar to the starting room except for the uniform lighting requirements. The copious amounts of fecal pellets are dried by fans that move the air at 30m/min beneath the trays when the larvae have reached the 5th instar. Unless the pellets are dried the wandering prepupae not only become mired in them but various contaminants may develop. The pupation room is held at 28 ± 1°C and 80-85% RH. Lights are off except as needed. The oviposition room is also held at 28 ± 1°C and 80-85% RH. Daylight, 6:00 am to 6:00 pm, is simulated with 25W flourescent or incandescent lamp and night time (active period) is provided with 7 1/2 W bulbs in an aluminum pie pan reflector and dimmed to reflect 0.16 cd/m^2 from the feeders and artificial leaves. If ambient light outside the cage is brighter than inside, the moths will orient to it and refuse to feed or oviposit. All rooms have individual temperature and humidity controls to limit spread of contaminants from one area to another.

2.2 Equipment and materials required for insect handling

2.2.1 Adult oviposition

- Walk-in or reach-in cages made of burlap, muslin or fine household type screen. A 137 x 121 x 125 cm high cage will accomodate 50 pairs of moths.
- A feeding station containing six 30 ml vials filled with 40% sucrose solution topped with polyethylene thistle tubes, top diameter of 4 cm and tubes extending to the bottom of the vials (Fig. 1).
- Four artificial leaves made from a 0.3 cm thick and 15 cm diameter piece of indoor-outdoor carpeting sandwiched between 0.2 cm polypropylene. The upper sheet is 12.5 cm in diameter and the bottom 15 cm so that a 1.25 cm wide band of the carpet is exposed at the top outer edge. (similar to Sparks, 1970) (Fig. 1).
- Fresh or fresh-frozen tobacco leaves blended with tap water - 100 g of tobacco: 1 liter water.
- If crowding is a factor use a ratio of 1 male:3 females.

2.2.2. Egg collection and storage

- A 20 x 30 x 10 cm plastic pan to receive eggs.
- Screened bottom dish 2.5 cm deep and 150 cm in diameter.
- Quart jar or other containers for frozen storage.

2.2.3 Egg sterilization

- A screened bottom dish 2.5 cm deep and 150 cm in diamter (same as for non-frozen storage).
- 5.25% sodium hypochlorite diluted 1:200 in tap water (0.026% AI).
- Clean air bench.

2.2.4 Egg hatch and larval inoculation of diet

- Starting containers-transparent plastic shoe boxes with bottom cut leaving lip and replaced with a 70 micron polyethylene filter 0.15 cm thick glued to lip at bottom
- A 1 cm mesh 60 CDS 39 Vexar plastic screen 26 x 30 cm folded in the form of an inverted U to support diet in the lid and to provide a perch for the larvae. (Fig. 2).
- Balance for weighing eggs (1 g = 750), or a small scoop calibrated to hold 400 eggs.

Fig. 1. Interior of colony cage, showing small 7 1/2-W night light in pie pan reflector, large 60-W daytime bulb, thistle tube feeders, artificial leaves, and moths.

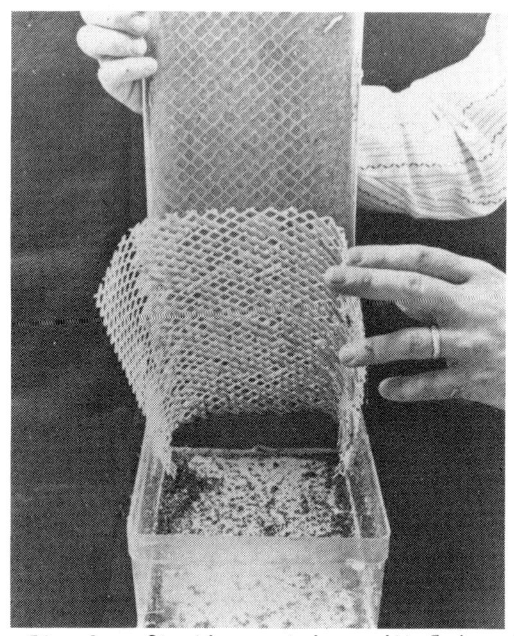

Fig. 2. Starting container with 5-day-old larvae.

2.2.5 Larval rearing containers

- Starting containers (see 2.2.4)
- Finishing containers, 0.6 cm thick plexiglas 62 x 44 x 11 cm with a 1.6 cm mesh hardware cloth cut to fit bottom and extend up 2 sides plus 2 V separate shaped extensions to support diet and provide a perch for the larvae. A 0.3 cm mesh 20 PDS 129 Vexar screen is positioned over the container and held in place with a frame made from a 5 cm wide T bar with a 1.6 cm projection (Fig. 3). A flap in the screen is opened to allow prepupae to drop into the collection tray below (Fig. 4).

Fig. 3. Finishing tray complete with diet support screen, 5-day-old larvae on starting screen and fine-mesh Vexar screen and T-bar (inverted).

Fig. 4. Flap opened to allow prepupae to drop into collection tray (one rearing tray removed to expose flap).

- Collection tray 62 x 44 x 11 cm (Fig. 4). Also see Fig. 5.
- Fans to provide air movement of 30 m/min under containers when larvae reach 5th instar.

Fig. 5. Finishing tray raised to show large 11-day-old fifth-instar larvae feeding underneath.

2.2.6 Pupation units, collection and storage
- Pupation cells 2.5 x 2.5 x 10 cm can be made either from 0.6 cm thick plyboard or polypropylene filter of the same thickness. A 0.6 cm mesh hardware cloth is stapled to one side of the unit and the other side consists of a 0.6 cm thick sheet of marine plywood or plexiglass held in place with large rubber bands cut from a truck inner tube (Fig. 6).
- Screened bottom storage trays 4.5 x 6.5 x 0.5 cm.

2.2.7. Insect holding
- The insects are held in the pupation cells for 7 days prior to removal to insure that they are well hardened and not subject to damage from premature handling. They are placed in single layers in screened bottom trays 4.5 x 6.5 x 0.5 cm and stored in the oviposition room, propped up on 0.5 cm wooden blocks to avoid saturation of the puparium with meconium. Too damp puparia will stretch rather than split when the moth attempts to emerge.

2.3 Diet preparation equipment
- 160 liter soup kettle with 5000 W heating element
- 1/3 horse power direct drive clamp on mixer with 2-9 cm diameter propellers
- Thermometer 0 - 100°C
- Balance 0 - 20 kg

Fig. 6. Prepupae in unit containing 160 cells.

3. ARTIFICIAL DIET

3.1 Composition - g/liter water except as noted

Nutrients	
CSM[1/]	120
Casein (sodium caseinate)	16
Torula yeast	64
Cholesterol	1
Vitamin pre-mix[2/]	10
Microbiocides	
Sorbic acid	2
Methyl-p-hydroxybenzoate	2
Formalin (40% formaldehyde)	2.5 (ml)
Streptomycin sulfate	.5
Propionic acid	1.0 (ml)
Gelling Agent	
Carrageenan HWG (gelcarin)	13

1/ 63.8% yellow cornmeal gelatinized; 24.2% soyflour, toasted; 5.0% nonfat dried milk; 5.0% soyoil, refined; 1.9% mineral premix; and 0.1% vitamin premix. If CSM is not available the following amounts (g) per liter of water may be used: yellow cornmeal 76; toasted soyflour 42; NaCl 0.65 and $CaCo_3$ 0.60.

2/ GUARANTEED ANALYSES

GUARANTEED ANALYSES		PER KILOGRAM
Vitamin A, I.U.	Minimum	22,046,000
Vitamin E, I.E.	Minimum	8,016
Vitamin B12, mg	Minimum	2.0
Riboflavin, mg	Minimum	500
Niacin, mg	Minimum	1,014
d-Pantothenic Acid, mg	Minimum	922
(Calcium d-Pantothenate, mg	Minimum	1,003)
Choline, mg	Minimum	43,490
(Choline Chloride, mg	Minimum	50,104)
Folic Acid, mg	Minimum	251
Pyridoxine, mg	Minimum	205
(Pyridoxine Hydrochloride, mg	Minimum	251)
Thiamine, mg	Minimum	101
(Thiamine Hydrochloride, mg	Minimum	251)
d-Biotin, mg	Minimum	20.0
Inositol, mg	Minimum	20,042

INGREDIENTS: Dextrose, Vitamin A Acetate in Gelatin-Sugar-Starch Beadlet, dl-alpha-Tocopheryl Acetate, Vitamin B12 Supplement, Ribo-flavin Supplement, Niacin Supplement, Calcium Pantothenate, Choline Chloride, Folic Acid, Pyridoxine Hydrochloride, Thiamine Hydro-chloride, Ascorbic Acid (501,046 mg/kg), added; d-Biotin, Inositol.

3.2 Brand names and sources of ingredients

CSM (Krause Milling Co., 4222 W. Burnham St., Milwaukee, WIS 53246; Lauhoff Grain Co., Danville, ILL 61832; Archer Daniels Midland Co., 4666 Faries Parkway, Decatur, ILL 62521); Casein (Erie Casein Co., P. O. Box 648, Erie, ILL 61250); Torula yeast (St. Regis Paper Co., Lake States Div., 515 W. Davenport St., Rhinelander, WIS 54501); yellow cornmeal (local grocery or mill); soyflour (Cargill, Inc. 1010 Tenth Ave., Cedar Rapids, IA 52402); cholesterol (Bio-Serv., Inc., P. O. Box B.S., Frenchtown, NJ 08825); vitamin premix (Hoffman La Roche, Inc., 601 Blam St., N. Wood Industrial Park, Salisbury, MD 21801); sorbic acid (Pfizer, Inc., 4360 NE Expressway, Doraville, GA 30340); methyl-hydroxy benzoate (Tenneco Chemical, Inc., P. O. Box 367, Piscataway, NJ 08854); formalin (Fisher Scientific, 3315 Winton Road, Raleigh, NC 27604); propionic acid (U. S. Biochemical Corp., P. O. Box 22400, Cleveland, OH 44122); streptomycin sulfate (Pfizer, Inc., 4360 NE Expressway, Doraville, GA 30340); Carageenan (gelcarin) Marine Colloids, 2 Edison Place, Springfield, NJ 07018.

Note: If only small quantities of diet are needed and not available from the firms listed try Bio-Serv., Inc., P. O. Box BS, Frenchtown, NJ 08825.

3.3 Diet preparation procedure

- Heat water to 82°C and turn off heating element.
- Weigh out dry ingredients.
- Measure liquid microbiocides.
- Start mixer and slowly add all ingredients.
- Mix 5 minutes prior to withdrawing diet into rearing containers. Continue mixing until all diet is removed.
- Mixture will be cooled to 74°C when all ingredients are added. Higher temperatures will allow heavier ingredients to settle to the bottom before gelling occurs.
- Draw mixed diet from kettle into starting container lid (0.5 liter) and finishing container (8.0 liter).
- Gelling occurs at 57°C.
- Cover containers with wrapping paper and stack criss-crossed.
- Store at 12°C or lower but do <u>not</u> freeze. Can be stored 1 week or more prior to use.
- If starting container lids with diet are stored more than two days, place in sterile plastic bags to prevent drying. If diet shrinks from the edges of the lids, newly hatched larvae may escape.

- If only small quantities of diet are needed heat water on stove or hot plate and mix in household or commercial type blender.

PRECAUTIONS
(i) Wear non-toxic particle mask when mixing dry ingredients.
(ii) Avoid breathing formaldehyde as far as possible.
(iii) Exposure to streptomycin may cause sensitization should worker require this medication.

4. HOLDING INSECTS AT LOWER TEMPERATURES

4.1 Adults

Adults are not normally held at lower temperatures because they are bulkier and more easily damaged than in the pupal stage.

4.2 Eggs

These can be held up to 7 days in a refrigerator at 4.5°C but must be kept dry. If used for insect food they are sealed in airtight containers and held at -23°C until used. Freezing is necessary not only for preservation but to prevent hatch. Frozen eggs have been stored up to 2 years without loss of nutritional quality for stilt bugs.

4.3 Larvae

Larvae have been held at 24°C and a photoperiod of 12 h to induce diapause but lower temperatures have not been tested. The pupal stage is preferred for extended storage.

4.4 Pupae

Non-diapausing pupae can be stored up to 30 days at 13°C and 85% RH. Survival is best if begun in the early or late stages of development. Diapause can be induced at 24°C and a photoperiod of 12 h or less. Diapausing pupae can be stored at 13°C for 9 months or longer. In nature some diapausing pupae do not emerge until the second year after pupation.

5. REARING AND COLONY MAINTENANCE

The founder colony of *M. sexta* was established from moths collected in blacklight traps during 1965 and has been reared continuously for 170 generations (1965-1983).

5.1 Egg collection

- Prepare artificial leaves as described under 2.2.1.
- During late afternoon place leaves on holders and spray with a blended mixture of 100 g fresh or fresh frozen tobacco leaves: 1 liter tap water.
- Collect leaves during morning and using thumb, rub eggs off into containers. (Eggs are very durable and almost impossible to crush between finger and thumb.)

PRECAUTION

Leaves must be soaked weekly in a Clorox solution to remove residues from tobacco solution or egg production will drop.

5.2 Egg sterilization

- Use clean air bench.
- Transfer eggs to screen bottom container and immerse in a 5.25% sodium hypochlorite solution (Clorox) diluted 1:200 (0.026% AI) for 15 minutes.
- Break up clumps to completely separate eggs.
- Rinse several seconds in tap water.
- Air dry promptly on clean air bench.

5.3 Diet inoculation

- Weigh out 0.534 g eggs to obtain 400 eggs or fabricate a small scoop to contain this amount.
- Place eggs in bottom of starting container.

- Place containers on open racks in starting room.

PRECAUTIONS

(i) Bottom of container must be dry or hatch will be reduced.

(ii) Avoid condensation inside container or neonate larvae will become entrapped in water droplets.

(iii) Lighting must be uniform so neonate larvae will not orient to the brightest source and become entangled with each other. They are very tenacious and will cling to each other until dead from starvation.

(iv) Weigh eggs promptly after drying to obtain accurate numbers. Eggs lose 20% of the initial weight by the time hatching begins.

5.4 Transfer

- Five days after hatch transfer support screen with 3rd instar larvae to finishing container.

5.5 Prepupae collection

When prepupae have finished feeding and begin searching for a place to pupate open the exit flap so they can drop or crawl into the collection tray below (7 days after transfer).

- Collect prepupae at the beginning and end of the work day.

5.6 Pupation

- Place prepupae in individual cells in pupation unit.
- Place cover on unit and secure with large rubber bands.
- Place units on open racks allowing 5 cm space between units.

PRECAUTIONS

(i) Humidity must be controlled near 85%. If too high the prepupae will be lying on a wet surface and the larval skin will stretch rather than split preventing ecdysis. If RH is low, desiccation occurs and the insect's blood volume is too low to cause rupture of the larval skin.

(ii) If large numbers of insects are reared a dehumidifier may be required since each prepupa loses ca 4.0 ml of water by the time ecdysis occurs (Joesten et al, 1982).

(iii) Although we have not been able to demonstrate any harmful effects to prepupae exposed to light we do use lights only when needed. When the room is dark prepupae settle down and cease their attempts to escape.

(iv) The bottom of the cells must be level so the various sacs in which the wings, legs, etc. develop will follow the proper course as they are pumped out. Uncovered areas will not harden resulting in bleeding or desiccation. See Fig. 7 for sequence of events from prepupa to adult.

5.7 Pupal collection

- At 7 days following placement of prepupae in the cells the rubber bands are removed and the pupation unit gently tipped so the plywood or plexiglass cover is resting on the bench. The unit is lifted up leaving the pupae on the cover from which they are transferred to screened trays for storage and/or emergence.

PRECAUTION

Although pupation occurs at 3.5 to 4.5 days after placement in cells pupae are not harvested until 7 days to be sure that all have pupated and are well hardened. Teneral pupae are easily injured and will bleed to death if handled too early.

Fig. 7. Development of the tobacco hornworm from prepupa to adult. A, Prepupa prior to ecdysis. B, Beginning of ecdysis; white lateral strip appears when tracheal linings are withdrawn from interior (0 min). C, Ecdysis 50 percent complete (5 min). D, Ecdysis complete (10 min). E, Expansion of ventral sacs containing antennae, legs, proboscis sheath, and wings, 75 percent complete (15 min). F, Lateral view of E. G., Pupae tanned and hardened (17 h), male on left, female on right, arrows show location of sexual characters. H, Moth emerging (14 d). I, Moth expanding wings (0 min). J, Wings unfolded (20 min). K, Wings hardened (65 min).

5.8 Adults

The sex ratio is approximately 1:1.

5.8.1 Emergence period

Moths begin to emerge 12 days after pupation and most have emerged by the 15th day. Males comprise 65% of the first 20% to emerge and females about 65% of the last 20%.

5.8.2 Collection of adults

To avoid handling of moths, trays of pupae are placed in the colony cage shortly after collection each week.

5.8.3 Oviposition

To induce moth activity lighting is reduced to 0.16 cd/m^2 (moths and cage barely visible to dark adapted human eye) reflected from the feeders and artificial leaves. Within minutes the moths become active. Mating occurs 1 to 2 days after emergence; oviposition begins at 4 days, peaks at 8 and drops off rapidly by the 12th day. Floor of the cage is swept daily to remove scales and dead moths.

PRECAUTIONS

(i) If handling newly emerged moths point the abdomen away from you or you may be sprayed with a stream of meconium.

(ii) If smell is detectable from meconium and other fecal material an exhaust fan is needed to prevent death of moths and discomfort to workers.

(iii) If the RH is kept at 85% moth scales will not pollute the air unless the moths are flying about. However, a simple mask should be used when entering the cage.

(iv) Avoid opening the cage door or entering the cage when moths are active or many will escape either through the door or on your clothing.

5.9 Sex determination

Females are slightly heavier than males and males have broader antenna than females but these are not accurate enough for 100% separation. The claspers of the males or lack of them in the females are more reliable characters for sex determination. Sex can be determined in the larval stage. Males have a characteristic spot on the midventral surface of abdominal segment 9 next to the junction with segment 8. The female lacks this character (Stewart el al, 1970). Pupae can also be separated; the 8th segment of the female pupa is divided by the 9t segment on the ventral median and the opening of the bursa copulatrix is visible in this segment. The 8th segment of the male is not divided and the 9th segment has 2 rounded pads. (Fig. 7g).

5.10 Rearing schedule

a. Daily
- Collect eggs and weigh.
- Feed moths.
- Sweep up dead and moribund moths and moth scales in colony room.
- Check temperature, RH and lighting in all areas.

b. Sunday
- Check 5th instar development. Turn fans on high speed if necessary to dry fecal pellets.

c. Monday
- Transfer 3rd instar larvae from starting containers to finishing tray.
- Clean and sterilize starting containers.
- Surface sterilize eggs.
- Collect prepupae and place in cells.
- Collect pupae.
- Clean and sterilize pupation cells and lids.
- Sex pupae for shipment and colony.
- Set up starting containers.

d. Tuesday
- Sterilize finishing trays and starting container lids and drain prior to adding diet.
- Mix diet and draw into rearing containers.
- Clean mixer and surrounding area.
- Collect remaining pupae.
- Collect prepupae and place in cells.
- Clean, sterilize and dry pupation cells and lids.

e. Wednesday
- Ship eggs.
- Collect prepupae and place in cells.

f. Thursday
- Collect prepupae and place in cells.
- Clean and sterilize finishing trays.

g. Friday
- Bring records up-to-date.
- See 5.10a.

i. Weekly
   - Clean moth feeders.
   - Clean artificial leaves.
   - Mop floors and working surfaces.

5.11 Insect quality

Insect quality is checked visually for rate of growth and size. Eggs are weighed to determine fecundity and % recovery from egg to adult is recorded. Field tests releasing sterile males showed that they were almost 100% competitive with native males in mating with native females. See 6.3 for nominal values.

5.12 Special problems

a. Diseases
   Pathogens such as *Serratia marcescens* and cytoplasmic polyhedrosis have occasionally infected the colony. Any containers with diseased larvae are promptly removed from the rearing area, the contents discarded and the rearing units sterilized in 0.5% (0.026% AI) Clorox before further use. *S. marcescens* can usually be recognized by the red pigments produced in the diseased larvae. Cytoplasmic polyhedrosis usually shows up in the 5th instars. The skin becomes very fragile and breaks if an attempt is made to handle the larvae. Eventually the larvae turn black. Diseased specimens should be submitted to a recognized insect pathologist for positive identification.

b. Microbial control of diet
   The various microbiocides listed are usually effective in controlling growth of yeast and molds, however the copious amounts of fecal pellets produced by 5th instars must be dried by holding the RH to 50-60% and providing an air flow beneath the finishing containers. Both the starting and finishing containers are designed to provide ventilation beneath the containers to minimize contamination. If contamination does occur the affected units are handled as in (a) above.

c. Human health
   (i) Adding propionic acid to the diet has allowed a reduction in the amount of formalin needed but since it has not been completely eliminated the mixing room should be well ventilated. This would also apply to streptomycin sulfate since workers may become sensitized to it causing a reaction should it be required for medical treatment.
   (ii) If RH is held at the prescribed 80-85%, moth scales tend to settle out when the moths become inactive. However, a simple mask should be worn to prevent inhalation. Ammonia may be a problem when cage populations are high and should be removed with an exhaust fan.
   (iii) If an allergic reaction is suspected either from diet ingredients or conditions in the moth colony, consult a doctor immediately.

5.13 General sanitation

All rearing containers are cleaned and soaked a minimum of 4 h in 0.5% Clorox (0.026% AI) and floors are mopped with the same solution at least once or more weekly. If fumigation of the room is necessary a 90 mm x 30 mm high dish containing 50 ml of 40% formaldehyde can be allowed to evaporate overnight when personnel are absent. Thoroughly air out the room the next morning.

6. LIFE CYCLE DATA

Life cycle and mortality data during development of various stages under optimum conditions are as follows:

6.1 Development data

Stage	% RH	Temp (C)	Number of Days		
			Min	Mean	Max
Eggs	50-60	28 ± 1	3.5	4.0	4.5
Larvae	50-60	28 ± 1	12.0	13.0	14.0
Pupae	80-85	28 ± 1	3.5	4.0	4.5
Adult emergence	80-85	28 ± 1	12.0	13.0	14.0
Pre-oviposition	80-85	28 ± 1	3.0	3.5	4.0
Total days			34.0	37.5	40.0
Egg to egg			35.0	38.5	41.0

6.2 Mortality data (%)

Non-viable eggs	5
Failure to establish in starting containers	6
Failure to pupate	3
Failure to emerge as sound adult	1
Total (accumulated)	14.25

6.3 Other data

Egg production peaks 8 days after eclosion and drops sharply by the 12th day.

Mean fecundity - 1300 eggs/female

Mean adult longevity - males 8 days, females 12 days

No. of larval instars - 5

Mean pupal weights (g) males 5.0 (min 3.8, max 5.5), females 5.4 (min 3.6, max 6.3)

Mean consumption of 40% sucrose solution - 2.9 ml (lifetime).

7. PROCEDURES FOR SUPPLYING INSECTS

Since most of the requests are for eggs and small quantities of pupae these are handled informally, usually by telephone. Orders are shipped early in the week to avoid weekend delays in receipt. Eggs are placed in small, sturdy plastic dishes. Pupae are placed on the bottom 6 layers of absorbent paper cushion, covered with the top 6 layers, rolled up and secured with rubber bands (Fig. 8). Hundred of shipments have been made with less than 3% loss.

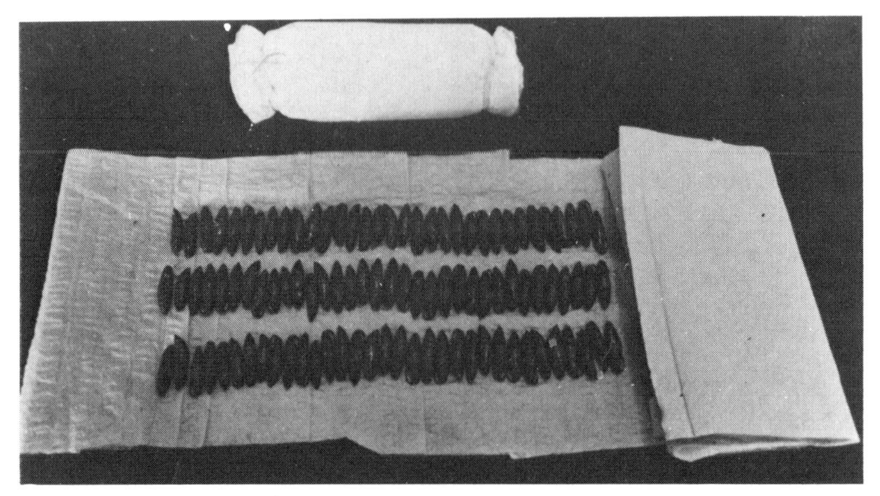

Fig. 8. Pupae (100) placed on bottom of paper cushions before being covered with top 6 layers rolled up and fastened with rubber bands (top of photo).

8. OTHER

When only small quantities of TBH are needed, they can be reared individually in 9 cm diameter cups 7 cm deep with a 40 g block of diet and a 6 cm x 4 cm (2 cm mesh)plastic screen on which the larvae can perch when molting. The lid must be porous to allow ventilation. Pupation cells may be made by drilling 2.5 cm diameter holes to a depth of 12 cm in wooden blocks. The cell walls must be smooth to prevent puncture of the prepupae or newly molted pupae. Pupation is also successful in 9 cm diameter cups 7 cm deep filled with fine, dampened vermiculite tamped firmly and leaving a 2.5 cm space at the top. (Also, see Bell and Joachim, 1976).

8.1 Miscellaneous insects

The diet has been found suitable for most foliar insects found feeding on tobacco including the tobacco budworm, *Heliothis virescens* (Fabricius); the corn earworm, *H. zea* (Boddie); the cabbage looper, *Trichoplusia ni* (Hubner); the salt marsh caterpillar, *Estigmene acrea* (Drury); the vegetable weevil, *Listroderes costirostris obliquus* (Klug); the corn root webworm, *Crambus caliginosellus* (Clemens); the velvet bean caterpillar, *Anticarsia gemmatalis* (Hubner); the fall armyworm, *Spodoptera frugiperda* (J. E. Smith); and various species of cutworms.

9. REFERENCES

Allen, N. and Kinard, W. S., 1969. Production of tobacco hornworm moths from field-collected larvae. J. Econ. Entomol. 62: 1068-1071.

Baumhover, A. H.; Cantelo, W. W.; Hobgood, J. M., Jr.; Knot, C. M.; and Lam, J. J., Jr., 1977. An improved method for mass rearing the tobacco hornworm. ARS-S-167, 13 pp.

Bell, R. A. and Joachim, F. G., 1976. Techniques for rearing laboratory colonies of tobacco hornworms and pink bollworms. Ann. Entomol. Soc. Am. 69: 365-373.

Hoffman, J. D., and Lawson, F. R., 1964. Preliminary studies on mass rearing of the tobacco hornworm. J. Econ. Entomol. 67: 354-355.

Joesten, M. E.; Royston, M. E.; Jiminez, M.; Wadewitz, A.; Melian, D.; and Lockship, R. A., 1982. Gain and loss of fluid in metamorphosing larvae of *Manduca sexta*. J. Insect Physiol. 28: 589-599.

Lam, J. J., Jr., Elsey, K. D., and Baumhover, A. H., Mass rearing of a predaceous stilt bug, *Jalysus wickhami* (Van Duzee) = *spinosus*. (In press)

Madden, A. H. and Chamberlin, F. S., 1945. Biology of the tobacco hornworm in the southern cigar-tobacco district. U. S. Dept. of Agric. Tech. Bull. No. 986. 51 pp.

Sparks, M. R. 1970. A surrogate leaf for oviposition by the tobacco hornworm. J. Econ. Entomol. 63: 537-540.

Stewart, P. A.; Baumhover, A. H.; Bennett, L. S.; and Hobgood, James M., Jr., 1970. A method of sexing larvae of tobacco and tomato hornworms. J. Econ. Entomol. 63: 994-995.

Snow, J. Wendell, Haile, D. G., Baumhover, A. H., and others, 1976. Control of the tobacco hornworm on St. Croix, U. S. Virgin Islands by the sterile male technique. U. S. Dept. of Agric. ARS-S-98. 6 pp.

Yamamoto, R. T., 1968. Mass rearing of the tobacco hornworm. I. Egg production. J. Econ. Entomol. 61: 170-174.

Yamamoto, R. T., 1969. Mass rearing of the tobacco hornworm. II. Larval rearing and pupation. J. Econ. Entomol. 62: 1427-1431.

ORGYIA PSEUDOTSUGATA

JACQUELINE L. ROBERTSON

USDA Forest Service, Pacific Southwest Forest and Range Experiment Station,
P. O. Box 245, Berkeley, CA 94701, USA

THE INSECT

Scientific Name: Orgyia pseudotsugata (McDunnough)
Common Name: Douglas-fir tussock moth
Order: Lepidoptera
Family: Lymantriidae

1. INTRODUCTION

The Douglas-fir tussock moth periodically defoliates coniferous forests in interior western North America (Furniss and Carolin 1977, Beckwith 1978). Primary hosts are Douglas fir, grand fir, white fir, and subalpine fir. Other conifers such as ponderosa pine are defoliated when they occur in stands mixed with firs. Shrubs such as bitterbrush adjacent to outbreaks are also attacked.

Newly hatched larvae feed on new needles. Death of these needles gives the tree a red cast; the tree tip often bears a conspicuous cap of silk deposited by migrating larvae. During further development, larvae feed on both old and new foliage. In heavy infestations, the forest becomes brown and purplish as more bare twigs become exposed. Top kill and tree death are the 2 most serious consequences of defoliation by the Douglas-fir tussock moth.

There is 1 generation a year. Adults appear from mid-August through October (Beckwith 1978). The flightless females remain on their cocoons, mate (males are active fliers), and lay their eggs. Diapause occurs in the egg stage; larvae begin to hatch late in May through early June of the following year at the same time the new foliage appears (Furniss and Carolin 1977). Five to 7 instars occur depending on environmental conditions and sex (in laboratory rearing, males have 6 and females 7 instars). Pupae appear late in the summer.

The Douglas-fir tussock moth has been reared continuously on several artificial diets (Lyon and Flake 1966, Peterson 1978, Thompson and Peterson 1978), including that used for rearing Choristoneura occidentalis (Robertson, Vol. II, p.227). Our laboratory colony has been reared through 24 generations over 12 years. Anyone rearing O. pseudotsugata is warned that the larval urticating hairs, particularly those of the later instars, can cause allergic reactions ranging from mild irritations to medical emergencies.

2. FACILITIES AND EQUIPMENT

2.1 Facilities

Rearing of larvae is done in an environmentally controlled room maintained at 23-26°C and 30-50% RH with a LD 16:8 photoperiod. Fluorescent light intensity is 100-120 footcandles in the photophase and 10 footcandles in the scotophase.

2.2 Equipment and materials required for insect handling

2.2.1 Adult oviposition

Two methods may be used

a. Individual mating

- Waxed, Dixie® cup with clear or translucent plastic lid (200 ml)

- Sterile, disposable gloves
- Dissecting forceps, curved, fine points
- Soft-nosed forceps
- Sheets of newsprint

b. Group mating (25-30 pairs)
- Ice cream carton, paper, 0.95 or 1.89 liter, unwaxed, with lid
- Waxed paper cut to fit half of inner carton surface
- Dissecting forceps, curved, fine points
- Soft-nosed forceps
- Sheets of newsprint
- Sterile, disposable gloves

2.2.2 Egg collection and storage
- Waxed Dixie cup with clear or translucent plastic lid, 200 ml
- Sterile, disposable gloves
- Sheets of newsprint
- Dissecting forceps, curved, fine points
- Refrigerator set at 5-10°C

2.2.3 Egg sterilization
- 600-ml (or larger) glass beaker
- Mechanical stirrer
- Cheesecloth
- Rubber band
- Tissue paper
- Stainless steel spatula
- Sterile, disposable gloves
- No. 1 filter paper, 15 cm diam, sterilized
- Transparent tape
- Dissecting forceps, curved, fine points

2.2.4 Larval rearing
- No. 3 camel-hair brushes, sterilized
- Sterile plastic petri dishes, 150 x 15 mm
- No. 1 filter paper, 15 cm diam, sterilized
- Sterile plastic petri dishes, 100 x 20 mm
- Grinding tool or wheel
- Soft-nosed forceps
- Sheets of newsprint
- Sterile, disposable gloves

2.2.5 Pupal collection
- Dissecting forceps, curved, fine points
- Sterile plastic petri dishes, 150 x 15 mm
- No. 1 filter paper, 15 cm diam, sterilized
- Sterile, disposable gloves

2.2.6 Insect holding
- Refrigerator set at 5-10°C

2.2.7 Anesthesia
- CO_2 source

<u>2.3 Diet preparation equipment</u> (See 2.3 <u>C.</u> <u>occidentalis</u>, Vol. II, p. 229)

3. ARTIFICIAL DIET: See 3.0 <u>Choristoneura</u> <u>occidentalis</u>, Vol. II, p. 229

4. HOLDING INSECTS AT LOWER TEMPERATURES

<u>4.1 Pupae and adults should not be stored</u>

<u>4.2 Eggs</u>
- Unsterilized eggs are stored for 4-6 months at 5-10°C

<u>4.3 Larvae</u>
- Larvae can be stored in petri dishes without food for 1-3 days at 10°C

5. REARING AND COLONY MAINTENANCE

The laboratory colony of O. pseudotsugata, established with egg masses collected in the field, has been reared through 24 generations over 12 years.

5.1 Egg collection

- Remove waxed paper from ice cream carton, carefully shaking off dead adults
- With dissecting forceps, pull off any adult bodies which remained after the shaking process
- Remove egg masses from waxed paper with dissecting forceps
- Gently separate the eggs by rolling the masses between 2 pieces of sterile filter paper
- Pour eggs into Dixie cup, close lid, and write the date that will be 4 months later on the lid with a marking pen
- Place in refrigerator set at 5-10°C for 4-6 months (discard unused eggs 6 months after placement in refrigerator)

PRECAUTION

Wear sterile, disposable gloves while collecting eggs.

5.2 Egg sterilization

- Remove eggs from refrigerator
- Pour the number desired into the bottom of a 600-ml glass beaker
- Pour 490 ml of distilled water mixed with 10 ml of 5.25% sodium hypochlorite over the eggs
- Add 4 drops of a wetting agent such as Tween 20® to the diluted sodium hypochlorite
- Wash for 15 min at high speed with a mechanical stirrer; cover the beaker with cheesecloth or gauze and fasten with a rubber band
- Decant the sodium hypochlorite
- Remove and retain the cheesecloth
- Repeat the 15 min sodium hypochlorite wash with fresh solution
- Again place the cheesecloth over the beaker, fasten with the rubber band, and decant the sodium hypochlorite
- Leave the cheesecloth over the beaker and rinse the eggs under running, distilled water for 15 min
- Drain off the water through the cheesecloth by inverting the beaker
- Remove the rubber band and carefully remove the cheesecloth with the beaker still inverted
- Blot the eggs, wrapped in wet cheesecloth, on tissue paper
- Spread out the cheesecloth on a sheet of newsprint
- With a spatula and dissecting forceps, transfer the eggs to sterile filter paper
- Air dry for 2 h
- Transfer 100-150 eggs to each 150 x 15-mm sterile, plastic petri dish lined with sterile filter paper
- Grind off lugs from inside of each petri dish top before taping or banding the dish closed
- After 10 days, open each dish and insert a 1-cm cube of artificial diet
- Tape or band the dish
- Check dishes daily for larval emergence

5.3 Larval rearing

- Grind lugs from the top of each sterile, plastic, 100 x 20-mm petri dish
- Within each dish, place 2 pieces of artificial diet approx. 1 x 1 x 6 cm in size
- With a sterile, no. 3 camel-hair brush, transfer 10 first instars to each petri dish
- Every 7 days, remove old diet and replace it with fresh

PRECAUTIONS

(i) Wear sterile, disposable gloves while handling larval rearing dishes, particularly those containing older instars.

(ii) As an added protection from urticating hairs, secure lab coat sleeves to wrists with rubber bands or by some other method.

(iii) Minimize air movements around opened dishes to prevent hairs from blowing onto table or into the air.

(iv) Wear a disposable particle mask to prevent breathing hairs.

5.4 Pupal collection

- With dissecting forceps, collect cocoons containing fully tanned pupae as they appear in rearing dishes
- Remove and dispose of cocoons
- Place 25-30 pupae of each sex in each sterile, filter-paper lined, 150 x 15-mm plastic petri dish

PRECAUTIONS

(i) Pupae are light green when newly formed and must be allowed to tan to a black-brown color before they can be handled.

(ii) Use the same precautions as noted for larval rearing to prevent contact with urticating hairs and cocoons, into which the hairs become mixed.

5.5 Adults

The sex ratio is approximately 1:1.

5.5.1 Emergence period

The first adult female will emerge about 9 days after pupation. Males will emerge about 12 days after pupation. Adult survival from first instars is about 50%.

5.5.2 Collection

This is done on the day of emergence using dissecting forceps. Females should be grasped by a rudimentary wing. Males should be anesthetized with CO_2, then grasped with forceps at the base of a forewing.

5.5.3 Oviposition

This is best done at 23-26°C in the dark. The preoviposition period is less than 24 h. Peak egg laying occurs within 2-3 days of pairing. Adults live for 7 days.

- Autoclave an unwaxed, 946 ml (qt) or 1892 ml (1/2 gal) ice cream carton
- Cut waxed paper to fit 1/2 of the inner linear surface of the carton
- Line half the inner linear surface of the carton with the waxed paper
- Place 25-30 pairs of adults in the carton
- Close the carton
- Lay the carton on the side lined with paper within a cabinet and close the doors
- After 14 days, open the carton and remove the eggs

5.6 Sex determination

- Female pupae are much larger than males
- Female pupae have very short wing pads, males have long wing pads
- Female moths have rudimentary wings and are flightless
- Male moths have normal wings and are strong fliers
- Females generally have 7 instars, males 6

5.7 Rearing schedule

5.7.1 Daily

- Collect pupae and place in adult emergence dishes
- Collect newly emerged moths and place in mating containers
- Refeed larvae as necessary
- Check previously sterilized eggs for larval emergence and transfer larvae to rearing

5.7.2 Weekly (or on alternate days)

- Collect eggs and store in refrigerator at 5-10°C
- Remove 4- to 6-month old eggs from 5-10°C refrigerator, sterilize eggs, and place them at room temperature for larval emergence

5.8 Insect quality

Insect quality is checked by continuously keeping records of egg hatch, larval survival, pupal weights, and adult emergence. Usual values for these parameters are: egg hatch-90%; larval survival-80%; male pupal weight-240 mg; female pupal weight-600 mg; adult emergence-90%.

5.9 Special problems

5.9.1 Diseases

Douglas-fir tussock moths are especially susceptible to nuclear polyhedrosis virus (NPV). Larvae become flaccid, inactive, and darker than usual. As the disease progresses, a larvae will virtually dissolve when touched. Once established, this virus spreads rapidly and can decimate an entire colony. All dishes containing larvae suspected of NPV infection must not be opened and should be autoclaved at once.

5.9.2 Microbial control

Bacterial and fungal infections will occur on the diet if proper sanitary conditions are not maintained. Escherichia coli, Aspergillus, and Penicillium may be particularly troublesome. Diet should not be left unwrapped or unrefrigerated for extensive periods; it should be handled only with sterile gloves and sterile utensils such as spatulas. All equipment used to handle insects, including brushes and forceps, should be sterilized and stored in sterile, closed containers such as enamel pans with lids. When in use, forceps should be wiped frequently with a surface sterilant.

5.9.3 Human health

If an allergic reaction occurs, discontinue handling the insects in any stage. If necessary, consult a doctor immediately.

6. LIFE CYCLE DATA

6.1 Developmental data

Stage	Weight (mg) on first day of stage		Duration (days)	
	Mean±SD	Range	Mean±SD	Range
Instar 1	0.6±0.1	0.3-0.8	7.1±0.4	7-9
Instar 2	4.3±1.6	1.2-6.6	5.6±1.9	4-8
Instar 3	18.7±9.1	10.0-46.2	5.1±1.5	3-7
Instar 4	37.7±8.1	19.3-56.2	7.5±1.5	6-13
Instar 5	101.2±25.5	38.5-33.1	9.7±1.9	8-13
Instar 6	275.7±79.7	183.6-420.0	16.0±3.6	11-24
Instar 7	566.3±113.2	438.4-765.2	11.0±1.5	11-14
Pupae				
male	239.3±8.7	229.3-244.8	14.0±0	--
female	603.5±93.3	537.5-669.5	10.0±0	--
Adult				
male	210.5±36.0	169.3-227.0	--	--
female	573.9±139.8	436.9-873.9	--	--

6.2 Survival data %

- Egg viability, 80-90
- Neonate larva to pupa, 83-96
- Neonate larva to adult, 80-92

7. REFERENCES

Beckwith, R. C., 1978. Biology of the insect. pp.25-29. Introduction In: M. H. Brookes, R. W. Stark, and R. W. Campbell (Editors), The Douglas-fir tussock moth: a synthesis, USFS-SEA Tech. Bull. 1585.

Furniss, R. L. and Carolin, V. M., 1977. Western Forest Insects. USDA, Forest Service. Miscell. Publ. 1339. pp. 222-227.

Lyon, R. L. and Flake, H. W., Jr., 1966. Rearing Douglas-fir tussock moth larvae on synthetic media. J. Econ. Entomol. 59: 696-698.
Peterson, L. J., 1978. Rearing the western tussock moth on artificial diet with application to related species. USDA Forest Service Res. Paper PNW-239. 5 p. illus.
Thompson, C. G. and Peterson, L. J., 1978. Rearing the Douglas-fir tussock moth. USDA Agric. Handbook 520. 17 p. illus.

OSTRINIA NUBILALIS

W. D. GUTHRIE, J. C. ROBBINS, and J. L. JARVIS

Corn Insects Research Unit, USDA-ARS, and Department of Entomology, Iowa State University, Ankeny (50021) and Ames, Iowa (50011).

THE INSECT

Scientific Name: *Ostrinia nubilalis* (Hübner)
Common Name: European corn borer
Order: Lepidoptera
Family: Pyralidae

1. INTRODUCTION

The European corn borer (ECB) is distributed in several countries in Africa, Europe and Asia, and was first discovered in the United States (Everett, Massachusettes in 1916). At present, the ECB is found in most states east of the Rocky Mountains, in several Canadian Provinces, including Prince Edward Island, and has 1 to 4 generations per year.

Although there are many species of plants in North America in which the ECB can complete its life cycle, maize is its preferred host, and this exotic species has become one of the most destructive insect pests of maize throughout the corn belt of the United States. First-generation borers cause damage primarily to leaf tissue (as plants grow out of the whorl stage the larvae invade sheathes, collars, and stalks, but most larvae pupate before much stalk damage occurs). Second-, 3rd-, and 4th-generation borers cause damage primarily to sheath, collar, shank, and stalk tissue.

The ECB has 5 larval instars and overwinters as diapausing larvae primarily in maize stalks. Moth emergence (from the overwintering population) in the corn belt of the United States usually occurs for 4 weeks, starting the last week of May or the first week of June. Moth emergence from these 1st-generation borers usually occurs over a period of 4 to 5 weeks, starting the last week of July or first week of August.

The ECB can easily be reared on a meridic diet. In 1982, ca 24 million egg masses (about 700 million eggs) were produced by the public and private sectors in North America, primarily for host-plant-resistance research.

2. FACILITIES AND EQUIPMENT REQUIRED

2.1 Facilities

- Room for diet preparations and cooking
- Larval rearing room (216 x 254 x 216 cm high) maintained at 27-28°C, 75-80% RH and constant light (four 120-cm cool white fluorescent bulbs per incubator)
- Oviposition room, temperature and lights controlled through a series of time clocks operating a heater, air conditioner, and lights (see Section 5.4.3)
- Egg storage room
- Room for cleaning and sterilizing insect rearing containers
- Autoclave for sterilizing vials, etc.

2.2 Equipment and materials required for insect handling

2.2.1 Adult oviposition; 2 methods are used

a. Individual mating

- Cage 6 cm high, 8.5 cm diam, bottom and sides made of 16- or 18-mesh copper or bronze wire, top made of 5-mesh hardware wire. Cage placed in petri dish containing moistened absorbent cotton for moth's drinking water.

- Waxed paper placed on top of cage for oviposition

b. Group mating

- A cage 60 x 31 x 31 cm high is of sufficient size to accomodate 200 to 300 pairs of moths. The sides, ends, and bottom consist of wooden frames covered with 16- or 18-mesh copper or bronze wire, fastened together with the screened surface inside. It is imperative that all wood surfaces are covered with screen because moths oviposit on smooth wood surfaces, but not on rough screen surfaces. The top is made of 5-mesh hardware wire. Two sheets of waxed paper (15 cm wide, 60 cm long) or 2 plastic sheets are placed on top of the cage for oviposition and are held in place with a rubber pad.

2.2.2 Egg collection and storage

- Waxed paper sheets (30 lb basis, waxed to 45 lb, Sealtite brand, 7-1/2 lb wax each side, 15 x 60 cm)
- Or plastic sheets (0.00125 gauge, 15 cm wide, D. W. white) from Fairmont Film, Inc., Fairmont, MN, USA
- Egg punching machine (punches out 1.3-cm discs of waxed paper containing 1 egg mass)
- Celotex boards (20 x 25 cm)
- Pins, trupoint chromonic stainless for pinning egg masses on celotex boards
- Pinning board (screen covered)

2.2.3 Larval rearing

a. Individual rearing

- Cooker (16,000 g capacity) Groen® TDB/7 model steam-jacketed kettle
- Motorized stirring rod attached to kettle
- Waring® blender (5 liter) with lid
- Variable autotransformer (need 2, 1 to control speed of stirrer in cooker and 1 to control speed of Waring blender)
- Thermometer to check temperature of diet
- Clock with timer to check length of time to cook diet at certain temperature
- Rubber spatula for scraping diet from kettle
- Water jugs (5 liter) for holding water for diet
- Balance (Mettler®, type R10, 0-10,000 g) for weighing diet
- Jars (screw cap), 59 ml (2 oz) for hatching larvae
- Vials, 3 dram (5.3 g)
- Small artist brush
- Plugging cotton, nonabsorbent
- Racks for holding vials containing larvae

b. Group rearing

- Need first 9 items listed under individual rearing
- Dishes, plastic (25 cm diam, 9 cm deep)
- Jelly cups, 30 ml (1 oz) for holding egg masses in dishes
- Corrugated paper for pupation sites
- Machine for cutting corrugated strips of paper to 2.3 cm wide
- Hood for waxing corrugated strips
- Pins, trupoint chromonic stainless for fastening 1 end of corrugated strips before waxing
- Small sewing machine motor mounted beneath portable typewriter table with adapter for wrapping corrugated strips of paper around apparatus
- Apparatus for wrapping corrugated strip for waxing
- Hot plate inside hood for keeping wax hot
- Container for holding wax
- Container with cold water for dipping waxed strips
- Plastic bags for storing waxed strips

3. ARTIFICIAL DIET

3.1 Composition

Ingredients	Batch amounts	%
- Water	12,900 ml	85.5
- Agar[a]	280 g	1.9
- Wheat germ [b]	520 g	3.4
- Dextrose[c]	400 g	2.7
- Casein[c]	440 g	2.9
- Beta-sitosterol[c]	32 g	0.2
- Salt mixture #2[c]	144 g	1.0
- Vitamin supplement[c]	92 g	0.6
- Ascorbic acid[c]	120 g	0.8
- Aureomycin[d]	27 g	0.2
- Fumidil B[e]	7 g	0.05
- Methyl-p-hydroxybenzoate[c]	21 g	0.1
- Sorbic acid[c]	8 g	0.05
- Propionic acid-phosphoric acid solution[b]	86 ml	0.6
- Formaldehyde[f]	7 ml	0.05
	15,084 ml	100.0%

[a] Moorehead & Co., Inc., 14801 Oxnard St., P.O. Box 2728, Van Nuys, California 91401, USA
[b] Mennel Milling Co., 128 W. Crocker St., Fostoria, Ohio 44830, USA
[c] Nutritional Biochemicals, 26201 Miles Road, Cleveland, Ohio 44128, USA
[d] Iowa Veterinary Supply Co., Box 616, Iowa Falls, Iowa 50126, USA
[e] Dadant & Sons, Hamilton, Illinois 62346, USA
[f] Fisher Scientific Co., 1241 Ambassador Blvd., P.O. Box 12405, St. Louis, Missouri 63132, USA

3.2 Directions for cooking meridic diet

- Place 6650 ml cold tap water into a preheated, steam-jacketed kettle
- Add 280 g agar to water and heat until melted; stir constantly with motorized stirring rod attached to kettle while heating
- When agar is at 90°C, turn off heating element and continue to stir
- Pour in 4000 ml water that has been cooled to 3°C
- Add 520 g wheat germ and stir until temperature of medium decreased to 58°C (this takes 20-60 min, depending on room temperature)
- Pour into a blender these ingredients: 400 g dextrose, 440 g casein, 32 g beta-sitosterol, 144 g salt mixture #2, 92 g vitamin supplement, 120 g ascorbic acid, 27 g aureomycin, 7 g Fumidil B, 21 g methyl-p-hydroxybenzoate, and 8 g sorbic acid
- Prepare mold inhibitor solution as follows:
- Mix 418 ml propionic acid with 82 ml distilled water
- Mix 42 ml phosphoric acid with 458 ml distilled water; combine with propionic acid solution
- Pour 2250 ml tap water into blender containing dry ingredients, then add 86 ml of the combined mold inhibitor solutions and 7 ml of 35% formaldehyde; blend for 3 min

PRECAUTION

Wear rubber gloves and protective mask while working with mold inhibitors.

- Pour contents of blender into kettle when temperature of agar-wheat germ water mixture reaches 58°C
- Stir until well mixed (2 to 3 min)
- Pour about 1/4 of this diet into a blender and blend for 1 min (this distributes ingredients evenly)
- Pour 950 g of diet into each dish
- Repeat previous two steps until all diet has been reblended and poured into dishes (will fill 15 dishes)

The procedure used in preparing the diet is very important. Agar and wheat germ mixture must be 58°C before adding any of the other ingredients because high temperature reduces the nutritional value of some of the ingredients.

Once the diet is poured into dishes, it is allowed to solidify at room temperature (24°C); then packaged in heavy (47 x 63 cm, .004 mil) plastic bags and stored at 4°C until used. The diet can be stored for at least 14 days.

PRECAUTIONS

(i) The dishes should not be infested with egg masses on the same day that they are prepared because fumes from the mold inhibitors may kill the eggs.

(ii) The diet from cold storage should be warmed to room temperature before being infested and placed in a room with 75-80% RH.

4. HOLDING AT LOWER TEMPERATURES

4.1 Adults and larvae

Moths and larvae are not held at lower temperatures.

4.2 Eggs

Egg masses can be held at 15°C for 10 days if necessary. If discs of waxed paper containing egg masses are pinned on celotex boards, stack the boards in a room with 90-95% RH. If egg masses are taken off plastic sheets, place egg masses in a 18.5 ml (5/8 oz) plastic cup with a plastic-lined lid.

4.3 Pupae

Strips of corrugated paper containing pupae can be placed in a plastic bag for 7 days at 15°C.

5. REARING AND COLONY MAINTENANCE

In our laboratory, the ECB has been reared for 200 generations over 19 years. This culture has not deteriorated in egg fertility, number of eggs per female, weight of pupae, etc., but it cannot be used for field research because ECB cultures reared continuously on a meridic diet for many generations survive on maize plants at a very low level. Cultures reared for 1 to 14 generations on a meridic diet survive on maize plants at a high level. A new culture is started each spring; ca 6,000 wild borers are dissected from maize plants each fall. These diapausing larvae are treated with phenylmercuric nitrate; 1 g is dissolved in 10,000 ml of 105°C water mixed constantly with a stirring rod on a magnetic hot plate. The mature (5th instar) larvae are dipped in the cooled solution for only an instant and placed on filter paper to remove excess liquid and then allowed to crawl into waxed corrugated strips for storage at 4°C. At weekly intervals for 5 weeks (starting January 25) the larvae are isolated individually into screen-capped 3.7-ml (1 dram) shell vials containing a strip of blotter paper (moistened once each week until pupation) and placed in a 27°C room. These larvae pupate in February and adults emerge 1 week after pupation. The progeny from the moths are reared through 2 generations on a meridic diet containing 1,500 ppm Fumidil B (1 larva/vial) for the first generation in March; 40 egg masses per dish for 2nd generation in April. Larvae reared for 3 to 14 generations on a meridic diet are used for field infestations.

PRECAUTION

This procedure is essential to eliminate *Nosema pyrausta*; this microsporidium is present in most larvae collected from the field and must be eliminated from laboratory cultures.

5.1 Egg mass collection

Remove waxed paper or plastic sheets containing eggs and replace with new sheets each morning.

a. Field infestation with eggs

- Use waxed paper for oviposition sites

- Cut 1.25-cm diam discs containing 1 egg mass each for field infestation
- Place 2 discs per pin onto 20 x 25-cm celotex boards
- Wrap in moist paper and incubate in plastic-lined boxes at 27°C and 80% RH

b. Field infestation with larvae
- Use plastic sheets for oviposition sites
- Hang sheets in room to incubate eggs at 27°C and 75-80% RH for 24 h, then move to a room with < 50% RH and 21°C for 3-4 h
- Pop eggs off sheets by drawing sheets back and forth over a narrow blunt edge (egg side away from edge)
- Collect eggs in teflon-coated pan coated with cornstarch to prevent eggs from sticking to each other
- Remove excess cornstarch by sifting through screen
- Place 1.6 g egg masses in sterile 1-liter jars which have been moistened with a fine mist of distilled water
- Place a lid on the top of each jar, but do not tighten; shake jar to distribute masses onto sides of jar
- Four days after oviposition, tighten lids and place in boxes covered with black plastic bags
- Place boxes in dark room for hatching
- When larvae hatch, put 50 g sterile corncob grits (20-40 mesh) in each jar and mix with larvae
- Replace solid lid with 0.3-cm-mesh galvanized-screen top to pour grits-larval mixture into larval inoculator bottles for application to crop

5.2 Diet infestation with neonate larvae or egg masses

a. In 5.3-g (3 dram) sterilized vials
- Condition diet by allowing to stand at 21°C overnight
- Cut plugs of diet 2 cm long (with a stainless steel cylinder with inside diam 15.5 mm and 16 x 90 mm outside dimensions) and insert 1 plug/vial, using a 15 x 125-mm test tube to force plugs out of cylinder into vial
- Transfer neonate larvae to diet with small artist brush
- Plug vials with nonabsorbent plugging cotton
- Place vials in rack
- Place rack containing larvae in room (27°C, 75-80% RH)

b. In dishes (25 cm in diam, 9 cm deep) sterilized with 2 ml of 10% Roccal® per liter of water (200 ppm alkyl dimethylbenzylammonium chlorides)
- Pour 950 g of diet in dish
- Condition diet by allowing to stand at 21°C overnight
- Scratch diet with fork to allow newly hatched larvae to penetrate the surface for good larval establishment
- Infest each dish with 0.1 g of blackheaded egg masses (removed from plastic sheets, about 1000 eggs) placed in a 18-ml jelly cup
- Disinfect jelly cups with 95% ETOH before using
- Imbed jelly cup in the surface of diet
- Place 6.8-cm wide, 600-cm long strip of corrugated paper inside dish (fits tightly below the lid) (This strip had previously been wrapped around an apparatus and dipped in hot paraffin wax containing 5% by weight sorbic acid and allowed to dry.)
- Cut out 8 x 11 cm portion of lid and replace with 80-mesh copper wire (to allow air exchange)
- Paint top of lid and top 3.5 cm of outside of each dish black to reduce light penetration so that larvae will enter corrugated strip to pupate (Diapause is prevented by rearing larvae under constant light)

5.3 Pupal collection

a. From vials
- larvae pupate in vials

b. From dishes
- Larvae crawl into corrugated strips to pupate

5.4 Adults
- Sex ratio is 48% female, 52% male (determined from 5 million insects)

5.4.1 Emergence period
- Rearing in 5.3-g (3 dram) vials or dishes
- Adults emerge 21-28 days after infesting diet with larvae or blackhead egg masses

5.4.2 Collection
a. From vials
- Remove plug of cotton from each vial after pupation and before emergence
- Put vials in a wire basket and place basket in oviposition cage (moths emerge directly from the vials inside the cage)

b. From dishes
- Place 1 corrugated strip containing about 600 pupae inside each oviposition cage (moths emerge from the strips inside the cage)

5.4.3 Oviposition
- Maintain room temperature of 27°C 16 h each day (5AM-9PM or 0500-2100)
- Maintain room temperature of 18-20°C 8 h each day (9PM-5AM or 2100-0500) cycling temperature is required to insure adequate mating
- Operate lights ON 6AM-8PM or 0600-2000
- Operate lights OFF 8PM-6AM or 2000-0600
- Maintain RH at 80-85%
- Sprinkle oviposition cages twice each day to provide drinking water for moths
- Most female moths oviposit for 10 days (approx. 10 masses/female; egg masses avg 25-30 eggs/mass)
- Discard adults after 16 days (ca 1 week is required for all adults to emerge from vials or strips of corrugated paper)

5.5 Rearing schedule (daily)
- Prepare diet for cooking
- Place diet in vials or dishes
- Place vials or dishes in incubation chambers
- Place vials or corrugated strips of paper containing pupae in oviposition cages
- Replace waxed paper or plastic sheets containing egg masses with new sheets
- Water 1 end and both sides of screen cages twice a day for drinking water for moths
- Punch out waxed paper discs containing egg masses or remove egg masses from plastic sheets
- Pin waxed paper discs containing egg masses (2 discs/pin) on 20 x 25 cm celotex boards
- Wrap moist paper around boards and place in plastic-lined boxes for incubation
- If plastic sheets are used for oviposition sites, remove egg masses from plastic sheets and place in jars for hatching

5.6 Insect quality
- Maintain insect quality by starting new cultures each year

5.7 Special problems
a. Diseases
- Rear insects in clean environment; to keep disease problems to nearly zero, add the following to diet:
- Fumidil B to control protozoa
- Aureomycin to control bacteria
- Methyl-p-hydroxybenzoate, sorbic acid, propionic acid, and formaldehyde to control fungi

b. Human health
- Wear a protective mask when scales are present

6. LIFE CYCLE DATA

Life cycle and survival data of various stages under 27°C, 75-80% RH, and constant light

6.1 Developmental data

- Eggs, 80 h
- Larvae, 13-19 days
- Pupae, 14-21 days
- Adult emergence, 21-28 days

6.2 Survival data (%)

- Individual rearing
- Egg viability 85-95% under high humidity (90% RH) conditions
- Neonate larva to pupa, 90-100%
- Neonate larva to adult, 85-97%

6.3 Other data

- Preoviposition period, 2 days
- Peak oviposition period, 6-7 days after emergence
- Avg oviposition period, ca 10 days
- Avg fecundity, 300 eggs/female
- Avg adult longevity, 12 days
- No. larval instars, 5
- Range of pupal weights (mg), females 80-150, males 60-85

7. REFERENCES

Guthrie, W. D., Russell, W. A. and Jennings, C. W., 1971. Resistance of maize to second-brood European corn borers. Proc. Annu. Corn and Sorghum Res. Conf. 26: 165-179.

Mention of a commercial or proprietary product does not constitute endorsement by the USDA-ARS.

PECTINOPHORA GOSSYPIELLA

ALAN C. BARTLETT AND WAYNE W. WOLF

USDA, ARS, Western Cotton Research Laboratory, 4135 East Broadway, Phoenix, Arizona 85040, and USDA, ARS, Southern Grain Insects Research Lab, Tifton, Georgia 31793, USA

THE INSECT

Scientific Name: *Pectinophora gossypiella* (Saunders)
Common Name: Pink bollworm (PBW)
Order: Lepidoptera
Family: Gelechiidae

1. INTRODUCTION

The pink bollworm, *Pectinophora gossypiella* (Saunders), (PBW) is probably indigenous to India and was described in 1843 from specimens found in cotton in 1842. The insect reached Egypt in infested cottonseed about 1906-07 and the western hemisphere around 1911. It was detected in Texas cotton in 1917 and reached Arizona by 1926. Infestations of PBW were detected in California in the Imperial Valley in 1965. At present, it is a major pest of cotton in Asia, Egypt, India, Mexico, Australia, and Brazil as well as the United States of America. The common host of the PBW is cultivated cotton (*Gossypium* spp.), however, the insect has been reported on seven families, 24 genera, and 70 species of plants (Noble, 1969).

The female oviposits on squares, terminals, and stems of the cotton plant early in the season, while in mid- and late-season, the eggs are laid primarily around the base of the bolls and under the calyx. Eggs are laid singly or in small clusters at night. Newly-hatched larvae move into squares, where they feed on the anthers, or into bolls, where they feed on the immature seed, often damaging two or more seeds and ruining the lint surrounding those seeds. The larvae cause damage to the seed and lint of cotton and to seed pods of okra and other plants of this type. The larval period is around 2 weeks in the summer. After completing development, the larvae cut their way out of the square or boll, drop to the ground, form a cocoon in the dirt and litter under the plant, and pupate. Depending on temperature, a generation will take from 22-32 days during the cotton-growing season. Starting in late August and September in the U.S. and many other cooler cotton-growing areas, the mature larvae will form a very tight cocoon and go into diapause during the host-free period. In latitudes between 10° north and south (such as in southern India) no diapause stage has been recorded for the PBW.

The PBW has been reared on several artificial diets (Beckman et al., 1953, Vanderzant and Reiser, 1956, Vanderzant, 1957) with the most successful being modifications of wheat-germ-based diet first reported by Adkinson et al. (1960). In our laboratory, the PBW has been reared for over 12 years. The laboratory-reared insects are used locally for field release programs, laboratory pheromone research, genetics studies, virus production, and shipment to other laboratories. A laboratory culture can easily be established by the collection of infested cotton bolls and collection of cut-out larvae in a pupation tray or by obtaining insects from an established laboratory colony.

Fig. 1. Racks holding PBW rearing containers and trays which hold styrofoam beads for collection of cut-out larvae.

2. FACILITIES AND EQUIPMENT NEEDED

2.1 Facilities

Rearing of larvae and pupae is carried out in controlled environment rooms maintained at 29 ± 2°C, 30-40% RH, and an LD14:10 photoperiod (see Fig. 1). Oviposition is carried out in controlled-environment rooms maintained at 27 ± 2°C, relative humidity over 50%, and an LD14:10 photoperiod (see Fig. 2). All environmental chambers are lighted with cool white fluorescent tubes which have a luminous efficacy of 79 lumens/watt and output wavelengths from about 350-750 mμ, with maximum output around 575 mμ. Measurements of light intensity with a Li-Cor Quantum Radiometer® Model LI-185A showed an average of 7 microeinsteins/ m^2/sec in each of the rearing chambers. Humidity is maintained by portable Westbend room humidifers (Model 4017).

2.2 Equipment and materials required for insect handling

Equipment sources and sizes are given the first time mentioned or in section 3.2.

2.2.1 Adult oviposition

a. Individual crosses (see Fig. 3)
- 237 ml (8 oz) squat food cup (Sweetheart® #S-308)
- Coverall lid (Lily® #85-77) with 55 mm diameter hole covered by fiberglass screen (screen mesh US #14, 18 × 16 openings per inch) and 7 mm diameter hole for vial
- 2 ml glass vial for sugar water (VWR 66011-020)
- 6% AI sugar water (w/v)
- 30 mm egg pad (cut from Masslinn® Shop Towels)

b. Group crosses (see Fig. 4).
- Plastic box (34.4 cm × 26.5 cm × 8.5 cm) with lid modified for scale removal and bottom removed and replaced with fiberglass screen
- 4 - 2 ml vials/box
- 26 cm × 21 cm fiberglass screen (separate from box screen)

- 24 cm × 18 cm rubber pad (used for weight on egg pad)
- 26.7 cm × 20.3 cm ordinary bond typing paper
- 10% AI sugar water with 0.2% AI methyl parasept

2.2.2 Egg collection and sterilization

a. Individual crosses
- Stapler and staples
- 9.3 cm diameter paper lids (Sweetheart® DS-308)
- marker (tube type, broad felt tip)
- no sterilization used

b. Group crosses.
- 0.1% AI NaClO
- 26 cm × 21 cm screen separators (same as 2.2.1.1)
- Paper clips (size to suit number of pads to be treated)
- 30 cm × 25 cm × 15 cm plastic container
- Drying rack (1.22 m long × 1.52 m high × .46 m deep with 4 shelves composed of 1 cm diameter rods spaced at 3 mm intervals, see Fig. 1)
- Long metal forceps (18 cm)
- Fumehood
- Transparent polyethylene bag (25 cm × 37 cm)

2.2.3 Egg inoculation of diet

a. Individual crosses
- 473 ml tall drink waxed cup (Lily-Tulip #16)
- Paraffin (AMOCO Parowax)
- Electric paraffin heater (American Scientific Products # M7395)

b. Group crosses
- 0.1% AI NaClO
- 946 ml paper food tubs and lid (Sealrite Neststyle sidewall, half gallon)
- Shredded diet

Fig. 2. Oviposition room showing oviposition chambers in place over scale collection ducts. Egg pads are on some trays.

- Polystyrene foam beads (Arizona Diversified, shredded styrofoam)
- Scissors or paper cutter

2.2.4 Collection of mature (cut-out) larvae

a. Individual crosses
- 1893 ml plastic tubs (Lily #10M tub)
- Plastic lids with screen-covered opening
- 15 cm diameter hexcel circles

b. Group crosses.
- Polystyrene beads
- Trays

2.2.5 Pupal collection and storage

a. Individual crosses
- 237 ml waxed squat food cups
- Coverall lid with screen-covered hole

b. Group crosses
- Soft-nosed forceps
- Fiberglass screen bag (30 × 30 cm, made from same screen as in 2.2.1)
- 3% AI NaClO solution
- Plastic wash tanks
- Drying racks
- Seed separator (industrial model air-blown seed separator)
- Volumetric cylinder (size appropriate to number of pupae produced)

2.2.6 Equipment required for insect holding (other than oviposition)
- Waxed food containers (237-946 ml)
- Screen top lids (appropriate size)
- 6% AI sugar water
- Paper toweling strips 7 cm × 20 cm
- 2 ml vials
- Controlled temperature and humidity cabinets

Fig. 3. Oviposition chamber for individual crosses showing feeding vial, egg pad, screened lid, and holding cup.

2.3 Diet equipment and preparation

2.3.1 Dry ingredients

- Top loading balance (Mettler Model P-10)
- Plastic bags
- Waxed food containers(237-473 ml)
- Standard weights
- Plastic and metal scoops

2.3.2 Finished diet

- 3.8 liter blender (Waring Model 91-215)
- 1 liter graduated cylinder
- Rubber spatula
- Steam kettle (Legion Model M-195-LP40) (Fig. 5)
- Lightning mixer (Lightning Model NS1VM) (Fig. 5)
- Cleaning brushes
- Asbestos gloves
- Thermometer (0 - 100°C)
- Diet pump (Fig. 5)
- Metal pans (60 × 46 × 11 cm)
- Sterile terry cloth towels
- Imperial Food Mill (french fry blade) (Imperial Model DUL)
- Plastic wrap (Saranwrap® or equivalent)
- 0.1% AI NaClO
- Autoclave

3. ARTIFICIAL DIET

3.1 Composition

Diet ingredients are added in the proportions shown in Table 1. The formula for the vitamin solution is found in Table 2. Sorbic acid can be

Fig. 4. Oviposition chamber for group crosses showing plastic box, modified lid, screen botton, screen cover, egg pad, and rubber pad.

replaced by a 1.34X measure of Sorbistat-K (potassium sorbate). For example, 16 g Sorbistat-K is equivalent to 12 g sorbic acid.

3.1.1 Liquid ingredients used in the media are prepared as follows:

Acetic acid

A 25% concentration of acetic acid is prepared by adding 750 ml of 100% acetic acid to 2,250 ml of tap water.

KOH (potassium hydroxide)

A 22% concentration of potassium hydroxide is prepared by dissolving 440 gm of KOH pellets in 1,560 ml of distilled water. Safety precaution: Do not handle KOH pellets. Heat generated while dissolving pellets should be dissipated with a water bath. Wear goggles or safety shield while preparing KOH solution.

Formaldehyde

A 10% AI solution of formaldehyde is prepared by adding 1,081 ml of 37% formalin to 2,919 ml of tap water. This should be done in a hood.

Choline chloride

A 10% concentration of choline chloride is prepared by dissolving 200 g of choline chloride in 1,800 ml of tap water. Store in refrigerator.

Vitamins

The dry ingredients are weighed and stored in a clean paper cup. Stock solution is prepared by blending 1,000 ml of distilled water and dry vitamins in an autoclaved high-speed blender for 20 seconds. Solution is stored in an empty clean 4 liter plastic jug with tight lid and refrigerated. The solution must be well agitated before being used.

Fig. 5. Steam kettle with associated diet pump (bottom left) and dispenser handle (center). The long hose allows dispensing of diet up to 10 ft away from kettle.

TABLE 1
Proportions of diet ingredients for diet used to rear pink bollworm larvae.

Ingredient	Quantity/liter
Acetic acid (25%)	11.6 ml
Agar	20.0 g
Alphacel	5.8 g
Aureomycin	0.9 g
Casein	40.6 g
Choline chloride (10%)	11.6 ml
Formaldehyde (10%)	4.83 ml
KOH (22%)	5.8 ml
Methyl parasept	1.9 g
Salt W	11.6 g
Sorbic acid	1.9 g
Sugar	40.6 g
Vitamin mix	3.9 ml
Water	774.1 ml
Wheat germ	34.8 g

TABLE 2
Vitamin solutions.

Ingredient	Quantity
Vitamin B_{12} (crystaline)	0.012 g
or	
(0.1%)[a]	(12.0 g)[a]
Biotin	0.24 g
Folic acid	6.0 g
Pyridoxine	3.0 g
Thiamin	3.0 g
Riboflavin	6.0 g
Nicotinic acid amide	12.0 g
D-pantothenic acid	24.0 g
Water	1000 ml

[a]The 0.1% Vitamin B_{12} is preferred because 12.0 g can be weighed more accurately.

Weigh ingredients carefully. Add 1000 ml distilled water to autoclaved blender. Blend 15 seconds. Pour into autoclaved glass jug. Keep lid on jug and refrigerate.

3.2 Brand names and sources of ingredicnts

Agar, casein, wesson salts (U.S. Biochemical Corp., P.O. Box 22400, Cleveland OH 44122); alphacel (Nutritional Biochemicals Corp., 26201 Miles Road, Cleveland, OH 44128); wheat germ (Kretschmer Wheat Germ Products, Minneapolis, MN 55402); MPH methyl parasept (Tenneco Chemicals, Inc., Turner Place, P.O. Box 2, Piscataway, NJ 08854); sorbic acid (Sigma Chemical Co., P.O. Box 14508, St. Louis, MO 63178); sorbistat-K (McKesson Chemical Co., 4909 W. Pasadena Ave., Glendale, AZ 85301); KOH, formaldehyde, acetic acid (VWR Scientific, P.O. Box 29027, Phoenix, AZ 85038); choline chloride (Grand Island Biological Co., 3175 Staley Road, Grand Island, NY 14072); aureomycin (Elco Vet Supply, 6317 Manchester Blvd., Buena Park, CA 90621); Hexcel® (Hexcel Corp., Valley Industrial Park, Casa Grande, AZ 85222).

3.3 Diet preparation procedure (operational)

In order to produce healthy insects, all media preparation procedures and sanitary precautions _must_ be followed. Measuring errors or omission of one or more media ingredients may cause malformed insects or prevent development. Adding vitamins at too high a temperature can degrade the vitamins. Since the media is exposed to the open air in the kitchen, all efforts must be taken to prevent microbial or pathogenic contamination.

3.3.1 Daily work sequence

The following is a suggested guide for daily kitchen operation. Details of each step are described in this section.

- Wash hands and put on clean laboratory clothing.
- Start up boiler.
- Review media orders and plan for proper number and sequence of media batches.
- Wash counters and floors with disinfectant.
- Measure media ingredients.
- Prepare media trays or cups.
- Prepare media in kettle.
- Dispense media.
- Complete media orders as requested.
- Clean kettle and dispensing equipment.
- Autoclave pans and utensils.
- Blow down boiler.
- Clean counters and floors daily; walls and ceilings monthly.
- Prepare liquid media ingredients if needed.
- Shred media.
- Check supplies and reorder if necessary.
- Perform preventative maintenance.
- Report equipment failures or abnormal supplies of media to supervisor.

Several of the above operations may be done concurrently in order to save time and use the equipment more efficiently. When several batches of the same recipe are to be made in one day, it may save time to measure them all at once. Dry ingredients for each batch may be weighed and stored in sealed plastic containers several days until ready for use. Rotate containers using oldest first.

3.3.2 Number of cups per media batch

The requests for media should be arranged according to kind of diet and priority. Table 3 gives the approximate volume of media needed for filling various numbers of different sized cups. By adding the media volumes required for various requests, the size and number of batches may be determined. Make proportional adjustments if larger amounts are used per cup.

TABLE 3
Volume of media required for various sized containers.

Number of cups	3/4-oz. cup (10 ml/cup)	8-oz. cup (80 ml/cup)	16-oz. cup (150 ml/cup)	2-qt. tub (500 ml/tub)
100	1,000ml	8,000 ml	15,000 ml	50,000 ml

Note: 1,000 ml = 1 liter. Requests for hot (liquid) media are normally filled by dispensing diet into clean 4 liter plastic bottles.

3.3.3 Measuring media ingredients

A media check sheet (9.1) is filled out at the start of each weigh-in session and each ingredient checked as it is added to a bag. This sheet is kept in the bag with the weighed ingredients until used. The check sheet and bag of ingredients are placed into plastic tubs, sealed, and stored in the weighing room. Various dry ingredients are combined into appropriate plastic bags or paper cups as follows: bag 1, agar, bag 2, aureomycin, and bag 3, all other dry ingredients.

3.3.4 Media preparation

Each step is checked off the media check sheet as it is performed. Performing the steps as given should provide uniform batches. Since some ingredients make up only 0.1% of the diet, thorough mixing is essential to provide a balanced diet to small larvae. The ingredients should be slowly poured - NOT DUMPED - into the kettle. Although the mixing produced by the mixer appears adequate, adding the ingredients slowly produces a more homogeneous mixture.

Before making the first batch each day, the steam kettle is washed with a 0.1% AI NaClO solution (bleach). Rinse kettle and media hoses with one or two gallons of water to remove any traces of disinfectant. Add required amount of water (12,000 ml or 24,000 ml) at 29°C or less to the steam kettle, verify with dip stick, and add agar. Steam is turned on and the water-agar mixture is stirred while being heated to 93°C. After agar is dissolved, the steam is turned off. Dry ingredients (except aureomycin and ascorbic acid) and remaining water are blended in a 1-gallon Waring blender at highest speed for 20 seconds. These blended dry ingredients, along with liquids (KOH, choline chloride, formaldehyde), are added to the steam kettle while constantly stirring with lightning stirrer. When the diet cools to 70°C, add the vitamins, aureomycin, and ascorbic acid. Turn on media pump and discard water remaining in hose. Recirculate media into the kettle for at least 2 minutes while vitamins and aureomycin are mixing.

3.3.5 Steam kettle cleanup

After media is pumped from kettle, the dispensing hose must be flushed with water before the media gels in the hoses. After flushing with water, the kettle is filled with water and let set for 15-30 minutes. The media is then easily scraped from the sides and flushed out the bottom. If media does gel in the hose, the dispensing nozzle should be removed (to reduce friction) and the pressure relief valve closed. The increased pressure should force the media from the hose. The pressure relief valve must be opened before replacing nozzle to prevent high pressure from damaging the pump. A bottle brush is used to clean diet from the kettle drain.

After all media is cleaned from inside the kettle, it is rinsed out and is ready for the next batch.

After the last batch of the day, the kettle, pump, and hoses are cleaned with a strong liquid detergent solution. Remove kettle drain valve assembly and clean with liquid detergent and Tergisol disinfectant.

3.3.6 Shredding and drying of media

Wheat germ diet is normally held for 2 days in pans before shredding. Trays must be covered with sterile towels during this period and diet handling will be done only with gloved hands.

The Imperial Food Mill is wiped down with 0.1% AI NaClO and rinsed before use. Use the french fry blade that produces 1/4" square shreds.

The media is cut into 2" strips and fed into the shredding machine. A piece of plastic taped to the discharge spout deflects diet particles into a pan. The pans of shredded diet are covered with sterile towels and moved to the infesting room.

Wash removable parts of food mill in dishwasher after each use and autoclave at least once a week. Clean diet from food mill bowl and spout and wipe with 0.1% AI NaClO. DO NOT let water run down motor shaft.

3.3.7 Sterilizing pans and utensils

All pans and utensils that contact media are to be autoclaved at least once a week. When space in autoclave is available, pans and utensils may be autoclaved more frequently. Containers should be stacked in the autoclave on their sides or upside down so steam will displace air. Non-autoclavable brushes, parts, spatulas, or utensils are soaked in 0.1% AI NaClO before using.

3.3.8 Cleaning counters, floors, and walls

3.3.8.1 Kitchen

Clean counters with soap and water and then wipe with 0.1% AI NaClO. Floors are soaked with detergent and 0.5% AI NaClO solution, washed down with hose, and wiped dry with squeegee. Stainless steel brush or scraper will remove stubborn media. Avoid using a mop because media particles remaining on mop provide food for organisms to grow. Media or dry ingredients left on sponges support bacterial and mold growth. Walls may be cleaned with high pressure sprayer. Wet the walls using detergent, then rinse with sprayer. Walls, ceilings, cabinets, and counters are sprayed weekly with a 0.1% AI formaldehyde spray using a portable hand sprayer.

3.3.8.2 Work room and hall entrance

Cabinets and work areas are washed each day with 0.1% AI bleach. Floors are mopped each day with a solution of 0.5% AI bleach. Filters are cleaned every 2 weeks. Walls are washed each month with a bleach solution of 0.1% AI.

3.3.8.3 Laundry room

Lab coats and towels are laundered twice weekly with a solution of detergent and 0.1% AI bleach. Lab coats and towels are dryed in dryer, then autoclaved. Dishwasher, dryer, and washer are wiped each day with 0.1% AI bleach. Floor is mopped daily with 0.5% AI bleach.

3.3.8.4 Rest rooms

Clean all areas with 0.1% Tergisol disinfectant daily. Mop floor daily with 0.1% AI bleach. Wash walls monthly.

3.3.8.5 Infest room

Wipe work area (counter) each morning and evening with 0.1% AI bleach. Mop floor morning and evening using 0.5% AI bleach solution. Walls are washed each month with 0.1% AI bleach solution.

3.3.8.6 Egg handling

Wipe counters each morning and evening with 0.1% AI bleach solution. Wipe egg rack each morning. Soak tank in 0.1% AI bleach solution each day and rinse with water. Mop floor twice daily. Wash walls each month with 0.1% AI bleach solution.

4. HOLDING INSECTS AT LOWER TEMPERATURES

4.1 Pupae

These can be held at 18°C in closed waxed cups for up to 5 days without appreciably lowering adult fecundity or fertility. If pupae are held longer at 18°C, reproduction is lowered in the adult stage. Pupae held at 18°C will show a 1- to 2-day delay in emergence compared to pupae held at normal rearing temperatures.

4.2 Eggs

These can be held at 18°C in plastic bags or tightly closed waxed cups for 1 or 2 days without affecting hatchability. Humidity within the cups must remain above 60% to prevent egg desiccation.

4.3 Larvae

Neonate larvae quickly desiccate at humidities below 60% and therefore should be stored only in the presence of artificial diet. If larvae are placed with diet and held at 18°C, development will be delayed, but larvae will live and resume development when placed into normal rearing temperatures. If larvae remain in low temperatures for longer than 6 or 7 days, diapause will be induced in a percentage of the population even after resumption of development; the longer the larvae remain in the cool temperature, the higher the proportion of diapause induction.

5. REARING AND COLONY MAINTENANCE

The laboratory colony was established from a laboratory colony maintained at the USDA, ARS Insect Laboratory, Tucson, Arizona with introductions of feral insects from Maricopa County, Arizona, in 1971 and 1972. The colony has been in laboratory culture without new feral introductions for over 100 generations.

5.1 Egg collection

5.1.1 Egg pads

Whether paper sheets or Masslinn toweling, egg pads are removed from oviposition cages at the end of the egg collection period. For individual crosses, egg collection can be made at 3- and 4-day intervals if egg pads are placed immediately in diet cups. For mass crosses, egg collection is done daily. Egg pads must be handled carefully to avoid crushing eggs. Crushed eggs act as a focus for mold contamination on the egg pad and from there to the diet.

5.1.2 Egg sterilization

Egg pads from individual crosses are not surface sterilized because of heavy losses of eggs in the process. The following applies to group crosses only.

- Wipe counters and drying rack with 0.5% AI formaldehyde solution or 0.1% AI NaClO.
- Stack egg sheets between screen separators and clamp with paper clips.
- Soak egg sheets in 9.5% AI formaldehyde solution for 30 minutes. Gently rock tank to free entrapped air. Renew formaldehyde solution once a week by mixing 1 part formalin (37.5 - 40% AI) with 3 parts water.
- Transfer eggs to rinse tank and rinse in running water for 30 minutes.
- Lay egg sheets on drying rack to dry.
- Fill request for eggs and notify person that their eggs are ready.
- Wipe out rinse tank with 0.1% AI NaClO after each day's use.
- Place remaining egg sheets in plastic bags and store in incubator at 20°C. Mark egg collection date on bag.

5.2 Diet infestation

5.2.1 Individual crosses

Egg pads are carefully stapled to the bottom of a non-waxed paper lid and the mating designation (code) and Julian date are written on the top of this lid. The lid is placed on a 473 ml waxed food cup half-filled with shredded diet. The crack between the lid and the cup is then coated with melted paraffin. The sealed cup is placed in an 1893 ml plastic tub containing 2 layers of fiber hexcel on the bottom of the tub and a screened lid is placed on the tub to help exclude externally-produced (escaped) larvae and to contain larvae moving from the diet cup during the cut-out period. Holding conditions are the same as for group crosses.

5.2.2 Group crosses

- Wash hands.
- Wash counters with 0.1% AI formaldehyde or NaClO.
- Fill tubs (2.2.3) with shredded diet, using an equal proportion of polystyrene foam beads. For 20 tubs of diet use 10 tubs of foam beads and 10 tubs of diet.
- Cut egg sheets into proper sized pads.
- For 946 ml tubs, use approximately 3,000-4,000 eggs per tub.
- Place egg pad in tub and place lid on tub.
- Write date on lids.
- Stack tubs on shelves in holding room.

5.3 Larval holding and cut-out

5.3.1 Individual crosses

Larvae develop in ca. 13 days and begin to cut out from the paper cup. Those that cut out form a puparium in the hexcel sheets at the bottom of the plastic tub or inside the paper cup against the cup wall and in the diet. Larvae can occasionally escape from screen lids by cutting through the fiberglass screen so care must be taken to protect important crosses from contamination by placing such containers on islands in shallow water pans or by placing such crosses on the highest shelves in the rearing chamber. Clean-up procedures are given in instructions for group crosses.

5.3.2 Group crosses

Infested diet is put in a dark room on rolling racks. The room is at a temperature of 27°C and humidity is 32 ± 10%. Pink bollworms are transferred to cut-out room after 10 days in dark room.

The floors are washed twice weekly by pouring a detergent disinfectant solution around the perimeter of the room. A squeegee is used to push water to the floor drain. The walls are washed in a rotation sequence. The filters for the air conditioners are washed weekly. The dehumidifier pan is emptied on Monday, Wednesday, and Friday or when full.

5.3.3 Cut-out room

The infested ice cream tubs (946 ml) containing wheat germ diet, polystyrene beads, and pink bollworm larvae are transferred (rack rolled) from the dark room to the cut-out room (60 tubs in number). Three pans of polystyrene beads are placed on bottom runners to catch fifth-instar larvae as they cut out of tubs.

The lights are kept on continuously to help control larvae in pans.

Larvae grow in the room with a temperature of 28 ± 2°C and humidity of 40 ± 10%. The tubs are discarded 18-21 days after date of infestation.

Mold is the biggest problem in the PBW rearing tubs so if any musty odor is detected, steps must be taken to control mold growth. If growth of mold is too bad, diet tubs must be discarded. Light infestations can often be controlled by lowering room humidity.

The floor is washed daily by pouring a detergent disinfectant solution around the perimeter of the room. Avoid splashing water into pans of beads. A squeegee is used to push water to the floor drain. The entire room is cleaned monthly with air conditioner filters being cleaned weekly.

5.4 Pupal collection

5.4.1 Individual crosses

Cups and hexcel are held in developmental chambers for 21 days after egg collection then pupae are removed from the cup and hexcel. Pupae in the hexcel are easily harvested by pulling the two hexcel disks from the bottom of the plastic tub where the larvae have firmly attached to it by their silk threads. Those pupae in the bottom disk fall out of the hexcel into the plastic tub. The pupae still in the hexcel are removed by separating the two hexcel disks, gently tapping the hexcel against the inside of the tub to dislodge the pupae into the tub. Pupae remaining in the paper cup are carefully removed from the diet and placed with the pupae in the plastic tub or in a new paper cup. The few remaining larvae are separated from the pupae, counted, and discarded or returned to diet for pupation (depending on the experimental procedure in effect). Pupae are counted, sexed, and records kept according to the cross designation.

5.4.2 Group crosses.

- Hold trays with beads and larvae 5-7 days for pupation.
- Empty beads into fiberglass screen bag and wash with 3% AI NaClO solution for 30 minutes. The NaClO solution must be kept strong enough to dissolve the silk cocoons.
- Rinse with fresh water for 30 minutes.
- Dry beads 3-6 hours depending on quantity and air humidity.
- Run beads through seed separator to separate pupae from beads.
- Rewash beads in NaClO (same dilution as above) for 1 hour, rinse for 30 minutes with fresh water, and dry. Pour beads in storage bin for recycling.
- Measure volume of pupae, record date and volume, and sample 250 ml of pupae and count number of dead and live larvae.

5.5 Adult emergence and oviposition

The sex ratio is approximately 1:1. The developmental period from egg to adult is 27-30 days at 29°C. Females eclose slightly before males (within 12 hours). Adults are held under a photoperiod of 14L:10D for oviposition.

Oviposition starts on the second day after emergence and peaks 6-8 days after emergence.

5.5.1 Individual crosses

Pupae are separated by sex and as the adults emerge they are examined for normal eclosion and complete expansion of wings. Individual crosses are made as appropriate with one male and 3-5 females per cross.

Moth cages are prepared by placing adults in a 273 ml paper cup, placing a screened lid on the cup with a 2 ml vial of sugar water. An oviposition pad is placed on the screen when the cage is prepared. Vials are replaced two times weekly or sooner if sugar water is depleted or contaminated by dead moths.

Cages are kept for 14 days after set-up, living moths are then frozen or placed into hot soapy water. The paper cup is discarded, lids and vials are washed in hot soapy water and rinsed in a 0.1% NaClO solution for reuse.

5.5.2 Group crosses

Moth scales can cause allergic reactions in some people. Since all particles are not removed by the air filtering system, a protective mask is worn while working with adult moths. Floors are washed daily similarly to the cut-out room.

The moth room temperature is maintained at 27 ± 2°C and the humidity above 50%. One hundred fifty pupae are volumetrically measured for each new cage and adults are allowed to emerge in the egg-laying cage. Cages are dated the day they are prepared and again when approximately 1/2 the moths have emerged.

PBW moths are fed a solution containing 90% water, 10% sucrose, and 0.2% methyl parasept (MPH). The solution is prepared by blending 300 g of sugar, 6 g MPH, and 700 ml of hot water for 30 seconds. This solution is added to 2,000 ml of hot water and stored in plastic jugs in the refrigerator.

Vial feeders for pink bollworms are changed daily. The old vials are rinsed to remove dead moths and sugar water. The vials are then soaked in 0.1% AI NaClO solution for 15 minutes, rinsed at least 3 times, and autoclaved.

Pink bollworm cages are cleaned for reuse. Moths are killed by immersion in hot soapy water and the cages are washed with detergent and sterilized with 0.1% AI NaClO. Cages are removed from egg production 14 days after the date of 50% emergence.

Moth scale collection ducts are cleaned each time cages are removed. Ducts are vacuumed and washed with high pressure hot water. Spare ducts are installed while dirty ducts are being cleaned.

5.6 Sex determination

Pink bollworms can be readily sexed as larvae, pupae, or adults. Male PBW larvae have dark-colored testes which can be seen with the unaided eye through the larval integument as early as the third larval instar. This character is reliable up to the pre-pupal stage when changes in the larval integument mask the clear observation of the testes. In the pupal stage, males have the 8th and 9th abdominal segments fused. On the mid-ventral surface of a male pupa two rounded genital lobes can be seen between the last segmental marking and the caudal end of the pupa. Female pupae have a segmental division between the 8th and 9th segment and the genital pore is marked as a dark slit immediately below the segmental marking in the 9th segment. These pupal characters can best be distinguished by use of a dissecting microscope at 8X to 12X magnification.

Adult PBW are easily distinguished since males show a pointed abdomen (because of the hair pencils and other scales surrounding the genital opening), while females have a shorter, blunt appearance of the abdomen (because the scales and tissue form an opening for the ovipositor). After training, an operator can sex adult PBW with the unaided eye, but accurate and rapid sexing requires the use of magnifications up to 8X.

6. QUALITY CONTROL

Monitoring various stages of the rearing operations provides indications of disease problems, errors in procedures, and standard values to compare with new procedures.

6.1 Pupae produced and dead larvae

Measure the volume of pupae harvested. Since there are approximately 25 pupae/ml, the volume of pupae multiplied by 25 is the approximate number of pupae harvested. The following protocol will be followed:

- Sample 250 ml of pupae and gently pour into a large pan.
- Count, remove, and record the number of live larvae in the sample.
- Count, remove, and record the number of dead larvae in the sample.
- Calculate and record percent live larvae, dead larvae, and number of pupae in the sample.

6.2 Pupal weight

Select 20 pupae at random from each batch of pupae harvested, weigh sample of pupae, and calculate and record average pupal weight.

6.3 Emergence

Select 5 random samples of 20 pupae each and put into 10 ml cups. Hold samples in moth room for 14 days, then count the number of pupae that did not emerge and calculate and record percent emergence.

6.4 Egg hatch

Each day that eggs are infested, a sample of 100 eggs will be taken from 3 separate egg sheets. Place egg samples in 10 ml cups, record date on lid, and store in moth room for 7 days, then count and record the number of first-instar larvae in each cup. Calculate percent egg hatch.

6.5 Air contamination

Monitoring of bacteria in the air indicates whether there is a bacterial or fungal problem in the rearing facility. Agar plates are set out every two weeks to take samples of particles floating in the air. The nutrient agar solution (250 ml water and 5.75 g of nutrient agar) is prepared by mixing agar with water and sterilizing in autoclave. Agar is poured into 12 petri dishes (100 × 15 mm). Plates are incubated for 48 hours at 30°C and then opened for 30 minutes in each room as indicated below. At the end of the 30 minutes, the plates are closed and incubated for another 48 hours. The bacteria and fungi colonies are then counted and recorded. If a room has large quantities of bacteria or mold (more than 10 colonies per plate), the room must be thoroughly disinfected. Agar plates are placed in the following rooms: Kitchen, infesting, egg wash, dark room, cutout room, bead room, moth room, and work area. Two blank (unopened) plates are kept to measure internal contamination from the petri dishes and agar.

7. LIFE CYCLE DATA

7.1 Developmental data are given in Table 4. No significant differences in developmental times due to sex have been observed (Bartlett et al., 1980).

TABLE 4 Developmental data for the pink bollworm at 29°C, 30% RH and an LD14:10 photoperiod.

Stage	Number of days Mean (± SD)
Eggs	4.2 (0.1)
Larvae	17.1 (1.3)
Pupae	9.3 (0.8)
Total Development (egg to egg)	32.6 (1.6)

7.2 Survival data (%)

7.2.1 Individual rearing

- Egg viability 82.4 ± 4.1
- Neonate larva to pupa 85 ± 3.9
- Neonate larva to adult 80 ± 3.6

7.2.2 Group rearing
- Egg viability 87.6 ± 3.5
- Egg to pupa 68.6 ± 3.4
- Egg to adult 63.1 ± 2.9
- Emergence from pupae 95.5 ± 1.9

7.3 Other data (at rearing temperatures)
- Preoviposition period, 1-2 days
- Peak egg laying period, 4-6 days after emergence
- Mean fecundity 266.4 ± 108.4
- Mean adult longevity 14.4 ± 3.0 days
- Number of larval instars, 5
- Mean pupal weight (mg), 18.3 ± 0.08

8. PROCEDURES FOR SUPPLYING INSECTS OR DIET

8.1 Placing a insect order

Orders for pink bollworms are made on the following form. Requests for up to 1000 individuals can be handled with 1 week notice. Requests exceeding that number must be made 45 days in advance. Continuing orders can be made with a starting and stopping date and with 45 days advance notice.

INSECT ORDER FORM[1/]

Date requested ______________
Date required ______________
If this is a continuing order place stop date here ______________
How often? ______________
Instar ______________
Type container ______________
Number of containers ______________
Number of insects ______________
Special diet ingredients for rearing (requires 45 day advance notice) ______________

Signature ______________________________
Project ______________________________

8.2 Ordering diet

Insect diet can be supplied fully made or in dry form. Ingredients can be added or left out. The following form must be used for diet requests:

REQUEST FOR PINK BOLLWORM DIET[1/]

Date requested ______________
Date required ______________
Size container ______________
Number of containers ______________
Special handling: (list all special ingredients, etc. needed for diet)
Signature: ________________________
Project: ________________________

NOTEAllow 2 days for requests. Media needed on Monday must be ordered on previous Wednesday.

9.1 WHEAT GERM DIET WORK SHEET

Ingredients Weighed 21,700 X1	Check	43,400 X2	Check	Ingredient	Check	
12,000 ml	____	24,000 ml	____	Water (to kettle)	____	Ing. added
525 g	____	1,050 g	____	Agar	____	cold
4,800 ml	____	9,600 ml	____	Water (blend with dry ing.)		Ing. added to boiling agar
882 g	____	1,764 g	____	Casein		
882 g	____	1,764 g	____	Sugar		
756 g	____	1,512 g	____	Wheat germ		
126 g	____	252 g	____	Alphacel	____	
252 g	____	504 g	____	Salt W		
42 g	____	84 g	____	MPH		
42 g[1]	____	84 g[1]	____	Sorbic acid[1]		
126 ml	____	252 ml	____	KOH (22%)	____	
252 ml	____	504 ml	____	Choline chloride (10%)	____	
105 ml	____	210 ml	____	Formaldehyde (10%)	____	
252 ml	____	504 ml	____	Acetic acid (25%)	____	
84 ml	____	168 ml	____	Wheat germ vitamins	____	Ing. added
20 g	____	40 g	____	Aureomycin (25 gm/lb)	____	to 70° diet

[1] Sorbistat-K may be used. See section 3.1.

DATE ________________ NAME ____

THIS DIET WAS PREPARED FOR: ___________________________________

REFERENCES

Adkisson, P. L., Vanderzant, E. S., Bull, D. L., and Allison, W. E., 1960. A wheat germ medium for rearing the pink bollworm. J. Econ. Entomol. 53: 759-762.

Bartlett, A. C., Butler, G. D., Jr., and Hamilton, A. G., 1980. Developmental rate of the sooty strain of Pectinophora gossypiella. Ann. Entomol. Soc. Am. 73: 164-166.

Beckman, H. F., Brukart, S. M., and Reiser, R., 1953. Laboratory culture of the pink bollworm on chemically defined media. J. Econ. Entomol. 46: 627-630.

Noble, L. W., 1969. Fifty years of research on the pink bollworm in the United States. USDA Agric. Handbook No. 357. 62 pp.

Vanderzant, E. S., 1957. Growth and reproduction of the pink bollworm on an amino acid medium. J. Econ. Entomol. 50: 219-221.

Vanderzant, E. S. and Reiser, R., 1956. Aseptic rearing of the pink bollworm on synthetic media. J. Econ. Entomol. 49: 7-10.

FOOTNOTES

Mention of companies or commercial products does not imply recommendation or endorsement by the U.S. Department of Agriculture over others not mentioned.

1/ These order forms are for use only by scientists in the Western Cotton Research Laboratory. Orders for insects by other scientists or agencies should be by letter to the Laboratory Director. Orders for diet cannot be honored.

PHTHORIMAEA OPERCULELLA

L. K. ETZEL

Division of Biological Control, University of California, 1050 San Pablo Avenue, Albany, CA 94706, USA

THE INSECT

Scientific Name: *Phthorimaea operculella* (Zeller)
Common Name: Potato tuberworm
Order: Lepidoptera
Family: Gelechiidae

1. INTRODUCTION

The potato tuberworm occurs world-wide on potatoes in storage and can cause serious losses on field potatoes in warm climates. In addition to attacking the tubers, larvae can behave as leaf miners and stem borers. The tuberworm can sometimes also be a pest of tomatoes, tobacco, eggplant, and other solanaceous plants.

The larval stage consists of 4 instars. There may be 5 to 8 generations a year in the field in warm climates, with the minimum generation time being about 4 weeks. The primary overwintering stage is the pupa. In stored potatoes, however, the tuberworm can breed year-round at temperatures above 10°C. No diapause is known. An adult female can lay a maximum of about 200 eggs, although fecundity is usually in the range of 80 to 150 eggs.

The potato tuberworm has been insectary reared at the University of California, Berkeley, since the 1940's, first as a factitious host for the braconid, *Macrocentrus ancylivorus* (Rohwer), and then as a source of food for colonies of lacewings and certain coccinellids. The basic rearing procedure has been described by Finney *et al.* (1947). A modified method was described by Platner and Oatman (1968).

An insectary colony can be initiated from field-collected larvae, or from specimens obtained from another insectary colony. Field-collected material is the preferable source, as it is less likely to be diseased.

2. FACILITIES AND EQUIPMENT REQUIRED

2.1 Facilities

All stages of the tuberworm are maintained in our laboratory at about 22.2°C (72°F) and 70% RH. The rearing room is 4.1 x 3.7 m. It is kept dark, except during work periods.

2.2 Equipment and materials required for rearing

2.2.1 Adult anesthetization

- Two 22.7-kg (50 lb) CO_2 cylinders (1 full and 1 empty)
- Copper tubing, 79 mm (5/16 in.), to connect cylinders
- T-connector to attach copper tubing (from full cylinder) and CO_2 regulator to empty cylinder
- CO_2 reduction regulator, 141 kg/cm^2 (2000 lb/in.^{2}) capacity, with flow meter incorporated
- 2000-ml flask containing ethyl ether and long piece of heavy copper wire (Fig. 1)
- Rubber stopper with 2 copper tubing inserts and rubber tubing to connect flask to CO_2 regulator and to moth-holding container

- Bucket of water to hold ether flask (not illustrated)
- 2 square plywood pieces, 30.5 cm per side and 50.8 cm per side, with a hole drilled in the center of each (for rubber tubing to be inserted from ether flask) to cover moth-holding units during anesthetization (Fig. 1 illustrates the 30.5-cm plywood piece)

2.2.2 Adult oviposition and egg collection

- Wooden trays (50.8 x 41.9 x 5.7 cm), bottom is 64-mm plywood, sides are 5.1 cm high x 1.9 cm wide with 1.3 cm deep grooves sawed into the tops; painted with glossy white oil enamel (Fig. 2)
- Muslin cloths, 61 x 50.8 cm
- 4 pieces of 1.3-cm angle aluminum which fit into the grooves to attach muslin cloth on each tray, 2 pieces 48.3 cm and 2 pieces 37.5 cm long (Fig. 2)
- 100-ml graduated cylinder
- Funnel to fit cylinder
- Stiff scrub brush
- Dust mask
- 18.9-liter (5-gal) workshop vacuum cleaner
- Wooden box for stacking oviposition trays, 73.7 cm high x 60.6 cm wide x 45.7 cm deep; trays held on 2.5-cm aluminum angle pieces positioned 7.6 cm apart up the insides of the box; inside dimension between 2 opposing aluminum pieces is 51.8 cm; box holds 8 trays (Fig. 3)

2.2.3 Larval infestation of potatoes

- Rigid plastic bucket, 18.9 liters, with 2 rows each with 6 holes (3.8 cm diam) around sides near the bottom, and a series of 79-mm holes in the bottom
- Potato holding box, 55.9 cm long x 38.1 cm wide x 30.5 cm high with a frame made with 6.4 x 9.9 cm wood pieces; welded 14-gauge (approx. 2 mm) galvanized wire with mesh spaces of 2.5 x 1.3 cm is stapled securely with 1.3-cm (1/2 in.) staples around the inside and on the bottom of the frame (not illustrated)
- "Bed of nails" for puncturing potatoes: piece of plywood, 38.1 x 45.7 x 1.9 cm on which there is a 25.4 x 25.4-cm area where 2.5-cm nails (1.7 mm diam.) are nailed on 64-mm centers, and protrude 64 mm above the board; coated with 3 layers of polyurethane varnish (Fig. 4)
- Heavy leather gloves
- Wooden trays, as described in 2.2.2 (Fig. 5)
- Wooden boxes for stacking trays, as described in 2.2.2

2.2.4 Larval holding

- Metal trays for holding infested potatoes 96.5 cm long x 38.1 cm wide; sides are 1.9-cm galvanized-iron angle pieces bolted together; 2 flat galvanized-iron cross supports, 38.1 cm x 3.2 cm x 32 mm are bolted to a tray 33 cm from each end; tray bottom of welded 14-gauge galvanized wire with mesh spaces of 2.5 x 1.3 cm is bolted to the frame (Fig. 6)
- Open shelves for storing trays

2.2.5 Larval cocooning

- "Barrier" for tuberworm emergence, 91.5 cm wide x 1.5 m long; bottom is 1.9 cm (3/4 in.) plywood to which heavy casters are attached; side pieces are 1.9 cm wide x 6.4 cm high; triangular piece of wood, 7.6 x 7.6 x 10.8 cm (x 6.4 cm high) is attached in each inside corner to eliminate 90° angle; asbestos sheeting, 3.2 mm thick, is glued around inner sides of barrier; inside corner between bottom and sides is coved with wood putty; inside bottom of barrier is sanded smooth and painted with glossy white oil enamel; high resistance 18-gauge (approx. 1 mm) Nichrome wire is snugly attached around inside of asbestos lining of barrier with staples (tiny holes must be drilled through the asbestos in order for the staples to penetrate the wood beneath), wire is slightly zigzagged between staples to allow for expansion and contraction, wire is electrified with about 36 volts from a transformer (rheostat may be used instead) and is consequently heated to about 82°C, which prevents the tuberworms from escaping (Fig. 7)

Fig. 1. Plastic jar in which moths emerge from pupae held in 473-ml (pint) containers. Note the piece of plywood resting on the cloth top of the jar. CO_2 carries ether from flask at left through hose inserted into a hole in the plywood for anesthetization of the moths.

- Rack to hold metal trays of infested potatoes over barrier (constructed of 1.6 cm (5/8 in.) galvanized piping welded together; 7 metal trays held on rack, with tray supports separated by 7.6 cm; square pieces of sheet metal (5.1 cm per side) glued to barrier to provide support for the legs of the rack) (Fig. 7)
- 6 sheets, 25.4 x 30.5 cm of heavy waxed paper (see on bottom of barrier in Fig. 7)
- 1 sheet, 12.7 x 12.7 cm, of heavy waxed paper
- Sand (about 900 ml = 150 ml on each waxed sheet; #1/30 = 30-mesh screen; see on waxed sheets in Fig. 7)
- Simple screen trays made of the same galvanized 14-gauge wire used for metal holding trays: 30.5 x 38.1 cm bottom, with the 4 edges bent up 2.5 cm to form sides (Fig. 7, one of these trays is present on top of the barrier rack)
- 1 piece of galvanized 14-gauge wire, 20.3 x 5.2 cm with the 2 edges in the long dimension bent down to form a raised platform about 2.5 cm high
- Plastic bucket, 11.4 liter (3 gal)
- Scrub brush
- Counter brush

2.2.6 Larval and pupal harvesting
- Plastic bucket, 11.4 liter
- Large colander, 27.3 cm diam x 9.5 cm deep
- Stainless steel mixing bowl, 7.6 liter (2 gal)
- Household bleach (5.25% NaOCl)
- Screen tray (as described under 2.2.5)

2.2.7 Larval processing as predator food
- Plastic bucket, 7.6 liter
- Large colander
- Paper towels

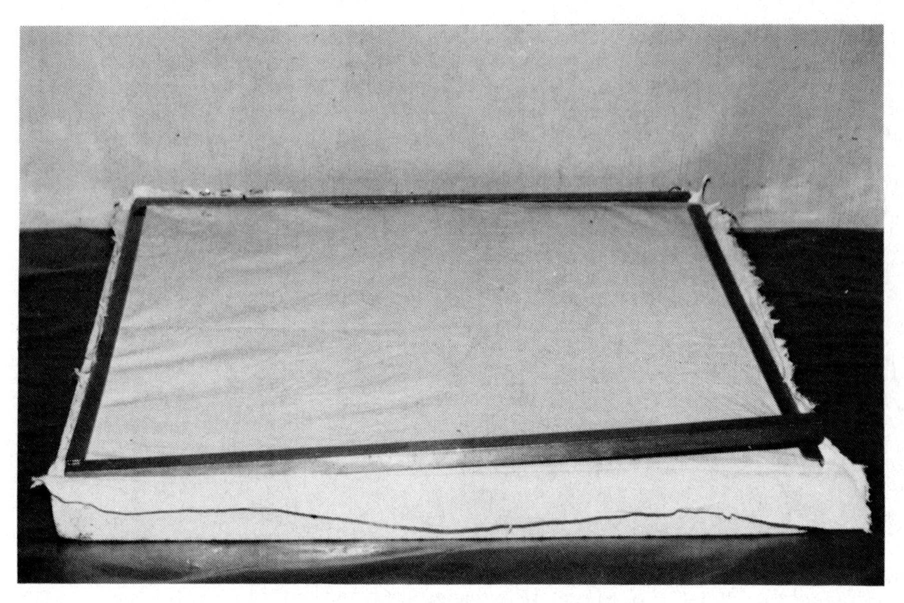

Fig. 2. Oviposition tray. One piece of angle aluminum is slightly raised out of groove in tray side to illustrate how oviposition cloth is held snugly in place.

- Electric household fan

2.2.8 Pupal/adult holding
- Capped, plastic, 473-ml (pint) containers (holes punched with a paper punch at 1.9-cm intervals near the container rim; numerous smaller holes of 1 mm diam made in the lower container sides and bottom (for aeration)
- Plastic jars, 25.4 cm deep x 21.6 cm diam (Fig. 2)
- Muslin cloths, 30.5 cm per side
- Large rubber bands

3. NATURAL DIET

3.1 Variety and source
- Grade B russet potatoes (approx. size range 3.2 x 4.4 cm to 5.1 x 8.9 cm are used in our insectary)
- "Mealy" varieties (reduced water content) provide the best substrate for rearing tuberworms
- Potatoes are obtained in 45.5-kg (100 lb) sacks either directly from a grower, or from a wholesale produce market

3.2 Storage
- Potatoes are purchased 2 or 3 times a year in large quantities
- Storage is in a refrigerated room at 4°C

4. REARING AND COLONY MAINTENANCE

4.1 Adults

4.1.1 Explanation of ether-CO_2 anesthetization equipment

The purpose for using a 22.7-kg CO_2 cylinder instead of a smaller one is to lengthen the time interval between cylinder replacements. However, a 22.7-kg cylinder has an internal pressure of 141 kg/cm^2, and a CO_2

Fig. 3. Boxes used to contain oviposition and infestation trays.

regulator attached to a cylinder with such a high pressure will "freeze" rapidly when CO_2 is allowed to flow through. One reason for connecting two 22.7-kg cylinders (1 empty) together is to eliminate the freezing of the regulator when CO_2 is allowed to flow for a lengthy period of time. The cylinder with the regulator attached is initially empty. CO_2 is introduced into the empty cylinder to a maximum pressure of 35.2 kg/cm^2. When CO_2 from this cylinder is allowed to pass over the ether in the flask, the regulator does not freeze.

The ether flask is held in a bucket of water to help keep the ether at room temperature when the cold CO_2 passes over it (Finney *et al.*, 1947). The bucket also serves the purpose of preventing the ether flask from breaking.

4.1.2 Collection

- Place the smaller piece of plywood (30.5 cm per side) over the pupal/moth holding jar, and insert the rubber tubing from the ether jar into the hole in the plywood (Fig. 1)
- Release CO_2 from the cylinder at a pressure of about 3.5 to 4.2

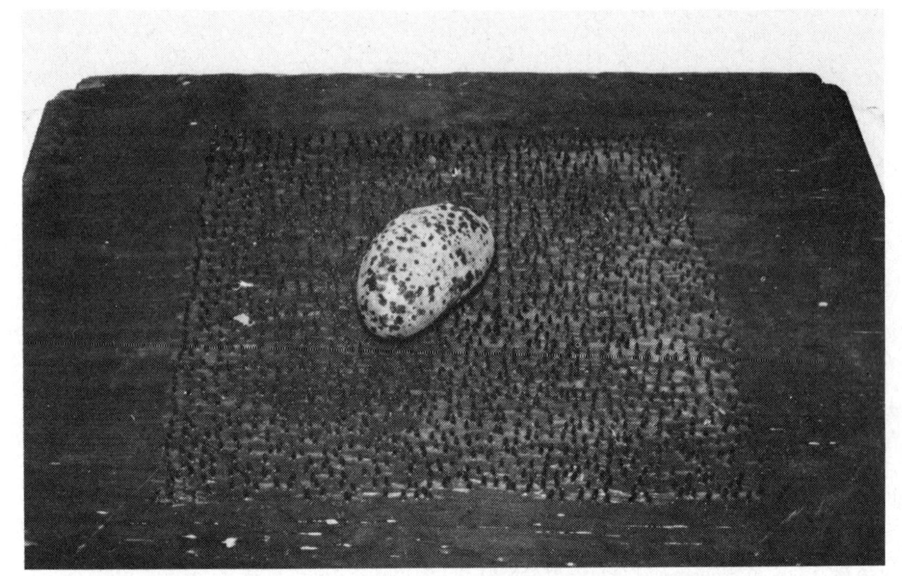

Fig. 4. "Bed of nails" used to puncture potatoes prior to infestation with potato tuberworms.

kg/cm^2 (50 to 60 lb/in.^{2}) for a period of 15 or 20 sec. (The CO^2-ether combination will anesthetize the moths for several minutes, whereas CO_2 alone will only anesthetize them for a few seconds, and therefore does not allow enough handling time.)
- Remove the cloth top of the jar, as well as the pupal containers within
- Pour the anesthetized moths into the graduated cylinder by use of the funnel
- Vacuum out the moth scales in the jar
- Return the pupal containers to the jar and place a new cloth on top of the jar

PRECAUTIONS

i) Moth anesthetization should be performed in a well-ventilated area, preferably in a fume hood with a sealed fan motor. This reduces or eliminates the exposure of the worker to ether and CO_2. Also, it is a precaution against the development of explosive atmospheres, although according to Finney et al. (1947), the use of CO_2 as a carrier gas for ether in the anesthetization device eliminates this danger.

ii) A long piece of heavy copper wire should be placed in the ether flask to prevent the formation of explosive peroxides. The flask should be thoroughly cleaned periodically.

iii) Moth scales are potentially allergenic for personnel rearing the tuberworm. Such allergenic reactions can be prevented or greatly alleviated if the worker uses a snugly fitting respiratory mask with adequate filter. A laboratory coat should be worn as well, since skin irritation can also occur.

4.1.3 Oviposition and egg collection

a. Setting up a clean oviposition unit
- Brush 1 side of a muslin cloth (61 x 50.8 cm) in 1 direction with a dry scrub brush; this raises the cloth nap and stimulates oviposition
- Place freshly emerged anesthetized moths in equal amounts (up to a

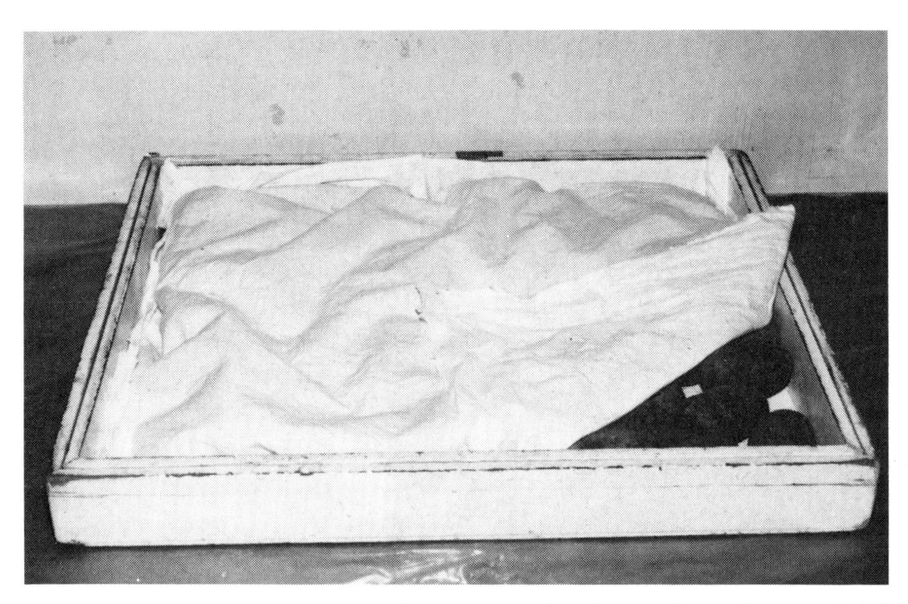

Fig. 5. Infestation tray holding punctured potatoes beneath eggs hatching on oviposition cloth.

Fig. 6. Close-up view of metal tray, showing construction details.

Fig. 7. Barrier and rack holding trays of potatoes. Tuberworms emerging from potatoes drop to barrier beneath and spin cocoons in the sand on the sheets of waxed paper. A heated wire around the inner sides of the barrier prevents larval escape.

maximum of 25 ml per unit) into the oviposition units (Fig. 2); we keep 4 oviposition units in operation for our level of production

- Lay the muslin cloth with the brushed side inward on top of the oviposition unit and fasten it in place by pressing the 4 pieces of angle aluminum on the cloth into the grooves in the tops of the unit (Fig. 2)

b. Servicing an oviposition unit already in use

- Hold the unit in a sloped position with 1 corner down, and tap 1 side with a heavy rubber hose or other suitable object; this knocks all the dead moths in the unit down to the corner
- Anesthetize the moths in the unit by using the anesthetization equipment with the larger piece of plywood (50.8 cm per side) to cover the unit
- Remove the egg-laden cloth and vacuum out the dead moths which were knocked down to 1 corner of the unit

- Knock anesthetized moths to 1 side of the unit and remove accumulated moth scales by vacuuming the bottom
- Place freshly emerged anesthetized moths in equal amounts (up to a maximum of 15 to 17 ml per unit) into the oviposition units
- Install a clean brushed cloth on top of the unit
- Return the oviposition units to the holding box in the rearing room (Fig. 3)
- Label the egg-laden cloths with the oviposition date and hang them in the rearing room for egg development to occur

4.2 Larvae

4.2.1 Infestation of potatoes

- Remove potatoes from the potato holding box and roll each one on the "bed of nails" in a continuous motion forward, back, and on each end. A heavy leather glove must be worn to protect the hand from the nails. The purpose for puncturing potatoes in this manner is to facilitate entry by hatching larvae. The potatoes are rolled just prior to larval hatch so that the puncture wounds will not scab over before larval entry (Fig. 4).
- Fill a wooden tray with punctured potatoes (Fig. 5)
- Take an egg-laden cloth, which has been hung in the rearing room for 4 days, and snugly place it, with the egg side down, over the tops of the potatoes in the tray (Fig. 5)
- Label the tray with a paper tag specifying the date of oviposition and the date the tray is set up
- Place the tray in a holding box in the rearing room (Fig. 3)
- Hold the tray in the box for 4 days until egg hatch and potato infestation are complete
- Remove sufficient potatoes from the cold room to be infested the next day; wash them in the 18.9-liter plastic bucket with a stream of water, and let dry for several hours; transfer the cleaned potatoes to the potato holding box in the rearing room

4.2.2 Holding of infested potatoes

- Move the infested potatoes from the wooden tray to a metal tray on an open shelf; 1 metal tray will hold the contents of 2 wooden trays
- Move the labelling tag from the wooden tray to the metal tray
- Hold the potatoes on the metal tray on the open shelf for 5 days

4.2.3 Emergence and cocooning

- Remove the 6 sheets of waxed paper (on which larvae are cocooned in sand) from the barrier (Fig. 7)
- Place the sheets with cocooned larvae on a small screen tray on the top rung of the barrier rack and put a collection date label with the sheets (Fig. 7)
- Remove the bottom metal potato-holding tray from the barrier rack and transfer the remaining trays down 1 level each
- Move the potatoes around on each tray to dislodge larvae that have cocooned between them, and shake the trays sharply
- Discard the potatoes from the removed bottom tray; later, thoroughly wash this tray with a soapy bleach solution and a scrub brush, and follow with a clear water rinse
- Transfer a metal tray of infested potatoes from the open shelf (after 5 days there) to the 2nd-from-top rung on the barrier rack
- Gently sweep up the larvae and debris on the barrier with a soft counter brush and place the material on a 12.7 cm^2 piece of heavy waxed paper on a small wire platform at 1 end of the barrier
- Place 6 pieces of heavy waxed paper beneath the rack of metal potato trays on the barrier
- Spread out about 150 ml of sand in a thin layer over each piece of waxed paper

4.2.4 Larval harvesting

- Make 7.6 liters of bleach solution (1 part bleach-3 parts water = 1.3% AI final solution) in bucket

- Transfer the sand mass from the waxed paper sheets into the bleach solution
- Agitate the bucket, let it sit for 5 to 10 min, and then repeat the agitation until the cocoons dissolve and release the bound sand
- Pour the bucket contents into the colander, which is resting on the stainless steel bowl or other suitable container (the colander will retain the larvae, while allowing the sand to pass into the receptacle)
- Rinse the larvae with a strong spray of tap water
- Discard the sand

4.2.5 Larval processing as predator food
- Place the colander with harvested larvae into the stainless steel bowl containing hot tap water (about 55°C) for 5 min to kill them
- Spread the dead larvae out on paper towels on a screen tray in front of a fan for drying
- Redistribute the larvae periodically to prevent them from sticking together (it is critical that overdrying does not occur, or the worms will not be suitable food for predatory insects)
- Weigh the larvae when dry, and record the production
- Store the larvae in cardboard container at 4°C (it is best to utilize the dead larvae as food within 3 days)

4.3 Pupae

4.3.1 Pupal harvesting
- Hold 1/4 of the daily collection of sand sheets for 7 days until the larvae have all transformed to pupae
- Follow the same procedure as for larval harvesting (section 4.2.4)
- Spread the harvested pupae out on paper towels on a screen tray for drying for 24 h
- Place the label specifying the date of the sand-sheet collection with the drying pupae

4.3.2 Pupal holding
- Measure the total volume of pupae harvested the previous day with a 100-ml graduated cylinder, and record the production
- Divide the pupae into 2 or 3 lots (with a maximum of 35 ml per lot) and place each lot into a capped, plastic 473-ml container
- Place 2 or 3 containers (473 ml) into a single plastic jar (Fig. 1) (moths later emerge from the pupae in the 473-ml containers, crawl through the larger holes in the same and move about the plastic jar)
- Cover the jar with a square of muslin cloth (30.5 cm per side) held in place by 2 large rubber bands

4.4 Rearing schedule

a. Daily (morning)
- Collect newly emerged moths from 3 holding jars of pupae/adults
- Service the 4 oviposition units and add the fresh moths
- Collect the eggs
- Dismantle the oldest pupal/adult holding jar, from which moths have been collected for 3 days (each jar is in use for 4 days)
- Set up 4 wooden trays with freshly punctured potatoes and with mature eggs (collected 4 days previously)
- Transfer the infested potatoes from the 4 wooden trays set up 4 days before to 2 metal trays which are placed on holding shelves
- Collect the 6 sand sheets from each of 2 barriers
- Save 3 of the sand sheets as a source of pupae for the colony, and place them in a screen tray on top of a barrier rack
- Use 9 of the sand sheets for larval harvesting; these sheets can be held at about 4°C until the harvesting procedure is performed
- Transfer the 2 metal trays which have been on the open shelves the longest (5 days) to the 2nd-from-top support, respectively, of each of the 2 barrier racks, after the bottom tray on each has been removed and the other trays moved down 1 space
- Set up 6 new sand sheets on each barrier
- Harvest the pupae from the 3 sand sheets collected 7 days before

- Set up a clean pupal/moth holding jar with pupae harvested the previous day
- Wash the "bed of nails", 4 wooden infestation trays, 2 metal potato holding trays, and 1 pupal/adult holding jar (with the pupal containers) with hot soapy bleach water and rinse

b. Monday, Wednesday, Friday
- Harvest the larvae
- Process the larvae as predator food

c. Weekly
- Replace the 4 oviposition trays with clean ones
- Carefully wash the bottoms of the 2 barriers with a soapy bleach solution and rinse with clear water
- Wash all muslin cloths
- Clean the rearing room

4.5 Special problems

Diseases are occasionally a problem in the culture. A treatment developed by Allen and Brunson (1947) is useful for controlling a protozoan disease caused by *Nosema*. The procedure consists of treating the eggs in hot water at 48.3°C for 20 min.

Bacterial diseases also sometimes occur in the rearing of potato tuberworms (Steinhaus, 1945). They can be suppressed by preventing high humidities in the rearing room, and by not rearing the insects at temperatures above 30.6°C (Finney *et al.*, 1947). A granulosis virus of the tuberworm exists, but has never appeared in our colony.

PRECAUTIONS

i) Moth anesthetization should be performed in a well ventilated area, preferably in a fume hood.

ii) Moth scales can be allegenic. A snugly fitting respiratory mask and a laboratory coat should be worn by the worker when performing the rearing procedures. A doctor should be consulted if an allergic reaction occurs.

iii) Ether is extremely flammable and must be handled accordingly.

4.6 Production capability

The following production figures apply to the level of activity specified in the previous narrative. Obviously, production can be increased or decreased according to the individual requirements of a laboratory.

About 1-1/2 h per day are spent maintaining the potato tuberworm colony. Harvesting and processing the larvae, which are used as food for lacewings and coccinellids, requires about 30 min 3 times a week. The total time spent per week is therefore approximately 12 h.

Production of potato tuberworms varies with the variety and age of the potatoes, the time they have been held in cold storage, and with the experience of the individuals performing the work. In this laboratory, maximum average production over a 7-month period was 409 ml (267 g) of pupae and 1169 g of larvae from approximately 120 kg of Grade B russet potatoes per week. The biomass yield of larvae and pupae from 45.4 kg of potatoes was therefore about 544 g (1.2% yield on a weight basis). With the mean weight of mature larvae being 12.6 mg, and the mean weight of pupae being 10.3 mg, it can be calculated that 1 potato tuberworm individual was produced for every 1 g of potato. Singh and Charles (1977) found in laboratory experiments that a maximum rearing density of 1 larva per 2 g of potato permitted optimum development, while Broodryk (1971) indicated that there is a rapid decline in survival and pupal weight of *P. operculella* at densitites greater than 1 larva per 1 g of potato.

Fecundity of the moths with this rearing technique (no provision of water) is 85 eggs per female, with 80% of these eggs being laid within 48 h after moth emergence. This corresponds to the egg production obtained by Fenemore (1979) for starved females (83.8 eggs). Fenemore (1979) was able to obtain egg production of 155.4 per female when water was provided.

5. LIFE CYCLE DATA

The tuberworm is currently reared in our laboratory at about 22.2°C and approximately 70% RH. At this temperature, the eggs hatch within 7 days, the larval stage (with 4 instars) lasts 12 or 13 days, the prepupal stage, 2 or 3 days, and the pupal stage 7 or 8 days. The total period from egg to adult is 28 to 31 days. The rearing room is kept dark, except during work periods.

6. REFERENCES

Allen, H. W. and Brunson, M. H., 1947. Control of *Nosema* disease of potato tuberworm, a host used in the mass production of *Macrocentrus ancylivorus*. Science 105: 394

Broodryk, W. W., 1971. Ecological investigations on the potato tuber moth, *Phthorimaea operculella* (Zeller) (Lepidoptera: Gelechiidae). Phytophylactica 3: 73-81.

Fenemore, P. G., 1979. Oviposition of potato tuber moth, *Phthorimaea operculella* (Lepidoptera: Gelechiidae): The influence of adult food, pupal weight and host-plant tissue on fecundity. N.Z.J. Zool. 6(2): 389-395.

Finney, G. L., Flanders, S. E. and Smith, H. S., 1947. Mass culture of *Macrocentrus ancylivorus* and its host, the potato tuber moth. Hilgardia 17: 437-483.

Platner, G. R. and Oatman, E. R., 1968. An improved technique for producing potato tuberworm eggs for mass production of natural enemies. J. Econ. Entomol. 61: 1054-1057.

Singh, P. and Charles, J. G., 1977. An artificial diet for larvae of the potato tuber moth. N.Z. J. Zool. 4: 449-451.

Steinhaus, E. A., 1945. Bacterial infections of potato tuber moth larvae in an insectary. J. Econ. Entomol. 38: 718-719.

ACKNOWLEDGEMENTS

The author is grateful to Dr. K. S. Hagen, Dr. L. E. Caltagirone, and Mr. John Andrews for their suggestions concerning this paper.

PHTHORIMAEA OPERCULELLA

G. W. RAHALKAR, M. R. HARWALKAR, H. D. RANANAVARE

Biology and Agriculture Division, Bhabha Atomic Research Centre, Trombay, Bombay - 400 085, India

THE INSECT

Scientific Name:	*Phthorimaea operculella* Zeller
Common Name:	Potato tuber worm
	Potato tuber moth
Order:	Lepidoptera
Family:	Gelechiidae

1. INTRODUCTION

The potato tuber worm, native of South America, is now an ubiquitous pest on all the continents, especially in subtropical regions. Besides potato (*Solanum tuberosum* L.) it is also a pest on tobacco, egg plant, tomato, capsicum pepper and other cultivated Solaneceous plants. In India, the potato tuber worm causes serious damage to the potato crop and stored potatoes.

The tuber worm female produces an average of 200 eggs and deposits them singly, either on the underside of the leaf or on an eye of the exposed tuber. The newly hatched larva usually makes a blotch mine in the leaf, but later works down into the stem. Some larva fasten several leaflets together with silk and form a protected place to feed. Foliage damage retards crop growth. Tuber injury is either in the form of a shallow burrow just under the skin or a deep tunnel is made into the tuber. At maturity the larva leaves the leaf or the tuber and forms a silken cocoon, which can generally be found among dead leaves or in trash on the ground. The nucleus infestation at storage time comes from tubers left exposed at harvest time.

Finney *et al.* (1947), developed a method for rearing the tuber worm on whole potatoes which yielded 770-900 larvae per kg of tubers. Ever since this method has been used with some modifications. Singh and Charles (1977) described an artificial diet on which it was reared for 3 generations. We have developed a potato slice method which is described here.

2. FACILITIES AND EQUIPMENT REQUIRED

2.1 Facilities

The rearing room is used for holding oviposition cages, larval feeding, pupation units, and adult emergence cages. The room (3.5 x 3.0 m) has double doors fitted with automatic door closers to prevent escape of insects. The space between the 2 doors is 1.5 m^2 and the walls are coated with white oil paint to facilitate cleaning and disinfection. The room is maintained at a constant environment of 29° $\pm$ 1°C and 60-70% RH and is provided with LD 12:12. Light intensity is at least 20 footcandles with cool white fluorescent tubes.

The media room (3.5 x 3.0 m) is used for slicing potatoes, wax coating of slices, egg sterilization, and adult collection. The walls of the room are coated with white oil paint. In the media room there should be 2 work benches (180 x 90 x 90 cm high) with adequate electrical outlets, a stainless steel sink with side drain boards, and a well ventilated fume hood.

2.2 Equipment and materials for insect handling

- Refrigerator (285 liter capacity)

- Carbon dioxide cylinder fitted with a 2-stage pressure regulator and indicators (1 for cylinder pressure and 1 for outlet pressure)
- Anthesthetizing unit consisting of a 500-ml Erlenmayer flask with side arm. A tightly fitting rubber stopper with a central hole in which a copper tube (1 cm diam) is inserted. When the cork is tightly fitted, 1 end of the tube should reach 1 cm from the bottom of the flask and the other end should protrude approximately 5 cm above the stopper, which is connected to the CO_2 cylinder through outlet of the pressure regulator with a rubber tubing (3-mm-inner diam). Another rubber tubing is connected to the side arm of the flask. The unit is kept in the fume hood.
- Autoclave

2.2.1 Adult oviposition

a. Oviposition cage
- Transparent plastic strips (30 x 8 x 1 cm thick) are screwed to form a square frame. All of the 1-cm-thick edges of frame on both the sides are grooved (12 mm deep, 3 mm wide) to accommodate 30-cm-long aluminium angle strips (1 cm).
- 3.5-g glass vial (2 dram)
- Cotton wicks
- Cane sugar
- Coarse black cotton cloth (40 cm^2)
- Hair brush (nylon bristles)
- Ethyl ether

b. Egg collection and sterilization
- Enamel tray (45 x 45 x 5 cm deep)
- Formaldehyde solution (10% active ingredient)
- Metal plate (40 x 40 x 0.2 cm) with a central hole (1 cm diam)
- Ethyl ether
- 1-liter measuring cylinder

2.2.2 Larval rearing

a. Potato slicing and wax coating
- Stainless steel knife (blade 20 cm long, 2.5 cm broad)
- Filter paper sheets or paper towelling
- Insect mounting pins
- Water bath set at 60°C
- Paraffin wax (M.P. 58-60°C)
- 500-ml beaker
- Slice holding tray (62 x 62 cm) cut from 2-mm-thick perforated aluminium sheets (5-mm-diam perforation)
- Potatoes

b. Description and operation of potato slicer (see Fig. 1)
- Sharp edged stainless steel blades (15 cm long and 0.5 cm wide) (A) are mounted in grooves (1 cm apart) between 2 metal blocks. One block is permanently screwed to the main frame made from 2 x 2 cm mild steel angle. Two screws (B) are provided on the second block to adjust tension on the blades. When hand operated metal plate plunger (C) is pressed over potato, it is sliced and the slices fall down into a tray.

c. Description of wax coating machine
- Roller hopper assembly (see Fig. 2)
  A tube (16 cm long) is cut from an aluminium pipe (7.4-cm-inner diam, 8 cm outer diam). Threads are cut on the inner surface at both ends of the tube. Two circular brass flanges (8 cm diam, 1 cm thick) (B) with a central hole (11 mm diam) are screwed over the open ends of the aluminium tube. A 40W heating element (E) is wound around a ceramic base (C) fitted over a hollow brass shaft (25 cm long, 1 cm outer diam, 5-mm-inner diam) (D). Threads are cut from outside at both the ends of the hollow shaft. This heater assembly is placed in the aluminium tube and the brass flanges are screwed in place. Two triangular brass flanges (7 cm base, 10 cm high, 1 cm thick) (G) with a hole (1 mm diam) are then fitted over the shaft at two ends of the roller. Two aluminium

plates (6 cm wide, 10 cm long, 2 mm thick) (F) are screwed on the edges of the triangular flanges. These aluminium plates, together with the

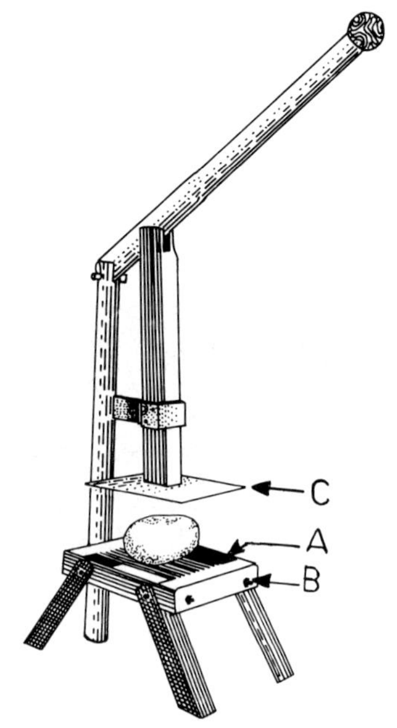

Fig. 1. Potato slicer: (A) stainless steel blades, (B) screws, (C) metal plate plunger.

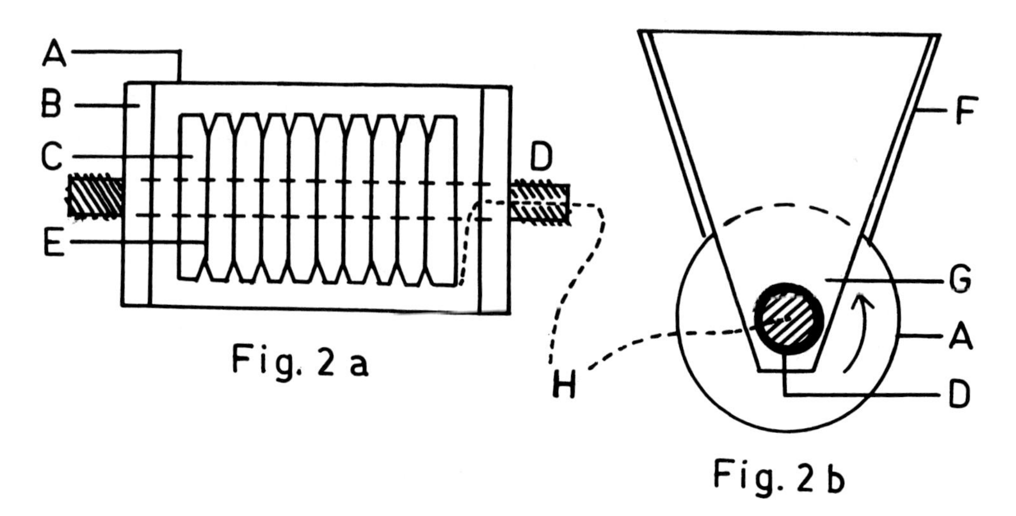

Fig. 2. Schematic drawing showing parts of the roller hopper assembly of the wax coating machine: (2a) roller with heater, (2b) roller hopper assembly. Legend: (A) hollow aluminium tube, (B) brass flanges, (C) ceramic base, (D), hollow metal shaft, (E) heating element, (F) aluminium plates, (G) triangular brass flanges, (H) wire to heating element.

triangular flanges, form a hopper over the roller and hold the paraffin wax.

- Base (see Fig. 3): a frame (45 x 90 cm) (B) is made from 2.5 x 2.5 cm perforated steel angles and an aluminium sheet (45 x 90 cm) (C) is fixed over the frame. Slots (2.5 x 1.3 cm) are cut (15 cm apart) in 3 rows on the aluminium sheet (5 slots in a row 15 cm apart). Brass rollers (2 cm diam, 1 cm wide) (D) are fixed in each slot from the bottom and protrude about 0.5 cm above the sheet. Three rollers in a line move around a common axle.

Fig. 3. Potato slice waxing machine: (A) roller hopper assembly, (B) stand, (C) base plate, (D) metal rollers, (E) energy regulator.

The roller hopper assembly (A) is mounted on 2 vertical metal supports fitted to the base frame. Height of the roller from the base is adjusted with the help of 2 screws fitted to the vertical supports. Energy regulator with indicator lamp is mounted on the vertical support (E). Heating element in the roller is connected to the main power supply through the energy regulator.

Operation:

- Potato slices are arranged on the perforated aluminium trays (slice holding tray) very close to each other
- Adjust the height of the roller from the base so that the roller surface touches the slices
- Paraffin wax is kept in the hopper
- Energy regulator is adjusted to heat the roller to the paraffin wax melting temperature
- Tray, holding slices, is pushed under the roller. The roller surface picks up a thin film of molten wax and simultaneously transfers the wax over the slices. The tray is moved sideways so that during 4 passes all the slices are coated.

Fig. 4. Pupation unit.

- A second tray is placed over the slices, the trays are inverted and the other surface of the slices is given a wax coating by repeating the process

2.2.3 Pupation (see Fig. 4)

a. Pupation unit

A cabinet (64 cm wide, 64 cm deep, and 70 cm high) with a removable door is made from 14-mm-thick plywood. Two windows (40 x 40 cm) are cut, 1 in the top of the cabinet and the other on the door. These windows are covered with 90-mesh nylon netting. Top and bottom of the cabinet protrude 2 cm in the front to accomodate the door. On the 2 sides of the cabinet from inside 2 strips of steel standards (65 cm long) are fixed 55 cm apart. Metal shelf clips are attached to each standard at 2.5-cm intervals. The unit is mounted on 4 revolving casters. Spring clamps are fitted on the cabinet body to hold the door tightly in place. The unit accommodates 24 trays.

b. Pupal collection

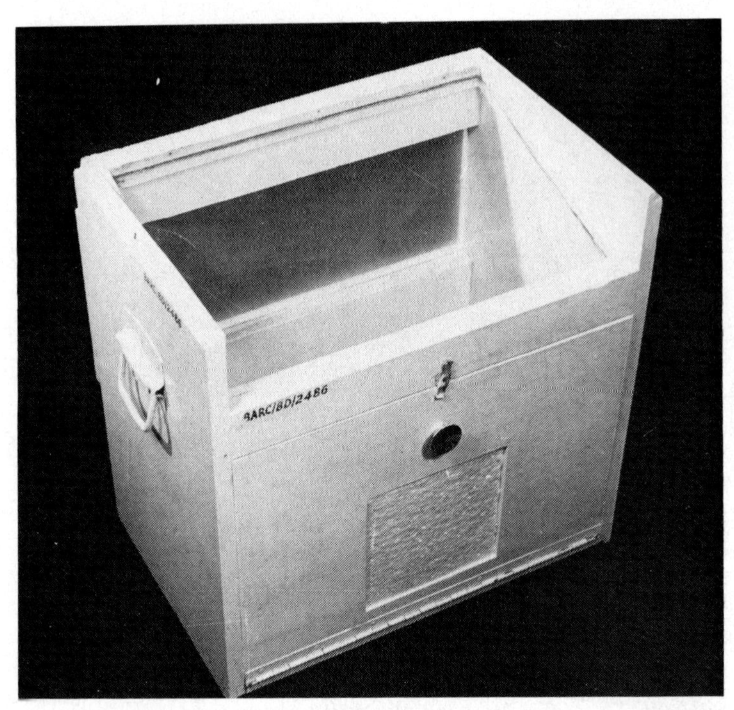

Fig. 5. Adult emergence cage.

- Pupation board - flat aluminium sheet (63 x 63 x 0.2 cm thick)
- Fine river sand
- Sieve (20 mesh)
- Flat piece of sheet metal (15 x 10 x 0.1 mm thick)
- Rectangular plastic or polythene cocoon holding tray (15 x 10 x 8 cm deep)

2.2.4 Adult collection
- Adult emergence cage (46 wide x 30 deep x 45 cm high) made from 2-cm-thick seasoned wooden planks (Fig.5). The cage has a slanting glass top, a hinged door, and a nylon-screen covered window (42 x 18 cm) at the back.
- Metal plate (45 x 20 cm) with a hole (1.5 cm diam) at the centre

2.2.5 Brand names and sources
- Ethyl ether (AR grade), formaldehyde solution (37-40%) and paraffin wax, Glaxo Laboratories (India) Ltd., Bombay-400 025, India

3. NATURAL HOST

Potatoes, any variety with 25% dry matter, 75% water, and Sp. Gr 1.05 to 1.1. We use variety 'kufri chandramukhi'.

4. INSECT HOLDING AT LOWER TEMPERATURES

4.1 Eggs

Egg sheets can be stored in polythene boxes with tight fitting lids. Eggs up to 48-h old can be stored up to 8 days at 10-12°C without affecting hatchability.

4.2 Adults

Newly emerged adults can be stored at 10°C for 2 days without affecting survival, fecundity, and egg viability.

5. REARING AND COLONY MAINTENANCE

The founder colony of P. operculella was established from field collected adults. Several generations were initially reared on whole potatoes and subsequently on potato slices.

5.1 Preparation of oviposition box

- Gently brush the black cotton cloth pieces with the hair brush to obtain a dense mat of cotton fibres. This can be seen by holding the cloth against light. Eggs are laid on the matted surface.
- Fill the 2-dram (3.5 g) vial with 10% sugar solution and insert the cotton wick until it reaches bottom of the vial. Dispense more sugar solution on the wick if required.
- Keep the oviposition cage on top of a table and place the cotton cloth over the frame with matted surface facing downward
- Press aluminium T-strips over the cloth so that it is wedged in the grooves stretching the cloth on all sides to hold it tight
- Anethestize moths
- Invert the oviposition cage and place 100 pairs of freshly emerged moths over the cloth surface
- Cover the cage with the second cloth piece as described
- Tightly fix the glass vial with cotton wick into the hole on the side of the frame
- Transfer the cage to the rearing room (if more cages are used they can be stacked one above another in a criss-cross fashion to allow aeration)

5.2 Egg collection

- Two days later transfer the oviposition cage(s) to the fume hood, anesthetize moths and remove the cloth piece; place a new cloth piece and fasten it as described
- Invert the cage and remove the second cloth piece
- Add about 20 pairs of newly emerged adult moths
- Fix a new cloth piece
- Check the glass vial; replenish sugar solution and change the wick if it is covered with too many scales or eggs
- Transfer the oviposition cages to the rearing room
- Repeat the process 2 days later
- Discard the moths and use fresh pairs
- The cloth pieces removed from the oviposition cage are referred to hereafter as egg sheets

5.3 Egg sterilization

Immediately after egg collection

- Place the enamel tray under the fume hood and pour 250 ml formaldehyde solution and add 750 ml distilled water
- Dip the egg cloth in the solution for 2 h with periodic shaking
- Remove the egg cloth, rinse several times in distilled water and air dry

5.4 Infestation of potato slices

- Thoroughly wash potatoes with tap water to remove all dirt and dry
- Place paraffin wax in the beaker and keep it in the water bath set at 60°C; allow the wax to melt
- Slice potatoes (1 cm thick) with the knife or use the slicer
- Press the slices between 2 sheets of filter paper or paper towelling to remove excess moisture
- Momentarily dip slices in the molten wax or use waxing machine and place them flat on the perforated aluminium tray very close to each other and

store for 3 or 4 days
- At the time of use, puncture the wax layer with the insect mounting pin (approx. 10 to 12 punctures per slice); periodically sterilize the pin by heating on a flame
- Place 3-day-old egg sheets on the slices facing down; use 4 egg sheets per tray; stack the tray in the larval feeding and pupation unit
- Cover approximately 3/4 area of the pupation board with 0.5-cm-thick layer of fine sand, thoroughly washed, dried, and passed through a 20-mesh sieve; introduce the board into the larval feeding unit and transfer the unit to the rearing room

PRECAUTION

Wax coated slices are stored for 3 or 4 days before use. Any latent fungal or bacterial infection or contamination during slicing would appear within this period. Discard such spoiled slices.

5.5 Collection of pupae
- Ten days after infestation of the slices, remove the pupation board and store in the rearing room for 3 days to complete pupation; replace with a new board
- Remove pupation board daily for 2 more days (3 pupal sheets are collected for a given set of infested slices and slices are discarded)
- Hold the pupation board on edge over a table top and scrape the pupae from the board with the flat piece of sheet metal
- Pour sand and cocoons into the woven wire sieve to separate excess sand from the cocoons
- Collect naked pupae and cocoons and place them in the shallow metal tray (some pupae are dislodged from their cocoons during scraping)
- Cover the tray with a steeply pitched cardboard roof and transfer to the adult emergence cage

5.6 Adult collection
- Transfer adult emergence cage to the fume hood, anesthetize emerging adults with CO_2-ether vapor mixture, collect and transfer to the oviposition cage
- Repeat adult collection daily for 6 days; discard cocoons

PRECAUTIONS

(i) Knife or insect mounting pin should be sterilized by autoclaving for 15 min at 1.05 kg/cm^2 before use.

(ii) Infested slices and cocoons are autoclaved before discarding them.

5.7 Quality of insects
- Mean pupal weight: males 6.25 to 6.95 mg; females 7.5 to 8.65 mg
- Avg fecundity: 155.2 eggs per female is acceptable

6.0 LIFE CYCLE DATA (29°C, 60-70% RH, LD 12:12)

- Egg hatch, 3 or 4 days
- Egg to prepupae, 13 to 15 days
- Egg to adult, 17 to 22 days

7. YIELD

When 2 oviposition cages with 100 pairs each are set daily and egg sheets are removed on alternate days, 4 egg sheets carrying approximately 15,000 eggs will be available each day. One tray holding approximately 240 slices (2.7 kg) is exposed every day to 4 egg sheets. With 90% egg hatch and 50% larvae developing into adults, yield would be approximately 2,500 adults/kg potatoes.

8. REFERENCES

Finney, G. L., Flanders, S. E., and Smith, H. S., 1947. Mass culture of *Macrocentrus ancylivorus* and its host, the potato tuber moth. Hilgardia 17: 437-483.

Singh, P., and Charles, J. G., 1977. An artificial diet for larvae of the potato tuber moth. N. Z. Jour. Zool. 4: 449-451.

PIERIS BRASSICAE

BRIAN O. C. GARDINER

ARC Unit of Insect Neurophysiology and Pharmacology, Department of Zoology, University of Cambridge, Downing Street, Cambridge, CB2 3EJ, England

THE INSECT

Scientific Name: *Pieris brassicae* Linnaeus
Common Name: Large white butterfly
Order: Lepidoptera
Family: Pieridae

1. INTRODUCTION

The large white butterfly, *Pieris brassicae* L., is a pest of *Brassica* crops and a migrant throughout the Palearctic region, has subspecies in the Himalayan region and on the Atlantic islands. Recently it has become established in the Neotropical region (Chile) (Gardiner 1974) where its range is expanding rapidly. It remains a serious pest of *Brassica* crops, particularly in the eastern half of its range and in Chile. Larvae can devour the entire plant but the majority of the damage is due to spoilage. Feltwell (1980) has quantified this damage at $200 million per annum with an additional potential of $20 million in the Neotropical region.

Depending on latitude, there are from 1 to 4 generations per year and diapause occurs in the pupal stage induced by a photoperiod of less than 16 h light per day on the larvae. The larvae are gregarious in all stages and may readily be reared on either their natural food plants or on artificial diet. Consequently this butterfly has been very extensively used by numerous workers for both physiological studies and the testing of insecticides. A comprehensive review of the literature on *P. brassicae* has recently been published by Feltwell (1982).

Anyone planning to establish a culture of *P. brassicae* should start from an existing stock which is already adapted to laboratory conditions. If local stock or race is desired, initial difficulties will be encountered as pointed out by David and Gardiner (1961). Briefly, wild adults do not feed readily at artificial flowers and batter their wings trying to escape. They will require hand-feeding for the first few generations and the larvae reared with more care to eliminate diseased and parasitized individuals.

The procedures described below are similar to those of David and Gardiner (1952) and David (1957), with some modifications. The rearing method is relatively simple since no specialized equipment or environment is required. It is also economical, requires a minimum of labor to keep it in production, and produces a large surplus of eggs (4000 per day) and immature larvae every week. Both eggs and pupae can be retained for several weeks under cool conditions and this enables constant supplies to be maintained.

2. FACILITIES AND MATERIALS

2.1 Facilities

- Glasshouse
- Constant temperature room set at 20°C
- Refrigerator

2.2 Facilities

a. Adults

- A large cage (1000 x 900 x 750 mm) of wood or metal frame with mosquito type netting on the top and 2 sides. The other 2 sides should be glass. On 1 side the netting should be fitted with fasteners (zip or velcro) designed for easy access for cleaning, introduction of butterflies, and placement and removal of cabbage leaves on which eggs have been oviposited.
- Two or more artificial flower feeders (Fig. 1). These are made of 10-mm thick perspex (leucite) sheet, 200 mm in diam around the rim of which 12.5-mm holes are drilled and into these 40-mm-long glass tubes are glued, flush with the top. Around these is painted a 40-mm white circle and on this circle 8 petals are then painted in a navy blue. (Cellulose paints supplied for touching-up car bodywork are used). One of these 'flowers' is positioned near the floor of the cage and the other about half-way up. The tubes are to be filled with 10% sucrose solution.

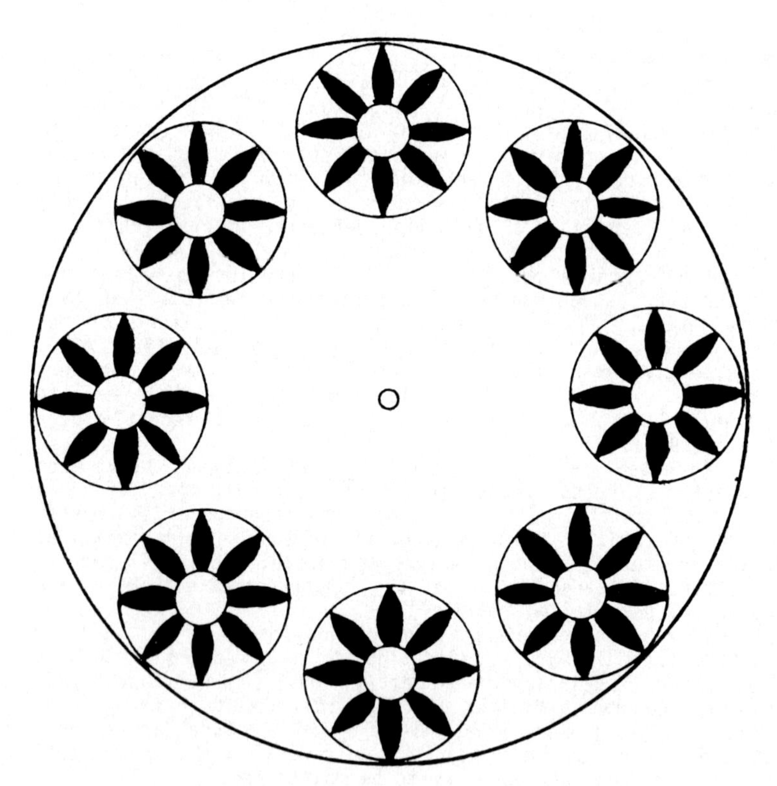

Fig. 1. Design of artificial flower.

- Plastic squeeze bottle for sugar solution
- 10% sucrose solution
- Supply of 200- to 300-mm high cabbages in 90-mm diam pots

b. Larvae

- Time switch and lights

- The most suitable cage is described in detail by Gardiner (1958, 1982). It has a wooden frame with a solid base and top with the sides covered in coarse synthetic-mesh fabric glued on with waterproof "cascemite" casein glue (obtain from hardware or do-it-yourself shops). It has a glass front held on by turn buttons. These cages can hold up to 500 larvae. Ten to 12 cages are needed.

3. MAINTENANCE OF THE CULTURE

3.1 Adults

- Keep cage in the glasshouse; maintain nighttime temperature at approx. 15°C, daytime temperature 20-30°C
- Fill the vials with sugar solution in the artificial flowers 2 or 3 times per day
- Supply a freshly potted cabbage daily (except weekends)
- Remove dead adults Monday through Friday
- Add 200 newly emerged adults weekly
- Wipe cage and wash flowers weekly
- Keep the cage at 15-20°C during weekends

3.2 Larval food

The larvae are fed throughout their life on various species of Cruciferae, particularly *Brassica* cultivars, as available in season. These plants may either be grown by the rearer or purchased from local growers. The best leaves are the large outer ones which, if picked regularly off cabbages, causes the inner leaves to constantly expand and so maintain the supply of large outer leaves. The cultivars 'January King' and 'Ellams Early' are excellent. Also ideal are 'Bolters' (plants running up to seed) and the tops of brussel sprouts after the sprouts have been harvested. When accepting a fresh supply from commercial sources it should be checked to ensure that it has not been treated recently with insecticide. If in doubt test it first against some small larvae. *Brassica* may be stored for up to a week under refrigeration. 'Primo', 'Dutch Pride', or similar hard white- or yellow-leaved cultivars are not suitable.

3.3 Larvae

These are kept in a constant temperature room at 20°C and LD 18:6 controlled by a time switch. One 100W tungsten or an 80W fluorescent bulb is sufficient.

Low intensity light should be supplied by a small 7.5W bulb (which is either on another time switch or else left permanently on) during the dark phase.

- Feed once daily
- On Fridays, feed twice the amount (for the weekend)
- Line floor of cages with several layers of old newspaper
- Put cabbage plant, upon which eggs are just hatching, into a clean cage, each week
- Start 1 or more further cages, as required
- Weekly, inspect older cages and discard any surplus larvae (aim at 200 final instar larvae per cubic foot of cage); cages may be cleaned by lifting up the cabbage on which they are feeding and sliding out the top layer of newspaper
- When 80-90% of larvae have gone up to the sides and roof of the cage to pupate, clean out and discard the remainder
- Wash cages when adults emerge

3.4 Chrysalids

- The larvae will pupate on the top and sides of their cages; nothing needs to be done to them
- Butterflies are transferred to the adult cage as they emerge, over 2-3 days; it may help to place the cage in the refrigerator for 30 min to decrease their activity

3.5 Eggs

- Are laid in batches on the potted cabbages

- Monthly, pot 30 cabbages in 90-mm plastic pots
- Monthly, sow some fresh cabbage seed
- Keep plants with eggs separate in the constant temperature room
- Discard (or use for supply) eggbearing, potted cabbage plants after the eggs have hatched

4. LIFE CYCLE DATA

TABLE 1. The length of various stages in the life cycle at differing temperatures.

Temperature (°C)	Eggs days to hatch	Larval instars 1	2	3	4	5	Total	Pupae
30.0	4.0	2.0	1.5	2.0	2.0	3.5	11.0	8
28.0		2.0	1.5	2.5	1.5	3.5	11.0	
25.0	4.5	2.8	1.7	1.8	2.2	3.5	12.0	
20.0	6.25	3.5	3.0	3.0	4.0	5.0	18.5	14
16.0	9.0							23
12.5	15.0	11.0	8.0	7.3	8.3	12.0	46.6	40

5. DIAPAUSE AND TEMPERATURE MANIPULATION

5.1 Diapause

- This is induced by rearing the larvae under a photoperiod LD 9:15
- Diapause chrysalids can be stored for 6 months at 20°C or 1 yr at 4°C
- Diapause is broken by exposing pupae to 4°C for 12 weeks and then emergence occurs after 3-4 weeks at 20-25°C

5.2 Temperature manipulation

- Eggs, when at least 48 h old, can be stored for up to 2 weeks at 4°C
- Larvae may be held back for 4 days at 4°C
- Chrysalids (nondiapause) should not be stored below 10°C
- Adults can be stored for several weeks at 4°C but gradually lose virility

6. PARASITES AND DISEASES

6.1 Parasites

- The Hymenopteran, Apanteles glomeratus L., has been known to attack the larvae
- The Chalcid, Pteromalis puparum L., attacks the chrysalids; they should be vacuumed up and destroyed and white-colored chrysalids should also be destroyed

6.2 Diseases

- Larvae may become infected with granulosis virus (symptoms: larvae turn yellow and then liquify); Infected larvae should be destroyed
- If the outbreak appears serious, set up numerous individual egg batches, rear them separately in large plastic vials, discard any as soon as the virus infection is detected
- Occasionally a bacterium, which seems to be introduced accidentally with the food, attacks the final instar larvae and the chrysalids (it can wipe out an entire cage while others placed alongside remain perfectly healthy); its spread is promoted by gross overcrowding and excessive humidity

7. SUPPLYING

The culture produces a large surplus of eggs, which are kept until hatching. The weekly thinning down of the larvae produces a potential surplus. Requests can usually be met from these 2 sources. If the request is for chrysalids or adults, then either extra stock cages are started or, for small numbers, the surplus larvae from the first thinning are reared through in a separate cage.

8. REFERENCES

David, W.A.L., 1957. Breeding *Pieris brassicae* (L.) and *Apanteles glomeratus* (L.) as experimental insects. Z. PflKrankh. Pflpath. PflSchutz., 64: 572-577.
David, W.A.L. and Gardiner, B.O.C., 1952. Laboratory breeding of *Pieris brassicae* L. and *Apanteles glomeratus* L. Proc. Roy. ent. Soc. Lond. (A), 27: 54-56.
David, W.A.L. and Gardiner, B.O.C., 1961. The mating behaviour of *Pieris brassicae* (L.) in a laboratory culture. Bull. ent. Res., 52: 263-280.
Feltwell, J., 1980. The depredations of the Large white butterfly (*Pieris brassicae*)(Pieridae). J. Res. Lepidopt., 17: 218-225.
Feltwell, J., 1982. Large white butterfly: the biology, biochemistry and physiology of *Pieris brassicae* (Linnaeus). Dr. W. Junk, The Hague. 535 p.
Gardiner, B.O.C., 1958. Making a moth cage. Hobbies Weekly, 126: 168.
Gardiner, B.O.C., 1974. *Pieris brassicae* L. established in Chile; another palearctic pest crosses the Atlantic. J. Lepidopt. Soc., 28: 269-277.
Gardiner, B.O.C., 1982. A Silkmoth Rearer's Handbook. Amateur Entomol. Soc., London. 256 p.

SPODOPTERA ERIDANIA

DONALD P. WRIGHT, JR.

Agricultural Research Center, American Cyanamid Co., Princeton, NJ 08540

THE INSECT

Scientific Name:	*Spodoptera eridania* (Cramer) Formerly *Prodenia*
Common Name:	Southern armyworm
Order:	Lepidoptera
Family:	Noctuidae

1. INTRODUCTION

The southern armyworm is a climbing cutworm found in the southern part of the United States. Of minor economic importance, it is perhaps the most commonly used laboratory caterpillar for insecticide research due to its omnivorous habits and its ease of rearing. The former characteristic not only makes it possible to rear it on many different host plants, but allows the evaluation of insecticides upon a variety of foliage with the maintenance of only one species of caterpillar. The eggs are laid in masses of about 100-150, primarily on the underside of the host plant leaves, and are covered with hairs from the female moth. They hatch in a few days into smooth, blackish-brown caterpillars bearing fine, yellow, longitudinal stripes dorsally and laterally. The larvae pass through five instars and reach a length of 30-40 mm after about 18 days before pupating in the soil. The pupal stage lasts about 10 days. The dull brownish moths have a wingspan of 40-45 mm and mate within a day of emergence. Following a pre-oviposition period of 2 days, the females lay eggs for several days. The moths live for 7 to 13 days.

This insect has been reared in American Cyanamid Laboratories since 1938, using procedures originally adapted from Waters (1937). It is probable that this and all other colonies of the southern armyworm were derived from that maintained by Waters at Ohio State University. Subsequent sharing of subcultures by the insecticide research groups of the major US chemical companies and many universities has led to a research insect that has given remarkably uniform responses from laboratory to laboratory and from year to year.

2. FACILITIES AND EQUIPMENT REQUIRED

2.1 Facilities

Rearing is carried out in a room maintained at 27° C and 40-60% rh. The room is continuously illuminated by fluorescent lamps.

2.2 Equipment and materials required for insect handling

2.2.1 Adult oviposition

- Screen cage 45x45x45 cm. Although these dimensions are not critical, such a cage will hold enough moths to produce > 1000 larvae per day. Cages with removable lids are more convenient than those with an access from the front.
- Black cloth to cover the top and sides of the cage.

2.2.2 Egg collection and storage

- Pots of bean plants, each containing about 6-8 plants with the primary leaves well expanded.

2.2.3 Egg hatch and larval innoculation of food
- 15 x 25 cm flats of bean plants in the 2-leaf stage.

2.2.4 Water pans
- Large pans capable of containing water 25 to 50 mm deep are useful for maintaining humidity in the moth cage and around the plants containing egg masses. They also make for ease of watering of plants and serve to help trap the larvae should they fall or migrate from the bean plants. These pans should be capable of being drained and flushed daily.

2.2.5 Larval rearing containers
a. Early instars
- 15 x 25 cm flats of bean plants in the 2 leaf stage.

b. Last instar
- Feeding enclosures as per Figure 1.
- Cardboard or blotting paper strips 5 cm x 50 cm.

Figure 1. Feeding enclosure

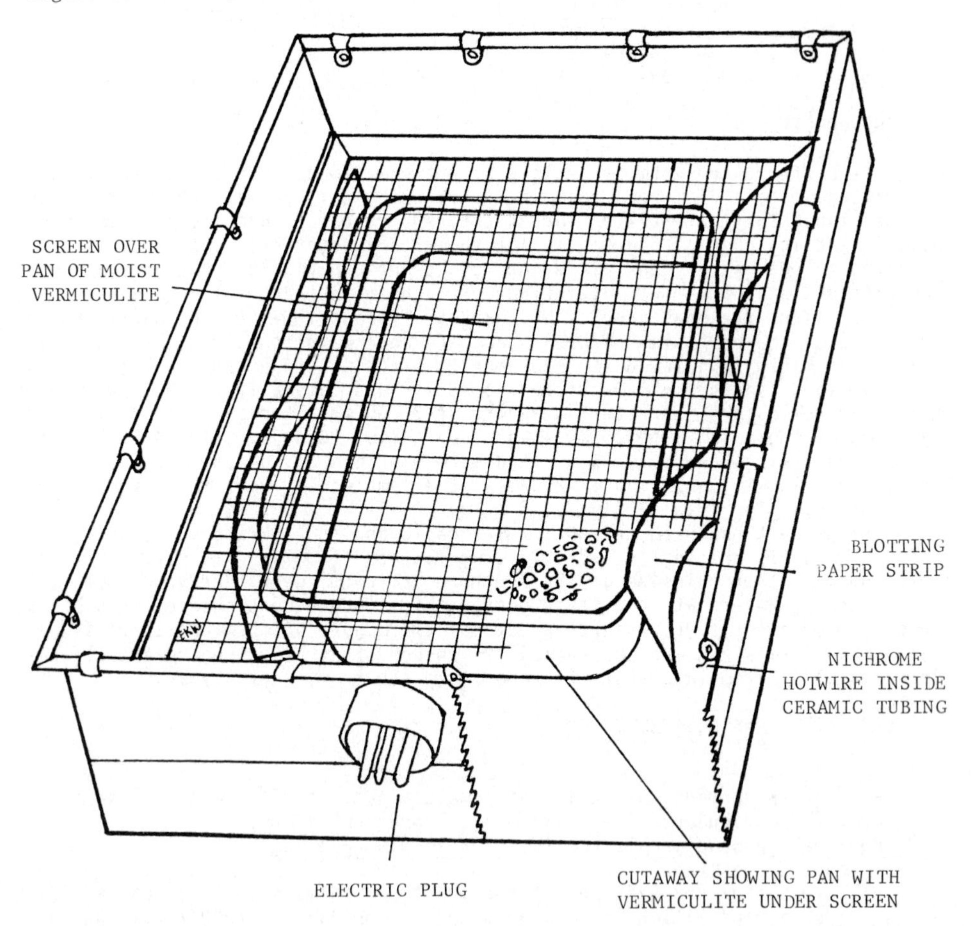

2.2.6 Pupation pans
- Stainless steel, enamel or plastic pans about 30x19x6 cm.
- Vermiculite. This must be fine enough to pass 6 mm mesh.
- Glass panes 21x32 cm.

2.3 Diet preparation equipment

- Flats to grow bean plants, about 15x25x6 cm.
- Pots to grow bean plants, about 9 cm diameter and 6 cm high.
- Bean seeds, any hairless-leaved variety of Phaseolus.
- Lettuce.

2.4 Brand names and sources

- Vermiculite. Zonolite Brand Industrial Insulation, net weight 10 lbs. Grace Construction Products, Cambridge, MA 02140. This must pass 6 mm mesh screen.

3. DIET PREPARATION

- Two weeks before use, bean seeds are planted in a greenhouse at 27°C. The plants are used when the primary leaves are well expanded, but before the trifoliates develop. There should be about 8-10 seeds per 9 cm pot and about 100 seeds per 15x25 cm flat. Fertile potting soil is used and the usual greenhouse practices are observed.
- Lettuce is purchased from local suppliers year-round. All of the outer leaves should be removed to insure that larvae are fed only insecticide free leaves from the interior of the head. It should be noted that this simple precaution is critical to the successful rearing of these caterpillars, due to almost universal use of insecticides in the commercial production of lettuce.

4. REARING AND COLONY MAINTENANCE

4.1 Egg collection

- Remove from the moth cage the four pots of bean plants on which the eggs from the previous day were laid. Most of the egg masses are on the undersides of the leaves. Set the pots aside until the eggs hatch in about 3 days.
- Remove dead moths from the cage.
- Crush any egg masses laid inside the cage.
- Place 4 fresh pots of bean plants in the cage.

4.2 Larval handling

a. Early instars

- Eggs turn black shortly before hatching. The leaves containing the dark eggs and/or newly emerged larvae are removed from the plants and distributed evenly over the foliage of a 15x25 cm flat of bean plants in the 2-leaf stage. Such a flat will support 1000 1st instar larvae, about 10-12 large egg masses.
- As the larvae consume the foliage, additional bean flats are placed adjacent to the older flats. After the larvae move to the new foliage, the old flats can be discarded.
- The caterpillars are usually used for testing in the third instar and those not needed for rearing can be discarded about a week after they emerged. To insure 1000 3rd instar larvae for daily tests, 200 larvae must be reared through to adults every other day.

b. Last instar

- After 15 days, larvae consume flats of bean plants so rapidly that it is necessary to transfer them to a feeding enclosure (Figure 1) where they can be fed lettuce.
- Break up a head of lettuce and distribute leaves over the screen to produce a layer about 25 mm thick. DO NOT USE THE OUTER LEAVES!
- Place 200 of the 15 day old larvae on the lettuce.
- Install the electric fence or a tight-fitting screen lid.
- Add lettuce twice daily as needed. Do not add more lettuce than the larvae will consume between feedings.

4.3 Pupation

- Prepare a pupation pan by putting 1200 cc of vermiculite and 400 ml of water in it and mixing thoroughly.
- Place the pan in the bottom of the feeding enclosure.
- Install cardboard or blotting paper strips to help larvae climb back into the pan of pupation medium when they fall into the bottom of the enclosure.
- Install the screen.
- Add enough lettuce to cover the screen.
- Install the electric fence or screen lid.
- The larval color pattern becomes more diffuse, and less food is consumed as the time for pupation approaches after about 8 days in the feeding enclosure. At this time decrease the amount of lettuce given in order to avoid a thick layer of partially-eaten food in which the larvae may pupate.
- When all larvae have descended into the pupation medium, the pan of vermiculite is removed from the feeding enclosure. Any food debris on the vermiculite is removed and the pan is covered with a pane of glass. The pan does not have to be kept in the dark.
- The feeding enclosure is cleaned thoroughly before re-use.

4.4 Adult handling

- The moths begin to emerge in about ten days. Each morning all newly-emerged moths are removed from the pupation pans and transferred to the moth cage. This cage is kept on blocks or legs in a pan of water to maintain humidity in the cage. Cover the cage with black cloth to encourage oviposition.
- The sex ratio is about 1:1.
- A new cage is set up at about monthly intervals and the old one is thoroughly cleaned before re-use.

4.5 Oviposition

- There is a 2 day pre-oviposition period.
- Moths from one pupation pan give good egg production for about two to three days.

4.6 Rearing schedule

a. Daily
- Remove the bean plants with eggs from the moth cage. Keep them in a pan of water until the eggs hatch.
- Remove dead moths from the bottom of the moth cage.
- Crush any egg masses laid inside the cage.
- Put 4 pots of bean plants in the moth cage for oviposition.
- Collect newly emerged moths and place all in the moth cage.
- Disperse newly hatched larvae on flats of bean plants.
- Discard larvae beyond test size if they are not scheduled for rearing through to adults.
- Place new flats of bean plants next to any flats needing replacement.
- Remove old flats no longer containing larvae.
- Put lettuce in feeding enclosures containing older larvae. This is best done early and late in the day rather than all at one time.
- Drain and flush water pans.
- Plant bean seeds for anticipated needs 2 weeks hence.

b. Alternate days
- Set up a feeding enclosure.
- Transfer about 200 15-day-old larvae into the enclosure.
- Set up a pupation pan in another feeding enclosure for larvae that are 20 days old.
- Remove pupation pans from enclosures when all larvae have descended into the pupation medium. Remove any food debris and cover the pans with a pane of glass.

4.7 Special problems

a. Parasites

Phorid flies and house flies (Musca domestica) can, on occasion, lay eggs in the pupation pans when the armyworms are in late stages of development. The subsequent maggots devour the prepupae and pupae, resulting in loss of all of the moths from that pan. Attention to general cleanliness in the rearing room, care to avoid over-feeding the larvae with lettuce, and the removal of all food debris from the pupation pans will help avoid this problem. In those cases where flies cannot be eliminated from the rearing room, it may be necessary to use a screen cover on the enclosures.

b. Food supply

i. Bean plants should be a smooth-leaved variety, as the first instar larvae do not feed readily on hairy leaves.

ii. Before using lettuce, it is essential to remove the outer leaves to avoid those that are contaminated with insecticides.

iii. Although this insect is most readily reared on bean plants and lettuce,its omnivorous habits allow the use of almost any locally available foliage or even slices of carrots or potatoes. It is also possible to rear this insect on artificial diets of the Vanderzant type.

c. Egg hatch

A common problem encountered in rearing this armyworm is poor hatch of eggs. Proper hatch depends on conditions under which the eggs and moths are kept. In addition, fewer than 25 moths per day added to the cage results in poor mating success and infertile eggs. Such eggs appear dessicated and yellowish in contrast with the normally greenish eggs. Many eggs from the first and last few days of oviposition are also yellowish and do not hatch. The eggs must be kept at > 50% rh in order to hatch. This is most readily accomplished by keeping pots of bean plants with the egg masses in a pan of water until the eggs hatch, rather than keeping detached leaves in humidity chambers.

d. Pupation medium

Vermiculite coarser than a 6mm mesh is not suitable for pupation. If such fine vermiculite is not available, sand can be used with some success, but more care must be used to maintain the proper moisture content. If either medium is too wet, it results in poor survival of the pupae and if too dry,it results in dead pupae or defective moths.

e. Moth cage

The humidity in the moth cage must be kept > 50% for good mating, oviposition and egg hatch. If this is difficult to accomplish, the cloth covering the cage can be kept wet by letting it hang into the water beneath the moth cage.

f. Human health

Avoid inhalation of moth wing scales.

5. LIFE CYCLE DATA

The following are typical data for various stages of the life cycle when the southern armyworm is reared in the laboratory at 27°C and 40-60% rh fluorescent illumination.

5.1 Developmental data

Stage	Number of Days		
	Min.	Mean	Max.
Eggs	2.5	3	3.5
Larvae	17	18	20
Pupae	8	10	14
Adults	7	9	13
Total development (egg to egg including pre-oviposition)	27	30	36

5.2 Survival data

- Egg viability: variable; poor on first day of oviposition, 90% on the 2nd and 3rd days, $<$ 50% on the 4th day and declining thereafter.
- Neonate larvae to pupae: 59%.
- Pupae to moths: 88%.
- Neonate larvae to moths: 52%.

REFERENCES

Waters, H. A. 1937. Methods and Equipment for Laboratory Studies of Insecticides. J. Econ. Entomology 30:179-203.

SPODOPTERA EXIGUA

RAYMOND PATANA

USDA-ARS , Biological Control of Insects Laboratory, 2000 East Allen Road, Tucson, AZ 85719, USA

THE INSECT

Scientific Name: *Spodoptera exigua* (Hubner)
Common Name: Beet armyworm
Order: Lepidoptera
Family: Noctuidae

1. INTRODUCTION

The beet armyworm is a foliage feeder of a wide range of field and vegetable crops around the world. The beet armyworm has been reared at the Tucson laboratory since the late 1950's, when it was used for insecticide screening tests. At that time and into the 1960's larvae were fed on cotton leaves during the summer and on chard during the winter months. Cultures were rarely maintained for any extended time periods due to virus disease. These frequent disease outbreaks were probably attributable to the use of natural food sources in most cases. Shorey's (1963) lima bean diet was first used at the Tucson laboratory in 1963. This diet was modified by Patana (1969) and a continuous beet armyworm laboratory culture has been maintained since 1965 without new introduction from the field. From the mid-1960's to the present time, the beet armyworm has been used as a host for predator and parasite research in our laboratory (Bryan et al. 1971, 1973, 1976). Stock for starting new cultures have been sent from this laboratory to numerous locations throughout the United States and to several foreign countries.

2. FACILITIES AND EQUIPMENT REQUIRED

2.1 Facilities - See *H. zea* 2.1 (Vol. II, p. 329)
2.2 Equipment and materials required for insect handling
2.2.1 Adult oviposition - See *H. zea* 2.2.1 (Vol. II, p. 329)
2.2.2 Egg collections and sterilization - See *H. zea* 2.2.2 (Vol. II, p. 330)
2.2.3 Egg hatch and larval inoculation of diet -See *H. zea* 2.2.3
2.2.4 Larval rearing

a. Parent culture
- Squat waxed containers, .177 liter (6 oz) No. 6S-BG Lily® (Owens-Illinois, Lily-Tulip Division, Toledo, OH 43601)
- Paper tab lids, N562A Lily®

b. Large-scale production
- Plastic utility boxes (UB-200), 8.9 x 25.4 x 34.3 cm (Sterling Products Co. Inc., St. Paul, MN 55118)
- Tissue, triple thickness wipes, Kaydry® (Kimberly-Clark Corporation, Roswell, GA 30076)

2.2.5 Pupal collection - See *H. zea* 2.2.5 (Vol. II, p. 330)
2.3 Diet preparation equipment
2.3.1 Pre-mix, mixing, and dispensing

a. Small batch - See *H. zea* 2.3.1 (Vol. II, p. 330)
b. Large-scale production - See *H. zea* 2.3.1

3. ARTIFICIAL DIET

3.1 Composition - See *H. zea* 3.1 (Vol. II, p. 330)
3.2 Brand names and sources of ingredients - See *H. zea* 3.2 (Vol. II, p. 331)
3.3 Diet preparation procedure
3.3.1 Dry mix (small and large scale) - See *H. zea* 3.3.1(Vol. II, p. 331)

4. REARING AND CULTURE MAINTENANCE

The original laboratory culture was established on diet in 1965 and has been reared continuously and for over 300 generations.

4.1 Larval stage
- Newly hatched larvae are sprinkled with a small brush on diet surface of .177 liter squat waxed cup, 20/cup, daily for parent colony.
- Cups are closed with a paper tab lid, dated and held at 27°C.
- Newly hatched larvae sprinkled on diet surface of plastic utility boxes with diet 150-300 larvae, depending on size wanted. A three-ply wiping tissue is placed over the box and it is covered with the lid.
- Larval rearing containers are held at 29°C.

PRECAUTION

Warm diet containers to room temperature and blot condensed water off surface before implanting.

4.2 Pupal stage
- Pupae are harvested after 12 days.
- Diet cups are washed and pupae are floated off into a screen sieve
- Pupae are surface sterilized with a 0.03 percent sodium hypochlorite solution, then rinsed with water.
- 75-150 pupae may be placed into 3.79 liter glass jars, lined with plastic bags (8.9 x 12.7 x 33 cm) blown up inside of the jars to fit the extremities of the jars. The bag is dated before placing it in the jar.
- One paper toweling strip 5 cm wide x 28 cm long is hung down inside the jar and the top of the jar is covered with a 12.7 cm^2 paper toweling held in place with a cutout jar lid.
- The jars are then held at 26°C for emergence.
- Sex ratio is 1:1 so sexing is not required.

4.3 Moth emergence
- Moth emergence begin in about 2 days.
- When first moth emerges, a feeding vial (2.5 ml screw top containing 10% honey) with the neck inserted through a hole in a 3 cm^2 piece of plastic screen over a hole in the toweling top of the jar.
- Mating occurs shortly after emergence.
- Oviposition begins 2 days after emergence.

4.4 Oviposition

a. Egg collection
- Oviposition jars are checked daily after emergence for eggs.
- When eggs are laid, toweling strips and top are changed and watering vials are filled daily.
- Moths are kept for 4 days after oviposition begins, the insert bag is then removed, tied and discarded.

b. Egg sterilization
- Egg-bearing toweling strip and tops are placed in Buchner funnel.
- The toweling is covered with a 0.03 % (AI) sodium hypochlorite solution for 5 minutes.
- The bleach solution is drawn off with a vacuum pump.
- The eggs are rinsed with water and this is then drawn off.
- The toweling strips are then spread on a wire tray, covered with tissue and allowed to dry overnight.

- Eggs from previous day are placed into 3.79 liter plastic containers held at 26°C and allowed to hatch.

4.5 Rearing schedule

a. Daily
- Implant newly hatched larvae on diet cups or boxes.
- Discard excess larvae.
- Feed moths.
- Collect and sterilize eggs.
- Discard old moths.

b. Monday, Wednesday, Friday
- Wash pupae (12, 13, 14 day old).
- Surface sterilize pupae.
- Put up pupae into jars for emergence.

4.6 Insect quality

In a long lasting insect culture, one maintained 1 to 5 years in the laboratory, the most meaningful indicator of insect quality is continued reproduction and survival. The current culture has been in the laboratory for over 18 years. When the culture was originally started, there were a number of "selections" made. The most obvious of these "selections" were the ability to adapt to a semi-artificial diet, and to adapt to the conditions of the laboratory. The rearing routine as used at this laboratory is such that it provides a limited mixing of genetic materials rather than a continuous program of inbreeding. This is accomplished by putting new larvae up daily from a mixed group of eggs. Essentially there is a random selection of larvae from parents reared over a 4-6 day period. Multiplying the number of eggs laid per female by the number of females laying per day, the chances of this mixing occurring are increased further. The entire rearing system's success is probably due to this mixing of genetic material which came more as a result of chance rather than design.

4.7 Special problems

Mold contamination

No moisture is added into the beet armyworm larval rearing spaces at the Tucson laboratory. All larval rearing containers are designed to "dry out" to a slight degree as the larvae develop. This system probably works better in a dry climate. This "drying" is thought to be the most important factor in controlling mold contamination.

5. LIFE CYCLE DATA

Life cycle data and survival data under normal rearing conditions, photoperiod LD 15:9. For additional development and oviposition data at different temperatures, see Fye and McAda (1972).

5.1 Developmental data

Stage (Temp. 25° 27°C)	Number of days
Egg (hatch)	3
Larvae to pupation	12
Pupal stage	2
Egg to adult emergence	17

5.2 Survival data (%)
- Egg hatch - 85-95
- Pupal yield (.177 liter cup) - 80
- Moth emergence - 95

5.3 Other data
- Preoviposition period - 1-2 days
- Peak oviposition period - 2-3 days after emergence

- Mean fecundity at 20°C - 1522/♀ (Fye and McAda, 1972)
- Longevity at 20°C - male 15.3 day, female 12.9 days

6. REFERENCES

Bryan, D. E.; Jackson, C. G.; Patana, Raymond; Neemann, E. G. 1971. Field cage and laboratory studies with Bracon kirkpatricki, a parasite of the pink bollworm. J. Econ. Entomol. 64: 1236-1241.

Bryan, D. E.; Fye, R. E.; Jackson, C. G.; Patana, Raymond. 1973. Releases of parasites for suppression of pink bollworm in Arizona. USDA, ARS, W-7. 8 p.

Bryan, D. E.; Fye, R. E.; Jackson, C. G.; Patana, Raymond. 1976. Non-chemical control of pink bollworms. USDA, ARS, W-39. 26 p.

Fye, R. E.; McAda, W. C. 1972. Laboratory studies on the development, longevity, and fecundity of six lepidopterous pests of cotton in Arizona. USDA, Tech. Bull. 1454. 73 p.

Patana, Raymond. 1967. A pressure paint tank modified for use as a dispenser of insect diets. J. Econ. Entomol. 60: 1755-1756.

Patana, Raymond. 1969. Rearing cotton insects in the laboratory. USDA, ARS, Prod. Res. Rep. 108. 6 p.

Patana, Raymond. 1977. Rearing selected western cotton insects in the laboratory. USDA, ARS, W-51. 8 p.

Shorey, H. H. 1963. A simple artificial rearing medium for the cabbage looper. J. Econ. Entomol. 56: 536-537.

SPODOPTERA LITTORALIS

A. NAVON

Division of Entomology, ARO, The Volcani Center, Bet Dagan, Israel

THE INSECT

Scientific Name: *Spodoptera littoralis* (Boisduval)
Common Name: Egyptian cotton leafworm
Order: Lepidoptera
Family: Noctuidae

1. INTRODUCTION

The Egyptian cotton leafworm is a major pest of economic crops, predominantly cotton, alfalfa, clover, sugarbeet and vegetables. The larva feeds on about 40 plant families containing not less than 87 host plants of economic importance (Avidov and Harpaz, 1969). It is widely distributed in Africa and the adjacent islands, Middle Asia and the circum Mediterranean region.

The species formerly known as *Prodenia litura* F. (Bishara, 1934) later became synonymous with *S. littoralis* in central and northern Africa (Viette, 1963). In the Persian Gulf and the Arabian Peninsula, the dividing line between *S. littoralis* and *S. litura*, a species found in the Far-East, is not clear and possibly they overlap. The identification of the two species (Mochida, 1973) would therefore be necessary prior to rearing of *S. littoralis*. In the main, the larva is a leaf-feeder, sometimes also damaging cotton buds, peanuts, tomato or green pepper. In the East Mediterranean region, up to 7 annual generations develop. Rapid migration occurs in the larva and adult stages. Under natural conditions the larval period lasts 15 days at 26°C, with minimum of 12 days at 30°C. The adults are polygamous and feed during their lifespan of 4-10 days. Preoviposition period is 1-2 days and neonates hatch from eggs after 3-4 days, at 25°C.

Several artificial diets have been composed to rear the species successfully (Moore and Navon, 1964, 1969; Levinson and Navon, 1969; Poitout *et al*. 1972; Kehat and Gordon, 1975; Navon and Keren, 1980).

2. FACILITIES AND EQUIPMENT

2.1 Facilities

Larvae are reared in a room 4 x 5 m illuminated by fluorescent and natural lights with a photoperiod of LD 16:8, temperature 24 ± 2°C, and RH 50-70%. Pupae are kept in an incubator at 20°C for moth emergence. Mating and oviposition are carried out in cages at 20-23°C.

2.2 Equipment and materials required for insect handling

2.2.1 Adult oviposition

a. Single-pair mating
- Plastic cup (200 ml)
- Paraffin paper sheets (24 x 35 cm)
- Rubber bands
- Cotton-wool plugs

b. Group mating
- Aluminum fly-net cage (40 cm diam x 120 cm)
- Water supply: test tube with 5% sucrose, upside down in solution pool

- Roll of paper towelling, 27 cm wide
- Scissors

2.2.2 Egg collection and storage
- Plastic cup
- Soft tissue paper
- Paper towelling
- Rubber bands

2.2.3 Egg sterilization
- Glass vial, 500 ml
- Scissors
- Short, metal forceps

2.2.4 Larval rearing containers

a. Individual rearing
- Plastic vials (4 cm diam x 5 cm) with plastic lid with hole (diam 1 cm) in center
- Squares of filter paper, Whatman No. 1 (2 x 2 cm)
- Plastic rack for vials

b. Group rearing
- Circular plastic box (14 cm diam x 7.5 cm) with tight-fitted plastic lid with hole (diam 10 cm)
- Plastic box (25 x 33 x 7.5 cm) with flat rim and a bottomless box of same size
- Roll of paper towelling, 32 cm wide
- Plastic grid (32 x 28 cm) mesh size 70 holes/m^2, equipped with 4 wooden legs at grid corners
- Vermiculite particle size 2-4 mm

2.2.5 Diet preparation equipment
- Waring® blender
- 15-liter mixer equipped with manual rotary scraper (A. Stephan u Sohne 325 Hameln, West Germany)
- 1-liter measuring cylinder
- 50-ml measuring cylinder
- Large, flat, plastic spatula
- Top loading balance (0.1-1000 g)

2.2.6 Finished diet
- Plastic tray (40 x 60 cm)
- Refrigerator (6°C)

3. ARTIFICIAL DIET

3.1 Composition

	Amounts	% (estimated)
a. Nutrients		
- Soybeans (autoclaved with water)	900 g	12.86
- Alfalfa meal	170 g	2.43
- Yeast powder	300 g	4.28
- L-ascorbic acid CP	26 g	0.37
b. Antimicrobials		
- Methyl-p-hydroxybenzoate	7.5 g	0.11
- Propyl-p-hydroxybenzoate	2.5 g	0.03
- Chloramphenicol	3.0 g	0.04
- Formaldehyde 37%	25.0 ml	0.36
c. Alginate-gel mixture		
- Sodium-alginate "Protanal SF"	130 g	1.86
- Calcium carbonate (precipitated)	6 g	0.09
- Citric acid CP	70 g	1.0
- Distilled water	5360 ml	76.57
Total diet	7000 g	100.0

3.1.1 Soybean treatment
- Add water to beans in equal weight
- Autoclave the beans-water mixture at 1.05kg/cm^2 (15 psi) for 20 min

- Cool to room temperature
- Weigh amount required for diet

3.2 Brand names and sources of ingredients

Soybeans (U.S.A.) (natural, raw); alfalfa meal (Granot, Israel) (natural, raw); whole milk powder (Nursia, de Bommelewaad-Heusden, Holland); yeast powder, *Saccharomyces cervisiae* (Fould-Springer, Maison-Alfort, France); L-ascorbic acid CP, methyl-p-hydroxybenzoate, propyl-p-hydroxybenzoate, citric acid, calcium carbonate (precipitated) (Merck, Darmstadt, West Germany); chloramphenicol (Abic, Israel); sodium-alginate "Protanal SF" (Protan, a/s Drammen, Norway).

3.3 Diet preparation

The diet is prepared at room temperature. Nutrients and antimicrobial fractions are mixed in Waring blender with 2000 ml distilled water. Ascorbic acid and formaldehyde are not included in the mixture.

- Add homogenate to industrial mixer
- Dissolve sodium-alginate while mixing gradually with 3300 ml distilled water in the Waring blender
- Add the alginate solution into the mixer
- Mix in the calcium carbonate and formaldehyde
- To set the diet, add dry citric and ascorbic acids to the mixer and mix for 10 sec at a speed of 1450 rpm and stop; let the diet gel
- Take out the gelled diet from the mixer after 30-60 min and use portions for breeding
- When not in use, store diet at 6°C

4. HOLDING INSECTS AT LOWER TEMPERATURES

- Moths can be stored at 17°C; below this temperature fertility drops
- Larvae can be stored at 6°C for 24 h
- Pupal emergence is affected at temperatures below 15°C

5. REARING AND COLONY MAINTENANCE

Egg clusters from the field were used to establish *S. littoralis* colony on the calcium-alginate diet. The colony has been reared continuously on the diet for over 70 generations at a production rate of 100-200 moths per day.

5.1 Egg collection

a. From plastic-paraffin paper cup (single-pair mating)

- Transfer moth pair to new container
- Add new moth diet of 5% sucrose absorbed on cotton-wool plug
- Cut egg clusters from the paraffin paper using a pair of scissors
- Transfer the egg clusters into plastic cup closed with polyethylene cloth
- Incubate the eggs at 17-20°C
- Collect eggs every 2nd day

b. From aluminum fly-net cage (group mating)

- Release paper towelling with egg clusters from the cage
- Cut egg clusters using scissors and incubate in plastic cups at 17-20°C
- Insert into the cage the new paper towelling sheet from the paper roll
- Remove dead moths from the bottom of the cage and add freshly emerged moths every 2nd day

5.2 Egg sterilization

Egg sterilization is optional. It is needed when epizootic is suspected in the insect colony.

- Dip egg clusters in 1.5% active chlorine solution containing a few drops of Triton X-100; dipping time is 10 sec
- Rinse the sterilized eggs in running water for 5 min
- Dry eggs at room temperature
- Put sterilized eggs in plastic vials containing wet tissue paper, close vial with polyethylene cloth and incubate at 17-20°C

5.3 Diet innoculation with neonate larvae

a. Individual rearing

- Condition diet by allowing to stand at room temperature for 2 h
- Put filter paper square on bottom of the plastic vial and a cube of diet (1 x 1 x 2 cm) placed on the filter paper
- Innoculate the diet with one neonate larva by a small camel-hair brush
- Close vial with paper towelling and lid
- Add fresh diet every 2nd day

b. Group rearing

- Put a double sheet of paper towelling on bottom of the circular box
- Place 100-g diet portion on top of the paper
- Innoculate the diet with 500 neonate larvae by a camel-hair brush
- Add fresh diet after 6 days

5.4 Late instar and pupal collection

- Transfer the larvae after 10-11 days to the rectangular box, containing 1 liter of vermiculite, the plastic grid and 2-3 portions, 100 g each, on top of the grid (Fig. 1)
- Add fresh diet every 2nd day
- Remove plastic grid
- Collect pupae from the vermiculite
- Wash pupae in running water and place them on paper towelling at room temperature to dry
- Transfer pupae to rectangular rearing box lined with paper towelling
- Incubate pupae at 17-20°C
- Female pupae have a greater distance between the genital and anal pores than the males; sex pupae accordingly

Fig. 1. Container for late instar larvae and pupation.

5.5 Adults

a. Single-pair rearing
- Moths emerging in small plastic vials are paired and transferred to plastic-paraffin paper vials
- Introduce adult solution absorbed on cotton-wool plug

b. Collective rearing
- Transfer moths from emerging boxes into aluminum fly-net cage
- Insert paper towelling used as an oviposition site inside the cage
- Fill the water supply device (essentially a test tube hung upside down) with moths' feeding solution: 5% sucrose in aqueous solution
- Fix the feeding device to the aluminum net
- Remove dead moths and add freshly emerged moths every 2nd day

5.6 Sex determination
- Males have white line in inner wing edges but females do not

5.7 Rearing schedule

a. Daily
- Innoculate diet with neonates in circular boxes
- Transfer 10-day-old larvae from circular to rectangular collective rearing boxes
- Change moth feeding solution
- Collect egg clusters and sterilize eggs if necessary

b. Alternate days
- Feed larvae in single and collective rearing containers
- Transfer moths to aluminum fly-net cage
- Remove dead moths from cage

c. Other rearing schedule
- Add fresh diet to circular box after 6 days of rearing
- Prepare fresh diet every 5-6 days at insect production rate of 100-200 moths per day

5.8 Insect quality

Check insects by recording reproductive capacity of moths, based on: spermatophore count, mating status of females, weight of eggs produced within 4 days, and egg fertility (Navon and Marcus, 1982); see 6.2, 6.3, and 6.4 for values.

5.9 Special problems

a. Diseases
- The insect is susceptible to nuclear polyhedrosis virus. Diseased larvae are recognized by the pinkish color on ventral parts of the body. Prophylactic routine consists of immersing rearing boxes in 0.4% aqueous KOH solution overnight, use of formaldehyde in the diet, and sterilizing the eggs. Formaldehyde level in the diet may be increased by 20% if virus-diseased larvae are observed. During the last 3 years no viruses were observed in the insect colony.

b. Nutrition
- The larvae require L-ascorbic acid in the diet for growth and development. Low moth fertility, locked pairs "permanent matings", and low mating capacity of males are symptoms of weak ascorbic acid deficiency in the larvae. The colony would recover from these symptoms by adding fresh diet every day, and by preparing new diet more frequently.

6. LIFE CYCLE DATA

At 24 ± 2°C; LD 18:6 photoperiod and 50-70% RH.

6.1 Development data

		Number of days (Mean ± SE)
Egg		3.5 ± 0.03
Larva		19.0 ± 0.03
Pupa	Female:	9.0 ± 0.2
	Male:	7.0 ± 0.2
Egg to egg		36.0 ± 0.2

6.2 Survival

	% (Means ± SE)
Neonate to pupa	73.0 ± 7.0
Pupa to adult	97.0 ± 1.6
Neonate to adult	70.0 ± 2.7

6.3 Adults reproductive capacity

Preoviposition period	1-2 days
Adult lifespan	Female: 8.0 ± 1.9 days
	Male: 7.0 ± 0.2 days
Oviposition period	4 days
Mean spermatophores/male	2.4 ± 0.2
Mean eggs/female/2 days	1088.0 ± 174.0
Mean eggs/female/4 days	2045.0 ± 518.0

6.4 Other data

Number of instars	6
Pupal weight	Female: 422 ± 17 mg
	Male: 398 ± 13 mg

7. PROCEDURES FOR SUPPLYING INSECTS

Eggs at a rate of 20,000 per week airmailed upon receipt of request. Pupae at a rate of 200 per week under specific arrangements with requesting laboratories.

8. REFERENCES

Avidov, Z. and Harpaz, I., 1969. Plant Pests of Israel. Israel Universities Press, Jerusalem, 549 p.

Bishara, I., 1934. The cotton worm Prodenia litura F. in Egypt. Bull. Soc. R. Ent. Egypte 18: 228-404.

Kehat, M. and Gordon, D., 1975. Mating, longevity, fertility and fecundity of the cotton leafworm, Spodoptera littoralis (Boisd.) (Lepidoptera: Noctuidae). Phytoparasitica 3: 87-102.

Levinson, H. Z. and Navon, A., 1969. Ascorbic acid and unsaturated fatty acids in the nutrition of the Egyptian cotton leafworm, Prodenia litura F. J. Insect Physiol. 15: 591-595.

Mochida, O., 1973. Two important insect pests, Spodoptera litura F. and S. littoralis (Boisd.) (Lepidoptera: Noctuidae), on various crops - Morphological descriptions of the adult, pupal and larval stages. Appl. Entomol. Zool. 8: 205-214.

Moore, I. and Navon, A., 1964. An artificial medium for rearing Prodenia litura F. and two other noctuids. Entomophaga 9: 181-185.

Moore, I. and Navon, A., 1969. Calcium alginate: A new approach in the artificial culturing of insects, applied to Spodoptera littoralis (Boisduval). Experientia 25: 221-222.

Navon, A. and Keren, S., 1980. Rearing the Egyptian cotton leafworm, Spodoptera littoralis, on a practical calcium-alginate diet. Phytoparasitica 8: 205-207.

Navon, A. and Marcus, R., 1982. D-isoascorbic acid fed to Spodoptera littoralis moths, induces sterility due to spermatophore malformation. J. Insect Physiol. 28: 823-828.

Poitout, S., Bues, R. and Le Rumeur, C., 1972. Elevage sur milieu artificial simple de deux nuctuelles parasites du cotton Earias insulana et Spodoptera littoralis. Ent. Exp. Appl. 15: 341-350.

Viette, P., 1963. Le complexe de "Prodenia litura (Fabricius)" dans la region Malgache (Lep. Noctuidae). Bull. Mens. Soc. Linn. Lyon 32: 145-148.

ACKNOWLEDGEMENT

The author thanks Dr. Isaac Moore of ARO, The Volcani Center, Bet Dagan, Israel, for useful suggestions in the preparation of this paper.

SYNANTHEDON PICTIPES

D. K. REED and N. J. TROMLEY

USDA-ARS , Fruit and Vegetable Insect Research Laboratory, P. O. Box 944, Vincennes, IN 47591

THE INSECT

Scientific Name: Synanthedon pictipes (Grote and Robinson)
Common Name: Lesser peachtree borer (LPTB)
Order: Lepidoptera
Family: Sesiidae

1. INTRODUCTION

The lesser peachtree borer (LPTB), a native American insect, was first described in 1868 and early workers reported it as a pest on plum and cherry trees. Since then, however, LPTB has become a major pest of peaches in all peach growing areas in the U.S. and cherry in the northern states. It is also hosted by the wild varieties, including wild blackcherry, wild red cherry, beach plum, wild plum and june berry. The larvae works at the margins of injured areas and feeds on actively growing tissue that is protected by overhanging bark. As the larvae tunnel, they provide openings to new wood for invasion by the canker organisms which may have been previously closed off by callus tissues (Hildebrand, 1947). The larvae are not antagonistic and one canker may harbor 50 or more. The LPTB overwinters as an immature larvae and can survive through most temperatures which do not harm the tree. Adults emerge in the spring and lay their eggs on roughened bark, normally near wounds or cankers, which attract them. In central and southern areas, 2 generations occur, while there may be only one generation in more northern areas (Wong et al., 1971).

A laboratory colony may be initiated by cutting peach limbs or trees in the early spring which have cankers that indicate active larval infestations and caging them. Adults are collected as they emerge in the cage. Pupae may be obtained from an already established laboratory colony if desired. The insects are reared on Golden Delicious apples in a manner similar to that described by Cleveland et al., 1968.

2. FACILITIES AND EQUIPMENT

2.1 Facilities

Rearing of larvae is conducted in controlled environment rooms or cabinets maintained at 26.6 ± 2 ° C and 55 ± 5% RH in continuous fluorescent light. Oviposition is carried out in a well lighted room maintained at the same conditions but with a 16:8 LD photoperiod. Adult emergence occurs in a cabinet or room maintained at similar conditions but with facilities to darken the location except for a lighted collection area.

Mating can be difficult without a proper environment. During warm weather, it may be accomplished in outdoor cages. In cold or cloudy weather, indoor conditions similar to rearing environment with adequate lighting and ventilation is suitable. Pheromone concentrations may build up in unventilated areas.

2.2. Equipment and supplies

2.2.1 Adult oviposition
- Plastic 15 ml jelly cups (Fill Rite Corp. 45-55 Liberty St., Newark, NJ 07052 or Premium Plastics, 465 W. Cermak Rd., Chicago, IL 60616)
- Paper lid with sawed (1.2x0.2 cm) slit (Same)
- Non-sterile cotton balls - Medium size

2.2.2 Egg Collection
- Sodium hypochlorite 0.1% A.I.
- Organdy cloth 6 cm^2 discs
- Plastic funnel 100-120 mm diam
- Sterile plastic petri dishes 100x15 mm
- Sterile filter paper 90 mm diam

2.2.3 Larval rearing
- Board with small nails protruding as apple and plastic bag puncher
- Captan (0.25%) in water
- Ascorbic acid (5.5%) in distilled water
- Handiwipes
- Plastic bags (2 mil) 40x36x94 cm (Bradley's Plastic Bag Co., 9130 Firestone Blvd., Downey, CA 90241
- Corrugated cardboard trays lined with aluminum foil (61x36x10 cm) (Packaging Corp. of America, 408 E. St. Clair, Vincennes, IN 47591)

2.2.4 Pupal collection
- Rough wood chips (5-10 cm) and Grit-O-Cobs® in a 2:1 ratio (The Andersons, Cob Division, P. O. Box 119, Maumee, OH 43537)
- Shop vacuum cleaner
- Shaker made of coarse hardware cloth (6-mm-mesh) or a mechanical shaker

Fig. 1. Adult collection device for lesser peachtree borer.

2.2.5 Adult collection
- Collector made of a vacuum cleaner and a 29x16.5x13 cm box with a plexiglass lid (Fig. 1)
- Collecting and holding cages made from quart (946 ml) ice-cream cartons modified with screen wire middles (24x9 cm), (Fig. 1), (Sealright Co., Inc., 605 W. 47th St, Kansas City, MO 64112)

2.2.6 Adult holding
- Metal or plastic tray containing 1.5-2 cm of moist sand with a plywood cover

2.2.7 Mating
- Cages 82 cm^2 or larger

3. DIET

3.1 Natural diet

Unsprayed golden delicious apples, are collected in the spring when they are ca 1.5-4.0 cm diameter. These are placed immediately into plastic bags, sealed and put under refrigeration at 4.4°C until needed. Other cultivars may be used but it is important that the apples are dry when stored.

4. INSECT HOLDING

4.1 Environmental conditions

4.1.1 Adults
- Moths may be stored for 3-4 days within the holding cages if placed on wet sand and held at 10°C in the dark.

4.1.2 Eggs
- Eggs can be stored up to 2 weeks in sterile petri dishes with moistened filter paper bottoms at 10°C.

4.1.3 Larvae
- These are not normally stored but neonate larvae may retain viability for 24 h or more if maintained in a dish with slightly moistened filter paper at 15°C. Too much water is detrimental.

4.1.4 Pupae
- Maintain at 7.5-10°C in the dark for up to 2 weeks.

5. REARING AND COLONY MAINTENANCE

The LPTB colony was established from field collected adults and has either had new insects introduced or has been completely renewed each year.

5.1 Egg collection

- Remove cotton balls from jelly cups and discard females after 5 days
- Soak cotton balls in 0.1% AI sodium hypochlorite for 5 min (Davis et al, 1979)
- Agitate and then remove cotton
- Pour eggs and liquid through organdy disc, fixed in the funnel
- Rinse 2X with distilled water
- Place discs in petri dishes on filter paper for storage or use immediately

5.2 Apple innoculation for larval rearing

- Wash apples thoroughly in 0.25% Captan
- Drain
- Punch 10-20 small holes in each apple to facilitate larval penetration
- Layer apples (ca 200) in bottom of tray
- Spray with Ascorbic acid solution (antioxidant)
- Cover with paper towels or napkins to dry overnight
- After 15-24 h, cover apples with Handiwipe® towels
- Brush eggs (adequate number can usually be obtained from 4-5♀/tray) onto towelling evenly after atomizing with water

- Place infested trays in plastic bags, tie end and punch 200-300 small holes for ventilation (provides 80-90% RH in trays)
- Place trays into rearing environment

5.3 Pupae collection

- Remove plastic bags and napkins after 23 days
- Sprinkle 25-30 cm layer of wood chip/Grit-O-Cob® mixture over apples (Burnside and Wong, 1973)
- After 22 days, vacuum chips and LPTB cocoons from trays with a shop vacuum cleaner and dispose of apples
- Shake wood chips through 6-mm-mesh hardware cloth which retains cocoons

5.4 Adult collection

The sex ratio is approximately 1:1

- Place trays containing cocoons in room with light source against far wall (collect only adults which fly to light)
- Collect male and female moths separately with vials or a vacuum collecting device
- Store until needed or mate immediately

5.5 Adult mating

Moths will mate the day of eclosion - Females will indicate mating receptiveness by extruding the last 3 abdominal segments into a "calling position" within 1 h after emergence. Males do not become sexually active for 2-4 h.

- Place 25-30 females into cage 1/2-1 h prior to introducing 30-35 males
- Observe hourly for copulation. Mated pairs will remain in copula for 1-1.5 h
- Remove pairs to small holding cages until they part, then males can be disposed of and females used for oviposition

5.6 Oviposition

This should be done at room temperature with supplemental fluorescent lighting.

Pre-oviposition period is very short, 2-3 h

- Place individually into oviposition cups with 2 moist cotton balls in bottom (saturate, then press water out)
- Cover with slitted lid which provides aeriation
- Maintain in oviposition chamber or room
- After 5-6 days, discard females

5.7 Sex determination

- Adult female is usually more robust with a simple filiform antennae and straight anal tuft.
- Adult male has a finely tufted antennae and the anal tuft is hastate in form.

5.8 Rearing schedule (Fig. 2)

Monday	- Collect adults and place in cooler
	- Infest trays with eggs from previous Tuesday's oviposition
Tuesday	- Mate adults from Monday and place females in oviposition cups
Wednesday	- Collect adults and place in cooler
	- Infest trays with eggs from previous Thursday's oviposition
Thursday	- Mate adults from Wednesday and place females in oviposition cups
Friday	- Screen all trays 23 days from date of infestation
	- Extract pupae 46 days from date of infestation

5.9 Insect quality

Quality of LPTB may be monitored by observing adult behavior (flight mating etc), pupal weights, fecundity and egg viability. Non-flying insects should be discarded. Wild insects should be periodically introduced into the colony to maintain quality.

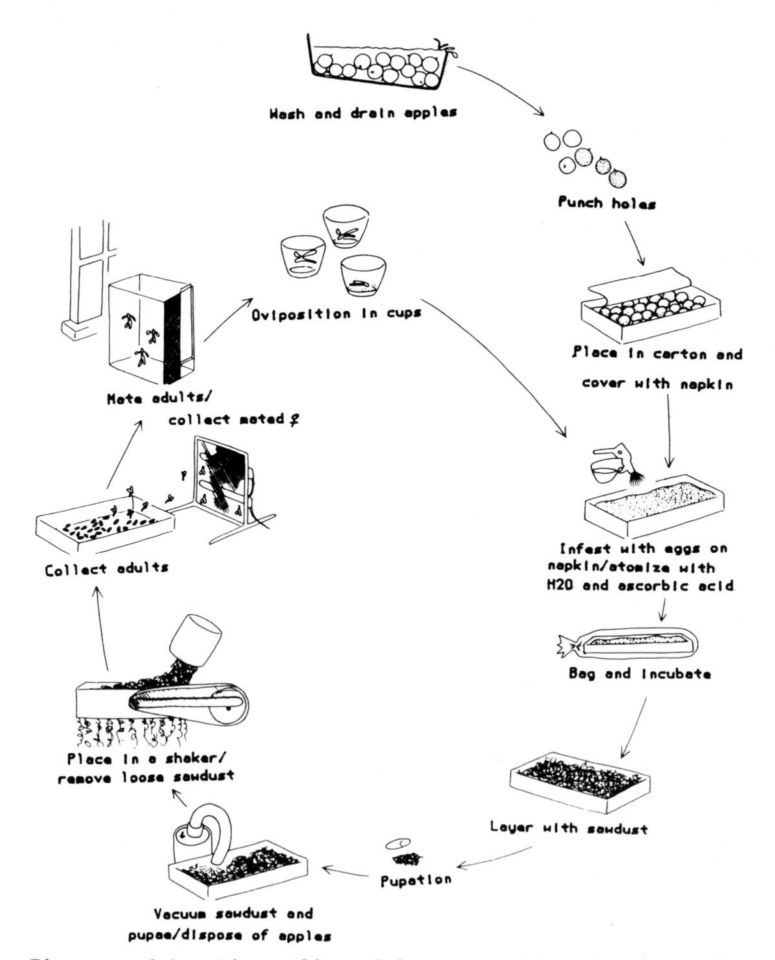

Fig. 2. Schematic outline of lesser peachtree borer rearing.

5.10 Special problems

No problems have been encountered with disease. Major rearing problems are mites and *Drosophila* flies. The flies are attracted to decaying apples. They can be controlled by using good sanitation and rooms as fly proof as possible. Traps using bait formulations can also be used. Mites are not normally a problem but if present, they can build up to damaging populations in a short time. They are controlled by strict sanitation and observation.

6. LIFE CYCLE DATA

6.1 Developmental data

Stage	Numbers of Days Min	Mean	Max
Eggs	7	9	12
Larvae	24	26	37
Pupae	12	14	29
Adult emergence		50	

Egg to Egg 50-51 days

6.2 Survival data (%)
Egg viability - 89% (85-96)
Neonate larvae to pupa - 48
Neonate larva to adult - 45

6.3 Other data
Pre-oviposition period - 2-3 h
Peak egg laying period - Day 1 and 2 after eclosion
Mean fecundity - 236 eggs/ ♀
Mean adult longevity - 9 days for ♂, 8 days for ♀
No. of larval instars - 7 (Head capsule of 7th instar is cast off within the pupal case)

7. PROCEDURES FOR SUPPLYING INSECTS

No special procedures have been developed to supply insects since the colony is essentially for research. Small numbers of insects may be supplied upon request with adequate lead time.

8. REFERENCES

Burnside, J. A. and Wong, T. T. Y., 1973. Lesser peachtree borer: A method of extracting pupae from rearing trays. J. Econ. Entomol. 66:247-48.

Cleveland, M. L., Wong, T. T. Y. and Lamansky, K. W., 1968. Rearing methods and biology of the lesser peachtree borer, Synanthedon pictipes, in the laboratory. Ann. Entomol. Soc. Am. 61:809-14.

Davis, D. G., Tromley, N. J., Wong, T. T. Y. and Reed, D. K., 1979. Lesser peachtree borer (Lepidoptera:Sesiidae): Influence of water and chemical washes on collection and hatchability of eggs. Proc. Ind. Acad. Sci. 89:225-30.

Hildebrand, E. M., 1947. Perennial peach canker and the canker complex in New York, with methods of control. Cornell Univ. Agric. Exp. Stn. Mem. 276:1-61.

Wong, T. T. Y., Kamasaki, H., Dolphin, R. E., Cleveland, M. L., Ralston, D. F., Davis, D. G., and Mouzin, T. E., 1971. Distribution and abundance of the lesser peachtree borer on Washington Island, Wisconsin. J. Econ. Entomol. 6:879-82.

Mention of a commercial product in this paper does not constitute on endorsement of this product by the USDA.

TINEOLA BISSELLIELLA

ROY E. BRY and ROBERT DAVIS

Stored-Product Insects Research and Development Laboratory, ARS/USDA, P. O. Box 22909, Savannah, GA 31403

THE INSECT

Scientific Name: *Tineola bisselliella* (Hummel)
Common Name: Webbing clothes moth or common clothes moth
Order: Lepidoptera
Family: Tineidae

1. INTRODUCTION

The webbing clothes moth is probably of African origin and is generally distributed world-wide. It is one of the most common of the several species of clothes moths responsible for damage in the U.S. The larvae feed on wool, hair, feathers, fur and on wool- or mohair-upholstered furniture. Occasionally the larvae feed on dead insects, dry dead animals, animal and fish meal, casein and other animal products such as bristles and leather. Adult moths do not feed. In nature, the life cycle will vary considerably according to temperature, relative humidity and diet available; Moncrieff (1950) indicates that the limits are ca. 48 days minimum and 4 years maximum. The life cycle under the rearing conditions in our laboratory is ca. 45-47 days.

The adult moth is covered with shiny golden scales and the top of the head has a pompadour of reddish golden hairs. The wings are without spots and those of the female have a span of ca. 13 mm; those of the male are slightly smaller. When at rest, the expanse of the folded wings ranges from 6.5 to 8.5 mm, with the former size being that of the female. The larva upon emerging from the egg is an active, white translucent insect ca. 1 mm long. The fully mature larva usually attains a maximum size of ca. 12.5 mm and average weight of ca. 3 mg. The larva may spin a feeding tunnel of silk that incorporates its own excrement and some of the fibers on which it is feeding. The larva may feed for some distance in this tube or else abandon it and build another.

In our laboratory, the webbing clothes moth has been reared continuously for over 35 years. The laboratory-reared insects are used primarily in a mothproofing research program. A laboratory colony can be started from either field-collected specimens or by obtaining starter cultures from an established laboratory colony.

2. FACILITIES AND EQUIPMENT REQUIRED

2.1 Facilities

Rearing is carried out in a controlled environment room maintained at 27±1°C and 60±5% RH. Lighting is with fluorescent lights (no less than 30-40 lux) with a 12:12 h L:D cycle. Oviposition is carried out under the same conditions but in nearly total darkness.

2.2 Equipment and materials required for insect handling

2.2.1 Adult oviposition

- A stainless steel chamber 0.305 m^3 is used for oviposition. The center of each side has a port 7 cm in diameter over which is soldered a ring for a 1-qt Mason jar (946 ml). The top of the chamber has a port 10 cm in diameter through which a 5 x 50-cm strip of moth test cloth is hung on a stainless steel rod positioned ca. 2.5 cm from the top of the chamber. The port in the top of the chamber is closed by a darkened plexiglass lid.

2.2.2 Rearing containers

- 1-qt (946 ml) Mason jars with lids having a 4 cm hole covered with 40-mesh screen and a No. 1 filter paper insert in the retaining rings

2.3 Diet preparation equipment

- Household type automatic clothes washer
- Household type clothes dryer
- Neutral soap (Ivory flakes)
- Scissors

3. DIET

For many years, the Savannah Laboratory has used the ingredients specified in the rearing procedures published by the American Association of Textile Chemists and Colorists (AATCC) (Anon. 1977) and by the Chemical Specialties Manufacturers Association (CSMA) (Anon. 1971) to rear the webbing clothes moth.

3.1 Composition

Moth test cloth
Brewer's Yeast

3.2 Brand names and sources of ingredients

- Moth test cloth, Testfabrics, Inc. P. O. Drawer O Middlesex, NJ 08846
- Vita-Food Red Label Brewer's Yeast, Vitamin Food Co., Inc., Newark, NJ 07104

3.3 Diet preparation

- Wash blanket-size pieces of moth test cloth in an automatic home washer using the wool cycle and dry in home dryer.
- Cut fabric into 5 x 50-cm strips for oviposition and into 14 x 50-cm strips for rearing.
- Store in large plastic shoe box in refrigerator at 3-5°C until required.

4. REARING AND COLONY MAINTENANCE

4.1 Egg collection

- Select rearing jars containing adults that have been emerging for 2-3 days. Attach these to the side ports of the rearing chamber. For subsequent seedings, always remove the two oldest cultures and replace with two cultures containing freshly emerging adults. In this manner, individuals from several rearing jars will be mixed to help minimize inbreeding. Once a month, the oviposition chamber should be thoroughly cleaned.
- Remove oviposition strip from oviposition chamber after 3 or 4 days and brush eggs into an enameled or pyrex pan with a stiff brush.
- Approximately 0.1 ml of eggs will equal ca. 2000 eggs which is sufficient to seed a 1-qt rearing container. A 0.1 ml volumetric scoop is recommended. The above number of eggs will produce 1000-1500 insects per jar.

4.2 Culture preparation

- Cultures are started by sprinkling 0.1 ml of eggs on a 14 x 50-cm strip of moth test cloth generously sprinkled with brewer's yeast.
- Roll the strip into a cigar-shaped roll and place in a 1-qt (946 ml) Mason jar.

PRECAUTIONS

Webbing clothes moth larvae are highly susceptible to the infectious polyhedral viruses and to some protozoan infections; therefore, cultures should be examined frequently and those having sick insects should be destroyed immediately. Symptoms of the polyhedral viruses consist of sluggishness and a dull white appearance of the larvae as opposed to the semi-translucent appearance of healthy larvae. Symptoms of protozoan infections also include sluggishness and slow response to external stimuli. Color of the larvae may also be altered.

5. LIFE CYCLE DATA

At 27+1°C, 60+5% RH and a 12:12 h L:D cycle, the normal time for a generation is ca 5 to 6 wk.

6. PROCEDURES FOR SUPPLYING INSECTS

All life stages may be shipped in the rearing container which is prepared for shipping by placing a 40-mesh screen lid over the filter paper disc in the retaining ring. The culture jars are placed in corrugated cardboard boxes and are cushioned with a suitable packing material such as expanded polyethylene, polystyrene, styrofoam or shredded newspaper to prevent breakage.

7. REFERENCES

Anon. 1971. Textile resistance test. pp. 168-71. Soap Chem. Spec.

Anon. 1977. Insects, resistance of textiles to. Standard Test Method 24-1977, pp. 287-91. Amer. Assoc. Text. Chem. and Color. Tech. Manual 53. 386 pp. Blue Book. 47(4A). 192 pp.

Moncrieff. 1950. Mothproofing. Leonard Hill, Ltd., London. 200 pp.

TRICHOPLUSIA NI

RICHARD H. GUY, N. C. LEPPLA, and J. R. RYE

Insect Attractants, Behavior, and Basic Biology Research Laboratory,
ARS-USDA, Gainesville, Florida 32604 USA.

C. W. GREEN, S. L. BARRETTE, AND K. A. HOLLIEN

Insect Attractants, Behavior, and Basic Biology Research Laboratory, and
Department of Entomology & Nematology, University of Florida,
Gainesville, Florida 32611 USA

THE INSECT

Scientific Name: *Trichoplusia ni* (Hübner)
Common Name: Cabbage looper
Order: Lepidoptera
Family: Noctuidae

1. INTRODUCTION

The cabbage looper is a cosmopolitan insect that originated in tropical or subtropical Africa and radiated to Europe, Asia, North and South America and several Pacific islands (Sutherland, 1965). The species is polyphagous, feeding primarily on the leaves of a wide variety of vegetables, field crops, flowers, ornamentals, and numerous kinds of wild hosts. In North America, it is particularly damaging to cole crops and tomatoes.

Cabbage looper larvae, due to their wide host range, have quite variable feeding habits. They not only defoliate plants but also may damage vegetative tissue such as celery tips, scar the surface of tomatoes and melons, and destroy cotton squares and flower buds. In many cases, they feed on plants that will not support their complete development.

Females deposit eggs singly on any part of the plant but leaves are preferred. Larvae emerge in ca three days and feed for another 14 at 27°C. After five instars, pupation occurs in a silken chamber often attached to the underside of a leaf near its base. Adults emerge seven days later and may produce another generation within three days. Development is continuous if the temperature is adequate.

The cabbage looper has been reared for more than two decades on a variety of diets (McEwan and Hervey, 1960; Ignoffo, 1963; Shorey, 1963; Shorey and Hale, 1965; Patana, 1969; Poitout and Bues, 1974), and has been mass reared for the production of insect pathogens (Vail et al., 1973) and for use in sterile insect technique (Henneberry and Kishaba, 1966). In our laboratory, it has been reared for more than 10 years for all kinds of basic research (Leppla et al., 1984). Thus, it is not difficult to rear but it is essential to establish and maintain a colony that is free of pathogens, such as nuclear-polyhedrosis virus.

2. FACILITIES AND EQUIPMENT REQUIRED

2.1 Facilities (Leppla et al., 1978; Leppla et al., 1982)

Larvae are reared in holding rooms maintained at 27°C, 50% RH, and an LD 14:10 photoperiod (300-750 nm spectrum, 180-310 lux) with continuously recirculating air. Moths are held at 28°C and 80% RH with a 14-h photophase (same light quality as larvae plus a 0.25-W night light). An air filtration system used to control scales is composed of 2.0 x 0.7 x 1.4 m high cabinets with fiberglass furnace filters connected to a self contained dust arrestor (Arrestall® size 40, American Air Filter Co., Inc., 215 Central Ave., Louisville, KY 40277).

2.2 Equipment and materials required for insect handling

2.2.1 Adult oviposition

- Holding cage, 30 x 21 cm diam cylindrical open bottom cage of 3 mm hardware cloth soldered at the seams.
- Cage base, 23.75 cm diam aluminum cake pan.
- Coarse vermiculite (Terra-lite Horticultural Products, W. R. Grace & Co., Cambridge, MA 02140).
- Feeder cups, 59 ml paper soufflé cups, no. 200, Sweetheart®.
- Absorbent cotton balls, non-sterile (Surgical Supply, 305 S. W. 7th Terrace, Gainesville, FL 32604).
- Sugar.
- Honey, unprocessed.
- Deionized water.
- Organdy cloth, white, 35.6 x 76.2-cm. of sufficient quality to withstand repeated agitation in 5% commercial bleach (5.25% sodium hypochlorite) solution for 5 min per wash.
- Paper clips, binders.

2.2.2 Egg collection and storage

- Manual operation.

2.2.3 Egg sterilization

- Portable washing machine (Rival 15-liter, Sears Wash-O-Matic®, No. 3400) modified by adding a deionized water supply port to the bottom, an internal strainer basket, and a vertical U-shaped 3 cm diam O.D. glass siphon tube to the outlet and by reducing the size of the agitator blades (Leppla et al. 1978).
- Commercial bleach.
- Egg collection net fabricated by attaching fine nylon organdy to a ringstand support ring.
- Beaker, 250 to 500 ml.
- Medicine dropper, 7.6-cm glass pipet with 2.5-cm suction cap (Bio-Quip Products, P. O. Box 61, Santa Monica. CA 90406).
- Rubber gloves.
- Sodium thiosulfate solution, laboratory grade (100 g/liter deionized water) (Fisher Scientific Co., 711 Forbes Ave., Pittsburg, PA 15219).

2.2.4 Egg hatch and larval inoculation of diet

- Paper toweling for surface-sterilized eggs.
- Casein glue.
- Drying table, 1.9 x 0.9 x 1.5 m high laminar-flow hood, Pure Aire 720B (Pure Air Corp., 8441 Canoga Ave, Canoga Park, CA 91304).

2.2.5 Larval rearing containers

- Plastic container, 30.5 x 30.5 x 12.7 cm deep I.D. Tupperware® #G-40, 6-liter.
- Plastic lid, 30.5 x 30.5 cm with 30% removed and replaced with a 28 x 28 cm sheet of 3 mm thick polypropylene 120 micron pore filter (Perox Materials Corp., P. O. Box 5481, Fairburn, GA 30213).
- Plastic insert, 28 x 28 x 1.1 cm thick translucent white acrylic 1.2 cm cube louver no. TA 225 (Scientific Lighting Products, 10545 Baur Blvd., St. Louis, MO 63132).

2.2.6 Pupal collection, sexing, and storage

- Soft-nose forceps, spring steel (Bio-Quip Products, P. O. Box 61, Santa Monica, CA 90406.)
- Collander, 3 liter aluminum (Wear-ever® No. 3123).
- Beaker, 500 ml.
- Deionized water.
- Commercial bleach.
- Portable washing machine, Rival 15 liter, Sears Wash-O-Matic®, No. 3400.
- Paper toweling.
- Paper cups, 0.47-liter, unwaxed.
- Analytical balance, 0 to 30-g, No. A-30 (Mettler Instrument Corp., Hightstown, NJ 08520).
- Stereozoom microscope, 10X-70X (Bausch and Lomb, Rochester, NY 14602).

2.2.7 Insect holding

- Controlled-temperature room, 27°C, 80% RH, and L:D 14:10 photoperiod (see Facilities 2.1).

2.3 Diet preparation equipment

2.3.1 Dry mix

- Clean tunnel, high efficiency particulate air-filtration system, No. FM-48 (Pure Air Corp., 8441 Canoga Ave., Canoga Park, CA 91304.
- Freezer, double-door, 1.2 m^3, No. T1-4-AD (Foster Refrigerator Corp., Thermodynamics, Inc., Parsons, TN 39363).
- Freezer, walk-in, 16 m^3, No. CPE-72, modular aluminum box with add-on coils (Larkin Coils, Inc., Atlanta, GA 30371).
- Top loading analytical balance, 0 to 8200-g, No. PC-8200 (Mettler Instrument Corp., P. O. Box 71, Hightstown, NJ 08520).
- Weighing pans, stainless steel pie pans.
- Graduated cylinder, 100 ml.
- Spatulas.
- Magnetic stirrer, No. PC-353 (Corning Glass Works, Corning, NY 14830).
- Fume hood.
- Diet dispensing table.

2.3.2 Finished diet

- Cooker, steam-jacketed kettle, 18.9 liter, No. TDB/4-20 (Groen Division, Dover Corp., 1900 Pratt Blvd., Elk Grove, IL 60007).
- Blender, 3.8 liter, No. CB-6 (Waring Products Division, Dynamics Corp. of America, New Hartford, CT 06057).
- Rheostat, Powerstat® variable autotransformer, No. BP57515, 120 V input, 0-140 V output.
- Pitchers, 2 liter plastic
- Sponges
- Kraft paper, 91.4 cm, 30 lb (13.6 kg) test (local supply).

3. ARTIFICIAL DIET

3.1 Composition

a. Dry mix to prepare 15.2 liters finished diet.

Ingredients	Amount	% (Estimated)
Gelcarin HWG	184 g	6.2
Casein	400 g	13.4
Pinto beans, ground	1100 g	36.8
Torula yeast	500 g	16.8
Wheat germ	800 g	26.8
Total	2984	100%

b. Finished diet

Dry mix	2984 g
Deionized water	10,800 ml
Vitamin mixture	120 ml
Ascorbic acid	52 g
Methyl paraben (MPH)	32 g
Sorbic acid	16 g
Formalin	60 ml
Tetracycline	4 (250 mg capsules)

c. Preparation of vitamin mixture (400 g Hoffman-LaRoche insect vitamin mix/1000 ml H_2O).

Vitamin	Amount/1000 gram (QS with dextrose)
Vitamin A	20,246,000 IU
Vitamin E	7,361.4 IU
Vitamin B_{12}	1.8 g
Riboflavin	459.0 mg
Niacinamide	921.1 mg
Calcium d-Pantothenate	921.1 mg
Choline chloride	46.0 mg
Folic acid	230.8 mg
Pyridoxine HCL	230.8 mg
Thiamine HCL	230.8 mg
d-Biotin	18.4 mg
Inositol	18.4 g

3.2 Brand names and sources of ingredients

Gelcarin HWG (Marine Colloids, Inc., 2 Edison Place, Springfield, NJ 07081); casein (National Casein Co., 601 W. 80th St., Chicago, IL 60620); pinto beans (local source, Dixie Lilly Co., Williston, FL 32696); torula yeast (St. Regis Paper Co., Lake States Division, 603 W. Davenport St., Rhinelander, WI 54501); wheat germ (Earthwonder, 1735 E. Trafficway, Springfield, MO 65802); vitamin mixture, ascorbic acid (Roach Chemical Division, Hoffman-LaRoche, Inc., Nutley, NJ 07110); methyl paraben (Tenneco Chemical, Turner Place, P. O. Box 365, Piscataway, NJ 08854); sorbic acid (American Hoechst Corp., Chemicals and Plastics Division, Rt. 202-206, Somerville, NJ 08876); formalin (Fisher Scientific Co., 711 Forbes Ave., Pittsburgh, PA 15219); tetracycline (Rugby Laboratories, Inc., Rockville Center, LI, NY 11570).

3.3 Diet preparation procedure

a. Preliminary
- Place 1-liter bottle of vitamin mixture on magnetic stirrer in clean tunnel.
- Cover diet dispensing table with kraft paper and arrange rearing containers to receive diet.
- Add deionized water to cooker.
- Weigh dietary ingredients and thoroughly mix the Gelcarin HWG, casein, pinto beans, torula yeast, wheat germ, methyl paraben, and sorbic acid.
- Place beaters on cooker, carefully lower into position, set speed at 6 for a 15-liter batch (slower for a smaller batch) and actuate.
- Set thermostat on cooker at 6 1/2 and switch on.

b. Preparation
- Slowly sprinkle dry ingredients into cooker.
- Cook until red thermostat light goes off, ca 15 min
- Add the formalin and cook 5 min, turn off heat for 2 min
- Sprinkle in the ascorbic acid and stir without heat for 3 min
- Turn off motor, remove beaters from cooker, and place in hot water to soak.

c. Pouring
- Transfer first 4 liters of diet from cooker into a pan.
- Dispense second 4 liters into blender.
- Return first 4 liters to cooker.
- Add 30 ml vitamin mix and 1 tetracycline capsule to blender.
- Blend for 30 sec with rheostat set at 100.
- Pour 1000 g of finished diet from blender into each rearing container.
- Repeat second through sixth steps for each 4 liters of diet.
- Cool diet in rearing containers at room temperature for at least 1 h before use.

d. Clean up
- Immediately use 4 liters of water to clean the empty cooker.
- Wash all utensils and clean room.
- Return all ingredients to freezers, restock supplies, and record diet output.

4. HANDLING INSECTS AT LOWER TEMPERATURES

Rearing is continuous so no storage is required.

5. REARING AND COLONY MAINTENANCE

The cabbage looper colony was established in 1973 with field-collected moths from north central Florida and has been maintained continuously on artificial diet. Eggs from ca 50 wild-type females were added to this colony in 1981 and 1982.

5.1 Egg collection
- Manually remove paper toweling substrate containing eggs from adult oviposition cages.

5.2 Egg sterilization
- Fill washer with 15 liters of deionized water, add 70 ml commercial bleach and mix.
- Place egg sheets (cloth substrate) into washer and wash for 5 min
- After washing, rinse sheets repeatedly to dislodge any remaining eggs.
- Squeeze excess bleach solution from sheets and dry them.
- Drain bleach solution containing eggs through siphon into net.
- Rinse washer and add remaining eggs to net.
- Invert net and rinse eggs into 500 ml beaker with ca 300 ml deionized water.
- Neutralize with ca 150 ml sodium thiosulfate solution.
- Soak eggs in solution for 2 min and decant upper layer containing unwanted eggs, larvae, and debris floating on surface.
- Pour mixture containing usable eggs into net, rinse with deionized water for ca 30 sec
- Place eggs into empty beaker with enough deionized water (ca 60 ml) to stir them freely.

5.3 Placement of eggs in rearing containers
- Apply eggs to 4 x 4-cm pieces of paper toweling with a medicine dropper (50-60 eggs per spot, 6 spots per larval rearing container).
- Attach a square of toweling with dry egg spots to the inside center of each container lid by using a few drops of casein glue.

5.4 Collection of pupae
- Transfer containers with 1 to 2-day-old pupae from development room to harvest area, remove lid, and carefully pull silk containing pupae from lid and sides of container.
- Use soft-nose forceps to pick remaining pupae from diet and frass in bottom of container.
- Fill pupal washer with 15 liters of deionized water and add 400 ml of commercial bleach.
- Place pupae into washer, wash 10 min, and drain into a ca 3 liter colander.

- Rinse pupae 2 min, spread to dry on paper toweling.
- After drying, remove all debris with forceps, count pupae, and weigh a random sample of 25 (pupal count = total weight of harvested pupae per average weight of sample pupae).

5.5 Adults

5.5.1 Emergence period

Females begin to emerge 18-19 days after larval rearing containers are established (4-5 day pupal stage). Males begin to appear 1 day later, after as many as 50% of the females have emerged. The emergence period lasts about 3 days with a peak on the second. The sex ratio is ca 1:1.

5.5.2 Collection

Pupae are placed into oviposition cages before emergence of adults.

5.5.3 Oviposition

Moths are maintained in an oviposition room at 28°C, 80% RH, and a 14 h photophase (see 2.1, Facilities). The preoviposition period is 2 to 3 days, peak egg production occurs after the moths are 4 to 5 days old, and they are discarded after ca 8 days.

Procedures

- Pour coarse vermiculite into cake pan base to a depth of ca 1.5 cm.
- Place a feeder cup containing cotton saturated with 5% honey and sugar solution (54 g sucrose + 35 ml unprocessed honey per liter of deionized water) in the center of the pan.
- Distribute 150 unsexed pupae on vermiculite around feeder cup.
- Wrap sides of hardware cloth oviposition cage with one layer of cloth substrate held in place with paper clip binders.
- Place top over pupae in base.
- After oviposition begins, remove egg sheets daily and replace with new substrate.

5.6 Sex determination

- Male pupae and adults usually are larger than females.
- Male pupae have paired elevated pads on the ventral side of the ninth abdominal segment whereas females lack these structures but have an orifice to the bursa copulatrix on the eighth (Butt and Cantu, 1962).
- Male moths have paired yellow tufts of setae that extend over several segments on the dorsal side of the thorax and also have the usual lepidopteran claspers.
- Female moths have an eversible ovipositor.

5.7 Rearing schedule (5-day work week)

a. Daily

- Prepare diet.
- Collect and surface-sterilize eggs.
- Place eggs on lids of larval rearing containers.
- Harvest pupae.
- Discard old and set up new oviposition cages.
- Check temperature, RH, and photoperiod in adult holding and larval rearing rooms.

b. Alternate days

- Clean and sanitize used oviposition cages.
- Replace old feeder cups with new ones in oviposition cages.

5.8 Insect quality

Records are kept on inclusion of ingredients to larval diet and use of new batches of ingredients. Rates of oviposition, egg fertility, and adult longevity are noted. Average yields and weights of pupae are determined. All rearing environments are monitored continuously.

5.9 Special problems

a. Diseases

Nuclear polyhedrosis virus infection is prevented by visually screening insects used to establish and maintain the colony and by surface-sterilizing eggs. Bacteria are controlled with tetracycline and fungi are impeded by dietary "inhibitors," cleanliness, and consistent environmental control.

b. Laboratory adaptation

Each year the eggs from ca 50 wild-type females are added to the colony. Efforts are made to prevent bottlenecks, particularly reductions in the proportions of ovipositing females.

c. Human health

Body parts of insects and microorganisms associated with rearing can be hazardous, so contact is avoided and protective masks and clothing are worn. Adequate ventilation is essential.

6. LIFE CYCLE DATA

Life cycle and survival data are given for development of eggs, adults and pupae at 28°C and 80% RH, and of larvae at 27°C and 50% RH.

6.1 Developmental data

Stage		Number of days		
		Min	Mean	Max
Eggs		2	3	4
Larvae	♂	13	14	15
	♀	12	13	14
Pupae	♂	3	4	5
	♀			
Adult emergence	♂	20	21	22
(from establishment)	♀	19	20	21
Total development (egg to egg, inclusive preoviposition)		23	24	25

6.2 Survival data (%)

- Egg viability ca 90.
- Neonate larva to pupa >90.
- Neonate larva to adult >90.

6.3 Other data

- Preoviposition period 3-4 days.
- Peak egg laying period 4-5 days after emergence.
- Mean fecundity ca 400 eggs per female.
- Mean adult longevity (discarded after ca 8 days).
- No. larval instars 5.
- Mean pupal weight 230 mg (min 190, max 250).

7. REFERENCES

Butt, B. A. and Cantu, E., 1962. Sex determination of lepidopterous pupae. U. S. Dept. Agric., Agric. Res. Serv. ARS-33-75. 7 pp.

Hennebery, T. J. and Kishaba, A. N., 1966. Mass rearing cabbage loopers, pp. 461-478 In C. N. Smith (ed.), Insect Colonization and Mass Production. Academic Press, New York.

Ignoffo, C. M., 1963. A successful technique for mass rearing cabbage loopers on a semisynthetic diet. Ann. Entomol. Soc. Am. 56: 178-182.

Leppla, N. C.; Carlyle, S. L.; Green, C. W.; and Pons, W. J., 1978. A custom insect rearing facility. In N. C. Lepla and T. R. Ashley (eds.), Facilities for Insect Research and Production. U. S. Dept. Agric. Tech. Bull. No. 1576, 86 pp.

Leppla, N. C.; Fisher, W. R.; Rye, J. R.; and Green, C. W., 1982. Lepidopteran mass rearing: An inside view, pp. 123-133 In Sterile Insect Technique and Radiation in Insect Control. International Atomic Energy Agency IAEA-SM-255/14, Vienna, Austria.

Leppla, N. C.; Vail, P. V.; and Rye, J. R., 1984. Mass rearing the cabbage looper. In E. G. King and N. C. Leppla (eds.), Advances and Challenges in Insect Rearing. U. S. Dept. Agric. Tech. Bull. pp. 248-254.

McEwen, F. L. and Hervey, G. E. R., 1960. Mass-rearing the cabbage looper, Trichoplusia ni, with notes on its biology in the laboratory. Ann. Entomol. Soc. Am. 53: 229-234.

Patana, R., 1969. Rearing cotton insects in the laboratory. U. S. Dept. Agric., Agric. Res. Ser., Production Research Report No. 108, 6 pp.

Poitout, S. and Bues, R., 1974. Elevage de chenilles de vingt-huit espèces de lépidoptères Noctuidae et de deux espèces D'Arctiidae sur milieu artificiel simple. Paticularités de L'elevage selon les espèces (Rearing larvae of 28 species of Noctuidae and 2 species of Arctiidae (Lepidoptera) on a simple artificial diet and breeding perculiarities according to the different species). Ann. Zool. Ecol. Anim. 6: 431-441.

Shorey, H. H., 1963. A simple artificial rearing medium for the cabbage looper. J. Econ. Entomol. 56: 536-537.

Shorey, H. H. and Hale, R. L., 1965. Mass-rearing of the larvae of nine noctuid species on a simple artificial medium. J. Econ. Entomol. 58: 522-524.

Sutherland, D. W. S., 1965. Biological investigations of Trichoplusia ni (Hübner) (Lepidoptera, Noctuidae), and related and associated species damaging cruciferous crops on Long Island, New York 1960-1963. Diss. Abstr. 25: 7434-7435.

Vail, P. V.; Anderson, S. J.; and Jay, D. L., 1973. New procedures for rearing cabbage loopers and other lepidopterous larvae for propagation of nuclear polyhedrosis viruses. Environ. Entomol. 2: 339-344.

Mention of a commercial or proprietary product does not constitute an endorsement by the USDA.

INDEXES

1. These indexes include only those species for which rearing methods have been described.

2. Page numbers refer to the first mention of the species in each chapter of the text.

3. Species are located by volume and page number.

4. The page numbers for those species for which detailed instructions are given are identified by **bold** Gothic type.

5. The page numbers for those species referred to in the chapters on multiple species rearing and entomophagous insects are listed in regular Gothic type.

6. Three catagories of index are provided:

   INDEX OF TAXONOMIC ORDER - This provides a rapid method of locating groups of genera within their representative families.

   INDEX OF COMMON NAMES - Species are listed alphabetically by their common names where these exist.

   INDEX OF SCIENTIFIC NAMES - Species reared are arranged in alphabetical order of scientific names.

INDEX OF TAXONOMIC ORDER

INDEX OF COMMON NAMES

INDEX OF SCIENTIFIC NAMES